ZOONOTIC VIRUSES OF NORTHERN EURASIA

ZOONOTIC VIRUSES OF NORTHERN EURASIA
Taxonomy and Ecology

DIMITRY KONSTANTINOVICH LVOV

MIKHAIL YURIEVICH SHCHELKANOV

SERGEY VLADIMIROVICH ALKHOVSKY

PETR GRIGORIEVICH DERYABIN

*D.I. Ivanovsky Institute of Virology of Russian Ministry of Public Health,
Moscow, Russian Federation*

ELSEVIER

AMSTERDAM • BOSTON • HEIDELBERG • LONDON
NEW YORK • OXFORD • PARIS • SAN DIEGO
SAN FRANCISCO • SINGAPORE • SYDNEY • TOKYO
Academic Press is an imprint of Elsevier

Academic Press is an imprint of Elsevier
125, London Wall, EC2Y 5AS
525 B Street, Suite 1800, San Diego, CA 92101-4495, USA
225 Wyman Street, Waltham, MA 02451, USA
The Boulevard, Langford Lane, Kidlington, Oxford OX5 1GB, UK

Notices
Knowledge and best practice in this field are constantly changing. As new research and experience broaden our understanding,
changes in research methods, professional practices, or medical treatment may become necessary.

Practitioners and researchers must always rely on their own experience and knowledge in evaluating and using any information,
methods, compounds, or experiments described herein. In using such information or methods they should be mindful
of their own safety and the safety of others, including parties for whom they have a professional responsibility.

To the fullest extent of the law, neither the Publisher nor the authors, contributors, or editors, assume any liability for any
injury and/or damage to persons or property as a matter of products liability, negligence or otherwise, or from any use
or operation of any methods, products, instructions, or ideas contained in the material herein.

ISBN: 978-0-12-801742-5

British Library Cataloguing-in-Publication Data
A catalogue record for this book is available from the British Library

Library of Congress Cataloging-in-Publication Data
A catalog record for this book is available from the Library of Congress

For information on all Academic Press publications
visit our website at http://store.elsevier.com/

Typeset by MPS Limited, Chennai, India
www.adi-mps.com

Working together
to grow libraries in
developing countries

www.elsevier.com • www.bookaid.org

Publisher: Janice Audet
Acquisition Editor: Jill Leonard
Editorial Project Manager: Elizabeth Gibson & Pat Gonzalez
Production Project Manager: Lucía Pérez
Designer: Vicky Pearson

Contents

List of Abbreviations

A	Asian (genotype)
aa	amino acid
AAV	Alma-Arasan virus
ABLV	Australian bat lyssavirus
ADRV	Adria virus
AGUV	Aguacate virus
AHFV	Alkhurma fever virus
AHSV	African horse sickness virus
AIDS	Acquired immune deficiency syndrome
ALTV	Altai virus
AMRV	Amur (Soochong) virus
ANADV	Anadyr virus
ANHV	Anhanga virus
ANIV	Aniva virus
APMV-1	avian paramyxovirus 1
ARAV	Aravan virus
ARGV	Araguari virus
ARTSV	Artashat virus
ARTV	Artybash virus
ASF	African swine fever
ASFV	African swine fever virus
AURAV	Aura virus
AVAV	Avalon virus
BAKV	Baku virus
BATV	Batai virus
BAUV	Bauline virus
BBKV	Babanki virus
BBV	Bukalasa bat virus
BDAV	Bandia virus
BEBV	Bebaru virus
BEFV	Bovine ephemeral fever virus
BHAV	Bhanja virus
BIRV	Birao virus
BKNV	Batken virus
BOZOV	Bozo virus
bp	base pairs
BSE	bovine spongiform encephalopathy
BTV	Bluetongue virus

BUJV	Bujaru virus
BUNV	Bunyamwera virus
BURV	Burana virus
CASV	Caspiy virus
CCHF	Crimean−Congo hemorrhagic fever
CCHFV	Crimean−Congo hemorrhagic fever virus
CD	circular dichroism
CDC	Centers for Disease Control and Prevention
cDNA	complementary DNA
CDV	canine distemper virus
CDUV	Candiru virus
CE	California encephalitis
CEV	California encephalitis virus
CeMV	cetacean morbillivirus
CESA	Centre, East and South African (genotype)
CEV	California encephalitis virus
CFT	complement-fixation test
CGLV	Changuinola virus
CHGV	Chagres virus
CHIKV	Chikungunya virus
CHIMV	Chim virus
CHPV	Chandipura virus
CHUV	Chenuda virus
CIV	Carey Island virus
CJD	Creutzfeldt-Jakob disease
CJSV	Carajas virus
CMV	Clo Mor virus
CNS	central nervous system
CNUV	Chenuda virus
COCV	Cocal virus
CORV	Corriparta virus
CPE	cytopathogenic effect
CPXV	Cowpox virus
CVV	Cache Valley virus
CWD	chronic wasting disease
CWV	Cape Wrath virus

D	Dalton (atomic mass unit)
DBV	Dakar bat virus
DENF	Dengue fever
DENV	Dengue virus
DGKV	Dera Ghazi Khan virus
DHOV	Dhori virus
DNA	deoxyribonucleic acid
DOBV	Dobrava−Belgrade virus
dsDNA	double-stranded DNA
dsRNA	double-stranded RNA
DTV	Deer tick virus
DUGV	Dugbe virus
DUVV	Duvenhage virus
EBLV	European bat lyssavirus;
EEEV	Eastern equine encephalitis virus
EHDV	epizootic hemorrhagic disease virus
ELISA	enzyme-linked immunosorbent assay
EMCV	encephalomyocarditis virus
emPCR	emulsion PCR
ENTV	Entebbe bat virus
ERVEV	Erve virus
Fab	fragments antigen binding
FFI	fatal familial insomnia
FMDV	Foot and mouth disease virus
FORV	Forécariah virus
FSV	Fort Sherman virus
GAV	Grand Arbaud virus
GERV	Geran virus
GETV	Getah virus
GGEV	Greek goat encephalitis virus
GGYV	Gadget's Gully virus
GIV	Great Island virus
GORV	Gordil virus
GROV	Guaroa virus
GSRV	Gissar virus
GSS	Gerstmann−Sträussler−Scheinker syndrome
HA	hemagglutinin
HAZV	Hazara virus
HBsAg	Hepatitis B s-antigen
HERV	Heartland virus
HFRS	hemorrhagic fever with renal syndrome
HFRSV	hemorrhagic fever with renal syndrome virus
HIT	hemagglutination inhibition test
HIV	human immunodeficiency virus
HPAI	highly pathogenic avian influenza
HTLV	human T-cell leukemia virus
HTNV	Hantaan virus
HUGV	Hughes virus
HYSV	Huaiyangshan virus
IACOV	Iaco virus
ICOV	Icoaraci virus
ICPI	intracerebral pathogenicity index
IERIV	Ieri virus
IFN	interferon
IHC	immunohistochemistry
IHNV	Infectious hematopoietic necrosis virus
IIV	D.I. Ivanovsky Institute of Virology (Moscow)
ILEV	Ilesha virus
INKV	Inkoo virus
IPNV	infectious pancreatic necrosis virus
IRKV	Irkut virus
ISAV	infectious salmon anemia virus
ISFV	Isfahan virus
ISKV	Issyk-Kul virus
ITPV	Itaporanga virus
IUMS	International Union of Microbiological Societies
JAV	Johnston Atoll virus
JCV	Jamestown Canyon virus
JE	Japanese encephalitis
JEV	Japanese encephalitis virus
JOSV	Jos virus
KAMV	Kama virus
KARV	Karimabad virus
kbp	kilobase pairs
kD	kilodalton
KEMV	Kemerovo virus
KEYV	Keystone virus
KFDV	Kyasanur Forest disease virus
KFV	Karelian fever virus
KHABV	Khabarovsk virus
KHAV	Khasan virus
KHTV	Khatanga virus
KHURV	Khurdun virus

KHUV	Khujand virus	**NA**	neuraminidase
KISV	Kismayo virus	**nAChR**	nicotinic acetylcholine receptor
KOMV	Komandory virus	**NDV**	Newcastle disease virus
KSIV	Karshi virus	**NEGV**	Negishi virus
KUNV	Kunjin virus	**NGS**	next-generation sequencing
KYZV	Kyzylagach virus	**nm**	nanometer
		NORV	Northway virus
LACV	La Cross virus	**NRIV**	Ngari virus
LBV	Lagos bat virus	**NSD**	Nairobi sheep disease
LD$_{50}$	50%-th lethal dose	**NSDV**	Nairobi sheep disease virus
LEIV	Laboratory of Virus Ecology (D.I. Ivanovsky Institute of Virology, Moscow)	**NSNJV**	vesicular stomatitis New Jersey virus
		nt	nucleotides
LGTV	Langat virus	**NT**	neutralization test
LIV	Louping ill virus	**NTR**	nontranslated region
LIV-Brit	LIV−British subtype	**nvCJD**	new-variant Creutzfeldt−Jakob disease
LIV-I	LIV−Irish subtype		
LIV-Spain	LIV−Spanish subtype		
LNAV	Lena river virus	**OCKV**	Ockelbo virus
LOKV	Lokern virus	**OHF**	Omsk hemorrhagic fever
LPAI	low-pathogenic avian influenza	**OHFV**	Omsk hemorrhagic fever virus
		OIE	World Organisation for Animal Health
M	mol (substance per liter)	**OKHV**	Okhotskiy virus
MABV	marine birnavirus	**ONNV**	O'nyong-nyong virus
MAC-ELISA	monoclonal antibody capture ELISA	**ORF**	open reading frame
		ORFV	Orf virus
MAGV	Maguari virus	**OZERV**	Ozernoe virus
MANV	Manawa virus		
MARAV	Maraba virus	**PALV**	Palma virus
MARBV	Marboi virus	**PBA**	pathogenic biological agent
MAYV	Mayaro virus	**PCR**	polymerase chain reaction
MBOV	Mboke virus	**PDV**	Phocine distemper virus
mcM	micromol (substance per liter)	**PFU**	plaque-forming unit
MD	megaDalton (atomic mass unit)	**PHSV**	Peruvian horse sickness virus
MEAV	Meaban virus	**PIRYV**	Piry virus
MELV	Melao virus	**PLAV**	Playas virus
MGBV	Magboi virus	**PMRV**	Paramushir virus
mM	millimol (substance per liter)	**poly(A)**	polyadenine sequence
MMLV	Montana Myotis leukoencephalitis virus	**POTV**	Potosi virus
		POWV	Powassan virus
MODV	Modoc virus	**PPBV**	Phnom Penh bat virus
MOKV	Mokola virus	**PPV**	Precarious Point virus
mRNA	matrix RNA	*PRNP*	prion protein (gene)
mtDNA	mitochondrial DNA	**PrP**	prion protein
MURMV	Murman virus	**PrP^{BSE}**	PrP−bovine spongiform encephalopathy
MURV	Murray virus		
MWAV	Manawa virus	**PrP^{c}**	prion protein (normal form)
MYKV	Mykines virus		

PrP^{CJD}	PrP—Creutzfeld-Jacob disease		**SHIBV**	Shimoni bat virus
PrP^{CWD}	PrP—chronic wasting disease		**SHOV**	Shokwe virus
PrP^d	PrP—disease-associated		**SHRV**	snakehead rhabdovirus
PrP^{RES}	PrP—proteinase K resistant		**SINV**	Sindbis virus
PrP^{SC}	PrP—scrapie		**sm**	santimeter
PTV	Punta Toro virus		**SNV**	Sin Nombre virus
PUNV	Punique virus		**SOKV**	Sokuluk virus
PUUV	Puumala virus		**SREV**	Saumarez Reef virus
			ssDNA	single-stranded DNA
QRFV	Quaranfil virus		**SSHV**	snowshoe hare virus
QYBV	Qalyub virus		**ssRNA**	single-stranded RNA
			SVCV	spring viremia of carp virus
RABV	rabies virus			
RAZV	Razdan virus		**TAGV**	Taggert virus
RBV	Rio Bravo virus		**TAHV**	Tahyna virus
RdRp	RNA-dependent RNA polymerase		**TAMV**	Tamdy virus
RFV	Royal Farm virus		**TBE**	tick-borne encephalitis
RNA	ribonucleic acid		**TBEV**	tick-borne encephalitis virus
rpm	revolutions per minute		**TBEV-Eur**	TBEV—European subtype
rRNA	ribosome ribonucleic acid		**TBEV-FE**	TBEV—Far Eastern subtype
RRV	Ross River virus		**TBEV-Mal**	TBEV—Malyshevo subtype
RT	reverse transcription		**TBEV-Sib**	TBEV—Siberian subtype
RT-PCR	reverse transcription polymerase chain reaction		**TCID$_{50}$**	50%-th tissue culture infection dose
RUKV	Rukutama virus		**TENV**	Tensaw virus
RVFV	Rift Valley fever virus		**TFAV**	Thiafora virus
			THAIV	Thailand virus
S	Svedberg unit (sedimentation)		**THOV**	Thogoto virus
s$_{20,w}$	sedimentation constant		**ThV**	theilovirus
SAFV	Saffold virus		**TILLV**	Tillamook virus
SAGV	Sagiyama virus		**TINDV**	Tindhólmur virus
SAKV	Sakhalin virus		**TLAV**	Tlacotalpan virus
SALV	Salehabad virus		**TLKV**	Tyulek virus
SARS	Severe acute respiratory syndrome		**TMEV**	Theiler's murine encephalomyelitis virus
SARV	Santa Rosa virus			
SAV	Sikhote-Alin virus		**TMV**	Tobacco mosaic virus
SCRV	St. Croix river virus		**TOPV**	Topografov virus
SCV	State Collection of Viruses		**TOSV**	Toscana virus
SDNV	Serra do Navio virus		**TPMV**	Thottapalayam virus
SDVFV	Syr-Darya valley fever virus		**TRBV**	Tribeč virus
SEOV	Seoul virus		**tRNA**	transport ribonucleic acid
SEPV	Sepik virus		**TRV**	Theravirus
SeV	Sendai virus		**TSE**	transmissible spongiform encephalopathy
SFCV	Sandfly fever Cyprus virus			
SFNV	Sandfly fever Naples virus		**TSEV**	Turkish sheep encephalitis virus
SFSV	Sandfly fever Sicilian virus		**TUCV**	Tucunduba virus
SFTSV	severe fever with thrombocytopenia syndrome virus		**TULV**	Tula virus
			TV	tellina virus
SFV	Semliki Forest virus		**TVTV**	trivittatus virus
SH	small hydrophobic		**TYUV**	Tyuleniy virus

UFAV	Ufa virus	**WGRV**	Wongorr virus
UNAV	Una virus	**WHAV**	Whataroa virus
USUV	Usutu virus	**WHO**	World Health Organization
UUKV	Uukuniemi virus	**WMV**	Wad Medani virus
UV	ultraviolet	**WNF**	West Nile fever
UZAV	Uzun-Agach virus	**WNV**	West Nile virus
VHEV	Vilyuisk human encephalomyelitis virus	**WYOV**	Wyeomyia virus
VHSV	Viral hemorrhagic septicemia virus	**XINV**	Xingu virus
VLAV	Vladivostok virus	**YFV**	yellow fever virus
VSAV	vesicular stomatitis Alagoas virus	**YOKV**	Yokose virus
		YTAV	yellowtail ascites virus
VSIV	vesicular stomatitis Indiana virus	**ZTV**	Zaliv Terpeniya virus
WA	West African (genotype)	ρ_{Sac}	floating density in the gradient saccharose solution
WCBV	West Caucasian bat virus		
WEEV	Western equine encephalitis virus		

HISTORY AND METHODOLOGY

Introduction

In 2014, Russian virologists celebrated the 150th birthday of Dmitry Ivanovsky and the 100th birthday of Victor Zhdanov, both of whom made seminal contributions to the field of virology. This book is dedicated to these pioneers. As W.M. Stanley, winner of the 1946 Nobel Prize in Chemistry for his isolation and crystallization of the tobacco mosaic virus, said,

> I believe that D.I. Ivanovsky's relationship to viruses should be viewed in much the same light as we view Pasteur's and Koch's relationship to bacteriology. There is considerable justification for regarding Ivanovsky as the father of the new science of virology, a field of endeavor which today is of great importance not only in medicine but in several closely allied fields of study. **W.M. Stanley, 1944**

The main goal of the book is to review modern data pertaining to the taxonomy, distribution, and ecology of zoonotic viruses in the ecosystems of Northern Eurasia. Included in the chapters that follow are the results of long-term investigations into biosafety and biodiversity in the different climatic belts (from subarctic to subtropical) that affect an area of more than 15 million km^2.

Climatic changes; increasing population density of arthropod vectors and vertebrate hosts; the development of virgin lands; the transference of viruses by birds, bats, and domestic animals to humans (by chance or through criminal action); and vectors allowing populations of viruses to adapt to a new environment, sometimes including humans, can lead to emerging or reemerging infections. The book presents data about the circulation, distribution, and evolution of more than 80 viruses, including the viruses that cause influenza, tick-borne encephalitis, and West Nile, Omsk, and Crimean—Congo hemorrhagic fevers; the hantaviruses; the Sindbis virus; viruses of the California encephalitis group; and other pathogenic viruses, as well as novel viruses classified for the first time with the use of the next-generation sequencing approach.

Antigenic diversity, the wide global distribution of viruses, the scale of epidemic or epizootic outbreaks, sometimes with a high mortality, and the absence of a specific treatment (and, often, a specific prophylaxis) for viral diseases together render the zoonotic virus problem significant for human and animal health all over the world.

The majority of zoonotic viruses circulate in nature independently of humans, but the appearance of large nonimmune populations of humans endemic to certain unexplored territories often results in wide epidemics of viral infections transmitted by bloodsucking vectors with natural foci. In this situation, an important biological principle becomes

Zoonotic Viruses of Northern Eurasia.
DOI: http://dx.doi.org/10.1016/B978-0-12-801742-5.00001-5

apparent: The more uncommon the pathogen is to the area, the more danger it presents. This danger often appears during intensive economic development of a previously uninhabited territory. On the one hand, knowledge of the basic laws of viral distribution is necessary for the purposeful exploration of such a territory and the exploitation of its natural resources without exposing those involved to dangerous diseases. On the other hand, the circulation intensity of some viruses could increase as the result of unreasoned or uncontrolled human activity. For example, irrigation and extended rice sowing could lead to rapid increases in populations of mosquito vectors and an intensification of epizootic processes involving domestic animals. The adoption of prophylactic measures could minimize such an undesirable succession of events.

The ability of many viruses to reproduce in both vertebrate hosts and arthropod vectors, depending on their circulation with regard to different biotic and abiotic factors in the environment, makes viruses a universal model for ecological studies. That same ability calls forth the total methodological arsenal of modern biology. Ecological investigations into levels of virus and host populations are needed in molecular, mathematical, medicogeographic, virological, epidemiological, and parasitological approaches. In addition, virus ecology investigations carried out in laboratories today involve genetic, biochemical, and electron microscopy methods as a matter of course.

The study of the ecology of viruses deals with relationships between pathogens and potential hosts on the population level and with complex factors on the environmental level. Because the human body is also an environment for viruses, its "ecology" is often discussed as epidemiology. Some epidemiological problems have significant practical importance: drift routes of the virus, transmission mechanisms, preservation during interepidemiological periods, etc. Every year, the growth

of knowledge leads to a decreasing number of strict anthroponoses. Practically all human viruses have their analogues, or just homologues, circulating among animal ones. That is why the investigation of extrahuman pathways of virus circulation has obvious epidemiological significance. Another direction for theoretical investigations is studying chronic and latent forms of virus infections, both of which could promote virus preservation during interepidemic periods. The evolution of viral gene pools is of extreme importance as the basis of the ecological plasticity of viruses and in forecasting the emergence or reemergence of dangerous viruses.

To simplify the presentation that follows, we state some common ecological definitions:

- *Ecosystem (biogeocenosis)*. A stable natural system with closed-cycle circulation of substances and energy between its biotic and abiotic components.
- *Biome (biocenosis)*. A set of populations of organisms in a given territory.
- *Species*. A set of populations of individuals that are similar in their morphological and physiological features, have a common origin, and interbreed under natural conditions.
- *Population*. A set of individuals of a given species living together in the same territory.
- *Genetic population*. A population occupying a limited territory.
- *Ecological niche*. The place of a species in an ecosystem.
- *Ecological plasticity*. The level of adaptability of species to changes in environmental factors.

This book presents the main results of long-term investigations into ensuring biosafety and biodiversity in different ecosystems of Northern Eurasia. The book describes the main ecosystems of Northern Eurasia in the context of the ecology of viruses, with an emphasis on the environmental factors necessary for the

TABLE 1.1 Zoonotic Viruses of Northern Eurasia

	Taxonomy					Ecology								Pathogenicity		
							Biom					Distribution			Animals	
							Shelter			Nesting of birds						
Order (if introduced)/ genome type	Family	Genus	Group	Virus	GenBank ID	Ecosystem/ vector	Pasture	Bats	Rodents	High latitudes	Temperate, subtropic belts	Northern Eurasia	Outside of Northern Eurasia	Human	Wild	Domestic
1	2	3	4	5	6	7	8	9	10	11	12	13	14	15	16	17
Mononegavirales / ssRNA (–)	Paramyxoviridae (Paramyxovirinae)	Avulavirus	NDV	Newcastle diseases virus (NDV)	AF077761	Wild and domestic birds	–	–	–	+	+	Everywhere	Worldwide	–	–	+
		Morbillivirus	PDV	Phocine distemper virus (PDV)	LY409630	Seals (Phocidae)	–	–	–	–	–	Baikal lake	North Europe	–	+	+
	Rhabdoviridae	Ephemerovirus	BEFV	Bovine ephemeral fever virus (BEFV)	AF234533	Cattle, buffalo/ Ceratopogoniidae	+	–	–	–	–	Central Asia (Turkmenia)	South Asia, Africa, Australia	–	–	+
		Lyssavirus	RABV	Rabies virus (RABV)	M13215	Wild and domestic Carnivora	–	–	–	–	–	Everywhere	Worldwide	+	+	+
			ARAV	Aravan virus (ARAV)	EF614259	Bats populations in human habitats	–	+	–	–	–	Central Asia (Kirgizia)	Unknown	Unknown	Unknown	Unknown
			EBLV-1	European bat lyssavirus 1 (EBLV-1)	EF157976	Bats populations in human habitats, caves	–	+	–	–	–	Central Asia (Kirgizia)	Europe	+	Unknown	Unknown
			IRKV	Irkut virus (IRKV)	EF614260	Bats populations in human habitats, caves	–	+	–	–	–	North-Eastern Siberia	Unknown	Unknown	Unknown	Unknown
				Ozernoe virus (OZERV)	FJ905105	Bats	–	+	–	–	–	South of the Far East	Unknown	+	Unknown	Unknown
			KHUV	Khujand virus (KHUV)	EF614261	Bats populations in human habitats, caves	–	+	–	–	–	Central Asia (Tajikistan)	Unknown	Unknown	Unknown	Unknown
			WCBV	West Caucasian bat virus (WCBV)	EF614258	Bats populations in human habitats, caves	–	+	–	–	–	North-Western Caucasus (Russia), Georgia	North-Eastern Iran	Unknown	Unknown	Unknown
		Novirhabdovirus	IHNV	Infectious hematopoietic necrosis virus (IHNV)	M16023	Aquaculture	–	–	–	–	–	Far East, Karelia	Europe, North America (Pacific coast)	–	+	+

(Continued)

TABLE 1.1 (Continued)

Order (if introduced)/ genome type	Taxonomy					Ecology									Pathogenecity		
	Family	Genus	Group	Virus	GenBank ID	Ecosystem/ vector	Biom					Distribution		Human	Animals		
							Shelter				Nesting of birds						
							Pasture	Bats	Rodents	High latitudes	Temperate, subtropic belts	Northern Eurasia	Outside of Northern Eurasia		Wild	Domestic
1	2	3	4	5	6	7	8	9	10	11	12	13	14	15	16	17
Mononegavirales / ssRNA (−)			VHSV	Viral hemorrhagic septicemia virus (VHSV)	EU481506	Aquaculture	−	−	−	−	−	Far East, Karelia	Europe, North America	−	+	+
		Vesiculovirus	SVCV	Spring viraemia of carp virus (SVCV)	AY527273	Aquaculture	−	−	−	−	−	Belarus, Ukraine, Moldova, Russia	Europe	−	+	+
Picornavirales / ssRNA (+)	Picornaviridae	Cardiovirus	Theilovirus (ThV)	Syr-Darya valley fever virus (SDVFV)	KJ191558	Desert river valleys/ Ixodidae	+	−	−	−	−	Kazakhstan, Turkmenistan	Unknown	+	Unknown	Unknown
				Sikhote-Alin virus (SAV)	In process	Taiga forest/Ixodidae	+	−	−	−	−	South of the Far East	Unknown	Unknown	Unknown	Unknown
dsRNA	Birnaviridae	Aquabirnavirus	IPNV	Infectious pancreatic necrosis virus (IPNV)	AF078668	Aquaculture of Salmon sp.	−	−	−	−	−	Baltic states	USA, Canada, Norway, Japan	−	+	+
	Reoviridae	Orbivirus	Kemerovo (KEMV)	Aniva virus (ANIV)	KJ191541-7	Subarctic sea costs/ Ixodidae	−	−	−	+	−	Okhotsk Sea basin	Unknown	−	−	−
				Kemerovo virus (KEMV)	KC288130-9	Taiga forest/Ixodidae	+	−	−	−	−	Western Siberia	Central part of Eastern Europe	+	+	+
				Okhotskiy virus (OKHV)	KF981623-32	Subarctic sea costs/ Ixodidae	−	−	−	+	−	Okhotsk, Bering, Barents Sea basins	Unknown	−	−	−
			BAKV	Baku virus (BAKV)	KJ191548-57	Subtropic sea costs/ Argasidae	−	−	−	+	−	Caspian Sea basin	Unknown	−	−	−
				Chenuda virus (CHUV)	KM311318-26	Arid Subtropic/ Argasidae, Mosquitoes	+	−	−	−	−	Central Asia	Egypt, SAR	−	−	−
			WMV	Wad-Medani virus (WMV)	KJ425426-35	Cattle, camels, sheep / Ixodidae	+	−	−	−	+	Central Asia (Turkmenistan, Kazakhstan)	Africa (Sudan, Egypt, Senegal), South Asia (India, Pakistan, Iran), Caribbean See basin (Jamaica)	Unknown	Unknown	Unknown

ssRNA (+/−) (ambisense) — Bunyaviridae — Hantavirus	Virus											
ARTV	Altai virus (ALTV)	EU424341	Insectivores habitats (*Sorex araneus*)	+	—	—	—	Altai (Western Siberia)	Unknown	Unknown	Unknown	Unknown
	Amur/Soochong virus (AMRV)	DQ056292 AY675353 AY675349	Rodents habitats (*Apodemus peninsulae*)	+	—	—	—	South of Far East	Unknown	Unknown	Unknown	Unknown
	Artybash virus (ARTV)	EU424339 EU424340	Insectivores habitats (*Sorex* spp.)	+	—	—	—	Altai (Western Siberia)	Unknown	Unknown	Unknown	Unknown
DOBV	Dobrava virus (DOBV)	L33685 L41916	Rodents habitats (*Apodemus agrarius, A. ponticus,* etc.)	+	—	—	—	Central Russia, Western Siberia	Europe	+	+	—
HTNV 1	Hantaan virus (HTNV)	X55901 M14627 M14626	Rodents habitats (*Apodemus agrarius*)	+	—	—	—	Far East	South Korea, China	+	—	—
	Khabarovsk virus (KHABV)	AJ011648 U35255	Rodents habitats (*Microtus fortis, Lemmus sibiricus*)	+	—	—	—	Far East	Unknown	Unknown	Unknown	Unknown
	Lena river virus (LNAV)	Unknown	Insectivores habitats (*Sorex calcutiens*)	+	—	—	—	Eastern Siberia	—	—	—	—
PUUV 2	Puumala virus (PUUV)	Z66548 X61034 X61035	Rodents habitats (*Clethrinomys glareolus*)	+	—	—	—	Central Russia, Western Siberia, Far East	Europe	+	—	—
	Ufa virus (UFAV)	AB297665–7	Rodents habitats (*Clethrinomys glareolus*)	+	—	—	—	Bashkiria	—	—	—	—
SEOV	Seoul virus (SEOV)	X56492 S47716 NC_005236	Rodents habitats (*Rattus norvegicus*)	+	—	—	—	Far East	South Korea, China	+	—	—
	Topografov virus (TOPV)	AJ011649 AJ011647 AJ011646	Rodents habitats (*Lemmus sibiricus*)	+	—	—	—	Eastern Siberia	Unknown	Unknown	—	
	Tula virus (TULV)	J005637 Z69993 ZC9931	Rodents habitats (*Microtus arvalis, M. rossiaemeridionales*)	+	—	—	—	Central Russia	—	—	—	—
	Vladivostok virus (VLAV)	AB011630	Rodents habitats (*Microtus fortis*)	+	—	—	—	Far East	—	—	—	—

(Continued)

TABLE 1.1 (Continued)

Taxonomy					**Ecology**									**Pathogenecity**		
							Biom					**Distribution**			**Animals**	
							Shelter			**Nesting of birds**							
Order (if introduced)/ genome type	Family	Genus	Group	Virus	GenBank ID	Ecosystem/ vector	Pasture	Bats	Rodents	High latitudes	Temperate, subtropic belts	Northern Eurasia	Outside of Northern Eurasia	Human	Wild	Domestic
1	**2**	**3**	**4**	**5**	**6**	**7**	**8**	**9**	**10**	**11**	**12**	**13**	**14**	**15**	**16**	**17**
ssRNA (+/−) (ambisense)		*Nairovirus*	CCHFV	Crimean-Congo hemorrhagic fever virus (CCHFV)	NC005300-2	Dry step, semi-desert, desert/*Ixodidae*	+	−	−	−	−	South Russia, Central Asia	Africa, South Europe, South Asia	+	−	−
			ARTV	Artashat virus (ARTSV)	KF7801650	Dry subtropics/ *Argasidae*	−	−	+	−	−	Armenia, Azerbaijan	−	−	−	−
			Hughes (HUGV)	Caspiy virus (CASV)	KF801656	Sea cost in dry subtropics	−	−	−	−	+	Caspian Sea basin	−	−	−	−
			Qalyub (QYBV)	Chim virus (CHIMV)	KF801656	Desert in dry subtropics/*Argasidae, Ixodidae*	−	−	−	−	−	Central Asia, Uzbekistan, Kazakhstan	−	−	−	−
				Geran virus (GRNV)	KF801649	Desert in dry subtropics/*Argasidae*	−	−	+	−	−	Transcaucasia	Unknown	−	−	−
			ISKV	Issyk-Kul virus (ISKV)	KF7801652 KF892055-7	Human habitats in dry subtropics/ *Argasidae, Ixodidae,* additional—mosquitoes	−	+	−	−	−	Central Asia	Malaysia	+	−	−
				Uzun-Agach (UZAV)	In process	Human habitats in dry subtropics/ *Argasidae*	−	+	−	−	−	Kazakhstan	Unknown	−	−	−
			SAKV	Sakhalin virus (SAKV)	KF801659	Sea costs subarctics/ *Ixodidae*	−	−	−	+	−	Okhotsk, Bering, Barents Sea basins	South Ocean basin (Australia)	−	−	−
				Paramushir virus (PMRV)	KF801656	Sea costs subarctics/ *Ixodidae*	−	−	−	+	−	Okhotsk Sea basin	Unknown	−	−	−
			TAMV	Tamdy virus (TAMV)	KF801653-5	Desert in dry subtropics/*Ixodidae*	+	−	−	−	−	Central Asia, (Uzbekistan, Turkmenia, Kirgizia, Kazakhstan, Tadzhikistan, Armenia, Azerbaijan)	−	+	−	−
				Burana virus (BURV)	KF801651	Low-mountain pasture of cattle/ *Ixodidae*	+	−	−	−	−	Central Asia (Kirgizia)	Unknown	−	−	−

Genus	Group	Virus (abbreviation)	GenBank accession	Habitat/vector					Distribution			
Orthobunyavirus	Bunyamwera (BUNV)	Batai virus (BATV)	JX846604–6	Forest steppe, steppe, temperate, subtropic belts, human, cattle habitats/mosquitoes	+	–	–	–	Southern Eurasia	+	+	+
		Anadyr virus (ANADV)	KM496335–37	Forest-tundra, tundra, human habitats/mosquitoes	+	–	–	–	North of the Far East	–	–	–
	California encephalitis virus (CEV)	Inkoo virus (INKV)	U47137 U88059 EU7895732	Forests/mosquitoes	+	–	–	–	North, Central USA, South Canada, North and Central Europe, China	+	+	+
		Khatanga virus (KHTV)	HQ734817–25	Forests/mosquitoes	+	–	–	–		+	+	+
		Tahyna virus (TAHV)	HM036208–10	Forests/mosquitoes	+	–	–	–		+	+	+
	KHURV	Khurdun virus (KHURV)	KF981633 KF981634 KF981635–8	Estuary of Volga river	–	–	–	+	South of European part of Russia	–	–	–
Phlebovirus	BHAV	Bhanja virus (BHAV)	JX961616–8	Cattle habitats/Ixodidae	+	–	–	–	European part of Russia, Caucasus	+	+	+
		Razdan virus (RAZV) (var. BHAV)	KC335496–8	Cattle habitats/Ixodidae	+	–	–	–	Transcaucasia (Armenia, Azerbaijan)	Unknown	Unknown	Unknown
	UUKV	Uukuniemi virus (UUKV)	D10759 M13417 M33551	Forests/Ixodidae additional—mosquitoes	+	–	–	+	Fennoscandia, Central Russia Transcaucasia (Azerbaijan)	+	Unknown	Unknown
		Zaliv Terpenia virus (ZTV)	KF89040–42 KF89243–5 KF767463–5 KF767460–2	Subarctic sea costs, subtropics/Ixodidae, additional—mosquitoes	–	–	–	+	Bering, Okhotsk, Barents Sea basins, Transcaucasia (Azerbaijan)	Unknown	Unknown	Unknown
		Komandory virus (KOMV)	KF892049–51	Subarctic sea cost/Ixodidae	–	+	–	–	Bering see basin	Unknown	Unknown	Unknown
		Khasan virus (KHAV)	KF892046–8	Forests/Ixodidae	+	–	–	–	S. Far East	Unknown	Unknown	Unknown
			KJ425423–5		–	–	–	+	France	Unknown	Unknown	Unknown

(Continued)

TABLE 1.1 (Continued)

Order (if introduced)/ genome type	Family	Genus	Group	Virus	GenBank ID	Ecosystem/ vector	Pasture	Bats	Rodents	High latitudes	Temperate, subtropic belts	Northern Eurasia	Outside of Northern Eurasia	Human	Wild	Domestic
Taxonomy						**Ecology**	**Biom** — Shelter			**Nesting of birds**		**Distribution**		**Pathogenicity**	**Animals**	
1	2	3	4	5	6	7	8	9	10	11	12	13	14	15	16	17
ssRNA (+/−) (ambisense)				Gissar virus (GISV)		Dry subtropics, human habitats (dovecotes)/ Argasidae	+	−	−	−	−	Central Asia (Tajikistan)	−			
			SFNV	Sandfly fever Naples virus (SFNV) and	HM566170–2	Semi-deserts, deserts/ Sandflies (Phlebotominae)	+	−	−	−	−	Central Asia, (Turkmenistan, South European part)	South Eurasia, Africa	+	−	−
				Sandfly fever Sicilian virus (SFSV)	NC_015411–3	Semi-deserts, deserts/ Sandflies (Phlebotominae)	+	−	−	−	−	Central Asia, (Turkmenistan, South European part)	South Eurasia, Africa	+	−	−
ssRNA (−)	Orthomyxoviridae	Influenzavirus A	Influenza A virus	H1 N[1, 3, 6]		Mostly habitats of aquatic birds	−	−	−	−	+	Everywhere	Worldwide	+	+	+
				H2 N[2, 3]		Mostly habitats of aquatic birds	−	−	−	−	+	Everywhere	Worldwide	+	+	+
				H3 N[1, 4, 6, 8]		Mostly habitats of aquatic birds	−	−	−	−	+	Everywhere	Worldwide	+	+	+
				H4 N[1, 4, 6, 8, 9]		Mostly habitats of aquatic birds	−	−	−	−	+	Everywhere	Worldwide	−	+	+
				H5 N[1, 2, 3]		Mostly habitats of aquatic birds	−	−	−	−	+	Everywhere	Worldwide	+	+	+
				H6 N[2, 4, 8]		Mostly habitats of aquatic birds	−	−	−	−	+	Everywhere	Worldwide	−	+	+
				H7 N[1, 3, 7, 8]		Mostly habitats of aquatic birds	−	−	−	−	+	Everywhere	Worldwide	+	+	+
				H8 N4		Mostly habitats of aquatic birds	−	−	−	−	+	Everywhere	Worldwide	−	+	+
				H9 N[2, 4]		Mostly habitats of aquatic birds	−	−	−	−	+	Everywhere	Worldwide	−	+	+
				H10 N[4, 5, 8]		Mostly habitats of aquatic birds	−	−	−	−	+	Everywhere	Worldwide	−	+	+
				H11 N[2, 6, 8, 9]		Mostly habitats of aquatic birds	−	−	−	−	+	Everywhere	Worldwide	−	+	+

Genome	Family	Genus	Species	Virus	GenBank			Habitat		Distribution (Russia)	Distribution (worldwide)			
				H12 N(2)		−	+	Mostly habitats of aquatic birds	−	Everywhere	Worldwide	−	+	+
				H13 N(1, 2, 3, 6, 8)		−	+	Mostly habitats of aquatic birds	−	Everywhere	Worldwide	−	+	+
				H14 N(5, 6)		−	+	Mostly habitats of aquatic birds	−	Everywhere	Worldwide	−	+	+
				H16		−	+	Mostly habitats of aquatic birds	−	Everywhere	Worldwide	−	+	+
		Quaranjivirus	Quaranfil (QRFV)	Tyulek virus (TLKV)	KJ438647–8	−	+	River lowlands in subtropics in burrow of birds/*Argasidae*	−	Central Asia, Kirgizia	Unknown	+	Unknown	Unknown
		Thogotovirus	DHOV	Dhori virus (DHOV)	GU962308–13	−	−	Semi-deserts, dry subtropics/*Ixodidae*	+	South European part of Russia, Transcaucasia	India, Egypt, Portugal	+	Unknown	Unknown
				Batken virus (BKNV) (var. DHOV)	KJ396672–4	−	−	Desert in dry subtropic/*Ixodidae*, additional—mosquitoes	+	Central Asia (Kirgizia)	Unknown	+	Unknown	Unknown
ssRNA (+)	*Togaviridae*	*Alphavirus*	Semliki Forest virus (SFV)	Chikungunya virus (CHIKV)	AY726732 KF872195	−	−	Imported cases	−	Moscow infectious hospital	Africa, South-Eastern Asia, Oceania	+	−	−
				Getah virus (GETV)	EF631998	−	−	Tundra, taiga, steppe/mosquitoes	+	Eastern Siberia, Far East	Mongolia, South-Eastern Asia, Australia	−	−	+
			SINV	Sindbis virus (SINV)	J02363	−	+	Human habitat (poultry) habitat of wild birds	−	South European part of Russia, Western Siberia	Africa, South Europe, South-Eastern Asia, Australia, Oceania	+	+	+
				Karelian fever virus (KFV) (var. SINV)	M69205	−	−	Middle taiga forests	−	Fennoscandia (Karelia)	Finland, Norway, Sweden	+	−	−
				Kyzylagach virus (KYZV) (var. SINV)	U946605 AF339478	−	+	Birds colonies in humid subtropics	+	Azerbaijan	Unknown	Unknown	Unknown	Unknown

(Continued)

TABLE 1.1 (Continued)

Order (if introduced)/genome type	Family	Genus	Group	Virus	GenBank ID	Ecosystem/vector	Biom: Pasture	Shelter: Bats	Shelter: Rodents	Nesting of birds: High latitudes	Nesting of birds: Temperate, subtropic belts	Distribution: Northern Eurasia	Distribution: Outside of Northern Eurasia	Pathogenicity: Human	Animals: Wild	Animals: Domestic
1	2	3	4	5	6	7	8	9	10	11	12	13	14	15	16	17
ssRNA (+)	Flaviviridae	Flavivirus	TBEV	Tick-borne encephalitis virus (TBEV):												
				European TBEV subtype (TBEV-Eur)	M27153 M33668	Forests/Ixodidae	+	−	−	−	−	European part of Russia	Europe	+	+	+
				Siberian TBEV subtype (TBEV-Sib)	L40361	Forests/Ixodidae	+	−	−	−	−	Ural, Western and Eastern Siberia	Mongolia	+	+	+
				Far Eastern TBEV subtype (TBEV-FE)	X07755	Forests/Ixodidae	+	−	−	−	−	Far East	Mongolia, Korea, China	+	+	+
				Malyshevo TBEV subtype (TBEV-Mal)	KJ744034	Lake valleys/mosquitoes	+	−	−	−	−	South of the Far East	Unknown	+	+	+
				Powassan virus (POWV)	L06436	Forests/Ixodidae, additional—mosquitoes	+	−	−	−	−	South of the Far East	Canada, North and Central USA	+	+	+
				Alma-Arasan virus (AAV)	KJ744033	Forest/Ixodidae	+	−	−	−	−	Kazakhstan	Unknown	+	Unknown	Unknown
				Omsk hemorrhagic fever virus (OHFV)	AY323489	Forest-steppe lowlands/Ixodidae, additional—mosquitoes	+	−	−	−	−	Western Siberia	Unknown	+	+	+
			JEV	Japanese encephalitis virus (JEV)	M18370	Forest lowlands/mosquitoes	+	−	−	−	−	South of the Far East	South-Eastern Asia	+	+	+
				West Nile virus (WNV)	M12294	Lowlands/mosquitoes, additional—Ixodidae	+	−	−	−	−	South European part of Russia	Africa, Southern Eurasia, Australia. North, Central and South America	+	+	+

			TYUV	Tyuleniy virus (TYUV)	DQ233748 KF815939	Sea costs in Subarctics	–	–	+	–	Bering, Okhotsk, Barents Sea basins	North America, South Ocean	+	+	–
				Kama (KAMV)	KF815940	Lake cost in Russian plain	–	–	–	+	Central Russia (Tatarstan)	Unknown	Unknown	Unknown	Unknown
			DENV	Dengue virus (DENV): DENV-1, DENV-2, DENV-3, DENV-4	U88536, M19197, M93130, AF320573	Imported cases	–	–	–	–	Moscow infectious hospital	Equatorial, Subequatorial, Tropical, Subtropical belts	+	–	–
				Entebbe bat virus (ENTV) Sokuluk virus (SOKV)	KP981619–22	Human habitat (bats) in subtropics/ *Argasidae*, additional – *Ixodidae*	+	–	–	–	Central Asia (Kirgizia)	Unknown	Unknown	Unknown	Unknown
dsDNA	*Asfarviridae*	*Asfivirus*	ASFV	African swine fever virus (ASFV)	AM712239	Human habitats (pigsty), habitats of wild boars	+	–	–	–	European part of Russia, Georgia, Armenia	Africa, Europe, Latin America	–	+	+
	Poxviridae	*Orthopoxvirus*	Cowpox	Murman virus (MURMV)	Unknown	Tundra	+	+	–	–	Kola peninsula	Unknown	–	–	–

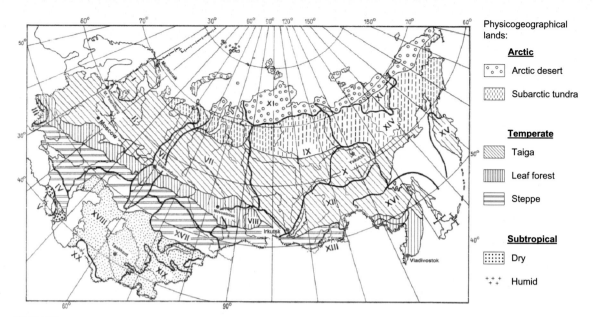

FIGURE 1.1 Physicogeographical zoning of Northern Eurasia (in the limits of the former USSR). Physicogeographical lands: I – Fennoscandia; II – Russian plain; III – Carpathians; IV – Crimean-Caucasian; V – Armenian Highland, VI – Ural, VII – West Siberian Lowland; VIII – Altai-Sayan;, IX – Middle Siberia; X – Yakutsk Hollow; XI – Northern Siberia; XII – Baikal, XIII – Mongolia-Dauria, XIV – North-Eastern Siberia, XV – Kamchatka-Kurils; XVI – Amur-Primorje, XVII – Eastern Kazakhstan; XVIII – Turan Lowland, XIX – Middle Asia; XX – Turkmen-Khurasan.

FIGURE 1.2 Federal districts of Russia (as of January 01, 2014).

TABLE 1.2   Federal districts of Russia (as of January 01, 2014)

N	Color on Figure 1.2	Name	Square (km²)	Population	Number of Federal subjects	Administrative center
I		Central	652,800	38,819,874	18	Moscow
II		South	416,840	13,963,874	6	Rostov-on-Don
III		North-Western	1,677,900	13,800,658	11	St. Petersburg
IV		Far Eastern	6,215,900	6,226,640	9	Khabarovsk
V		Siberian	5,114,800	19,292,740	12	Novosibirsk
VI		Ural	1,788,900	12,234,224	6	Yekaterinburg
VII		Volga	1,038,000	29,738,836	14	Nizhny Novgorod
VIII		North-Caucasian	172,360	9,590,085	7	Pyatigorsk

circulation of the viruses and the potential impact of climate change on the distribution of viruses. The book also summarizes data on the circulation of about 80 viruses from the *Paramyxoviridae*, *Rhabdoviridae*, *Picornaviridae*, *Birnaviridae*, *Reoviridae*, *Bunyaviridae*, *Flaviviridae*, *Orthomyxoviridae*, *Togaviridae*, and *Asfarviridae* families that were isolated in natural foci of Northern Eurasia. Discussed is their taxonomy, ecology, distribution, and pathogenicity for humans and animals, as well as the penetration of viruses outside of their usual areas, caused by migrating birds, bats, or human activity (Table 1.1). The sections on the systematization

of data on the taxonomy, ecology, and geographical distribution of viruses; their significance as regards human and animal pathology; and the role of landscape conditions and arthropod vectors for virus circulation may be used as a manual on those subjects. The book is addressed to human health, veterinary health, and environmental field virologists; infectious disease specialists; ecologists; parasitologists; epidemiologists; and graduate and postgraduate students.

To help readers, we list the physicogeographical zoning (Figure 1.1) and federal districts (Figure 1.2, Table 1.2) of Russia, which takes up the major part of Northern Eurasia.

The Development of Virology: The History of Emerging Viruses

The problem of emerging and reemerging infections has a high priority in human affairs because such infections are considered unpredictable and sometimes lead to epidemics through either natural factors or criminal actions. These infections know no borders and constitute a threat to national and global safety. They are like a dormant volcano, and we know very little about the pathogenic potential of the agents that carry them. The situation has not changed in the last 10,000 years, since the formation of human society.[1]

The development of virology as one of the natural sciences occurred in the last decade of the nineteenth century, and the leading role in that event belongs to three scientists who worked on the tobacco mosaic virus (family Virgaviridae, genus *Tobamovirus*): the German chemist Adolf Mayer (Figure 2.1), the Russian botanist Dmitry Ivanovsky (Figure 2.2), and the Dutch microbiologist Martinus Beijerinck (Figure 2.3).

Adolf Mayer named the tobacco disease and established its infectious nature.[2] Dmitry Ivanovsky, whose five-year studies began in 1887, first showed in 1892 that it was necessary to differentiate two diseases, one of which was of fungus etiology and the other unknown.[3,4]

Ivanovsky was 28 years old at the time. He demonstrated that the tobacco mosaic agent was neither a bacterium, nor an enzymes, nor a soluble substance, but rather was an "ultraparasite": a microorganism filterable through bacteria-proof filters, reproducing infection in healthy plants *in vivo*, and incapable of growing on artificial media.[4] This event was a decisive step in the birth of the biological science of virology.[1]

Fifty years later, in 1944, Wendell Stanley, writing in *Science*, stated, "there is considerable justification for regarding Ivanovsky as the father of the new science of virology."[5] He also recognized Ivanovsky's leading role in virus crystallization:[6] "Ivanovsky's right for the fame is growing year after year. I believe that his consideration for the viruses should be considered in the same light as the names of Pasteur and Koch in microbiology."[5] In recognition of Ivanovsky's services, the D.I. Ivanovsky Institute of Virology (IIV) in Moscow, established in 1944 as a part of the USSR Academy of Medical Sciences, was named after him in 1950. A nominal prize was established to be given once every five years for the best work in general virology.

Ivanovsky's importance was recognized by M. Beijerinck.[7] At the same time, it is scarcely

17

FIGURE 2.1   Adolf Eduard Mayer (1843–1942).

necessary to exaggerate Beijerinck's contribution to the birth of virology. In his independent studies published six years after Ivanovsky's seminal papers, Beijerinck was the first who used the term "virus,"[8] although virological terminology was not completely determined until the 1930–1940s. The discipline of virology itself was not fully formed until the end of the nineteenth century, when the German scientist Friedrich Loeffler (Figure 2.4)[9] described the foot-and-mouth disease virus (Picornaviridae, *Aphthovirus*; the first filterable infectious agent of mammalian animals), and the very beginning of the twentieth century, when American military doctors under the guidance of Walter Reed (Figure 2.5)

FIGURE 2.2   Dmitry Iosifovich Ivanovsky (1864–1920).

FIGURE 2.4   Friedrich Loeffler (1852–1915).

FIGURE 2.3   Martinus Willem Beijerinck (1851–1931).

FIGURE 2.5   Walter Reed (1851–1902).

determined the etiology of yellow fever, linked to the yellow fever virus [10,11] (Flaviviridae, *Flavivirus*; the first arbovirus to be explored).

After 1917 and the Russian Revolution, only embers of Russian virology smoldered on the ashes of the war. Nevertheless, beginning at the end of the 1920s and continuing into the 1930s, little by little, Russian virology began to be restored from the ashes.

During the 1920s, Russia suffered serious smallpox epidemics (up to 200,000 cases per year). With the aim of mass immunization, smallpox vaccine was manufactured and a compulsory system of smallpox vaccinations was introduced that eliminated morbidity from the disease in 1936. In 1958, at the 11th Assembly of the World Health Organization (WHO), the famous Russian scientist Viktor Zhdanov (Figure 2.6) from the IIV, spoke in support of a proposal for global smallpox eradication. More than 1.5 million doses of vaccine were transferred free of charge by Russia in this unprecedented program, which was conducted under the guidance of Donald Henderson (Figure 2.7). Many Russian virologists—first and foremost, the well-known virologist Svetlana Marennikova (Figure 2.8)—participated in the realization of the program.

Ilya Mechnikov and Nikolay Gamaleya were the pioneers in rabies studies in Russia. Pasteur's antirabies station—the second in the world after Paris—was founded in 1886 in Odessa, and 636 such stations producing vaccines, making diagnoses, and presiding over vaccinations were located in Russia by 1936. The effort led to a considerable reduction in rabies mortality in the country.[12]

In 1937–1938, Russian virologists carried out complex studies in the Far East of the etiology of the tick-borne encephalitis virus (Flaviviridae, *Flavivirus*); later, they studied the Japanese encephalitis virus (Flaviviridae,

FIGURE 2.7 Donald Ainslie Henderson (1928).

FIGURE 2.6 Viktor Mikhailovich Zhdanov (1914–1987).

FIGURE 2.8 Svetlana Sergeevna Marennikova (1923).

FIGURE 2.9   Lev Alexandrovich Zilber (1894−1966).

FIGURE 2.11   Anatoly Alexandrovich Smorodintsev (1901−1986).

FIGURE 2.10   Mikhail Petrovich Chumakov (1909−1993).

FIGURE 2.12 Yevgeny Nikanorovich Pavlovsky (1884−1965), among professors and graduate students of the general biology and parasitology department of the Military Medical Academy (1955). Upper row: Yu.V. Chicherin, A.A. Gorovenko, D.K. Lvov, K.F. Dobrovolsky, A.A. Karyakin; middle row: B.N. Nikolaev, G.G. Smirnov, E.N. Pavlovsky, A.V. Gutsevich; Lower row: V.N. Motorin, A.S. Nedelko, N.N. Ushakov, V.I. Shut.

Flavivirus).[13] Lev Zilber (Figure 2.9), Mikhail Chumakov (Figure 2.10), Anatoly Smorodintsev (Figure 2.11), Yevgeny Pavlovsky (Figure 2.12; founder of the natural foci doctrine), Valentin Soloviev (Figure 2.13), Elisaveta Levkovich (Figure 2.14), Antonina Shubladze (Figure 2.15), and a number of other young scientists from that time formed a cohort of researchers that determined the development of virology in the country for many years. Afterward, Omsk hemorrhagic fever virus (Flaviviridae, *Flavivirus*), Crimean−Congo hemorrhagic fever virus (CCHFV) (Bunyaviridae, *Nairovirus*), and hemorrhagic fever with renal syndrome (Bunyaviridae,

Hantavirus) were explored under the guidance of Chumakov.[14]

Active investigations of arboviruses are continuing up to today with new viruses isolated, descriptions of new viral infections, and the establishment of a geographical distribution of natural foci and population interaction mechanisms. An approach called ecological sounding has revealed the basic circulation characteristics of arboviruses belonging to different ecological complexes in large territories ranging from the Arctic to the Subtropics.[15−18] The approach has uncovered reasons for the worsening of epidemiological situations linked to CCHFV (Bunyaviridae, *Nairovirus*) and West Nile virus (Flaviviridae, *Flavivirus*) in the south of the European part of Russia during 1999−2005.[19]

Studies of influenza and other viral respiratory diseases began in the 1930s and 1940s. Conducted under the guidance of Viktor Zhdanov (Figure 2.6), Anatoly Smorodintsev (Figure 2.11), Anna Gorbunova (Figure 2.16), and Ludmila Zakstelskaya (Figure 2.17), these investigations were aimed at isolating viral

FIGURE 2.13 Valentin Dmitrievich Soloviev (1907−1986).

FIGURE 2.14 Elisaveta Nikolaevna Levkovich (1900−1982).

FIGURE 2.16 Anna Sergeevna Gorbunova (1900−1981).

FIGURE 2.15 Antonina Konstantinovna Shubladze (1909−1993).

FIGURE 2.17 Ludmila Yakovlevna Zakstelskaya (1918−1996).

strains, investigating their biological properties, and elaborating on methods for diagnosis and prophylaxis. Much attention was given to drift and shift changes in antigenic structure and to factors of pathogenesis. The issue was discussed by Nicolai Kaverin (Figure 2.18), and Anatoly Smorodintsev and Viktor Zhdanov came to different theoretically based general conclusions in classic papers, although they had different points of view to explain the phenomena.[20–22] A number of useful collaborations on influenza viruses and other virological problems, especially arbovirology, were carried out with foreign colleagues—for example, Frank Fenner (Figure 2.19), Robert Webster (Figure 2.20), David Suarez (Figure 2.21), Hans-Dieter Klenck (Figure 2.22), Yoshiro Kawaoka (Figure 2.23), Friedrich Deinhardt (Figure 2.24), Charles Calisher (Figure 2.25), Jordi Casals (Figure 2.26), Robert Tesh (Figure 2.27), Barry Beaty, (Figure 2.28) and many others.

In 1978, again under the guidance of Viktor Zhdanov, a newly appearing pandemic influenza A (H1N1)—the so-called Russian flu, which circulated in China for half a year afterward—was speedily identified, and immediate information about it was sent to WHO and to world and regional influenza centers.[23] Broad international cooperation in the area of

FIGURE 2.18 Nicolai Veniaminovich Kaverin (1933–2014).

FIGURE 2.20 Robert Gordon Webster (1932).

FIGURE 2.19 Frank John Fenner (1914–2010).

FIGURE 2.21 David L. Suarez (1964).

FIGURE 2.22 Hans-Dieter Klenk (1938).

FIGURE 2.25 Charles H. Calisher (1936).

FIGURE 2.23 Yoshihiro Kawaoka (1955).

FIGURE 2.26 Jordi Casals-Ariet (1912–2004).

FIGURE 2.24 Friedrich Deinhardt (1926–1992).

FIGURE 2.27 Robert B. Tesh (1936).

FIGURE 2.28   Barry Beaty (1944).

FIGURE 2.29   Albert Bruce Sabin (1906–1993).

influenza monitoring did not stop even in the most frozen periods of the Cold War, when Soviet–Western connections were almost completely broken off.

Today, studies of influenza are developing in different applied and theoretical directions, basically in two centers in Russia: the IIV in Moscow and the Institute of Influenza in St. Petersburg. Much attention is being given to the ecology of influenza viruses and their monitoring in human and animal populations on the territory of Northern Eurasia. Fifteen hemagglutinin-type influenza strains were isolated from natural sources and deposited into the Russian State Collection of Viruses (SCV) at the IIV. This permanent work enabled researchers to predict a sharp intensification of H5 subtype circulation in Southeast Asia after 2001, forming new H5 variants,[24,25] and a further penetration of the highly pathogenic H5N1 variant into Northern Eurasia and then Western Europe, Southwestern Asia, and Africa.[26–28]

One of the last examples of an emerging–reemerging situation is the appearance of the new pandemic influenza A (H1N1) pdm09. Complex studies by epidemiologists, virologists, biochemists, and molecular biologists revealed the mechanism of appearance of mutant virus variants with amino acid substitutions in receptor-binding sites of hemagglutinin, changing receptor specificity and provoking severe lethal viral pneumonia.[29–32]

Global poliomyelitis epidemics in the 1950s led to active studies of the mass production of live polio vaccine, chiefly in the former USSR, in close cooperation with the American virologist Albert Sabin (Figure 2.29). Particularly valuable were the contributions by Mikhail Chumakov (Figure 2.10) and Anatoly Smorodintsev (Figure 2.11). Mikhail Chumakov was head of the IIV from 1950 to 1955 and of the then newly organized Institute of Poliomyelitis and Viral Encephalitis from 1955 to 1972. By the end of 1959, all children in Russia were vaccinated, and that mass vaccination led to a considerable decrease in morbidity and the eradication of the epidemic in 1960.[33] In 1988, WHO passed the Global Polio Eradication Initiative, which has resulted in a 99.9% reduction in the incidence of the disease. Poliomyelitis has not been seen in Russia since July 1, 2002, except for some imported cases.

The role of academicians Vasily Syurin (Figure 2.30) and Mikhail Gulyukin (Figure 2.31) looms large in the investigation of viral infections in animals. Syurin's monograph *Viral diseases of animals* has become a necessary handbook for Russian veterinary doctors.[34]

FIGURE 2.30   Vasily Nikolayevich Syurin (1915–2004).

FIGURE   2.32   Anatoly   Timofeevich   Kravchenko (1905–1976).

FIGURE 2.31   Mikhail Ivanovich Gulyukin (1944).

FIGURE   2.33   Pavel   Nikolayevich   Kosyakov (1905–1993).

The history of Russian virology is indissolubly linked with the IIV. A year before the end of the Second World War, the USSR Academy of Medical Sciences was founded in accordance with resolution N 797, 30.06.1944 of the Council of People's Commissars and the IIV was founded as a part of the Academy. The first director of the IIV was Anatoly Kravchenko (Figure 2.32), the famous specialist in biosafety, who led the Institute from 1944 to 1950. Mikhail Chumakov (Figure 2.10) was the director from 1950 to 1955, when he founded the Institute of Poliomyelitis and Viral Encephalitis (now called the Chumakov Institute of Poliomyelitis and Viral Encephalitides in his honor). During 1955–1960, the IIV was led by Pavel Kosyakov (Figure 2.33). Viktor Zhdanov (Figure 2.6) was the director from 1961 to 1986, a quarter century in which the Institute staff was enlarged twice, the molecular biology of viruses began to develop as a scientific discipline, and six WHO collaborating centers (for research into the problems of influenza, arboviruses, the ecology of viruses, herpes, viral hepatitis, and AIDS) began operation.

During its 10th International Congress in 1970, spurred by an initiative of Zhdanov,

FIGURE 2.34   Joseph Louis Melnick (1914—2001).

FIGURE 2.36   President of the Russian Federation Vladimir Vladimirovich Putin awards the State Prize of the Russian Federation to Dmitry Konstantinovich Lvov (1931), director of the D.I. Ivanovsky Institute of Virology.

FIGURE 2.35   Nils Christian Oker-Blom (1919—1995).

together with Joseph Melnick (Figure 2.34) (USA) and Nils Oker-Blom (Figure 2.35) (Finland), the International Union of Microbiological Societies (IUMS) decided to create three sections respectively covering the fields of bacteriology, virology, and mycology. Subsequently, these sections became three separate divisions of the IUMS, possessing complete autonomy in the conduct of their affairs and in the International Congresses of the IUMS. Viktor Zhdanov was president of the IUMS in 1970—1974. Dmitry Lvov (Figure 2.36), who was vice-director of the IIV for 19 years (1968—1986), was elected head of

the Institute in 1987 and completed his tenure as head in 2014. The Center for the Ecology of Rare and Dangerous Disease Agents was founded on the basis of Institute-led methodological and scientific organization work in the area of emerging and reemerging infections. The Center qualified virological staff all over the USSR during multiple seminars, conferences, and joint expeditions. A system for monitoring new infectious agents was developed, and the theoretical basis for such monitoring in different ecosystems was introduced into practice. The methodological approach included the collection of field material in the entire territory of Northern Eurasia. More than 15 million km^2 were investigated with the help of 12 meridian zones crossing arctic, tundra, taiga, leaf forest, steppe, semidesert, and desert climatic belts within 18 physicogeographic lands with unique ecosystems. About 80 viruses were isolated, many of which turned out to be new to science. The etiological role of isolated viruses was revealed. New emerging infections—Issyk-Kul, Syr-Darya valley, Karshi, Tamdy, and Karelian fevers—were discovered. The risk of an epidemiological outbreak of these diseases was evaluated, and, on

FIGURE 2.37   Petr Grigorievich Deryabin (1947).

FIGURE 2.39   Sergey Minovich Klimenko (1929).

FIGURE   2.38   Sophia   Yanovna   Gaidamovich (1921–2003).

FIGURE   2.40   Valentin   Lvovich   Gromashevsky (1934–2010).

the basis of knowledge gained, a prognosis was given regarding the risk of viral infections in Northern Eurasia.[15,17,18]

On the basis of all this history, the *Atlas of Distribution of Natural Foci Infections on the Territory of [the] Russian Federation* was published in 2001.[15] The following IIV scientists were awarded the State Prize of the Russian Federation: Dmitry Lvov (Figure 2.36), Petr Deryabin (Figure 2.37), Sophia Gaidamovich (Figure 2.38), Sergey Klimenko (Figure 2.39), Valentin Gromashevsky (Figure 2.40), Alexander Butenko (Figure 2.41), Sergei Lvov (Figure 2.42), and Ludmila Kolobukhina (Figure 2.43).

The main goal of the IIV is to investigate the evolution of emerging and reemerging viruses that are able to provoke dangerous epidemiological situations that threaten the biosafety of Russia. The Russian SCV operates in accordance with the mission of the IIV. The SCV uses the following methods to carry out its task: (i) clinical, epidemiological, and zoologico-parasitological methods, for field material collection during epidemic outbreaks; (ii) virological and serological methods, for diagnosis, isolation, and identification of viruses; and (iii) molecular–genetic methods (virtual- or real-time polymerase chain reaction testing; sequencing, including next-generation

FIGURE 2.41   Alexander Mikhailovich Butenko (1940).

FIGURE 2.42   Sergei Dmitrievich Lvov (1962).

FIGURE 2.43   Ludmila Vasilievna Kolobukhina (1951).

sequencing; and the use of biological microchips), for phylogenetic analysis and taxonomy studies of viruses. Results obtained are used to investigate biodiversity, deposit data into the SCV and GenBank® (the NIH genetic sequence database), construct modern diagnostic testing systems, develop and approve antiviral drugs, and select candidate strains of viruses for future vaccines.[27]

References

1. Lvov DK. Birth and development of virology the history of emerging−reemerging viral infection investigations. *Vopr Virusol* 2012;(Suppl. 1):4−19 [in Russian].
2. Mayer A. Die Mosaikkrankheit der Tabakspflanze. In: *Landivertschaftlichen Versuchs-Stationen*, vol. 32; 1886. p. 451−567.
3. Ivanovsky DI. On two diseases of tobacco plants. *Agric Forestry* 1892;**CCIXX**:104−21 [in Russian].
4. Ivanovsky DI. Uber die Mosaikkranicheit der Tobakhflanze. *Bull Acad Imp Sci St Petersburg* 1892;**35**: 67−70.
5. Stanly W. Soviet studies of viruses. *Science* 1944;**99**: 137−9.
6. Stanley W. Some chemical, medical and philosophical aspects of viruses. *Science* 1941;**93**:143−5.
7. Beijerinck MW. Bemerkungen zu dem Aufsatz von Herrn Ivanovsky über die Mosaikkrankheit der Tabakpflanze. *Zbl Bakt Parasit Kund Abteilung* 1899;**28**:127−31 [in German].
8. Beijerinck MW. Über ein contagiun vivum fluidum als Ursache der Fleckenkrankheit der Tabaksblätter. In: *Verhandelungen Koninkl. Akademie van Wetenschappen te Amsterdam*; 1898. Pt. 11. Sect. D. No 6. S. 3−21 [in German].
9. Loeffler F. Frosch: Summarischer bericht über die Ergebnisse der Untersuchungen der Kommission zur Erforschung der mau und Klauenseuche bei dem Institute fur Infektionskrankheiten in Berlin. *Zbl Bakt Parasit Infekt* 1897;**1**(22):257−9.
10. Reed W, Carroll J, Agramonte A, Lazear J. Expedition report. In: *USA Senate Documents* 1901, vol. 66. p. 156−72.
11. Reed W, Carroll J. The etiology of yellow fever. *Ameyer Med* 1902;**3**:301−5.
12. Selimov MA. *Rabies*. Moscow: Meditsina;1976344: [in Russian].
13. Zilber LA, Levkovich EN, Shubladze AK, et al. Etiology of spring-summer epidemic encephalitis. *Archiv Biologicheskikh Nauk* 1938;**52**:162−83 [in Russian].

14. Chumakov M. *Viral hemorrhagic fevers.* Moscow: Meditsina;1979190: [in Russian].

15. Lvov DK, Deryabin PG, Aristova VA, et al. *Atlas of distribution of natural foci virus infections on the territory of [the] Russian Federation.* Moscow: SMC MPH RF Publ;2001192 [in Russian].

16. Lvov DK, Klimenko SM, Gaidamovich SYa. *Arboviruses and arbovirus infections.* Moscow: Meditsina;1989336 [in Russian].

17. *Organization of ecological–epidemiological monitoring in Russian Federation for anti-epidemic defense of the civilians and army.* In: Lvov DK, editor. Moscow: Russian Ministry of Public Health, Federal Office of Biomedical and Extreme Problems, D.I. Ivanovsky Institute of Virology; 1993. p. 128 [in Russian].

18. Shchelkanov MY, Gromashevsky VL, Lvov DK. The role of ecologo-virological zoning in prediction of the influence of climatic changes on arbovirus habitats. *Vestn Ross Akad Med Nauk* 2006;(2):22–5 [in Russian].

19. Lvov DK, Butenko AM, Gromashevsky VL, et al. West Nile and other zoonotic viruses as examples of emerging–reemerging situations in Russia. *Arch Virol* 2004;(Suppl. 18):85–96.

20. Smorodintsev AA, Luzianina TY, Aleksandrova GI, Taros L. Basis of the anthroponotic nature of human pandemic influenza A viruses. *Vopr Virusol* 1981;(2):250–4 [in Russian].

21. Zhdanov VM. Why is it so difficult to defeat influenza? In: Zhdanov VM, editor. *Soviet medical reviews. Section E. Virology reviews*, vol. 2. London: Harwood Academic Publisher GmbH;1987. p. 1–14.

22. Zhdanov VM, Lvov DK. *Evolution of agents of infectious diseases.* Moscow: Meditsina;1984. p.266 [in Russian].

23. Zhdanov VM, Lvov DK, Zakstelskaya LY, et al. Return of epidemic A1 (H1N1) influenza virus. *Lancet* 1978;**1** (8059):294–5.

24. Lvov DK, Yamnikova SS, Fedyakina IT, et al. Ecology and evolution of influenza viruses in Russia (1979–2002). *Vopr Virusol* 2004;**49**(3):17–24 [in Russian].

25. Lvov DK, Yamnikova SS, Lomakina NF, et al. Evolution of H4, H5 influenza A viruses in natural ecosystems in Northern Eurasia. *Int Congr Ser* 2004;**1263**:169–73.

26. Lvov DK. Populational interactions in biological system: influenza virus A wild and domestic animals humans; relations and consequences of introduction of high pathogenic influenza virus A/H5N1 on Russian territory. *J Microbiol Epidem Immunobiol* 2006;**3**:96–100 [in Russian].

27. Lvov DK. Evolution of emerging-reemerging viruses in Northern Eurasia global consequences. In: Lvov DK, Uryvaev LV, editors. *Investigation of evolution of viruses in the frame of the problems of biosafety and social infections.* Moscow; 2011. p. 5–16 [in Russian].

28. Lvov DK, Shchelkanov MY, Prilipov AG, et al. Evolution of HPAI H5N1 virus in Natural ecosystems of Northern Eurasia (2005–2008). *Avian Dis* 2010;**54**:483–95.

29. Lavrischeva VV, Burtseva EI, Khomyakov YN, et al. Etiology of fatal pneumonia cause by influenza A (H1N1) pdm09 virus during the pandemic in Russia. *Vopr Virusol* 2013;**58**(3):17–21.

30. Lvov DK, Burtseva EI, Prilipov AG, et al. A possible association of fatal pneumonia with mutations of pandemic influenza A/H1N1 swl virus in the receptor-binding site of HA1 subunit. *Vopr Virusol* 2010;**55**(4):4–9 [in Russian].

31. Lvov DK, Shchelkanov MY, Bovin NV, et al. Correlation between the receptor specificities of pandemic influenza A (H1N1) pdm09 virus strains isolated in 29–211 and the structure of the receptor-binding site and the probabilities of fatal primary virus pneumonia. *Vopr Virusol* 2012;**57**(1):14–20 [in Russian].

32. Lvov DK, Yashkulov KB, Prilipov AG, et al. Detection of amino acid substitutions of asparaginic acid for glycine and asparagine at the receptor-binding site of hemagglutinin in the variants of pandemic influenza A/H1N1 virus from patients with fatal outcome and moderate form of the disease. *Vopr Virusol* 2010;**55**(3):15–18 [in Russian].

33. Chumakov MP, Voroshilova MK, Drozdov SG. *Some results of mass immunization of the population of USSR against poliomyelitis by virion vaccine on the base of Albert Sabin strains. Virion peroral anti-poliomyelitis vaccine.* Moscow: USSR Ministry of Public Health;196112–26. [in Russian].

34. Syurin VM, Samuilenko AY, Soloviev BV, et al. *Viral diseases of animals.* Moscow: Institute of Biological Industry;1998 [in Russian].

I. HISTORY AND METHODOLOGY

Ecological Approach to Investigating Zoonotic Viruses

3.1 GENERAL CONCEPTS

The term "ecology" (from ancient Greek οκος—house, and λόγος—science) was coined by the German scientist Ernst Haeckel (Figure 3.1) in 1866 to denote the scientific discipline that studies populational interactions among biological species and between those species and the environment.[1] Later, the discipline was enriched by adding the concept of multispecies communities—ecosystems.[2,3]

It is not the purpose of this book to list general epidemiological statements or to go deep into details about the special epidemiology of single viral diseases. Such information is readily found in numerous handbooks. Instead, in what follows, we shall attempt to analyze general features and regularities peculiar to the viruses as biological species during their interactions with their hosts, of which humankind is one. The investigation of populational interactions between viruses and the cells of human tissues is the central task of modern epidemiology. Note, however, that the ecology of viruses is significantly wider than their epidemiology, which is a subdiscipline thereof. Thus, studying the complex sets of events that follow the interactions of viruses with permanently changing environmental conditions must be grounded in an ecological approach, one of the principal features of which is the discussion of species as communities of populations of individuals.[3] Special attention has to be paid to pathogen—host coevolution in unstable environmental conditions.[4,5]

Population is the unit of evolution. The evolution of a species should be regarded as the history of its ecology.[6] Viruses yield to the laws of evolution,[7] but although populations are similar in general, they are not identical, because they are isolated from each other by geographical and ecological factors. Each population has its unique gene pool that determines its features. Examining both gene pools themselves and the direction of their changes for the evolving population is key to understanding the causes of the appearance of epizooties and epidemics.[8]

The traditional directions of epidemiological investigations have resulted in a wealth of experience with the struggle with viral infections. The most demonstrative example is the global eradication of smallpox.[9] Nevertheless, the deep causes of the appearance of epidemics are often unknown. Where are the ecological niches of virus populations? How do virus

FIGURE 3.1   Ernst Heinrich Haeckel (1834–1919).

populations change? Why does a particular virus–host coevolution proceed in the way it does? Answers to these questions are necessary for the exploration and prediction of the conditions that lead to the development of epizootic outbreaks and emerging–reemerging epidemics. Exploring and predicting these conditions is the main practical purpose of the discipline of the ecology of viruses. As a theoretical discipline, the ecology of viruses seeks to reveal the mechanisms of the preservation of viruses as biological species in the biosphere, of virus evolution, and of gene pool transformations.

In this era of technological progress, numerous anthropogenic factors place pressure on ecosystems, significantly accelerating their evolution. The most well-known examples of these factors are environmental pollution by waste; the universal use of pesticides, antibiotics, vaccines, and other biopreparations; urbanization with extremely high human populations; the development of modern vehicles; the economic exploitation of new territories; and the development of industrial animal husbandry with extraordinary sizes and densities of domestic animals. Such factors lead to changing ecosystem structures, promote the introduction of new pathogens that cause epidemics, contribute to modification of the features of known viruses, and facilitate alterations in the reactivity and sensitivity of human populations.

The appearance, development, and recession of epizootic and epidemic waves is determined by properties of the interactions between populations of viruses and susceptible hosts. In accordance with evolution, only the most successful relationships among species form. Often, these relationships represent middle levels of the pathogenicity of the virus and susceptibility of the host. In some cases, the infection is persistent.[10,11] This type of virus–host relationship is the one which is most suitable for the period that is unfavorable to transmission of the virus and to stability of the host population. A virus's persistence and latency could lead to transformation of the gene pool and play a crucial role in preserving the virus during interepidemic periods. A virus's persistence in birds and bats could promote its dissemination over large territories. Nevertheless, both epizooties and epidemics represent only brief episodes in the evolution of viruses. One of the main causes of outbreaks of infectious diseases is the increasing density of the susceptible host population. There is a threshold population density permitting the development of epizooty. The existence of this threshold makes it clear that humankind plays an important role in the permanent modification of the structure of ecosystems.

It is difficult to find viral infections of humans for which there are no analogues among animal ones.[6,8] This parallelism has serious epidemiological consequences. Humans often have a great susceptibility to natural-foci viruses and are unable to adapt to new pathogens that appear in undeveloped territories. In some cases, previously unknown viruses could be the cause of outbreaks of an epidemic.[12]

On the one hand, the presence of natural foci of viruses makes it impossible—at least today—to eradicate these viruses and the threat of an epidemic caused by them. On the other hand, through rational modification of ecological factors, one could promote conditions that are incompatible with the existence of certain virus populations.

As human society evolves (about 400 generations in 10,000 years), humans come into contact with etiological agents of both wild and domestic animals. Some of these agents have adapted to human populations. Others have transformed into obligatory or facultative anthroponoses. Still others cause zoonoses that can provoking outbreaks of epidemics. Thus, the etiological agents of infectious diseases result from interactions between evolving populations of viruses and their potential hosts under the influence of natural and—nowadays—anthropogenic factors.[6,8]

For many biological species, large populations travel over significant distances. These migrations facilitate the transmission of adaptive pathogens. New natural-foci viruses could evolve, resulting in a dramatic worsening of epizootic or epidemic situations, as occurred, for example, after West Nile virus penetration into America in the autumn of 2001.[5,13] Natural-foci of viral infections could be ephemeral, seasonal, or permanent (i.e., existing for thousands or millions of years).

Seasonal migrations of birds (Aves) play an especially important role in global transmissions of virus populations. Birds are one of the most ancient (about 300 million years old) reservoirs for viruses. Some species' population densities are extremely high, one of the predisposing conditions of epizootic outbreaks. Some species of birds are synanthropic, living in close contact with human beings and domestic animals. The migrations of many millions of birds could be compared to a gigantic pendulum, whose swing over one continent and then another twice a year enables the birds to adapt to the viruses they encounter (Figure 3.2). During migrations, close interactions between different populations take place and therefore so do exchanges of adaptive

infectious agents in the stops, hubs, and overwintering places.[14]

3.2 VERTEBRATE HOSTS

Ecological links have been established between birds and more than 200 viruses belonging to 30 genera from 18 families (Table 3.1). In some cases birds are the main (or even single) host of the virus, while in other cases birds are virus amplifiers, necessary for the worsening of epizootic and other epidemic outbreaks. The role of birds in virus circulation is determined by the following factors: (i) the number and density of bird populations in nesting places (colonies of Alcidae birds and penguins in subarctic and subantarctic climatic belts, and colonies of herons, gulls, terns, and cormorants in temperate and subtropical climatic belts); (ii) high concentrations of populations in overwintering places (ducks, geese, coots, etc.); (iii) flyways of seasonal migrations (Passeriformes and Anseriformes); (iv) contacts of dendrophylic and terrestrial species (thrushes, Passeriformes, etc.) with vectors (mosquitoes, as well as Ixodidae, Argasidae, and Gamasidae ticks); (v) shelter habitats in burrows (wheatears, bee-eaters, rollers, swallows, and pigeons), where the birds readily make contact with Argasidae ticks; (vi) water and near-water habitats of birds (ducks, geese, sandpipers, gulls, terns, cormorants, and coots), which support the circulation of viruses through alimentary transmission.[14,15]

Six hundred twenty-two species of birds making up 231 genera from 60 families of 24 orders inhabit Northern Eurasia (Table 3.2). The main species are from the families Alcidae, Anatidae, Laridae, Sternidae, Ardeidae, Rallidae, Corvidae, and Phalacrocoracidae, the clade Columbidae, the order Passeriformes, and some others. (The ecological links among separate species of birds are discussed in later sections.)

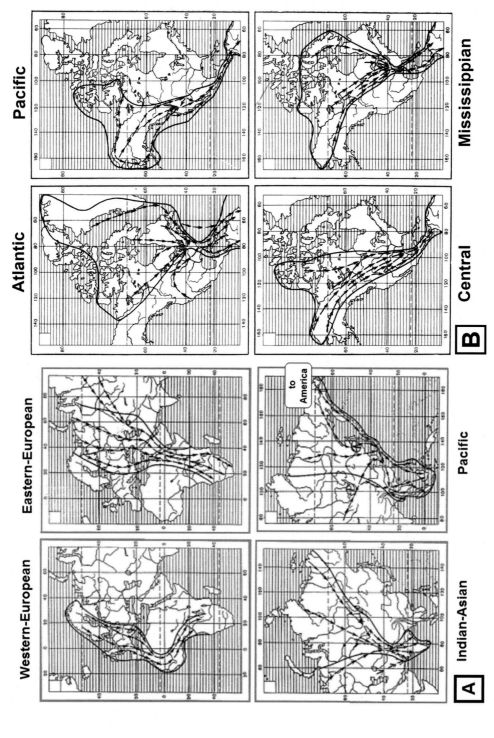

FIGURE 3.2 The main migration flyways of birds: (A) in the Old World and (B) in America.

TABLE 3.1 Distribution of Viruses of Different Taxa Among Vertebrate Hosts

Genome/Order	Family	Genus	Fishes (4)	Amphibians (5)	Reptiles (6)	Birds (7)	Humans (Hominidae) (8)	Monkeys (9)	Bats (10)	Dogs (11)	Cats (12)	Other [Carnivora] (13)	Horses (14)	Other [Perissodactyla] (15)	Pigs (Suidae) (16)	Cattle (Bovidae) (17)	Sheep, goats (Caprinae) (18)	Deers (Cervidae) (19)	Other [Artiodactyla] (20)	Rodents (Rodentia) (21)	Hares (Lagomorpha) (22)	Insectivorous (Disectivora) (23)	Pinnipeda (24)	Cetacea (25)	Other (26)
dsDNA / Herpesvirales	Alloherpesviridae	*Batrachovirus*	−	+	−	−	−	−	−	−	−	−	−	−	−	−	−	−	−	−	−	−	−	−	−
		Cyprinivirus	+	−	−	−	−	−	−	−	−	−	−	−	−	−	−	−	−	−	−	−	−	−	−
		Ictalurivirus	+	−	−	−	−	−	−	−	−	−	−	−	−	−	−	−	−	−	−	−	−	−	−
		Salmonivirus	+	−	−	−	−	−	−	−	−	−	−	−	−	−	−	−	−	−	−	−	−	−	−
	Herpesviridae	*Iltovirus*	−	−	−	+	−	−	−	−	−	−	−	−	−	−	−	−	−	−	−	−	−	−	−
		Mardivirus	−	−	−	+	−	−	−	−	−	−	−	−	−	−	−	−	−	−	−	−	−	−	−
		Simplexvirus	−	−	−	−	−	+	−	−	−	−	−	−	−	+	−	−	−	−	−	−	−	−	+
		Varicellovirus	−	−	−	−	+	+	−	+	+	−	+	−	+	+	+	+	−	−	−	−	−	−	−
		Cytomegalovirus	−	−	−	−	+	+	−	−	−	−	−	−	−	−	−	−	−	−	−	−	−	−	−
		Muromegalovirus	−	−	−	−	−	−	−	−	−	−	−	−	−	−	−	−	−	+	−	−	−	−	−
		Proboscivirus	−	−	−	−	−	−	−	−	−	−	−	−	−	−	−	−	−	−	−	−	−	−	+
		Roseolovirus	−	−	−	−	+	−	−	−	−	−	−	−	−	−	−	−	−	−	−	−	−	−	−
		Lymphocryptovirus	−	−	−	−	+	+	−	−	−	−	−	−	−	−	−	−	−	−	−	−	−	−	−
		Macavirus	−	−	−	−	−	−	−	−	−	−	−	−	+	+	−	−	+	−	−	−	−	+	−
		Percavirus	−	−	−	−	−	−	−	−	−	−	+	+	−	−	−	−	−	−	−	−	−	−	−
		Rhadinovirus	−	−	−	−	+	+	−	−	−	−	−	−	−	+	−	−	−	+	+	−	−	−	−
dsDNA	Adenoviridae	*Mastadenovirus*	−	−	−	−	+	+	+	+	−	−	+	−	+	+	+	−	+	+	−	+	−	−	+
		Aviadenovirus	−	−	−	+	−	−	−	−	−	−	−	−	−	−	−	−	−	−	−	−	−	−	−
		Atadenovirus	−	−	+	+	−	−	−	−	−	−	−	−	+	+	−	−	−	−	−	−	−	−	−
		Siadenovirus	−	+	+	+	−	−	−	−	−	−	−	−	−	−	−	−	−	−	−	−	−	−	−
		Ichtadenovirus	+	−	−	−	−	−	−	−	−	−	−	−	−	−	−	−	−	−	−	−	−	−	−
	Asfarviridae[a]	*Asfivirus*[a]	−	−	−	−	−	−	−	−	−	−	−	−	+	−	−	−	−	−	−	−	−	−	−

(Continued)

TABLE 3.1   (Continued)

Viruses			Hosts (classes of Vertebrata)																						
							Mammalia																		
							Primates			Carnivora			Perissodactyla		Artiodactyla										
Genome/Order	Family	Genus	Fishes	Amphibians	Reptiles	Birds	Humans (Hominidae)	Monkeys	Bats	Dogs	Cats	Other	Horses	Other	Pigs (Suidae)	Cattle (Bovidae)	Sheep, goats (Caprinae)	Deers (Cervidae)	Other	Rodents (Rodentia)	Hares (Lagomorpha)	Insectivorous (Disectivora)	Pinnipeda	Cetacea	Other
1	2	3	4	5	6	7	8	9	10	11	12	13	14	15	16	17	18	19	20	21	22	23	24	25	26
dsDNA (continued)	Iridoviridae	Ranavirus	–	+	+	–	–	–	–	–	–	–	–	–	–	–	–	–	–	–	–	–	–	–	–
		Lymphocystivirus	+	–	–	–	–	–	–	–	–	–	–	–	–	–	–	–	–	–	–	–	–	–	–
		Megalocytivirus	+	–	–	–	–	–	–	–	–	–	–	–	–	–	–	–	–	–	–	–	–	–	–
	Papillomaviridae	Alphapapillomavirus	–	–	–	–	+	+	–	–	–	–	–	–	–	–	–	–	–	–	–	–	–	–	–
		Betapapillomavirus	–	–	–	–	+	–	–	–	–	–	–	–	–	–	–	–	–	–	–	–	–	–	–
		Gammapapillomavirus	–	–	–	–	+	–	–	–	–	–	–	–	–	–	–	–	–	–	–	–	–	–	–
		Deltapapillomavirus	–	–	–	–	–	–	–	–	–	–	–	–	–	–	+	+	+	–	–	–	–	–	–
		Epsilonpapillomavirus	–	–	–	–	–	–	–	–	–	–	–	–	–	–	+	–	–	–	–	–	–	–	–
		Zetapapillomavirus	–	–	–	–	–	–	–	–	–	–	+	–	–	–	–	–	–	–	–	–	–	–	–
		Etapapillomavirus	–	–	–	+	–	–	–	–	–	–	–	–	–	–	–	–	–	–	–	–	–	–	–
		Thetapapillomavirus	–	–	–	+	–	–	–	–	–	–	–	–	–	–	–	–	–	–	–	–	–	–	–
		Iotapapillomavirus	–	–	–	–	–	–	–	–	–	–	–	–	–	–	–	–	–	–	+	–	–	–	–
		Kappapapillomavirus	–	–	–	–	–	–	–	–	–	–	–	–	–	–	–	–	–	–	–	+	–	–	–
		Lambdapapillomavirus	–	–	–	–	–	–	–	+	+	–	–	–	–	–	–	–	–	–	–	–	–	–	–
		Mupapillomavirus	–	–	–	–	+	–	–	–	–	–	–	–	–	–	–	–	–	–	–	–	–	–	–
		Nupapillomavirus	–	–	–	–	+	–	–	–	–	–	–	–	–	–	–	–	–	–	–	–	–	–	–
		Xipapillomavirus	–	–	–	–	–	–	–	–	–	–	–	–	–	–	+	–	–	–	–	–	–	–	–
		Omikronpapillomavirus	–	–	–	–	–	–	–	–	–	–	–	–	–	–	–	–	–	–	–	–	–	+	–
		Pipapillomavirus	–	–	–	–	–	–	–	–	–	–	–	–	–	–	–	–	–	–	+	–	–	–	–
	Polyomaviridae	Polyomavirus	–	–	–	–	+	+	–	–	–	–	–	–	–	–	+	–	–	–	+	+	–	–	–
	Poxviridae[a]	Avipoxvirus	–	–	–	+	–	–	–	–	–	–	–	–	–	–	–	–	–	–	–	–	–	–	–
		Capripoxvirus	–	–	–	–	–	–	–	–	–	–	–	–	–	–	–	+	–	–	–	–	–	–	–
		Cervidpoxvirus	–	–	–	–	–	–	–	–	–	–	–	–	–	–	–	–	+	–	–	–	–	–	–

(Continued)

Family	Genus																						
dsDNA (continued)	Leporipoxvirus	−	−	−	−	−	−	−	−	−	−	−	−	−	−	−	−	−	−	−	−	+	−	−
	Molluscipoxvirus	−	−	+	−	−	−	−	−	+	−	−	−	−	−	−	−	−	−	−	−	−	−	+
	Orthopoxvirus[a]	−	−	+	−	+	+	+	−	−	+	−	−	+	−	−	−	−	−	−	−	+	−	+
	Parapoxvirus	−	−	−	−	+	+	+	−	−	+	−	+	−	−	−	−	+	−	+	−	−	+	−
	Suipoxvirus	−	−	−	−	−	−	−	−	+	−	−	−	−	−	−	+	−	−	−	−	−	−	−
	Yatapoxvirus	−	−	+	−	+	+	−	−	−	−	−	−	−	−	−	−	−	−	−	−	−	−	−
ssDNA — Anelloviridae	Alphatorquevirus	−	−	−	+	−	+	−	−	−	−	−	−	−	−	+	−	−	−	−	−	−	−	−
	Betatorquevirus	−	−	−	+	−	+	−	−	−	−	−	−	−	−	+	−	−	−	−	−	−	−	−
	Gammatorquevirus	−	−	−	+	−	+	−	−	−	−	−	−	−	−	+	−	−	−	−	−	−	−	−
	Deltatorquevirus	−	−	−	−	−	−	−	−	−	−	−	−	−	−	−	−	−	−	−	−	−	−	−
	Epsilontorquevirus	−	−	−	−	−	+	−	−	−	−	−	−	−	−	−	+	−	−	−	−	−	−	−
	Zetatorquevirus	−	−	−	−	−	+	−	−	−	−	−	−	−	−	−	+	−	−	−	−	−	−	−
	Etatorquevirus	−	−	−	−	−	−	−	−	−	−	−	+	−	−	−	−	−	−	−	−	−	−	−
	Thetatorquevirus	−	−	−	−	−	−	−	−	−	−	−	−	−	+	−	−	−	−	−	−	−	−	−
	Iotatorquevirus	−	−	−	−	−	−	+	+	−	−	−	−	−	−	−	−	+	+	−	−	−	−	−
ssDNA — Circoviridae	Circovirus	−	−	−	+	−	−	+	−	−	−	−	−	−	−	−	+	+	−	−	−	−	−	+
	Gyrovirus	−	−	−	+	−	−	−	−	−	−	−	−	−	−	−	−	−	−	−	−	−	−	−
ssDNA — Parvoviridae	Parvovirus	−	−	−	+	−	+	−	−	+	−	−	+	−	−	+	+	+	−	+	−	−	+	−
	Erythrovirus	−	−	−	+	−	+	−	−	−	−	−	−	−	−	+	+	−	−	−	−	−	−	−
	Dependovirus	−	+	−	+	−	+	+	−	−	+	−	+	−	+	−	+	−	+	−	+	−	−	−
	Amdovirus	−	−	−	−	−	−	−	−	−	+	−	−	−	−	−	−	−	−	−	−	−	−	−
	Bocavirus	−	−	−	+	−	−	−	−	+	−	−	−	+	−	+	+	−	−	−	−	−	−	−
	Unclassified	−	−	−	+	−	+	+	+	+	−	−	+	−	+	+	+	+	−	−	−	−	−	−
DNA/RNA with RT — Hepadnaviridae	Orthohepadnavirus	−	−	−	−	−	+	−	−	−	−	−	−	−	−	+	−	−	−	−	−	+	−	−
	Avihepadnavirus	−	+	−	−	−	−	+	−	−	−	−	−	−	−	−	−	−	−	−	−	−	−	−
DNA/RNA with RT — Retroviridae	Alpharetrovirus	−	−	−	+	−	−	−	−	−	−	−	−	−	−	−	+	−	−	−	−	−	−	−
	Betaretrovirus	−	−	−	+	−	+	−	−	−	−	−	+	−	−	+	+	−	+	−	−	+	−	+
	Gammaretrovirus	−	+	−	+	−	+	−	−	+	−	−	+	−	+	+	+	−	+	−	+	+	−	+
	Deltaretrovirus	−	−	−	−	−	+	−	−	−	−	−	−	−	−	+	−	−	+	−	−	−	−	−
	Epsilonretrovirus	+	−	−	−	−	−	−	−	−	−	−	−	−	−	−	−	−	−	−	−	−	−	−
	Lentivirus	−	−	−	+	−	+	−	−	+	−	−	+	−	+	+	+	−	+	−	−	+	−	+
	Spumavirus	−	−	−	+	+	−	−	−	+	−	−	+	+	+	+	+	−	+	−	−	+	−	+

TABLE 3.1 (Continued)

	Viruses		Hosts (classes of Vertebrata)																						
							Primates		Bats	Carnivora			Perissodactyla		Mammalia		Artiodactyla								
Genome/Order	Family	Genus	Fishes	Amphibians	Reptiles	Birds	Humans (Hominidae)	Monkeys	Bats	Dogs	Cats	Other	Horses	Other	Pigs (Suidae)	Cattle (Bovidae)	Sheep, goats (Caprinae)	Deers (Cervidae)	Other	Rodents (Rodentia)	Hares (Lagomorpha)	Insectivorous (Disectivora)	Pinnipeda	Cetacea	Other
1	2	3	4	5	6	7	8	9	10	11	12	13	14	15	16	17	18	19	20	21	22	23	24	25	26
dsRNA	Birnaviridae[a]	Aquabirnavirus[a]	+	−	−	−	−	−	−	−	−	−	−	−	−	−	−	−	−	−	−	−	−	−	−
		Avibirnavirus	−	−	−	+	−	−	−	−	−	−	−	−	−	−	−	−	−	−	−	−	−	−	−
		Blosnavirus	+	−	−	−	−	−	−	−	−	−	−	−	−	−	−	−	−	−	−	−	−	−	−
	Picobirnaviridae	Picobirnavirus	−	−	+	+	+	−	−	+	−	−	−	−	+	+	−	−	−	+	−	−	−	−	+
	Reoviridae[a]	Orthoreovirus[a]	−	−	+	+	+	+	+	−	−	−	−	−	+	+	+	−	−	+	−	−	−	−	−
		Aquareovirus	+	−	−	−	−	−	−	−	−	−	−	−	−	−	−	−	−	−	−	−	−	−	−
		Coltivirus	−	−	−	−	+	−	−	−	−	−	−	−	−	−	−	−	−	+	−	−	−	−	−
		Orbivirus[a]	−	−	−	+	−	+	−	−	−	−	+	+	−	+	+	+	−	+	−	−	−	−	+
		Rotavirus	−	−	−	+	+	+	−	−	−	−	−	−	+	−	−	−	−	−	−	−	−	−	−
		Seadornavirus	−	−	−	−	+	−	−	−	−	−	−	−	−	−	−	−	−	+	−	−	−	−	−
ssRNA (−)/Mononegavirales	Bornaviridae	Bornavirus	−	−	−	−	+	−	−	−	−	−	−	−	−	−	−	−	−	−	−	−	−	−	−
	Filoviridae	Marburgvirus	−	−	−	−	+	−	−	−	−	−	−	−	+	−	−	−	−	+	−	−	−	−	−
		Ebolavirus	−	−	−	−	+	−	+	−	−	−	−	−	−	−	−	−	−	−	−	−	−	−	−
	Paramyxoviridae[a]	Avulavirus[a]	−	−	−	+	−	−	−	−	−	−	−	−	−	−	−	−	−	−	−	−	−	−	−
		Rubulavirus	−	−	−	−	+	+	+	−	−	−	−	−	+	−	−	−	−	−	−	−	−	−	−
		Henipavirus	−	−	−	−	+	−	+	−	−	−	+	−	+	−	−	−	−	−	−	−	−	−	−
		Respirovirus	−	−	−	−	+	+	−	−	−	−	−	−	−	−	−	−	−	−	−	−	−	−	−
		Metapneumovirus	−	−	−	+	+	−	−	−	−	−	−	−	−	−	−	−	−	−	−	−	−	−	−
		Morbillivirus[a]	−	−	−	−	+	−	−	+	−	−	−	−	−	+	+	−	−	−	−	−	+	+	+
		Pneumovirus	−	−	−	−	+	−	−	−	−	−	−	−	−	−	+	−	−	+	−	−	−	−	−
		Unclassified	+	−	−	−	−	−	−	−	−	−	−	−	−	−	−	−	−	−	−	−	−	−	−
	Rhabdoviridae[a]	Vesiculovirus	+	−	−	−	−	−	−	−	−	+	+	+	+	+	+	+	+	+	+	−	−	−	−
		Lyssavirus[a]	−	−	−	−	+	−	+	+	+	+	−	−	−	−	−	−	−	+	−	−	−	−	−

ssRNA (−)		*Ephemerovirus*[a]	−	−	−	−	−	−	−	−	−	−	−	+	−	−	−	−	−	−	−	−	−
		Novirhabdovirus	+	−	−	−	−	−	−	−	−	−	−	−	−	−	−	−	−	−	−	−	−
	Orthomyxoviridae[a]	*Influenzavirus A*[a]	−	+	+	−	−	+	+	+	+	+	+	−	+	−	+	−	−	−	−	+	+
		Influenzavirus B	−	−	+	−	−	−	−	−	−	−	−	−	−	−	−	−	−	−	−	−	−
		Influenzavirus C	−	−	+	−	−	−	−	−	−	−	−	−	−	−	−	−	−	−	−	−	−
		Quaranjavirus[a]	−	+	+	−	−	−	−	−	−	−	−	−	−	−	−	−	−	−	−	−	−
		Thogotovirus[a]	−	−	+	−	−	−	−	−	+	−	−	+	+	−	+	+	+	−	−	−	−
		Isavirus	+	−	−	−	−	−	−	−	−	−	−	−	−	−	−	−	−	−	−	−	−
Without family		*Deltavirus*	−	−	+	−	−	−	−	−	−	−	−	−	−	−	−	−	−	−	−	−	−
ssRNA (+/−)	*Arenaviridae*	*Arenavirus*	−	−	+	−	−	−	−	−	−	−	−	−	−	−	−	+	−	−	−	−	−
	Bunyaviridae[a]	*Orthobunyavirus*[a]	−	−	+	−	−	+	+	−	+	−	+	−	+	+	+	+	+	−	−	−	+
		Hantavirus[a]	−	+	−	−	+	−	−	+	−	−	−	−	−	−	−	+	−	−	−	+	−
		Nairovirus[a]	−	+	+	−	−	−	−	−	+	−	+	−	+	+	+	+	+	−	−	+	+
		Phlebovirus[a]	−	−	+	−	−	−	−	−	+	−	−	−	+	−	+	+	+	−	−	−	−
ssRNA (+)/Nidovirales	*Arteriviridae*	*Arterivirus*	−	−	−	+	−	−	−	−	+	+	−	−	−	−	−	+	−	−	−	−	−
	Coronaviridae	*Alphacoronavirus*	−	−	−	−	+	+	+	+	+	+	+	−	−	−	−	−	−	−	−	−	−
		Betacoronavirus	−	−	−	−	−	−	+	+	+	−	+	+	−	−	−	−	−	−	−	−	−
		Gammacoronavirus	−	+	−	−	−	−	−	+	−	−	−	−	−	−	−	−	−	−	−	−	−
		Torovirus	−	−	+	−	−	−	−	−	+	+	+	−	−	−	+	−	−	−	−	−	−
		Bafinivirus	+	−	−	−	−	−	−	−	−	−	−	−	−	−	−	−	−	−	−	−	−
ssRNA (+)/Picornavirales	*Picornaviridae*[a]	*Enterovirus*	−	−	+	+	−	−	−	−	+	+	−	−	−	−	−	−	−	−	−	−	−
		Cardiovirus[a]	−	−	+	−	−	−	−	−	−	−	+	−	−	−	−	−	−	−	−	−	−
		Aphthovirus	−	−	−	−	−	−	−	+	+	−	−	−	−	−	−	−	−	−	−	−	−
		Hepatovirus	−	−	+	−	−	−	−	−	−	−	−	−	−	−	−	−	−	−	−	−	−
		Parechovirus	−	−	+	−	−	−	−	−	−	−	−	−	−	−	−	−	−	−	−	−	−
		Erbovirus	−	−	−	−	−	−	−	−	+	−	−	+	−	−	−	−	−	−	−	−	−
		Kobuvirus	−	−	+	−	−	−	−	−	−	−	−	−	−	−	−	−	−	−	−	−	−
		Teschovirus	−	−	−	−	−	−	−	−	−	−	+	−	−	−	−	−	−	−	−	−	−
		Sapelovirus	−	+	−	−	−	−	−	−	−	+	−	−	−	−	−	−	−	−	−	−	−

(Continued)

TABLE 3.1 (Continued)

Genome/Order	Family	Genus	Fishes (4)	Amphibians (5)	Reptiles (6)	Birds (7)	Humans (Hominidae) (8)	Monkeys (9)	Bats (10)	Dogs (11)	Cats (12)	Other (13)	Horses (14)	Other (15)	Pigs (Suidae) (16)	Cattle (Bovidae) (17)	Sheep, goats (Caprinae) (18)	Deers (Cervidae) (19)	Other (20)	Rodents (Rodentia) (21)	Hares (Lagomorpha) (22)	Insectivorous (Disectivora) (23)	Pinnipeda (24)	Cetacea (25)	Other (26)
ssRNA (+)		Senecavirus	–	–	–	–	–	–	–	–	–	–	–	–	+	–	–	–	–	–	–	–	–	–	–
		Tremovirus	–	–	–	+	–	–	–	–	–	–	–	–	–	–	–	–	–	–	–	–	–	–	–
		Avihepatovirus	–	–	–	+	–	–	–	–	–	–	–	–	–	–	–	–	–	–	–	–	–	–	–
	Astroviridae	Avastrovirus	–	–	–	+	–	–	–	–	–	–	–	–	–	–	–	–	–	–	–	–	–	–	–
		Mamastrovirus	–	–	–	–	+	–	+	+	–	+	–	–	+	+	+	–	–	+	–	–	+	–	–
	Caliciviridae	Vesivirus	–	–	–	–	–	–	–	+	+	–	–	+	+	–	–	–	–	–	–	–	–	–	–
		Lagovirus	–	–	–	–	–	–	–	–	–	–	–	–	–	–	–	–	–	–	+	–	–	–	–
		Norovirus	–	–	–	–	+	–	–	–	–	–	–	–	+	–	+	–	–	+	–	–	–	–	–
		Sapovirus	–	–	–	–	+	–	–	–	–	–	–	–	–	–	–	–	–	–	–	–	–	–	–
		Nebovirus	–	–	–	–	–	–	–	–	–	–	–	–	–	+	–	–	–	–	–	–	–	–	–
	Flaviviridae[a]	Flavivirus[a]	–	+	–	+	+	+	+	+	+	+	+	+	+	+	+	+	+	+	+	+	–	–	+
		Pestivirus	–	–	–	–	–	–	–	–	–	–	–	–	+	–	–	–	+	–	–	–	–	–	+
		Hepacivirus	–	–	–	–	+	+	–	–	–	–	–	–	–	–	–	–	–	–	–	–	–	–	–
	Hepeviridae	Hepevirus	–	–	–	+	+	–	–	–	–	–	–	+	+	–	–	+	–	+	–	–	–	–	–
	Nodaviridae	Betanodavirus	+	–	–	–	–	–	–	–	–	–	–	–	–	–	–	–	–	–	–	–	–	–	–
	Togaviridae[a]	Alphavirus[a]	–	–	–	–	+	+	–	–	–	+	+	+	+	–	+	+	+	+	+	+	+	–	+
		Rubivirus	–	–	–	–	+	–	–	–	–	–	–	–	–	–	–	–	–	–	–	–	–	–	–
		Unclassified	+	+	+	+	–	–	–	–	–	–	–	–	–	–	–	–	–	–	–	–	–	–	–
Unclassified viruses of Vertebrata			–	–	+	+	+	+	–	–	–	–	–	–	–	–	–	–	–	–	–	–	–	–	–
Prions			–	–	–	–	–	–	–	–	+	+	–	–	–	+	+	+	+	–	–	–	–	–	–
Total 33	**33**	**143**	**16**	**4**	**8**	**37**	**69**	**33**	**11**	**15**	**15**	**12**	**20**	**12**	**35**	**42**	**25**	**11**	**11**	**36**	**12**	**3**	**4**	**4**	**10**

[a]Taxa of viruses described in this book.

TABLE 3.2 List of Potential Vertebrate Hosts for Zoonotic Viruses in Northern Eurasia

Class	Order	Family	Genus	Species	Known significance
Amphibian (*Amphibia*)	Caudate (Caudata)	2	6	9	+
	Salientia (Anura)	5	6	23	+
Total	**2**	**7**	**12**	**32**	**+**
Reptilian (*Reptilia*)	*Testudines*	5	6	7	+
	Sauria	6	18	74	+
	Serpentes	7	22	55	+
Total	**3**	**18**	**46**	**136**	**+**
Birds (*Aves*)	Geese, ducks, swans (*Anseriformes*)	1	14	48	+ + +
	Herons, storks (*Ciconiiformes*)	3	12	18	+ + +
	Coots, rails, cranes, bustards (*Gruiformes*)	2	8	17	+ + +
	Gulls, terns, guillemots, tufted puffins, shorebirds (*Charadriiformes*)	8	40	111	+ + +
	Crows, rooks, daws, magpies, sparrows, larks, starlings (*Passeriformes*)	25	96	286	+ + +
	Cormorants, pelicans, solan-geese, frigate-birds (*Pelecaniformes*)	2	2	8	+ + +
	Pigeons, doves (*Columbiformes*)	1	2	10	+ +
	Hens, pheasants, partridges (*Galliformes*)	2	12	20	+ +
	Grebes (*Podicipediformes*)	1	1	5	+ +
	Albatrosses, petrels (*Procellariiformes*)	1	2	4	+ +
	Diurnal birds of prey (*Falconiformes*)	3	15	44	+
	Halcyon, rollers (*Coraciiformes*)	4	5	6	+
	Woodpeckers (*Piciformes*)	1	5	14	+
	Nightjars (*Caprimulgiformes*)	1	1	3	+
	Sandgrouses (*Pteroclidiformes*)	1	2	4	+
	Owls (*Strigiformes*)	2	12	18	+
	Cuckoos (*Cuculiformes*)	1	1	5	+
	Flamingo (*Phoenicopteriformes*)	1	1	1	
	Loons (*Gaviiformes*)	1	1	3	
Total	**24**	**60**	**231**	**622**	**+ + +**

(*Continued*)

TABLE 3.2   (Continued)

Class	Order	Family	Genus	Species	Known significance
Mammalian (*Mammalia*)	*Artiodactyla*	5	12	20	+
	Carnivora	5	14	37	++
	Bats (*Chiroptera*)	3	12	40	+++
	Insectivora	3	11	45	+++
	Perissodactyla	1	1	2	+
	Pinnipedia	3	7	12	+
	Rodentia	10	47	127	+++
Total	**7**	**30**	**104**	**283**	

Viruses of 33 families and 143 genera produce Vertebrata pathology (Table 3.1). The periodical intrusion of vertebrate animals into the virus circulation scheme creates conditions under which virus populations are enriched by new variants. Viruses with a segmented genome (e.g., Bunyaviridae, Reoviridae, Orthomyxoviridae, and more) have an additional opportunity for recombination. In some cases, penetration into vertebrate populations led to the mastering of a new ecological niche by the virus populations. Animals of the Vertebrata subphylum have special significance for viruses that haven't yet gained or have already lost the capability of vector transmission. With these viruses, persistent infections in vertebrates (more often in rodents or insectivorous bats) usually develop.

The main factors determining the epizootic and epidemic significance of the *Vertebrata* are (i) the number of contacts of humans with bloodsucking arthropods; (ii) susceptibility to viruses; (iii) the development of high levels of viremia (more than $2.0-2.5$ $\log_{10}(LD_{50})/$ 10 mcL for newborn mice infected intracerebrally); (iv) the density of human populations; (v) seasonal migrations; (vi) the presence of persistent infections; and (vii) the number

of contacts between domestic animals and humans.[15]

Members of the order Chiroptera play a special role among the Vertebrata, harboring more than 50 viruses from the families Bunyaviridae, Coronaviridae, Arenaviridae, Flaviviridae, Herpesviridae, Orthomyxoviridae, Paramyxoviridae, Filoviridae, Rhabdoviridae, Reoviridae, and Togaviridae (Table 3.1). The fauna of Northern Eurasia includes 40 species of Chiroptera from 12 genera of three families (Table 3.2). Some species of bats participate in seasonal migrations over thousands of kilometers twice a year, as birds do. The Chiroptera are distributed everywhere except the polar zones and some ocean islands. Bat populations with high densities are a common component of anthropogenic biocenoses. Experimental data show that bats support a high level of long-term viremia. Brown fat is the main tissue for virus accumulation. There is regular transplacental transmission among vertebrates. Viruses could be preserved in bats for months during the overwintering period in temperatures of $5-10°C$. The factors listed in the previous paragraph determine the epizootic and epidemic significance of the Chiroptera.[15,16]

More than 100 viruses are linked with rodents (order Rodentia), which have a wide distribution over all landscape and vertical zones, exhibit a high biomass and high densities of populations, interact closely with bloodsucking arthropods, and often develop persistent infections with vertical transmission of viruses. Also, many species of rodents have an anthropogenic relationship with humans. Populations of Rodentia as well as Insectivora are the usual components of natural foci and the most common reservoir of viruses from the families Arenaviridae, Bunyaviridae, Flaviviridae, Paramyxoviridae, Poxviridae, Rhabdoviridae, and Togaviridae (Table 3.1). The fauna of Northern Eurasia includes 127 species of Rodentia from 17 genera of 10 families (Table 3.2).[15]

Aside from human beings (order Primates), orders of Vertebrata that are important for virus circulation in Northern Eurasia are the wild and domestic perissodactyls (Perissodactyla) and artiodactyls (Artiodactyla), among which the most representative are domestic horses (*Equus ferus caballus*) as well as domestic swine (*Sus scrofa domesticus*) and wild boars (*Sus scrofa*).

3.2.1 Arthropod Vectors

Bloodsucking arthropods (phylum Arthropoda) often are specific vectors for zoonotic viruses (arboviruses), transmitting them by biting into susceptible vertebrates. Arthropods are also the natural reservoir for viruses that are transmitted transovarially or transstadially.[7] Long-term surveillance of viruses has found that, in arthropods, they could be transmitted vertically during metamorphosis, through a sexual pathway or by the alimentary inoculation of mosquito larva with further infection of the imago. In such cases, arboviruses could be categorized as symbionts of arthropods or their parasites. In the evolutionary context, many viruses (e.g., from the families Togaviridae and Flaviviridae) appeared as arthropod symbionts in equatorial and subequatorial climates.[7,15]

Confirming the symbiont hypothesis are a number of factors: the presence of virus symbionts in arthropods, the persistence of infection during the life of the imago, the transmission of viruses through the stages of life and the transovarial transmission during metamorphosis, the cyclic recurrence of virus amplification in arthropods with necessary reproduction of the virus in the intestine, and the possibility of reproduction under relatively low temperatures (which is usual for plant viruses but unique for arthropod viruses). Taking into account the mechanism of inoculation, one may categorize typical arboviruses as blood parasites. This form of parasitism is secondary and emerged from intestinal parasites of the arthropods. Later, viruses acquired the ability to penetrate the intestine, enabling them to accumulate in salivary glands and to be transmitted by the host biting into hematothermal animals (including birds with temperatures up to 42°C), yet keeping the ability to reproduce under the low temperatures in arthropods (20–25°C).

The majority of arboviruses can reproduce in arthropods when the temperature of the environment is greater than 16–18°C but less than about 28–30°C, at which point the incubation period decreases significantly. At the latter temperatures, the virus reproduces in the epithelium of the intestine, penetrates into the host's salivary glands, and accumulates there at a level sufficient for infection when the host bites. The presence of microfilariae enhances the ability of arbovirus transmission (e.g., the microfilaria *Brugia malayi* promotes the dissemination of dengue virus in *Aedes aegypti* mosquitoes).[3] That is why the most favorable conditions for arboviruses form in the equatorial climatic belt (Figure 3.3, Table 3.3), where permanently high temperatures (about 27°C) and humidity prevail. The role of ticks in the equatorial and subequatorial climatic belts is not significant. For example, the modern life cycle of ticks from the family Ixodidae in the equatorial

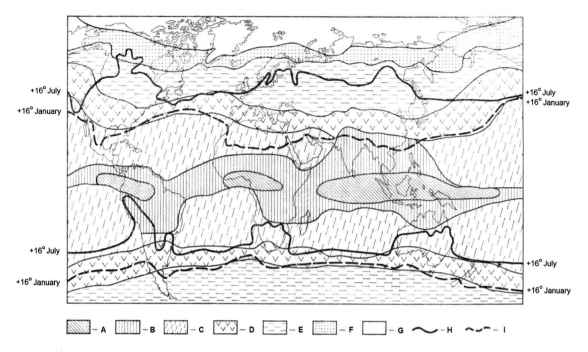

FIGURE 3.3 Climatic belts of the world. A − Equatorial; B − Sub-equatorial; C − Tropical; D − Subtropical; E − Temperate; F − Subarctic; G − Arctic.

TABLE 3.3 The number of viruses belonging to the Togaviridae, Flaviviridae, and Bunyaviridae families and transmitted by mosquitoes and ticks in different climatic belts

		Climatic belt				
The main vector	Family	Equatorial	Subequatorial	Tropical	Subtropical	Temperate
Mosquitoes	*Togaviridae*	15	13	9	10	4
	Flaviviridae	20	19	11	6	6
	Bunyaviridae	62	62	26	21	13
	Total	97	94	46	37	23
Ticks	*Togaviridae*	0	0	0	0	0
	Flaviviridae	2	2	3	4	11
	Bunyaviridae	3	7	13	5	9
	Total	5	9	16	9	20
Sum	*Togaviridae*	15	13	9	10	4
	Flaviviridae	22	21	14	10	17
	Bunyaviridae	65	69	39	26	22
	Total	102	103	62	46	43

climatic belt usually includes just one host feeding.

The subequatorial climatic belt is also suitable for arbovirus reproduction practically all year long. Nevertheless, the subequatorial climate has a dry period, which is differently expressed in different regions. Mosquito numbers are suppressed during the dry period, reducing the number of mosquito-borne arboviruses that do not have a sufficient level of ecological plasticity. Some arboviruses circulating in ticks have adapted to the dry period. Such adaptation is seen to become more important as one moves northward.

In the tropical climatic belt, environmental conditions change dramatically because there is a "winter period" with low temperatures that are not sufficient for the reproduction of the majority of arboviruses. These conditions explain the decreasing number of arboviruses linked with mosquitoes in the tropics with the simultaneously increasing number of tick-borne arboviruses.

In the subtropical climatic belt, the number of mosquito-borne arboviruses continues to decrease. Overwintering is more difficult for arboviruses. Again, the role of ticks as a natural reservoir for arboviruses continues to increase, especially in arid zones, which are common in the subtropics.

In the temperate climatic belt, there is a very short period that is suitable for the reproduction of arboviruses in mosquitoes. Arboviruses transmitted by mosquitoes are distributed mainly in the southern part of this belt. A prolonged and difficult overwintering period significantly impedes the surveillance of arboviruses. What is likely is viral penetration of birds and bats in spring from overwintering places or of mosquitoes carried by the airflow. The latter pathway may be usual in regions with a monsoonal climate (e.g., the Far East, where Japanese encephalitis (family Flaviviridae, genus *Flavivirus*) is seen). Seasonal natural foci are quite often observed.

Nevertheless, the opportunity to form stable natural foci exists when local populations of mosquitoes and ticks are included in virus circulation. Ixodidae ticks adapted to the more difficult conditions of the temperate climatic belt by shifting to two to three host feedings. Thereby, this family of ticks—mainly *Ixodes uriae*—ensured the adaptation of their parasitic arboviruses in bird colonies on the shelf zone in the northern part of the temperate climatic belt. Such adaptation occurred as the result of a decrease in the already low threshold of virus reproduction (to 0−10°C) in arthropod vectors.

Whether there are vectors to transmit viruses depends on the gene pools of the vector and the host. The ability of a virus to reproduce in arthropod tissues under a given temperature after the minimal inoculation dose, to penetrate into the hemolymph through the side of the intestine and reach salivary gland tissue, and to accumulate there in an amount sufficient to infect the vertebrate host during a bite, as well as a high density of vertebrate hosts, are the main criteria determining the significance of a species (or even a population) for the preservation and transmission of arboviruses.

The majority of arboviruses have been isolated from mosquitoes. Mosquito larvae can be easily infected via the alimentary canal. In addition to mosquitoes (family Culicidae), midges (family Ceratopogonidae) and sandflies (family Psychodidae, genus *Phlebotomus*) are the most effective vectors of arboviruses. These families of bloodsucking insects belong to the thread-horns (suborder Nematocera, order Diptera). There are no known effective vectors of arboviruses among the Brachycera suborder of Diptera, although some families of this suborder—Tabanidae, Simulidae, Hippoboscidae, etc.—contain bloodsucking insects. Nevertheless, some peculiarities of midge ecology permit us to attribute a high significance to these insects as vectors of arboviruses. Another bloodsucking insect is the American swallow bug (*Oeciacus vicarius*), which is the vector of at least two viruses of the family *Togaviridae*.[17,18] Under experimental conditions, many arboviruses can reproduce and increase their numbers in nonbloodsucking insects following parenteral inoculation.

Because of their long-term metamorphosis cycle (up to 5–7 years) and ability to be transmitted transstadially and transovarially, ixodid ticks (taxon Acari, family Ixodidae) are effective vectors and a natural reservoir for more than 70 arboviruses. Therefore, they are not only vectors, but also hosts, of many viruses.

The family Ixodidae comprises about 700 species grouped into 14 genera (Table 3.4). The main significance of ixodid ticks is in the

TABLE 3.4   Superfamily *Ixodoidea*

Family	Subfamily	Genus	Number of species
Ixodidae	*Amblyomminae*	*Amblyomma*	130
	Bothriocrotoninae	*Bothriocroton*	7
	Haemaphysalinae	*Haemaphysalis*	166
	Hyalomminae	*Hyalomma*	27
		Nosomma	2
	Ixodinae	*Cornupalpatum*	1
		Ixodes	243
	Rhipicephalinae	*Anomalohimalaya*	3
		Cosmiomma	1
		Dermacentor	34
		Margaropus	3
		Rhipicentor	2
		Rhipicephalus[a]	82
	Unclassified	*Compluriscutula*	1
Total	**6**	**14**	**702**
Argasidae		*Antricola*	17
		Argas	61
		Nothoaspis	1
		Ornithodoros	112
		Otobius	2
Total		**5**	**193**
Total sum	**6**	**19**	**895**

[a]*Includes as a subgenus the previously separate genus* Boophilus.

temperate climatic belt. Members of the genus *Ixodes* penetrate far in the high (subarctic and subantarctic) latitudes.

Members of the genus *Ixodes* have worldwide distribution. This genus is the most ancient and primitive in the Ixodidae family. The most primitive forms are distributed over Australia and South America. All *Ixodes* species have a three-host type of metamorphosis. The range of *I. uriae* comprises territories outside of the polar (Arctic and Antarctic) circles. The *Dermacentor* genus of ixodid ticks is distributed over Europe, Asia, Africa, and North America. The northern boundary of the range of *Dermacentor* lies more south than the northern boundary of *Ixodes*. Some species are usual habitants of leaf and mixed forests, as well as arid zone and rain forests, in Africa and southeastern Asia.[19] The life cycle of *Dermacentor* includes three host feedings. The range of the *Haemaphysalis* genus of ixodid ticks extends over all the continents except Antarctica. The majority of *Haemaphysalis* species are in Asia, especially southeastern Asia. Almost all species are parasites whose life cycle includes three host feedings. Species of the genus *Boophilus* (now categorized as a subgenus of the genus *Rhipicephalus*) are most common in tropical and subtropical steppes and in forest landscapes elsewhere. They are distributed over arid zones and have a life cycle of two or three host feedings.[20] Species of the genus *Hyalomma* of the Ixodidae family are distributed over arid landscapes of Africa, South and Central Asia, and the Mediterranean region of Europe. The life cycle includes one to three host feedings. The genus *Amblyomma* is common in the tropical, subequatorial, and equatorial belts of Africa, America, and Asia but is absent in Northern Eurasia.[15,21,22]

About 10% of arboviruses are isolated argas ticks (taxon Acari, family Argasidae). The family comprises five genera (Table 3.4), among which *Argas* and *Ornithodoros* are the most important. The northern border of Argasidae distribution in Eurasia is 45°N, while in North America it is 50°N. North America is the continent with the richest fauna of Argasidae; most species are xerophilous and therefore are distributed mainly over arid regions. In Northern Eurasia, argas ticks are distributed over a strip of landscape between arid Central-Asian and desert Mediterranean provinces.[16,23]

Argas ticks are shelter habitants, are polyphagous, and can attack a wide set of Vertebrata taxa from reptiles to humans. Their life cycle could be as long as 20–25 years, and they can survive without feeding for up to nine years.[22,23] These characteristics lend themselves to the long-term preservation of arboviruses in argas ticks, forming extremely stable natural foci.[14,15] Many species (especially those of the *Argas* genus) are closely associated with birds that enable the transcontinental transmission of viruses adapted to Argasidae ticks.

Gamasidae ticks (taxon Acari, superfamily Gamasoidea) have worldwide distribution. These ticks are in one of the most ancient phyla: the Arthropoda. Their role in virus circulation has not yet been investigated in detail, although about 10 arboviruses were isolated from natural sources and a set of laboratory experiments confirms the ability of Gamasidae ticks to preserve and transmit viruses.[15,16]

References

1. Haeckel E. *Generelle morphologie der organismen*. Berlin: Georg Reimer;18661: Allgemeine Anatomie der Organismen; 2: Allgemeine Entwickelungsgeschichte der Organismen [in German].
2. Fisher RA. *The genetical theory of natural selection*. Oxford: Oxford University Press;2000.
3. Turell MJ, Mather TN, Spielman A, Bailey C. Increased dissemination of dengue 2 virus in *Aedes aegypti* associated with concurrent ingestion of microfilariae of *Brugia malayi. Am J Trop Med Hyg* 1987;37:197–201.
4. Mettler LE, Gregg TG. *Population genetics and evolution*. Trenton, NJ: Prentice Hall;1969.

5. Theophilides CN, Ahearn SC, Grady S, et al. Identifying West Nile virus risk areas: the dynamic continuous-area space—time system. *Am J Epidem* 2003;**157**(9):843—54.

6. Lvov DK. Ecology of viruses. In: Lvov DK, editor. *Handbook of virology. Viruses and viral infections of human and animals*. Moscow: MIA;2013. p. 66—86 [in Russian].

7. Williamson M. *The analysis of biological populations*. London: Arnold;1972.

8. Lvov DK. Ecology of viruses. *Vestnik Academii Meditsinskikh Nauk SSSR* 1983;(12):71—82 [in Russian].

9. Henderson DA. Smallpox eradication—the final battle. *J Clin Pathol* 1975;**28**:843—9.

10. Kuno G. Transmission of arboviruses without involvement of arthropod vectors. *Acta Virol* 2001;**45**:139—50.

11. Lvov DK. The role of midges (Ceratopogonidae) in the circulation of arboviruses. *Parasitologiya* 1982;**16**(4):293—9 [in Russian].

12. Lvov DK, Deryabin PG, Aristova VA, et al. *Atlas of distribution of natural foci virus infections on the territory of Russian Federation*. Moscow: SMC MPH RF Publ;2001 [in Russian].

13. Murray KO, Mertens E, Despres P. West Nile virus and its emergence in the United States of America. *Vet Res* 2010;**41**(6):67.

14. Lvov DK, Il'ichev VD. *Migration of birds and the transfer of the infectious agents*. Moscow: Nauka;1979 [in Russian].

15. Lvov DK, Klimenko SM, Gaidamovich SY. *Arboviruses and arboviral infections*. Moscow: Meditsina;1989 [in Russian].

16. Lvov DK, Lebedev AD. *Ecology of arboviruses*. Moscow: Meditsina;1974, p.184. [in Russian].

17. Chamberlain RW. Epidemiology of arthropod-borne Togaviruses: the role of arthropods as hosts and vectors and of vertebrate hosts in natural transmission cycles. In: Schlesinger RW, editor. *The Togaviruses*. NY—London—Toronto—Sydney—San Francisco: Academy Press;1980, p. 175—228. Chapter 6.

18. Monath TP, Lazuick JS, Cropp CB, et al. Recovery of Tonate virus ("Bijou Bridge" strain), a member of the Venezuelan equine encephalomyelitis virus complex, from Cliff Swallow nest bugs (*Oeciacus vicarius*) and nestling birds in North America. *Am J Trop Med Hyg* 1980;**29**(5):969—83.

19. Kolonin GV. *World distribution of Ixodidae ticks: Dermacentor, Anocentor, Cosmiomma, Dermacentonomma, Nosomma, Rhipicentor, Rhipicephalus, Boophilus, Margaropus, Anomalohimalaya*. Moscow: Nauka;1984 [in Russian].

20. Kolonin GV. *World distribution of Ixodidae ticks: Haemaphysalis*. Moscow: Nauka;1978 [in Russian].

21. Pomerantsev BI. Fauna of USSR 1950;**4**(2): Ixodidae ticks [in Russian].

22. Pavlovsky EN. *Natural foci of transmission infections in connection with landscape epidemiology of zooanthroponoses*. Moscow—Leningrad: Nauka;1964 [in Russian].

23. Filippova NA. Fauna of USSR 1966;**4**(3): Argas ticks (Argasidae) [in Russian].

CHAPTER

4

Methods

4.1 ECOLOGO-VIROLOGICAL MONITORING OF NORTHERN EURASIA TERRITORIES

4.1.1 Purpose of Investigation

In the former USSR, there was a functioning and vigorous antiterrorism system similar in purpose to the CDC's Epidemiological Intelligence Service.[1–3] Its investigations, coordinated by the Center of Virus Ecology, were aimed at providing both a permanent means of predicting and uncovering emerging–reemerging infections and a way of minimizing the consequences of those infections, be they of natural or artificial origin. Results obtained from the investigations were also used for the study of biodiversity in different ecosystems of Northern Eurasia.[4–9] The system produced ecological evaluations of various regions and simultaneously collected field materials in along-meridian sections about 2,000 km long through the entire territory of Northern Eurasia. As a result, a landmass of more than 15 million km^2 was studied. Twelve zones were distinguished, each passing through unique ecosystems, including arctic, tundra, northern taiga, middle taiga, southern taiga, mixed forests, leaf-bearing forests, steppe, and desert, within 18 physicogeographical lands (Figure 4.1, Table 4.1). Natural preconditions for the existence of virus foci in various environments were investigated.

About 3 million mosquitoes, 100,000 ticks, 5,000 samples from the Vertebrata, and 4,000 human samples with unrecognized diagnoses were collected for virological investigation. The distribution of the main ecological complexes of viruses is presented in Table 4.2. The routes of transcontinental transfer of viruses associated with birds are shown in Figure 4.2. About 90 different viruses were isolated, of which 24 were described for the very first time. The etiological role of these viruses in human and animal diseases was evaluated. The potential for the emergence of epidemics in different landscape belts was determined. Also, areas ripe for the emergence of new viruses were predicted,[5–7,10,11] and all this activity was a part of the system for providing a permanent means of preventing bioterrorism events and consequences for the foreseeable future.[12]

4.1.2 Ecosystems in High Latitudes

Usually "high latitudes" means the arctic and antarctic), as well as the subarctic and subantarctic, tundra zones: the so-called ice zones (with July temperatures up to +5°C).[13,14] We would add to that list the wooded tundra and northern and middle taiga subzones. The reason for this addition is that the isotherm $\Sigma_{T \geq 10°C}$ ≥10°C reaches within the middle

49

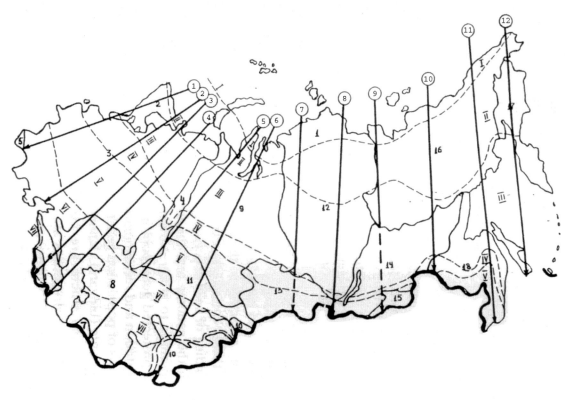

FIGURE 4.1   Ecological sounding for zoonotic viruses in the former USSR through different physicogeographical lands.

taiga, which corresponds approximately to the northern boundary of the range of *Ixodes persulcatus*, the main vector of the tick-borne encephalitis virus in the Eastern Hemisphere, as well as of other, ecologically related, *Ixodes* ticks found in the Western Hemisphere. In North America (Canada and Alaska) and in the Far East of Russia, the middle taiga region lies mostly to the north of latitude 50°N, whereas in the rest of Russia it lies mostly to the north of latitude 55°N (Figures 4.1 and 4.2, Tables 4.1 and 4.3).

We next discuss the principal high-latitude zones, keeping in mind that, of all abiotic factors, the organic world is the most sensitive to temperature, precipitation, and light. In respect of arboviruses, the greatest limiting factor of

these zones, compared with other climatic zones, is the temperature deficit that is more or less manifested. In some cases, the lower temperatures may limit the replication of viruses in bloodsucking Arthropoda; in others—for instance, in arctic conditions—the Arthropoda, which are both vectors and reservoirs of arboviruses, cannot themselves live. In its zoogeographic aspect, the territory discussed lies mainly within the arctic and circumboreal subregions of the Holarctic region. The existence of the same (or related) species in both the Eastern and the Western Hemisphere of this territory is typical. In Eurasia, the high-latitude territories to the west and to the east of the Yenisei River are within the European–Ob and Angara provinces, respectively; in America,

TABLE 4.1 The Main Landscape-Climatic Complexes of the North of Eurasia Included in Various Physicogeographical Lands (see Designations in Figure 1.1)

Zone, Subzone Landscape belts	In physicogeographical lands	Days per year with $T \geq 20°C$	$\Sigma_{T \geq 10°C}$	July isotherm, °C
Arctic	1, 9, 12, 16, 17	0	0	0–2
Central Arctic	1	0	0	0
Arctic desert	9, 12, 16, 17	0	0	0–2
Subarctic	2–4, 9, 12, 16, 17	0–10	0–600	2–13
Arctic tundra	2–4, 9, 12, 16, 17	0	0	2–5
Moss–lichen tundra	2–4, 9, 12, 16, 17	0	0–200	5–8
Bush tundra	2–4, 9, 12, 16, 17	0	200–400	8–10
Forest-tundra	2–4, 9, 12, 16, 17	10	400–600	10–13
Taiga	2–4, 9, 12, 16, 17	10–15	600–1,600	13–18
Northern taiga	2–4, 9, 12, 16–18	10	600–1,200	13–16
Middle taiga	2–4, 9, 12, 16, 17	10–15	1,200–1,600	16–18
Southern taiga	3–5, 9, 12,14,18	15–20	1,600–1,800	18–19
Mixed forest	3–6, 9, 11, 13, 15, 18	20–30	1,000–2,400	19–20
Leaf-bearing forest	3–6, 9, 11, 13, 15, 18	20–60	1,800–3,000	19–21
Steppe	3, 6–9, 10–13, 19	30–120	2,400–3,800	20–26
Wooded steppe	3, 6, 7, 9–11, 13, 19	30–60	2,400–3,000	20–21
Dry steppe	3, 6–8, 10, 11	60–90	3,000–3,200	21–22
Semi-desert	3, 6–8, 10	90–120	3,200–3,800	23–26
Subtropics (dry, humid)	6, 8, 10	120–150	3,800–5,000	26–30

they range across the Canadian provinces. The territories examined are situated in three climatic belts: the arctic and subarctic zones and the northern part of the temperate zone.

In its physicogeographical aspect, the territory in question is situated mostly within the following lands: (1) the Arctic, (2) the eastern part of Fennoscandia, (3) the northern part of the Russian Plain, (4) the northern part of the West Siberian Plain, (5) the northern part of the Central Siberian Plateau, (6) northeastern Siberia, and (7) the North Pacific part of Asia (Figure 4.1).

The arctic belt is subdivided into the central (inner) arctic and the arctic desert. The central arctic, with July temperatures about 0°C and with the period with isotherm $\Sigma_{T \geq 10°C}$ ≥10°C absent, has no dry land. The land fauna is represented by birds, mostly Laridae and Alcidae families. Among birds of the order Procellariiformes (Tubinares) are found those of the genus *Fulmarus* and, sporadically, birds of the order Anseriformes and suborder Limicolae. Among birds in the subfamily Passerinae, the lapland bunting (*Calcarius lapponicus*) is the most frequent. In July–August, the central

TABLE 4.2 Distribution of the Main Ecological Complexes of Zoonotic Viruses in Different Landscape Belts of Northern Eurasia

| | Abiotic markers | | | Complexes of viruses transmitted by | | | | | | | | | | | | | | | | | |
| | | | | Mosquitoes | | | | | Ticks | | | | | | | | | Sandflies | Without vectors | | |
Landscape belt	Number of days per year with T ≥ 20°C	Isotherm of July, °C	Sum of effective temperatures, $\Sigma T_{\geq 10^\circ C}$	California 1	Batai 2	Sindbis 3	Japanese encephalitis 4	West Nile 5	Tick-born encephalitis 6	Tyuleniy 7	Crimean-Congo hemorrhagic fever 8	Issyk-Kul 9	Tamdy 10	Sakhalin 11	Zaliv Terpeniya 12	Okhotskiy 13	Wad Medani 14	Sandfly fever Naples 15	Rabies 16	Hantaviruses 17	Influenza A 18
Arctic	0	0–2	0	−	−	−	−	−	−	−	−	−	−	−	−	−	−	−	−	−	−
Subarctic (tundra)	0	2–10	0–400	+	+	−	−	−	−	−	−	−	−	−	−	−	−	−	+	+	+
Northern taiga	0–10	10–16	400–1,200	+++	+	+	−	−	−	+++	−	−	−	+++	+++	+++	−	−	++	++	++
Middle taiga	10–15	16–18	1,200–1,600	+++	+	++	−	−	+	++	−	−	−	+++	+++	+++	−	−	++	++	++
Southern taiga	15–20	18–19	1,600–1,800	+++	+	+	−	−	+++	+	−	−	−	++	++	++	−	−	++	++	++
Leaf-bearing forest	20–60	19–21	1,800–3,000	++	++	+	+	+	++	−	−	−	−	+	+	+	−	−	++	+++	+++
Steppe	60–120	21–26	3,000–3,800	+	+++	+	+	+++	−	−	+++	+	−	−	−	−	−	++	++	+++	+++
Subtropic (wet, dry)	120–150	26–30	3,800–5,000	+	+++	+++	++	+++	−	−	+++	++	++	−	−	−	++	++	++	+	+++

1. *Bunyaviridae, Orthobunyavirus*, antigenic complex of California encephalitis virus.
2. *Bunyaviridae, Orthobunyavirus*, antigenic complex of Bunyamwera virus.
3. *Togaviridae, Alphavirus*, antigenic complex of Western equine encephalitis virus.
4. *Flaviviridae, Flavivirus*, antigenic complex of Japanese encephalitis virus.
5. *Flaviviridae, Flavivirus*, antigenic complex of Japanese encephalitis virus.
6. *Flaviviridae, Flavivirus*, antigenic complex of tick-borne encephalitis virus.
7. *Flaviviridae, Flavivirus*, antigenic complex of Tyuleniy virus.
8. *Bunyaviridae, Nairovirus*, antigenic complex of Crimean—Congo hemorrhagic fever virus.
9. *Bunyaviridae, Nairovirus*, antigenic complex of Issyk-Kul virus.
10. *Bunyaviridae, Nairovirus*, antigenic complex of Tamdy virus.
11. *Bunyaviridae, Nairovirus*, antigenic complex of Sakhalin virus.
12. *Bunyaviridae, Phlebovirus*, antigenic complex of Uukuniemi virus.
13. *Reoviridae, Orbivirus*, antigenic complex of Kemerovo virus.
14. *Reoviridae, Orbivirus*, antigenic complex of Kemerovo virus.
15. *Bunyaviridae, Phlebovirus*, antigenic complex of sandfly fever Naples virus.
16. *Rhabdoviridae, Lyssavirus*, antigenic complex of rabies virus.
17. *Bunyaviridae, Hantavirus*, antigenic complex of hemorrhagic fever with renal syndrome virus.
18. *Orthomyxoviridae, Influenza A virus*, antigenic complex of Influenza A Virus.

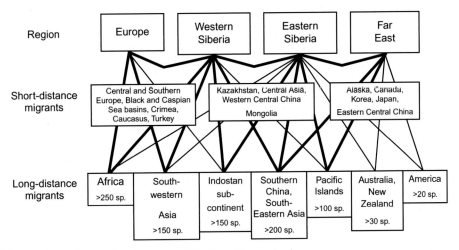

FIGURE 4.2 Nesting and overwintering areas of Northern Eurasia birds.

arctic becomes the arena of regular migrations of birds inhabiting arctic islands and coastal regions. Transcontinental migrations between Eurasia and America also occur. Bloodsucking Arthropoda vectors (*Ixodes* ticks and mosquitoes) are absent; therefore, there are no conditions favorable to the emergence of arbovirus natural foci.

The arctic desert subzone includes the terminal northern parts of Asia and America. In Russia (and in the former USSR as well), this territory covers 150,000 km². Mean July temperatures range from 0°C to 2°C; the conditions $T \geq 10°C$ is never reached. Glaciers occupy up to 80% of land territories; the rest—rock fields—is arctic desert and spotted polygonal semideserts.

About 30 bird species—mostly colonial seabirds—nest on the arctic coast. There is a regular exchange between avian faunas of the Northern and Southern Hemispheres. Besides birds, such vertebrates as lemmings (tribe Lemmini), polar foxes (*Alopex lagopus*), and polar bears (*Ursus maritimus*) are occasionally found. Natural infectious foci may be related to both bird colonies and lemmings. However, because there are no bloodsucking Arthropoda (although the role of Gamasidae ticks was not

investigated), there can be no foci of transmissible arbovirus infections. Of course, the absence of such foci does not exclude the adaptation of some viruses to other ways of transmission (e.g., contact, aerogenic, alimentary). It is important to note the same circumpolar character of species dissemination in the arctic belts of America and Asia, with possible interpopulation exchanges between animals and agents.[7,15-17]

Within the subarctic belt (tundra), one may distinguish the arctic, moss—lichen, shrub, and wooded tundra.

The arctic tundra in Russia (and the former USSR) occupies about 160,000 km². The condition $T \geq 10°C$ is never reached; the July—August isotherm is from 2°C to 5°C. The frost-free period is less than 110 days. Swamps and lakes occupy up to 60% of the total area, with polygonal tundra accounting for 20% and lichen tundra 15%. The dissemination of most species is circumpolar. The fauna is represented mainly by birds (about 50 species). The great number of bird colonies are scattered along the seacoast. Sparrows and snipes may bring viruses from places of hibernation. Mammals are represented by lemmings, polar foxes, and bears. It is important to note that

TABLE 4.3 Distribution of Arboviruses to the North from 55°N and to the South from 50°S

Family	Genus	Serological group	Virus	Place of isolation and distribution	Biotope	Isolation source
1	2	3	4	5	6	7
Togaviridae	*Alphavirus*	Western equine encephalitis	Karelian fever virus (Ockelbo, Pogosta disease)	Russia (Karelia), Sweden, Norway, Finland	Northern and middle taiga	*Ae. communis*, other *Aedes* mosquitoes, humans
Flaviviridae	*Flavivirus*	Tick-born encephalitis	Tick-born encephalitis virus	Eurasia to the South from 60–65 N	Middle taiga	*Ixodes persulcatus* and *I. ricinus* ticks, humans
		Tyuleniy	Gadgets Gully virus	Macquorie Islands (54°S, 159°E)	Coastal colony of *Penguin rookeries*	*Ixodes uriae* ticks
			Tyuleniy virus	Russia (Far East, Kola Peninsular), USA	*Alcidae* birds colony	*Ixodes uriae* ticks
Bunyaviridae	*Orthobunyavirus*	Bunyamwera	Batai virus	Norway (60°N)	Middle taiga	*Aedes, Anopheles* mosquitoes
			Northway virus	Alaska, Canada	Tundra, Northern taiga	*Aedes* mosquitoes, rodents
		California	Snowshoe hare virus	USA, Canada (to the north from 50°N)	Northern and middle taiga, tundra	Mostly *Aedes* and *Culiseta* mosquitoes
			Khatanga virus	North of Russia	Northern and middle taiga, tundra	Mostly *Aedes* and *Culiseta* mosquitoes
			Inkoo virus	Finland	Middle taiga	*Aedes* mosquitoes
			Unidentified virus	Norway	Tundra	*Aedes* mosquitoes
	Phlebovirus	Uukuniemi	Precarious Point virus	Macquorie Islands (54°S, 159°E)	Coastal colony of *Penguin rookeries*	*Ixodes uriae* ticks
			Uukuniemi virus	European part of Russia, Finland (to the south of 61°N)	Middle taiga	*Ixodes ricinus* ticks
			Zaliv Terpeniya virus	North of Eurasia, North of America, Iceland	*Alcidae* birds colony	*Ixodes uriae* ticks
			Komandory virus	Bering Sea basin	*Alcidae* birds colony	*Ixodes uriae* ticks

Family	Genus	Serogroup	Virus	Location	Host	Vector
	Nairovirus	Sakhalin	Sakhalin virus	Okhotsk, Bering, Barents Sea basins, South Ocean	*Alcidae* birds colony, penguins	*Ixodes uriae* ticks
			Paramushir virus	Okhotsk Sea basin	*Alcidae* birds colony	*Ixodes uriae* ticks
			Taggert virus	Macquorie Islands (54°S, 159°E)	Coastal colony of Penguin rookeries	*Ixodes uriae* ticks
			Klo-Mor virus	Russia (north of the Far East), UK	*Alcidae* birds colony	*Ixodes uriae* ticks
Reoviridae	*Orbivirus*	Kemerovo (Great Island)	Aniva virus	Okhotsk Sea basin	*Alcidae* birds colony	*Ixodes uriae* ticks
			Bauline virus	Canada (Newfoundland)	*Alcidae* birds colony	*Ixodes uriae* ticks
			Cape Wrath virus	Canada (Newfoundland)	*Alcidae* birds colony	*Ixodes uriae* ticks
			Great Island virus	UK (Scotland), Canada (Newfoundland)	*Alcidae* birds colony	*Ixodes uriae* ticks
			Mikines virus	Faeroes Islands (62°N, 07°W)	*Alcidae* birds colony	*Ixodes uriae* ticks
			Nugget virus	Macquorie Islands (54°S, 159°E)	Coastal colony of Penguin rookeries	*Ixodes uriae* ticks
			Okhotskiy virus	Russia (north of the Far East, Kola Peninsular	*Alcidae* birds colony	*Ixodes uriae* ticks
			Tindholmur virus	Faeroes Islands (62°N, 07°W)	*Alcidae* birds colony	*Ixodes uriae* ticks
			Yaquina Head virus	USA (Oregon, Alaska)	*Alcidae* birds colony	*Ixodes uriae* ticks
Rhabdoviridae	*Lyssavirus*	Sawgass	New Minto virus	USA (Alaska)	Northern taiga	*Haemaphysalis leporispalustris* ticks
Coronaviridae	Unclassified	Unclassified	Runde virus	North of Norway	*Alcidae* birds colony	*Ixodes uriae* ticks

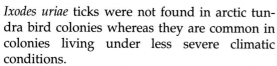

FIGURE 4.3　*I. uriae* ticks on the stones of Tyuleniy Island in the Sea of Okhotsk.

FIGURE 4.4　Nesting colony of guillemots (*Uria aalge*) on the rocks of the Commander Islands in the Bering Sea.

Ixodes uriae ticks were not found in arctic tundra bird colonies whereas they are common in colonies living under less severe climatic conditions.

The moss—lichen (typical) tundra in Russia occupies about 150,000 km^2; The sum of daily temperatures higher than 10°C over the year ($\Sigma_{T \geq 10°C}$) is about 200°C; July temperatures are 5—10°C. The frost-free period lasts 110—150 days. A large part of the area (30%) is under moss—lichen associations, while 25% are lakes and swamps and about 10% are shrubs. The ornithofauna includes 65—70 species, 50% to 80% of which are nesting within river valleys along river and lake banks. Avian faunas of arctic and moss—lichen tundra are similar in respect of the species involved. *I. uriae* ticks (Figure 4.3) have already been found in coastal and island bird colonies within the subzone (Figure 4.4). Therefore, the moss—lichen tundra subzone may be regarded as a potential northern boundary of arbovirus areas. Among mammals besides lemmings and polar foxes, the reindeer is widespread and, to the east from the Lena River, marmots as well. Mosquitoes can be found in this subzone, especially on the coast, although their population is not very high. Nevertheless, their presence determines in principle a possible existence of some arboviruses adapted to replication in mosquitoes.

Mosquito activity is observed in July and early August. Absolutely dominating species are circumpolar *Aedes* members: *Ae. communis*, *Ae. punctor*, *Ae. hexodontus*, *Ae. impiger*, and *Ae. nigripes*.

A permanent, though not numerous, representative of the mosquito fauna is *Culiseta alaskaensis*, which is very aggressive toward humans. Multiple small stagnant water bodies emerging in tundra during springtime serve as mosquito breeding sites. A short period with above-zero temperatures, low summer isotherms, possible frosts at a time detrimental for larvae—all these factors support an ecologically cold-adapted group of mosquito species in tundra. To the extent of moving down from the moss—lichen tundra to shrub and wooded tundra, mosquito populations increase greatly. The number of species found also expands somewhat: there appear *Ae. pullatus*, *Ae. excrucians*, *Ae. cantans*, *Ae. diantaeus*, *Ae. cinereus*, and *Ae. riparius*, all typical for the taiga zone. The population peak occurs in the middle of July. The active period of mass species in tundra does not exceed 1.0—1.5 months. All species attach to both animals and humans, but animals are preferred. Daily activity is regulated by temperature conditions. In years with a warm summer, activity lasts from 7—8 p.m. to 8—10 a.m. and then is somewhat suppressed by the bright light. In cold summers, the night flight of the mosquitoes is stopped at the morning maximum. Activity starts from the temperature $\pm(2.8-3.8)°C$.

In tundra, because shelter is absent, activity is suppressed by the wind. Suppression begins with a wind speed of 1.5−3.5 m/s and stops at 5.0 m/s.

Most fauna species in tundra are characterized by circumpolar dissemination in Eurasia and America, with possible interpopulation exchange of species and adapted agents.

There is no consensus as to the origin of the tundra landscape. Probably, its main elements appeared in the early Glacial Age, during the Neocene epoch, in the northern part of eastern Siberia and in the Bering Sea region of North America. To the west (western Siberia, the Russian Plain, Fennoscandia) the tundra landscape came from the east.

In the shrub tundra, taiga species prevail over tundra ones. This zone covers a part of the Kola Peninsula, as well as northern segments of the Russian Plain, of the West Siberian Plain, of Middle Siberia, and of Kamchatka−Kuril Land. Among mammals are snowshoe hares and some species of mouselike rodents. The population (and species composition) of mosquitoes increases.

The wooded tundra subzone stretches over a narrow strip from 20−30 to 200 km wide. $\Sigma_{T \geq 10°C}$ is about 600°C (from 400°C to 800°C). Mean July temperature is 11−14°C. There is a short period with mean daily temperatures above 15°C. Annual precipitation is 300−400 mm, which greatly exceeds the rate of evaporation. This proportionality presupposes an excessive wetting of the territory and its bogging up. Taiga species in this zone are even more prevalent than in the shrub tundra zone. Very large populations of mosquitoes with small numbers of basic species are typical for the zone. Valley forests of the wooded tundra directly turning into taiga masses are ecological routes for dissemination of the taiga

species to the north. Birds are represented here by the capercaillie, hazel grouse, and others, mammals by the snowshoe hare and others. Tundra fauna, however, is also found in watershed areas, swamps, and other water bodies, although in small numbers. The wooded tundra is the main grazing territory for domestic reindeer in winter, spring, and autumn.[7,15,17]

The taiga forest belt includes subzones of northern, middle, and southern taiga. In the northern subzone, $\Sigma_{T \geq 10°C}$ varies from 800°C in the north to 1,200°C in the south. The mean July temperature is from 12−13°C to 16−18°C. Except at the center of the Yakut hollow, the mean humidification index is over 1.0. Dark coniferous forests prevail in the western and eastern parts of the belt, which have a moderately continental climate, and light coniferous and leaf-bearing forests predominate in the inner part, which has an extreme continental climate. Like the wooded tundra, the taiga is very much bogged up.

The mosquito fauna in the northern taiga is much more diverse than its subarctic counterpart. Over 25 species are found in the northern taiga, over 30 in the middle taiga. Aedes mosquitoes, such as Ae. punctor, Ae. communis, and, in some places, Ae. pullatus, Ae. excrucians, Ae. intrudens, Ae. sticticus, and Ae. pionips, prevail. The mosquito flight starts in late May to early June. The first species to appear, having hibernated at the imago stage, are Culiseta alaskaensis and Cs. bergrothi. However, their populations are small. During the first 10 days of June on which the temperature is under 10°C, Ae. communis and Ae. cataphylla fly out, and 5−6 days later Ae. punctor, Ae. dorsalis, and Ae. diantaeus; in the middle of June, Ae. excrucians, Ae. cinereus, Ae. pionips, and Ae. pullatus appear. A mass flight of Ae. communis and Ae. punctor (the dominant species) is observed from the middle of June to late August, with maximum numbers appearing from late June to late July. In this period, 4,000−6,000 mosquitoes per hour may attack a human, a figure higher than in other zones. Larvae start to hatch in water temperatures of 4–5°C, and hatching is optimum in temperatures of 12–14°C. In spring, with snow thawing, a large number of small bodies of water are formed, yielding larvae that hatch in late May through the middle of June.

It is important to note that, in the northern taiga, all species have time for one gonotrophic cycle only. As in the subarctic zone, natural foci may exist only for arboviruses that are capable of transovarial transmission, the only way to preserve viruses during winter periods and to ensure the flight of virusophoric mosquitoes in the year after infection.[7,14,15,17]

In the middle taiga subzone, $\Sigma_{T \geq 10°C}$ is increased up to 1,400–1,600°C and the isotherm of July is 16–18°C. The subzone occupies nearly 40% of the taiga, with a 43% share of dark coniferous forests. The number of mosquito species rises to 30. The dominant species remain *Aedes* members: *Ae. communis*, *Ae. punctor*, and, in some places, *Ae. excrucians*, *Ae. intrudens*, *Ae. cyprius*, *Ae. pionips*, *Ae. diantaeus*, *Ae. pullatus*, *Ae. cataphylla*, *Ae. leucomelas*, and *Ae. cinereus*. Some species of *Anopheles* (*An. beklemishevi*, *An. messeae*) and *Culex* (*C. pipiens*) also appear.

Note that *Ae. communis* and *Ae. punctor* dominate all over the north, from the tundra to the middle taiga. The time for flight and mass activity of the dominant species in the middle taiga is 1–2 weeks ahead of that in the northern taiga, and the active season lasts up to 4 months, from late May to middle September. The attack density per person–hour may reach 3,000 specimens, lower than in the northern taiga or wooded tundra. Under middle taiga conditions, *Ae. punctor* can make a second gonotrophic cycle, greatly enhancing its role as a vector. Nevertheless, the monocycle prevails in the middle taiga as well. Therefore, the diverse seasonal population manifests one maximum. Both forest water bodies (in which 17 species are found) and open ones (19 species found) may serve as breeding sites.

The concentration of cattle and reindeer in the middle taiga stimulate mosquitoes to breed, thus raising their population.

It is known that *Ixodes persulcatus* ticks are the main vectors of the tick-borne encephalitis virus, and the northern boundary of their range is within the middle taiga. Probably, all the subzone should be regarded as dangerous in respect of tick-borne encephalitis. The ranges of other *Ixodes* members (e.g., *I. lividus*, *I. apronophorus*, *I. trianguliceps*, and *I. angustus*) are a bit further to the north from that of *I. persulcatus*. The relationship between these tick species and arboviruses has not been studied yet been studied. At bird nesting places in the shelf subzone of the Pacific and Atlantic Oceans, temperatures correspond approximately to those of the middle taiga. The subzone is the southern border of the range of *I. uriae*.

The taiga fauna is much richer than in the subarctic one. The number of migrating bird species increases sharply in moving successively from the subtropical, to the tropical, subequatorial, and equatorial zones, in the latter of which a year-round circulation of arboviruses is possible. In the springtime, some of these agents may be brought into the taiga zone, leading to the formation of seasonal and, in certain cases, permanent natural foci. A typical example is Karelian fever.

The *passeridae* are predominant. The greatest number (over 400) of typical taiga species was found in eastern Siberia. Probably, this geographical region was the center where many bird species were formed during the postglacial period. Later, they disseminated to the west as far as eastern Europe.

Among mammals in the taiga are snowshoe hares, many species of mouselike rodents, and insectivorous animals serving as blood meals for both Ixodidae ticks and mosquitoes.

The fauna of eastern Siberia takes its origin in the fauna of the ancient Angara continent, which in turn came out of the Oligocene–Miocene Chinese–Manchurian fauna influenced by North American fauna through Beringia. The reverse enrichment of the American fauna by eastern Siberian species took place through Beringia as well. The glacial period did not affect the eastern fauna of the northern Palearctic ecozone very much compared with the western one. Eastern Siberia's xerothermic climate in the inter- and postglacial periods determined the population of the taiga (and even the tundra) by steppe elements. For instance, in Yakutia and to the east of the Lena River, one may detect links between taiga and tundra fauna, on the one hand, and Dauria steppes, on the other: Altai pika, marmots, and ground squirrels inhabit the steppes of both biomes.

The taiga landscape emerged out of a cold snap in the Neocene epoch before the glacial period. In North America, dark coniferous forests were common during the Eocene. In Eurasia, the homeland of taiga landscapes are mountain ridges of eastern and northeastern Siberia. In the period of the tugai forests, the taiga was a vertical landscape, but after the cold snap, it proliferated further to the Siberian plains and then to the Russian Plain and Fennoscandia. The formation of forest-free moss and grass swamps in places of excessive humidity typical of the taiga also dates from the beginning of the glacial period of the Neocene.[7,14,15,17]

4.1.3 Ecosystems of the Central Part of the Temperate Climatic Belt

The temperate climatic belt includes the southern taiga subzone of the taiga belt in the north, mixed forests in the west, leaf-bearing forests in the east, and a forest–steppe belt in the south. The territory covers the Baltic coast, the Western European lowland, the center of the Russian Plain, the middle and southern Ural Mountains, southern Siberia, the valleys in the low stream of the Amur River, Primorsky Krai (the Ussuri Taiga), and the southern part of Sakhalin Island. Evaporability in the belt is slightly less than the annual sum of precipitation for practically all of the regions (except the southern forest–steppe strip). Swamps predominate (especially in the southern taiga), producing temporary and constant water reservoirs suitable for the development of mosquito larvae with several gonotrophic cycles in the warm period of the year. $\Sigma_{T \geq 10°C}$ is 1,600–3,800°C, insufficient for the reproduction of arboviruses distributed over the equatorial, subequatorial, and tropical climatic belts. A short (<60 days) period with $T \geq 20°C$ is a limiting factor for the circulation of many arboviruses as well (Table 4.4).

The southern taiga climatic belt is characterized by $\Sigma_{T \geq 10°C}$ 1,600–1,$\approx$800°C and a July isotherm of 18–19°C.

The species composition of bloodsucking mosquitoes is sufficiently depleted in the zone, but the density of the mosquito population is much higher in the southern part of the temperate and subtropical climatic belts. In the northern parts of the range of bats (*Myotis mystacinus*, *Myotis nattereri*, *Myotis dasycneme*, *Myotis daubentonii*, *Plecotus auritus*, *Nyctalus lasiopterus*, *Nyctalus leisleri*, *Nyctalus noctula*, *Pipistrellus pipistrellus*, *Pipistrellus nathusii*, *Eptesicus nilssonii*, *Eptesicus bobrinskoi*, *Eptesicus nilssonii*, *Eptesicus serotinus*, *Murina leucogaster*, and *Murina ussuriensis*), the number of these animals is insufficient for the maintenance of virus circulation. The ornithofauna is enriched by both migrating and settled species. There is a possibility of virus penetration by migrating birds from overwintering places in the springtime, with the further formation of seasonal or stable natural foci (Figures 4.5–4.6). In the southern part of the temperate climatic zone,

TABLE 4.4   Antibacterial Preparations[a], Used for Decontamination of Collection Strains

Antibiotics	Sensitive microbes	Type of action	Optimal concentration for cell cultures U/mL, μg/mL	Optimal concentration for virus infected cell cultures un/mL, μg/mL
Penicillin	B+	B/c	100.0	50.0
Streptomycin	B−, B+	B/c	100.0	50.0
Monomicin	B−, B+, M	B/c	100.0	50.0
Neomycin	B−, B+, M	B/c	50.0	25.0
Kanamycin	B−, B+, M	B/c	200.0	50.0–100.0
Gentamicin	B−, B+, M	B/c	200.0	30.0
Tetracycline	B−, B+	B/c	100.0	50.0
	M	B/st	30.0	50.0
Nystatin	mf	F/st	50.0	25.0
Thiamuline	B+, B−, M	B/c	10.0	5.0
Tilmicosin	B+, B−, M, mf	B/c	5.0–10.0	2.5–5.0

[a]Designations: B+, Gram-positive bacteria; B−, Gram-negative bacteria; mf, microscopic fungi; M, mycoplasma; B/st, bacteriostatic action of the preparation; B/c, bactericidal effect of the drug; F/st, fungistatic action of the preparation.

up to isoline $\Sigma_{T \geq 10°C}$, equaling 2,000–2,400°C, there is a possibility of outbreaks that are etiologically linked to viruses (Japanese encephalitis virus, West Nile virus, Sindbis virus, etc.) transmitted by mosquitoes. In the more northern parts, only those viruses can circulate which are capable of transovarial transmission and are adapted to temperatures <15°C (Bunyamwera serogroup, California serogroup, etc.). Pasture Ixodidae ticks are widely distributed, with optimum distribution in the southern taiga (*I. persulcatus*) and mixed forests (*I. ricinus*).

The leaf-bearing forest climatic belt is characterized by $\Sigma_{T \geq 10°C} \approx 1,800–2,400°C$ and July isotherm of 19–20°C. In this landscape belt, one can obtain the number and peak species variability for the *Haemaphysalis* genus of Ixodidae ticks, a genus that also can be found to the south, in the forest–steppe zone, but in that zone environmental conditions are too droughty for hygrophilous *Haemaphysalis* ticks.

The members of the *Dermacentor* genus of Ixodidae ticks are also distributed over the leaf-bearing forest climatic belt, but their optimum shifts into the forest–steppe zone.

The forest–steppe climatic belt is characterized by $\Sigma_{T \geq 10°C} \approx 2,400–3,000°C$ and a July isotherm of 20–21°C. The zone is divided into north and south subzones. The latter reaches near the steppe and is enriched by cereals in the plains and valleys, as well as by separate islets of wood in the watersheds. Before anthropogenic modification, motley grass steppe alternated with the leaf-bearing wood. Human activity led to a reduction in the area of forests.[7,16]

4.1.4 Ecosystems of the Southern Part of the Temperate and Subtropical Climatic Belts

The southern part of the temperate and subtropical climatic belts includes the southern

FIGURE 4.5 Migratory roots of continental and transcontinental seasonal flights of birds in Eurasia: West European and Indian–Asian pathways.

regions of the European part of the former USSR, as well as the Caucasus Mountains and Middle Asia, within steppe, semidesert, and subtropical (arid and wet) belts. Evaporability is more than the annual sum of precipitation in all belts except the wet subtropics. $\Sigma_{T \geq 10^{\circ}C} \approx 3{,}800{-}5{,}000^{\circ}C$, a temperature range that is quite sufficient for many viruses that reproduce in Arthropoda (including mosquitoes and midges) within the temperature range 16–24°C. The number of days with $T \geq 20^{\circ}C$ exceeds 60. The diverse species composition of arthropod vectors and vertebrate hosts results in a large potential for the circulation of viruses associated with mosquitoes, midges, and Argasidae and Ixodidae ticks. A high number and population density of bats, as well as *A. vespertilionis* and *I. vespertilionis* linked with bats, promote the formation of stable natural foci in the system bats–virus–ticks. Also, during bird migrations, viruses could form both seasonal and stable natural fociin the system birds–virus–(ticks, mosquitoes). The most suitable environmental

FIGURE 4.6 Migratory roots of continental and transcontinental seasonal flights of birds in Eurasia: East European and East Asian pathways.

conditions for such natural foci are in the subtropics, especially the wet regions (Georgia, Azerbaijan) with $\Sigma_{T \geq 10°C} \approx 3,800-5,000°C$, a period of about 120–150 days when $T \geq 20°C$, and a mean July temperature of 26–30°C. To the north, abiotic factors of the environment that influence arbovirus circulation become less favorable.

In the steppes, $\Sigma_{T \geq 10°C}$ is 3,000–3,200°C, $T \geq 20°C$ for 60–90 days, mean July temperature is 21–22°C. In the semidesert belt, $\Sigma_{T \geq 10°C}$ is 3,200–3,800°C, $T \geq 20°C$ for 90–120 days, and mean July temperature is 24–26°C. Within

the steppe belt lies the northern boundary of the Argasidae tick's range (the boundary corresponds approximately to the isoline, where the frost-free period is 150–180 days and $T \geq 20°C$ is 90–100 days long), as well as the ranges of the subfamilies of ticks Haemaphysalinae, Hyalomminae, and Rhipicephalinae. A set of viruses having significance for human pathology is distributed in this region. The subtropical zone lends itself naturally to the circulation of practically all viruses. Taking into account the presence of local populations of *Ae. aegypti* and *Ae. albopictus* in the wet subtropics on the Black

Sea coast of the Caucasus region makes it clear that at least seasonal circulation of the yellow fever, dengue, and chikungunya viruses is possible in the future.[7,14,17,18]

4.2 LONG-TIME STORAGE OF ISOLATED STRAINS

The main goal for the collection of viruses that are pathogenic or conditionally pathogenic to humans is to preserve the gene pool of the viruses isolated during the 20th and 21st centuries for present and future generations of researchers. Collecting strains of viruses (reference strains, strains of viruses suitable for the production of diagnostic preparations and vaccines, and strains selected in the laboratory or viruses isolated from nature that are of human, animal, or vector origin) is only one part of the task involved. As important, or even more important, is the necessity to store the original strains of the viruses for decades with no change to their initial properties: their biological, antigenic, and genetic structure, and their immunogenic potency. We know how important it is to keep the viruses intact without passages for many decades. Storing viral strains without passages ensures that they remain safe from contamination in passage.[1] Suitable methods of preserving viruses were developed at the D.I. Ivanovsky Institute of Virology of the Russian Ministry of Public Health.

The Russian State Collection of Viruses (SCV) has been housed in the D.I. Ivanovsky Institute of Virology of the Russian Ministry of Public Health since the Institute was founded in 1944. Order N 205a of the Minister of Public Health of the USSR approved the SCV in 1956. The status of the Russian SCV was confirmed in 1993, and the agency issued "regulations on the State collection of viruses belonging to I–III pathogenic groups." The status also was confirmed and recorded in decision 24.06 of the Government of the Russian Federation in 1996. In the decision, the Russian SCV was recognized as a national treasure, and the government acknowledged the necessity to create conditions for the maintenance of the viruses. According to the program of the Ministry of Public Health of the Russian Federation on the maintenance of the SCV in 2011–2012, work was undertaken to provide the SCV with modern facilities, affording reliable conditions for the storage of viruses and the safety of the Russian SCV personnel handling them. A system of monitoring and warning systems for emergencies arising in warehouses of viruses was established, as were a secure local area network of the Russian SCV and the automation of processes for depositing, accounting for, storing, and delivering material. Electronic forms of documents and an electronic document management system were developed. The Russian SCV now has new low-temperature freezers with increased energy conservation, including freezer Sanyo MDF-S2156 VAN ($-150°C$) for the storage of frozen virus suspensions. All freezers are equipped with reserve cooling based on liquid carbonic acid, as well as with recorders to record and monitor their temperature.

A programmable freezer IceCube14S-A, SY-LAB Gerate GmbH (Austria), was purchased for the storage of samples of virus strains and cell cultures in liquid nitrogen. Local servers were installed to save data collected, and software was installed that allows low-temperature freezers and thermostats to connect up to the same network in order to monitor the condition of equipment and preserve the data on the server.

The Russian SCV has numerous systems to ensure its safety: an alarm system, an access control system for the delimitation of zones, the primary and backup camera systems, and an automatic fire extinguishing system.

A negative pressure was created in the isolation areas of the lab, eliminating the potential for the release of pathogens into the environment during use of the lab equipment. Twenty-eight air conditioners regulate both the low-temperature freezers in the Russian SCV and an array of backup freezers, as well as maintain the required temperature for processing equipment and scientific instruments. The Russian SCV has a new-generation sequencer—the SU 410-1001 PRE—and a high-performance sequencing system—the Illumina MiSeq.

The Russian SCV is registered in World Health Organization (WHO) Collection Centers as an information center that allows scientists to deposit and issue strains of viruses, including strains that can be used for production. So, the Russian SCV both is the national property of Russia and has international importance.

According to the latest report from the International Committee on Taxonomy of Viruses (2012), 6 orders, 87 families, 19 subfamilies, and 349 genera comprising 2,284 viruses have been classified (37 strains of viruses remain unclassified).[2] Only 32 families include viruses related to groups I–III, which are pathogenic to humans, and the Federal Service for Surveillance on Consumer Rights Protection and Human Wellbeing of the Russian Federation has granted permission for scientists to work with these viruses in biological and virological laboratories in large cities of Russia.[3–5] The Russian SCV has kept a collection of approximately 3,000 virus strains belonging to the following 18 families: the RNA viruses Arenaviridae, Bunyaviridae,[6] Caliciviridae, Coronaviridae, Flaviviridae,[7] Orthomyxoviridae, Paramyxoviridae, Picornaviridae, Rhabdoviridae, Togaviridae,[8] Retroviridae, and Reoviridae; and the DNA virus strains Adenoviridae, Hepadnaviridae, Herpesviridae, Papillomaviridae, Parvoviridae, and Poxviridae.

The Russian SCV also contains nine strains of prions and 59 strains from two *Chlamydia* genera: *Chlamydia* and *Chlamydophilia* of the family Chlamydiaceae.[9]

The Russian SCV is collecting and keeping alive model virus strains isolated in the Russian Federation and abroad in view of their potential value for the development of theoretical bases for the medical and biological sciences, as well as their potential value in studying the etiology and epidemiology of human diseases and in preparing diagnostic and prophylacticmedications.[10]

It should be stressed that many laboratories in Russia that work with viruses or with suspicious material on infectious viruses have their own working collections of virus strains. They deposit the most notable strains of these collections into the Russian SCV with the purpose of patenting a procedure to obtain a certificate acknowledging the deposit of original author's strain. Such a certificate is necessary for the subsequent patenting by authors of different kinds of preparations created from that strain. At present, most viruses, chlamydiae, and prions are stored long term in liquid nitrogen at $-196°C$, and are preserved by freeze-drying (i.e., drying in a vacuum while in a frozen state).[11–13]

The material presented in this section is a summary of the results of a multiyear (up to 60 years) experience of Russian SCV work aimed at depositing, collecting, and preserving viruses, prions, and chlamydiae for long-term storage by means of freeze-drying them. All studies were conducted on dozens of samples of each virus strain deposited from 18 families.[9,14–17] Preservation methods based on freeze-drying and on the selection of effective temperatures for storage in the Russian SCV have retained all viable pathogens without passages: human pathogenic viruses from 18 families, 59 strains of the *Chlamydia* family Chlamydiaceae, and 9 strains of prions. This record of preservation is most significant for the collection work of the SCV, allowing pathogens to retain their biologic properties for many decades.[12,18]

All national and state collections in many countries currently hold viruses, prions, and chlamydia in liquid nitrogen at −196°C or in the form of freeze-dried suspensions at a storage temperature of −20°C, −40°C, or −60°C. According to our research, it is possible to capture the original properties, structure, and characteristics of a strain for many decades by freeze-drying using optimal fillers and a definition of constant effective temperature storage of −20°C, −40°C, or −60°C. This technique has made it possible not only to preserve the gene pool of strains from the 20th and early 21st centuries, but also to provide model (reference) virus strains of the viral profile of the Russian SCV to industrial enterprises and scientific research institutions the world over.

The principal pattern of the cryostasis of freeze-dried viruses, chlamydiae, and prions is as follows: The deaths of infectious virions, chlamidiae, and prions are the same for all members of the same family and depend not only on the filler or the method of freeze-drying, but, above all, on the constant subzero temperatures and long-storage vacuum containers (vials or bottles). The inside container should not be conducive to the development of the processes of oxidation. Vacuum storage must be in a container ≤100 microns.

The freeze-drying technique comprises the following operations: (i) preparation of the pathogenic biological agent (PBA) as a suspension ready for freeze-drying; (ii) the choice of the protective environment; (iii) the choice of the drying mode; (iv) sealing of the vials with permissible residual moisture.[12]

PBAs—viruses, prions, and chlamydiae—should be cleaned only by ballast, cell detritus by centrifugation at 2,000 rpm × 10 min or 10,000 rpm × 5 min. A higher degree of virus purification by ultracentrifuging is not necessary, because virions can lose their infectiousness if they are not protected by proteins left over after lyophilization.

Protective components during lyophilization serve as a buffer, which provides uniform drying of all components in the drying process. These components also provide a protective effect against exposure to PBA extreme factors: freezing, drying, and storage. Protection is achieved by lyophilization of the virus with a compound that uses protective components, which do not form chemically strong ties with the PBA. The rigidity of the virus's protein structures is enhanced through the redeployment of the molecular environment of the PBA in the drying components and, above all, by increasing the energy of the water—that is, by reducing the stock of free water in the system.

Comparative evaluation of the protein-free protective components (lecithin, sodium sulfate, polyalcohol, polyethylene oxide, and the sugars sucrose and sorbitol) showed that they can have a positive impact on the process of PBA freeze-drying, but a negative effect on the subsequent storage of dry viruses, prions, and chlamydiae at +4°C and −20°C. The PBAs are killed within two to three years of storage under these conditions. The protein and protein−carbohydrate combination protective components have a positive effect not only on the process of lyophilization, but on subsequent storage, affording the ability to store freeze-dried culture viruses, prions, and chlamydiae dozens of years in subzero temperatures. Like most researchers, we recommend these protective protein or protein−carbohydrate components, obtained on the base components for virus cultivation, because they have a greater stabilizing effect. Suspensions of mouse brain in an allantois liquid of embryo chicken origin with a PBA can be successfully freeze-dried and stored without preservatives. Note the negative effect of embryo yolk as a preservative. Our research indicated that, as a rule, adding yolks from 2.5% to 25% into the PBA suspension hinders lyophilization and subsequent storage at subzero temperatures. Under these conditions, chlamydiae die off within one year. Of the many preservatives of protein–carbohydrate origin we have studied, we recommend the one most suitable for viruses, for prions, and for chlamydiae: a mixture of 1% gelatin +5–10% sucrose or sorbitol.

The mode of drying is also important in the process of freeze-drying PBAs. First of all, it is necessary to consider to which group of pathogenicity the PBA belongs. The drying of pathogens belonging to the second group of pathogenicity requires a special protective block and personnel who are knowledgeable about protection. A combination of protective measures should always be provided, and filters— HEPA filters and filters from the Petriyanov's tissue—should always be installed in the drying apparatus at the outlet on the vacuum pump. For the lyophilization of microorganisms with high biological risk, chamber drying machines are used, with clogging under vacuum. Such machines have extended capabilities, making it possible to achieve total decontamination of the drying chamber.

Generally, samples were dried in vials 1–3 mL in volume, with silicone cork under vacuum or under an argon atmosphere, and then were drained by aluminum cap compression. Distinctive features of the modern chamber dryer are maximal automation on all stages of drying; the programming of all parameters of freeze-drying (time, temperature of the samples, temperature of the collector, strength of the vacuum), from freeze samples up to the closing of vials; and full display of all parameters, both electronically and on paper. A great means of enhancing the long-term storage of viruses, chlamydiae, and prions is to get rid of any residual moisture left over after lyophilization. The optimal residual humidity of the dried PBA must be from 1% to 3%. No more than 0.5–1.0 mL of the PBA should be poured into ampoules, and no more than 0.2 mL into vials.

Because viruses, prions, and chlamydiae are subject to common rules of preservation by the method of lyophilization, the foregoing recommendations and others are detailed in the book *The Basic Principles of Preservation of Viruses by Lyophilization*.[12]

Achieving purity of the viral cultures and other PBAs, as well as eliminating contaminants from the systems and components used in the cultivation of strains, is one of the highlights of collection work, because it affects the assessment of the main biological characteristics of the strains. Cultures collected can be contaminated by bacteria, protozoa, fungi, mycoplasmas, chlamydiae, and viruses. By regulation of the Russian SCV, each strain shall be tested for the presence of contaminants and must be cleared of them before being, or at the moment they are, deposited into the Russian SCV (Table 4.4). Then the purified strain pool is preserved by lyophilization and, in accordance with copyright laws, the strain is ready for long-term storage. Details of the methods of diagnosing collection cultures for the presence of contaminants and decontaminating the cultures are described in the publication *Methodological Manual on Diagnostics of the Contaminants in the Viral Cultures and Methods of Decontamination*.[18–20] Table 4.4 lists the action of antibiotics on those contaminants which are most often found upon depositing strains into the Russian SCV.

Diagnosing and cleaning retroviruses, as well as other spontaneous viruses, that can contaminate virus strains and make them unsuitable for production is difficult and requires a great deal of responsibility.[1]

According to WHO requirements, vaccines and immunoglobulins, as well as strains of viruses used for production, must be free of exogenous agents. Modern design enables one to identify the viruses that cause changes in virus strains. Many human and animal viruses can cause latent or chronic infection both *in vitro* and *in vivo*. The persistence of the virus in the organism or in cell cultures can change many properties, including pathogenic properties of persisting viruses. One problem is that changes in viral contaminants can be so well hidden that they cannot be detected by conventional methods.

It is important to identify the viral contaminants or contaminants of other origin in samples of virus strains recommended for production and in probes of media used to grow viruses in cell cultures. To conduct the comprehensive studies required to identify those contaminants, it is necessary to know the history of obtaining cells and of obtaining the main ingredient of a culture medium, namely, serum.[1] For example, if bovine serum or calf serum will be used, it should be tested on parainfluenza type 3, rhinotracheitis, and diarrhea viruses; if fibroblasts of nine-day-old chick embryo are to be used, they should be tested on leukemia viruses, myeloblasts, and other oncogenic viruses of birds; if green monkey cell cultures are used, they should be tested on herpesviruses and cytomegalovirus of monkeys; if primary fibroblasts of human embryos will be used, they should be tested on HBsAg, human T-cell lymphotropic virus (HTLV), hepatitis C virus, and human immunodeficiency virus (HIV).

Cell lysis, the destruction of the cell monolayer's ability to form syncytia, is a cell transformation process that may take place after 5−10 serial passages of the virus. To test for contaminants that can cause or enhance cell lysis, cells are precipitated with the use of a centrifuge operating at 1,000 rpm for 10 min. The supernatant obtained is used as antigen-laden material with a set of hyperimmune sera for different model strains of viruses. In this manner, we can identify contaminants, depending on the nature and history of the cell culture. None of the strains of viruses used as models or in the manufacture of immunobiological products and that have been stored for decades in the Russian

SCV contain any contaminants acquired during passages in cell cultures or in chicken embryos. Viral strains in the lyophilized state are stored without passages.

4.3 NEXT-GENERATION SEQUENCING

4.3.1 Overview

Next-generation sequencing (NGS) technology has been developing since the late 1990s as a method aimed at dramatically improving the performance of genomic research, primarily for sequencing of the human and other genomes.[1] The first widely used method of DNA sequencing, developed by Frederick Sanger in 1977, employs synthesis with a chain-terminating inhibitor.[2] Frederick Sanger received a Nobel Prize in Chemistry for this invention in 1980, shared with Paul Berg and Walter Gilbert, the latter of whom developed a method of DNA sequencing by chemical degradation. Using his own method, Sanger sequenced the first DNA genome: that of the bacteriophage phiX174.[3] Following automation of the process of DNA electrophoresis and analysis, Sanger's method of sequencing has become routine and was used in all of the projects involving the sequencing of the complete genomes of various organisms, including *Haemophilus influenzae* (1995), *Mycoplasma genitalium* (1995), *Escherichia coli* (1996), *Bacillus subtillis* (1997), the tubercle bacillus *Mycobacterium tuberculosis* (1998), the yeast *Saccharomyces cerevisiae* (1996), the nematode *Caenorhabditis elegans*, arthropods, mammalians, and others.[4−9] Sanger's method played a crucial role in the celebrated Human Genome Project that was declared completed in 2003. The first pathogenic virus whose genome was completely sequenced was the Epstein−Barr virus (human herpesvirus 4

(*Herpesviridae*, *Lymphocryptovirus*)) (1984). Since then, the ever-accumulating viral genomic data have enabled scientists to detect viruses with hybridizing probes or by the specific amplification of DNA (through a polymerase chain reaction, or PCR). Currently, PCR and sequencing are the main universal tools for studying the distribution, diversity, evolution, and ecology of the viruses.

The main disadvantage of both the PCR technique and Sanger's sequencing of viruses is the necessity to use specific primers, making it difficult to detect and to sequence viruses that have significant genetic differences from known viruses. Also, many viruses are difficult or impossible to cultivate in the cell line or in laboratory animals, and they become "invisible" to the researcher. All these limitations can be resolved by using NGS technology, whose principal advantage is a far better performance. NGS technology is developing as a platform that allows the simultaneous sequencing of a plurality of individual DNA clones that are fixed to a solid surface (hence the name "massively parallel sequencing"). The number of the simultaneously sequencing clones can reach several hundred million, with a total sequenced length of up to 600 billion nucleotides. The length of the human genome is 3 billion base pairs (bp), the *E. coli* genome has about 5 million bp, the genome of the herpesviruses (Herpesviridae) is 150–200 kbp, and that of the flaviviruses (Flaviviridae) is about 10 kbp. The high performance of NGS makes it possible to find the viral sequences in the sample, even if they constitute just a hundredth of a percent of the total genomic material. Bioinformatic analysis of the sequences obtained enables the researcher to differentiate sequences belonging to different viruses and determine the totality and diversity of viruses in the sample. This approach opens new possibilities, both in medical virology and in research on the circulation of viruses in nature. Its chief advantage is that it affords rapid recognition of new pathogenic viruses and

timely interpretation of the outbreaks of emerging or reemerging viral infections. In recent years, many pathogenic viruses were discovered by the NGS approach: a new Ebola virus (of the family Filoviridae) in Uganda,[10] a novel phlebovirus (family Bunyaviridae) associated with severe fever with thrombocytopenia syndrome,[11] a new arenavirus (Arenaviridae) associated with deaths after organ transplantation,[12] new cosaviruses (*Picornaviridae*) causing child diarrhea,[13] and others.[14,15]

NGS technology gives researchers the opportunity to solve the important problem of the detection of minor variants in virus populations. The method's high performance allows the scientist to sequence virtually all copies of the viral genomes contained in a sample and to determine the presence of mutated viruses, even when they are few in number. Certain pathogenic mutations may alter the antigenic properties of the virus or change its specificity to cell receptors. Identifying these mutations in the virus population will more accurately predict the development of infection and enable physicians to adjust therapy. Other mutations can lead to the formation of viral resistance to antiviral drugs, so monitoring these mutations is essential for effective chemotherapy. This problem is particularly acute in such infections as HIV, influenza, hepatitis, and others.[16–19]

Viruses are the most abundant organisms on Earth. Up to 10^{10} virus particles can be found in 1 l of ocean waters.[20] A metagenomic approach based on NGS was used to research viral diversity in different natural reservoirs, including animal or bird feces, arthropods, plants, and the environment (ocean water, hot spring water, soil, etc.). Many new viruses belonging to more than 50 families were discovered in different sources.[21–27]

4.3.2 Sample Preparation

In the study of environmental samples (water, soil, feces, etc.), the first stage is homogenization and clarification of the material. This stage is followed by ultrafiltration through a 0.8-mcm pore filter and concentration of the viruses by ultracentrifugation prior to nucleic acid purification. In the analysis of biological samples (blood, organ tissues, insects, or plants), sample preparation involves the isolation of DNA or RNA. Because the virus genome is much smaller than the genomes of prokaryotes and eukaryotes, it is desirable to use techniques that get rid of high-molecular-weight DNA during the process of separation. Eliminating this DNA will reduce the contamination of the sample host DNA. If cDNA needs to be synthesized, it may be necessary to analyze the associated RNA reverse transcription reaction with a random primer.

Several platforms for NGS were developed that implement different strategies of sample preparation and registration of the signal, but all of them are based on the same general principles. The first step consists of the fragmentation of a DNA molecule to a certain size by ultrasonic or enzymatic means. This step is followed by ligating a DNA fragment with adapters and fixing it onto a solid surface (of a flow cell, or chip). Next, clonal amplification is performed, whereby physically isolated clusters of amplified DNA from individual DNA molecules are made to form on the chip's surface.

On an Illumina platform, clonal amplification and clustering are carried out on the total surface of the chip by the "bridge PCR." On a Roche or Life Technologies (Ion Torrent and Ion Proton) platform, clonal amplification is carried out on the surface of microspheres by the method of emulsion PCR (emPCR), thereby forming a mixture of microspheres, each of which carries its clone of the amplified DNA. Further, these microspheres are fixed in individual cells on the chip. Thus, the formation of isolated clusters of DNA is determined by the locations of cells on the chip and by filling them with microspheres. The number of clusters (cells) on the chip is between of millions and billions, depending on the performance of the system. Further sequencing is performed in parallel for each cluster. As a result, for each cluster (cell), a corresponding DNA sequence is determined. Usually, the length of a "read"—a short DNA sequence—is small, averaging from 36 to 250 bp. However, because of the large number of clusters, the total length of the sequenced DNA is from 40 million to 600 billion bp.

Different platforms use various enzymatic reactions to register the sequencing reaction. On an Illumina platform, sequencing is performed by synthesis, using fluorescently labeled deoxynucleotides. In 454 devices on the Roche platform, sequencing by synthesis is also applied, but registration of the reaction is based on the detection of pyrophosphate (pyrosequencing). In Sequencing by Oligonucleotide Ligation and Detection (SOLiD), sequencing is implemented through the ligation reaction with fluorescently labeled probes. In Ion Torrent and Ion Proton devices, protons (H +) that are released during the synthesis reaction are detected.

4.3.3 Bioinformatics

Row NGS data are represented by several million reads, corresponding to individual clusters. Processing is begun with the assembly of reads to a more extensive "contig" DNA sequence. This process is carried out with the help of special programs called assemblers and requires a high-performance computer. The assembly forms extensive consensus sequences of DNA—contigs, composed of many overlapping reads. Further analysis involves searching for homologous sequences of known viruses in the genomic database. The main tool is a service Basic Local Alignment Search Tool (BLAST, http://blast.ncbi.nlm.nih.gov), which uses different algorithms to search for homologues in the GenBank database (http://www.ncbi.nlm.nih.gov/genbank).

4.3.4 RNA Isolation

This section describes a standard protocol that is used for sequencing of RNA viruses at the D.I. Ivanovsky Institute of Virology in Moscow, Russia. The most appropriate method for the purification of RNA for subsequent sequencing is the use of a TRIzol® reagent. For RNA isolation from the tissues of infected animals, up to 30 mg of tissue should be homogenized. A pestle or ball mill in 1 mL of TRIzol reagent can purify 5–50 mcg of total RNA. After an incubation period of 10 min on the table, add 0.2 mL chloroform and mix thoroughly. Place the tubes into the centrifuge, which is precooled to 4°C, and centrifuge at 12,000g for 15 min. The result should be a separation into two phases: The lower phase contains phenol with fat-soluble contaminants, and the upper, clear water phase contains the RNA. The upper, water phase (approx. 0.5 mL) should be carefully transferred to a clear 1.5-mL tube.

For further purification, as well as to remove the low-molecular-weight fractions of rRNA (5S) and tRNA, RNA from the upper, aquatic phase can be cleaned by "RNeasy mini kit" (QIAGEN, Germany), according to the instructions.

To concentrate the virus from the culture medium, ultracentrifugation is used (30,000 rpm for 1 h is usually sufficient to precipitate virus). The virus pellet obtained may be dissolved in 1 mL TRIzol or in 350 mcL of RLT buffer for subsequent RNA purification by the "RNeasy mini kit."

To remove rRNA (18S and 28S), we used the "GenRead rRNA depletion Kit" (QIAGEN, Germany) according to the instructions. In general, about 1 μg of total RNA is sufficient. The efficacy of the depletion reaches 70–90%; thus, the amount of RNA obtained for further analysis is about 100 ng. For DNA-library preparation, 20–100 ng of the ribodepleted RNA is enough.

4.3.5 Preparation of DNA Libraries and Sequencing

In preparation for sequencing, 20–100 ng of ribodepleted RNA was fragmented in 15 mcL of reverse transcriptase buffer with random primer at 85°C for 5 min and then placed on ice. Next, 200 units of enzyme RevertAid Premium (Thermo Scientific, USA) and 20 units of RNase inhibitor RNasin (Promega, USA) were added to the fragmented RNA, and the reaction mixture was incubated at 25°C for 10 min and then at 42°C for 60 min. The reaction was stopped by heating at 70°C for 10 min. Second-strand cDNA synthesis was performed with the use of a "NEBNext® mRNA Second Strand Synthesis Module" (NEB, USA) in accordance with the instructions. The resulting dsDNA was purified with a "MinElute PCR Purification Kit" (QIAGEN, Germany) by the company's QIAcube automatic stations.

For the preparation of DNA libraries, a "TruSeq DNA Sample Prep Kits v2" (Illumina, USA) was used in accordance with the instructions. Alternatively, the NEBnext Ultra RNA Library Prep Kit (New England BioLabs Inc., USA) could be used, again in accordance with the manufacturer's recommendations. For the size selection of the DNA libraries, "AMpure XP" (Beckman Coulter, USA) beads were used. The size of the libraries was 270 nt, which corresponds to the size of the insert, 150 nt. The data requirements for the size of DNA libraries are associated with the use of a sequencing kit that produces pair-end sequences 150 nt in length (MiSeq Reagent Kit V2 300PE). The library that was obtained was visualized on the automatic electrophoresis station "QIAxcel Advanced System" (QIAGEN, Germany). Quantification of the libraries obtained was done by real-time PCR according to the recommendations outlined in the "Sequencing Library qPCR Quantification Guide" (Illumina, USA). Sequencing DNA libraries are held on the device MiSeq (Illumina, USA) with the use of a set of "MiSeq Reagent Kits V2 (300PE)" in accordance with the manufacturer's instructions.

Assembling contigs *de novo* and mapping reads to the reference sequence were performed with the program "CLC Genomics Workbench 6.0" (CLC bio, USA). Preliminary searching of homologous sequences was performed with the service BLASTX (http://blast.ncbi.nlm.nih.gov). The software package "Lasergene Core Suite" (DNAstar, USA) was used to analyze nucleotide and amino acid sequences. Sequence alignments were performed by means of the algorithm ClustalW. Phylogenetic analysis and dendrogram construction were carried out with the program MEGA5, a nearest-neighbor or maximum-likelihood algorithm with thousandfold bootstrap testing.

References

4.1 Ecologo-Virological Monitoring of Northern Eurasia Territories

1. Goodman RA, Bauman CF, Gregg MB, et al. Epidemiologic field investigations by the Centers for Disease Control and Epidemic Intelligence Service, 1946—1987. *Public Health Rep* 1990;**105**(6):604—10.
2. Langmuir AD. The Epidemic Intelligence Service of the Center for Disease Control. *Public Health Rep* 1980;**95**(5):470—7.
3. Langmuir AD, Andrews JM. Biological warfare defense. 2. The Epidemic Intelligence Service of the Communicable Disease Center. *Am J Public Health Nations Health* 1952;**42**(3):235—8.
4. Lvov DK, Aristova VA, Butenko AM, et al. *Viruses of Californian serogroup and etiologically linked diseases: clinic-epidemiological characteristics, geographical distribution, methods of virological and serological diagnostics. Methodological manual.* Moscow: RAMS;2003 [in Russian].
5. Lvov DK, Deryabin PG, Aristova VA, et al. *Atlas of distribution of natural foci virus infections on the territory of Russian Federation.* Moscow: SMC MPH RF Publ.;2001 [in Russian].
6. Lvov DK, Il'ichev VD. *Migration of birds and the transfer of the infectious agents.* Moscow: Nauka;1979 [in Russian].
7. Lvov DK, Klimenko SM, Gaidamovich SY. *Arboviruses and arbovirus infections.* Moscow: Meditsina;1989 [in Russian].
8. *Organization of ecological-epidemiological monitoring in Russian Federation for anti-epidemic defense of the civilians and army.* In: Lvov DK, editor. Moscow: Russian Ministry of Public Health, Federal Office of Biomedical and Extreme Problems, D.I. Ivanovsky Institute of Virology; 1993 [in Russian].
9. Shchelkanov MY, Gromashevsky VL, Lvov DK. The role of ecologo-virological zoning in prediction of the influence of climatic changes on arbovirus habitats. *Vestnik Rossiiskoi Akademii Meditsinskikh Nauk* 2006;**2**:22—5 [in Russian].
10. Lvov DK. Arboviral zoonoses of Northern Eurasia (Eastern Europe and the commonwealth of independent states). In: Beran GW, editor. *Handbook of zoonoses. Section B: Viral zoonoses.* London—Tokyo: CRC Press; 1994. p. 237—60.
11. Lvov DK, Butenko AM, Gromashevsky VL, et al. *West Nile and other emerging—reemerging viruses in Russia. Emerging Biological Threat. NATO Science Series. Series I. Life and Behavior Sciences*, vol. 370. Amsterdam: IOS Press;200533—42.
12. Lvov DK. Importance of emerging-reemerging infections for biosafety. *Vopr Virusol* 2002;**47**(5):4—7 [in Russian].
13. Chernov YI. *Environment and societies of tundra zone. Environment of Far North and man.* Moscow: Nauka;19858—22. [in Russian].
14. Puzachenko YG. *Climatic limit of southern border. Environment of Far North and man.* Moscow: Nauka;198522—56. [in Russian].
15. Lvov DK. *Ecological soundings of the former USSR territory for natural foci of arboviruses. Soviet Medical Reviews. Section E. Virology Reviews*, vol. 5. London: Harwood Academic Publisher GmbH.;19931—47.
16. Lvov DK. Ecology of viruses. In: Lvov DK, editor. *Handbook of Virology. Viruses and viral infection of human and animals.* Moscow: MIA;2013. p. 66—86 [in Russian].
17. Lvov SD. *Natural virus foci in high latitudes of Eurasia. Soviet Medical Reviews. Section E. Virology Reviews*, vol. 5. London: Harwood Academic Publisher GmbH; 1993137—85.
18. Shchelkanov MY, Lvov DK, Kolobukhina LV, et al. Isolation of Chikungunya virus in Moscow from Indonesian visitor (September, 2013). *Vopr Virusol* 2014;**59**(3):28—34 [in Russian].

4.2 Long-Time Storage of Isolated Strains

1. Shalunova NV. *The problem of contamination of medical biological products. A standardization of methods of control, the lack of which contaminate viruses* [Ph.D. thesis]. Moscow; 1991 [in Russian].

2. King AMQ, Adams MJ, Carstens EB, editors. *Virus Taxonomy: Ninth report of the international committee on taxonomy of viruses*. San Diego: Elsevier Academic Press;2012.

3. Lvov DK, Deryabin PG, Fadeeva LL. *Virology. Safe use of microorganisms in pathogenic groups I–II (endangered)*. SR 1.3.1285-03. Moscow: Ministry of Public Health; 2003 [in Russian].

4. Order of Russian Ministry of Public Health N 96 from 18-Feb-2004 "on safe storage of Wild Polioviruses" [in Russian].

5. Procedure for recording, storage, transmission and transportation of micro-organisms in pathogenic groups I–IV. SR 1.2.036-95. Moscow; 1996: 80p. [in Russian].

6. Fadeeva LL, Deryabin PG, Khaletskaya EV, et al. Catalogue of strains deposited and stored in the Russian State collection of viruses (SCV). 1. Bunyaviridae. Moscow; 2004 [in Russian].

7. Lvov DK, Fadeeva LL, Deryabin PG, et al. Catalogue of strains deposited and stored in the Russian State collection of viruses (SCV). 2. Flaviviridae. Moscow; 2005 [in Russian].

8. Lvov DK, Fadeeva LL, Deryabin PG, et al. Catalogue of strains deposited and stored in the Russian State collection of viruses (SCV). 3. Togaviridae. Moscow; 2006 [in Russian].

9. Fadeeva LL, Deryabin PG, Sukhno EO. Catalogue of strains deposited and stored in the Russian State collection of viruses (SCV). 4. Chlamydiaceae. Moscow; 2007 [in Russian].

10. Fadeeva LL, Khaletskaya EV, Melnichenko AV. The value of the Russian State collection of viruses in the identification and prevention of epidemic situations. In: Modern aspects of rehabilitation in medicine. Yerevan; 2003 [in Russian].

11. Dolinov KE. Technology basics of dry biological products. Moscow; 1996 [in Russian].

12. Fadeeva LL, Khaletskaya EV, Lvov ND, et al. The basic principles of preservation of viruses by lyophilization. Moscow; 2004 [in Russian].

13. Flint DM. *Some basic aspects of freeze drying of materials*. Moscow: Mir;19811–13 [in Russian].

14. Fadeeva LL, Khaletskaya EV, Lvov ND. Methodical recommendations. Conservation of Prions skreppi and conservation regimes. Moscow; 2002 [in Russian].

15. Fadeeva LL, Khaletskaya EV, Lvov ND. Methodological manual on the conservation of the viruses of the family Bunyaviridae and long-term storage modes. Moscow; 2003 [in Russian].

16. Lvov DK, Fadeeva LL, Deryabin PG. Manual for arboviruses stored in the Russian State collection of viruses. Moscow; 2006 [in Russian].

17. Nosik DN, Gushchin NV, Fadeeva LL. The preservation of the human immunodeficiency virus by lyophilization. *Vopr Virusol* 1990;3:243–5 [in Russian].

18. Fadeeva LL, Lvov ND, Khaletskaya EV, et al. Methodological manual on diagnostics of the contaminants in the viral cultures and methods of decontamination. Moscow; 2004 [in Russian].

19. Dyakonov LP, Citkova VI. Animal cell culture. Moscow; 2000: 3–398 [in Russian].

20. Granitov VM. Chlamydia. Moscow; 2002 [in Russian].

4.3 Next-Generation Sequencing

1. Ansorge WJ. Next-generation DNA sequencing techniques. *New Biotechnol* 2009;25(4):195–203.

2. Sanger F, Nicklen S, Coulson AR. DNA sequencing with chain-terminating inhibitors. *PNAS* 1977;**74**(12):5463–7.

3. Sanger F, Coulson AR, Friedmann T, et al. The nucleotide sequence of bacteriophage phiX174. *J Mol Biol* 1978;**125**(2):225–46.

4. C. elegans Sequencing Consortium. Genome sequence of the nematode *C. elegans*: a platform for investigating biology. *Science* 1998;**282**(5396):2012–18.

5. Casari G, Andrade MA, Bork P, et al. Challenging times for bioinformatics. *Nature* 1995;**376**(6542):647–8.

6. Fleischmann RD, Adams MD, White O, et al. Whole-genome random sequencing and assembly of *Haemophilus influenzae* Rd. *Science* 1995;**269**(5223):496–512.

7. Fraser CM, Gocayne JD, White O, et al. The minimal gene complement of *Mycoplasma genitalium*. *Science* 1995;**270**(235):397–403.

8. Goffeau A, Barrell BG, Bussey H, et al. Life with 6000 genes. *Science* 1996;**274**(5287):563–7.

9. Moszer I, Kunst F, Danchin A. The European *Bacillus subtilis* genome sequencing project: current status and accessibility of the data from a new World Wide Web site. *Microbiology* 1996;**142**(11):2987–91.

10. Towner JS, Sealy TK, Khristova ML, et al. Newly discovered ebola virus associated with hemorrhagic fever outbreak in Uganda. *PLoS Pathogens* 2008;**4**(11):e1000212.

11. Xu B, Liu L, Huang X, et al. Metagenomic analysis of fever, thrombocytopenia and leukopenia syndrome (FTLS) in Henan Province, China: discovery of a new bunyavirus. *PLoS Pathogens* 2011;**7**(11):e1002369.

12. Palacios G, Druce J, Du L, et al. A new arenavirus in a cluster of fatal transplant-associated diseases. *N Engl J Med* 2008;**358**(10):991–8.

13. Holtz LR, Finkbeiner SR, Kirkwood CD, et al. Identification of a novel picornavirus related to cosaviruses in a child with acute diarrhea. *Virol J* 2008;**5**:159.

14. Dacheux L, Cervantes-Gonzalez M, Guigon G, et al. A preliminary study of viral metagenomics of French bat species in contact with humans: identification of new mammalian viruses. *PloS One* 2014;**9**(1):e87194.

15. Lau SK, Li KS, Tsang AK, et al. Genetic characterization of Betacoronavirus lineage C viruses in bats reveals marked sequence divergence in the spike protein of pipistrellus bat coronavirus HKU5 in Japanese pipistrelle: implications for the origin of the novel Middle East respiratory syndrome coronavirus. *J Virol* 2013;**87**(15):8638−50.

16. Ghedin E, Laplante J, DePasse J, et al. Deep sequencing reveals mixed infection with 2009 pandemic influenza A (H1N1) virus strains and the emergence of oseltamivir resistance. *J Infect Dis* 2011;**203**(2):168−74.

17. Gong L, Han Y, Chen L, et al. Comparison of next-generation sequencing and clone-based sequencing in analysis of hepatitis B virus reverse transcriptase quasispecies heterogeneity. *J Clin Microbiol* 2013;**51**(12):4087−94.

18. Hoffmann C, Minkah N, Leipzig J, et al. DNA bar coding and pyrosequencing to identify rare HIV drug resistance mutations. *Nucleic Acids Res* 2007;**35**(13):e91.

19. Wang C, Mitsuya Y, Gharizadeh B, et al. Characterization of mutation spectra with ultra-deep pyrosequencing: application to HIV-1 drug resistance. *Genome Res* 2007;**17**(8):1195−201.

20. Bergh O, Borsheim KY, Bratbak G, et al. High abundance of viruses found in aquatic environments. *Nature* 1989;**340**(6233):467−8.

21. Breitbart M, Hewson I, Felts B, et al. Metagenomic analyses of an uncultured viral community from human feces. *J Bacteriol* 2003;**185**(20):6220−3.

22. Cann AJ, Fandrich SE, Heaphy S. Analysis of the virus population present in equine feces indicates the presence of hundreds of uncharacterized virus genomes. *Virus Genes* 2005;**30**(2):151−6.

23. Honkavuori KS, Shivaprasad HL, Briese T, et al. Novel picornavirus in Turkey poults with hepatitis, California, USA. *Emerg Infect Dis* 2011;**17**(3):480−7.

24. Li L, Victoria JG, Wang C, et al. Bat guano virome: predominance of dietary viruses from insects and plants plus novel mammalian viruses. *J Virol* 2010;**84**(14):6955−65.

25. Schoenfeld T, Patterson M, Richardson PM, et al. Assembly of viral metagenomes from yellowstone hot springs. *Appl Environ Microbiol* 2008;**74**(13):4164−74.

26. Shan T, Li L, Simmonds P, et al. The fecal virome of pigs on a high-density farm. *J Virol* 2011;**85**(22):11697−708.

27. Zhang T, Breitbart M, Lee WH, et al. RNA viral community in human feces: prevalence of plant pathogenic viruses. *PLoS Biol* 2006;**4**(1):e3.

ZOONOTIC VIRUSES OF NORTHERN EURASIA: TAXONOMY AND ECOLOGY

Order Mononegavirales

Order Mononegavirales includes four different viral families (Bornaviridae, Filoviridae, Paramyxoviridae, and Rhabdoviridae) possessing enveloped virions and a single-stranded, −negative-sense RNA genome that lacks a cap- and poly-A structure on the 5′ and 3′ termini.[1] The 3′ and 5′ ends of the order's genomic RNA contain untranslated sequences that possess some degree of terminal complementarity.[1−3] The length of the genomic RNA varies between 8,900 and 19,000 nt for different families. The number of encoded viral proteins also varies, from 5 to 11, but with the common structure 3′-N-P-M-G-L-5′. Genomic RNA is not infectious, and launching virus replication in an infected cell requires the presence of viral RNA polymerase, a few molecules of which are included in of the virus's nucleocapsid. All viral genes exist as separate transcriptional elements with their initiating and terminating structures. Genomic RNA contains, in its 3′-end region, a sequencing element that is necessary for viral RNA polymerase to bind and thereby initiate mRNA transcription. The genes are transcribed by sequential interrupted synthesis from 3′ to 5′, resulting in discrete mRNAs for each gene.[4−7] The transcribed mRNAs are capped and polyadenylated in the 5′ and 3′ termini, respectively. This manner of transcription leads to the accumulation of the viral mRNAs and proteins along a gradient that decreases from the 3′ terminus to the 5′ terminus.[5,7,8] In general, in the direction from 3′ to 5′, the nucleocapsid protein (N), the phosphoprotein (P), and the matrix protein (M) (all three encoded by the 3′ part) are more abundant in infected cells than are the attachment protein (glycoprotein (G), hemagglutinin (H), or hemagglutinin−neuraminidase (HM)) and the viral RNA polymerase (L), both encoded by the 5′ part.[5,8] Some of the members of the Mononegavirales order encode additional proteins, the open reading frames (ORFs) of may overlap. These additional proteins also may be transcribed by viral RNA polymerase, using alternative ORFs or RNA editing.[1,8]

5.1 FAMILY PARAMYXOVIRIDAE

The Paramyxoviridae family consists of seven genera divided into two subfamilies: the Paramyxovirinae (genera *Avulavirus*, *Henipavirus*, *Respirovirus*, *Morbillivirus*, and *Rubulavirus*) and the Pneumovirinae (genera *Pneumovirus* and *Metapneumovirus*).[1] Many members of the family Paramyxoviridae, such as human parainfluenza viruses, measles virus, human respiratory syncytial virus, mumps virus, Newcastle disease virus (NDV), Nipah

Zoonotic Viruses of Northern Eurasia.
DOI: http://dx.doi.org/10.1016/B978-0-12-801742-5.00005-2

and Hendra viruses, and other viruses causing mainly respiratory and neurological diseases, are important human and animal pathogens.[2–4]

Morphologically, the virions may vary in size (150 nm and greater) and take a filamentous, pleomorphic, or spherical shape. The nucleocapsid of the paramyxoviruses consists of a genomic RNA and three virion-associated proteins: N, P, and L. The lipid bilayer, surrounding the nucleocapsid, contains two viral transmembrane envelope glycoproteins: attachment protein and fusion protein (F). The attachment glycoprotein of the Paramyxovirinae subfamily possesses either hemagglutinating (H) (genera *Henipavirus* and *Morbillivirus*) or hemagglutinating–neuraminidase activities (HN) (*Avulovirus*, *Rubulavirus*, and *Respirovirus*).[5,6]

The length of the genomic RNA of the known members of the Paramyxoviridae family varies from 13,280 to 19,212 nt. The genome structure of the paramyxoviruses is the same as that of all members of the Mononegavirales, but, in addition to encoding the common proteins 3-N, P, M, G, and L-5′, it encodes a fusion protein (F) and two or more nonstructural proteins derived from overlapping ORFs of the P locus (P/C/V) (subfamily Paramyxovirinae) or in the M2/L locus (subfamily Pneumovirinae).[7–10]

5.1.1 Genus *Avulavirus*

The prototypical species of the *Avulavirus* genus is NDV—an important avian pathogen distributed worldwide and posing a constant threat to the poultry industry.[1,2] NDV (also named Avian parainfluenza virus 1) is one of the nine types of avian paramyxoviruses belonging to the *Avulavirus* genus. They have a common genomic structure and common phylogenetic and antigenic relationships.[3] The length of the genomic RNA of the avuloviruses is between 14,904 and 17,262 nt. Envelope proteins (HN) of the avulaviruses have hemagglutinating and neuraminidase activity. Phylogenetically, the avulaviruses are most closely related to the viruses of the *Rubulavirus* genus, but differ from them in the manner of transcription of the P locus, which leads the synthesis of P and V proteins and lacks the C protein ORF. Also, unlike some species of the *Rubulavirus* genus, none of the avulaviruses have an additional small hydrophobic (SH) gene locus (Figure 5.1).[3,4]

5.1.1.1 *Newcastle Disease Virus*

History. NDV, or avian paramyxovirus 1 (APMV-1) (order Mononegavirales, family Paramyxoviridae, subfamily Paramyxovirinae, genus *Avulavirus*), the etiological agent of Newcastle disease (pseudoplague of birds, fowl pseudopest, Asian plague of birds, Ranikhet disease, avian pneumoencephalitis, pigeon paramyxovirus 1), was first described during epizooty among poultry on Java (Indonesia) in 1926.[5] At the same time, an outbreak among poultry arose in Newcastle (UK) and was described as viral "Newcastle disease."[6] The disease was then discovered among

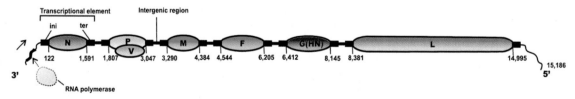

FIGURE 5.1   Scheme of the genome of the NDVs (genus *Avulavirus*, family Paramyxoviridae). The genome of NDV is a single-stranded negative-sense RNA genome that is 15,186 nt in length. *Drawn by Tanya Vishnevskaya.*

poultry in 1941—1945 in Ukraine, Belorussia, and Moldavia (USSR).[7]

Virion and genome. NDV has a virion envelope (140—150 nm) with a spheral nucleocapsid, is stable in the environment for pH 2—10, and keeps its biological activity under 8—20°C over several years—in frozen form, at least two to three years. NDV is sensitive to ultraviolet radiation (including sunshine), high temperatures, and disinfectants: 0.5—2.0% caustic rubbed, 1—2% formalin, or 5% phenol solution could inactivate NDV in 20—30 min.[3,7]

The NDV genome is ssRNA (−) ($1.5 \cdot 10^4$ nt) that encodes six structural proteins: RNA-dependent RNA polymerase (L), hemagglutinin—neuraminidase (HN), fusion protein (F), matrix protein (M), phosphoproteins (P), and nucleoprotein (NP).[3,8]

NDV strains are classified as velogenic (a high level of virulence), mezogenic (a middle level of virulence), and lentogenic (a low level of virulence). The last two are widely used as candidates for anti-NDV vaccines. It has been demonstrated that the main genetic marker of the level of NDV virulence is an amino acid sequence of the proteolytic site of envelope fusion protein F. Modern classification specifies 16 phylogenetic variants of six genotypes: 1, 2, 3a—d, 4a—d, 5a—e, and 6. On the basis of phylogenetic analysis of F protein sequences, circulating NDV strains may be divided into two main classes. NDV strains of class I can be isolated from wild aquatic birds and poultry and are generally avirulent for chickens. The majority of the virulent NDV strains belong to class II, which is subdivided into 11 genotypes (I—XI).[2] In Russian and Kazakhstan, several strains of NDV belonging to different genotypes were isolated from wild birds. These strains are represented in Figure 5.1. Different genotypes belong to different pathotypes. The highest frequencies of velogenic NDV strains isolated were established for genotypes 5a—e, 3c, and 4b.[8—10]

The severity of Newcastle disease depends on the level of virulence and the host's age,

immune status, and susceptibility. Birds of the Gallidae family are the most susceptible, waterfowl the least susceptible, among poultry. An infected bird excretes NDV by exhaled air, respiratory secretions, and excrement during the 24 h before the first clinical symptoms present and continues for up to 5—10 days after the last clinical symptoms. Under experimental conditions, NDV remains infectious for 10 months in *Argas persicus* (Argasidae ticks) and for 7.5 months in *Dermanyssus gallinae*, or the red poultry mite (Gamasidae ticks). Wild, synanthropic, decorative, and exotic birds all could be the source of NDV. The main avenue for the long-distance transfer of viruses is seasonal migrations of wild birds, contaminated stock, eggs, and poultry.[5,7,11—13]

Epizootiology. NDV has been detected in all continents and belongs to the especially dangerous infectious diseases. Velogenic NDV strains are endemic to the majority of countries in Asia, Africa, South and Central America, and Mexico. Canada, the United States, and Scandinavian countries are thought to be free of NDV as the result of a ban on import of poultry, strict quarantine actions, and the necessary depopulation of infected birds.[14]

About 30 species of wild birds are susceptible to NDV under both natural and experimental conditions.[7,8,11—13] Paresis of extremities, labored breath, and 20% mortality are often detected in pigeons during epizootics among poultry. Among starlings (*Sturnus vulgaris*), NDV is accompanied by tremor and coordination disorder. NDV infections of ravens (*Corvus corax*), horned owls (*Bubo virginianus*), white-tailed sea eagles (*Haliaeetus albicilla*), halcyons (*Alcedo atthis*), and the African ostrich (*Struthio camelus*) were described in zoos. NDV was isolated from nestlings of koel (*Eudynamys scolopaceus*) in India, from the horned owl (*Bubo virginianus*) in the United States, and from the fish hawk (*Pandion haliaetus*), cormorant (*Phalacrocorax carbo*) and swallow (*Delichon urbicum*) in the Netherlands.

The isolation of lentogenic strains from the booby (*Sula nebouxii*) on the coasts of the Orkney and Hebrides Islands, as well as from the Canada goose (*Branta canadensis*) in the United States, indicates the role of seabirds in the preservation of NDV in nature.

Surveillance of NDV in Northern Eurasia was started in 1971. Antibodies to this virus were found in 1% of colonial Alcidae seabirds in the north of the Far East.[15] Eighty-two NDV strains were isolated from wild birds on the Komandor Islands, in the south of the Far East,[16,17] in North Caucasia,[18] and in the Volga River estuary[7,19] (Table 5.1). We believe that colonies of seabirds are a natural reservoir of NDV. One source described an outbreak of the disease among pigeons in Kazakhstan in 2005.[20] was Another source demonstrated that, before 2004, only lentogenic NDV strains belonging to genotype 1 circulated among wild birds in the south of the Russian Far East, but velogenic strains of genotypes 3a and 5b appeared in 2004.[17] NDV strains were isolated from migrating birds in the United States,[20] in the Volga River estuary in Russia,[19,21] and in the Far East (Figure 5.2).[16]

Epidemiology. In humans, NDV appears after 1–5 days of incubation, with influenzalike symptoms and often with conjunctivitis. The person is fully recovered after 1–2 weeks.[22–24] Risk groups for NDV infection are the staffs of poultry farms, vaccination teams, and virological laboratories.

Pathogenesis. After inoculation, NDV reproduces in endothelium with consequent destruction of the walls of blood vessels and with the development of inflammatory and necrotic processes. In 24–36 h after infection, NDV localizes in the parenchymal organs of the bird; in the marrow, brain, and muscles; and in the respiratory, intestinal, and reproductive paths, resulting in hemodynamic and necrotic–dystrophic disorders.[25]

Clinical features. The incubation period in birds is 5–15 days. Clinical features vary from asymptomatic to lethal infection. Diarrhea with watery yellow-green excrement is an early and permanent clinical feature of NDV. The fulminant form of the infection leads to the death of the bird in 1–3 h. Infected turkeys (*Meleagris gallopavo*) demonstrate lesions of the central nervous system (CNS) followed by paresis of the extremities. Experimentally infected pigeons (*Columba livia*), siskins (*Carduelis spinus*), sparrows (*Passer domesticus*), crows (*Corvus brachyrhynchos*), jackdaws (*Corvus monedula*), and magpies (*Pica pica*) have the same symptoms.

The clinical classification of NDV includes five units:

- Doyle form, or velogenic viscerotropic NDV: severe course, hemorrhagic lesions of intestinal path, high mortality.[6]
- Beach form: severe course, respiratory and neurological symptoms, lethal outcome.[26]
- Beaudette form, or mesogenic NDV: respiratory and germinal lesions, mortality among young.[27]
- Hitchner form, or lentogenic NDV: poor respiratory symptoms.[28]
- Asymptomatic form: weak intestinal lesions[29]; diagnosis can be established only with laboratory methods.

Pathological changes depend on the level of virulence. Significant lesions are obvious in the presence of the Doyle and Beach forms of the disease. The most serious lesions are hemorrhagic lesions in the intestinal path as the result of necrosis of the mucous and lymph nodes as well as hypostasis and necrosis of the spleen and thymus. The germinal organs often are reduced in size, and yolk could spread into the abdominal cavity. Changes in the respiratory path occur with the Doyle, Beach, Beaudette, and Hitchner clinical forms, accompanied by petechia on the mucous membranes and the development of stagnation in the trachea as the result of complications during secondary infections.[12,30] There are no significant lesions in the CNS in all clinical forms of NDV.[29]

TABLE 5.1 Isolation of NDV from Wild Birds in Russia (1971–2010)

Far East		Europe		
North of the Far East (Komandory Islands, Barents Sea basin)	South of the Far East (Hanka lowland, the coast Peter the Great gulf)	Northern Caucasia	Volga river estuary	
1971	2001–2010	2008	1974	2001
410 samples	348 samples	147 samples	244 samples	336 samples
Guillmots (*Uria aalge*), tufted puffins (*Fratercula cirrhata*) rhinoceros auklets (*Cerorhinca monocerata*), glaucous-winged gulls (*Larus glaucescens*), red-faced cormorants (*Phalacrocorax urile*)	Mallards (*Anas platyrhynchos*), spot-billed duck (*Anas poecilorhyncha*), common teals (*Anas crecca*), garganeys (*Anas querquedula*), Baikal teals (*Anas formosa*), oriental turtle-doves (*Streptopelia orientalis*), common pheasants (*Phasianus colchicus*)	Mallards (*Anas platyrhynchos*)	Mallards (*Anas platyrhynchos*), common teals (*Anas crecca*), garganeys (*Anas querquedula*), common terns (*Sterna hirundo*), coots (*Fulica atra*), European herring gulls (*Larus argentatus*), black-headed gulls (*Larus ridibundus*), cormorants (*Phalacrocorax carbo*), grey herons (*Ardea cinerea*), purple herons (*Ardea purpurea*), white herons (*Ardea alba*), night herons (*Nycticorax nycticorax*), rock doves (*Columba livia*), tree sparrows (*Passer montanus*)	

Number of isolated strains

Tracheal swabs	Foul place swabs	Internals	Total	Foul place swabs	Total	Internals	Total	Foul place swabs	Internals	Total	Foul place swabs	Total
4	4	1	9 (2.2%)	3	3 (0.9%)	2	2 (1.4%)	7	25	32 (13.1%)	27	27 (8.0%)

Diagnostics. The variable nature of the clinical symptoms of NDV does precludes them from being reliable diagnostic criteria. Modern diagnostics are based on reverse transcription (RT)-PCR, isolating the virus, or serological investigations (hemagglutination inhibition test, or HIT; enzyme-linked immunosorbent assay, or ELISA; and neutralization assay, or NA). In reference laboratories, one could use monoclonal antibody panels,[12,13,31] oligonucleotide fingerprinting,[10] or sequencing F and NH genes,[16,17,24,27] or even the full genome[19] of NDV, together with further phylogenetic analysis.

Severe clinical forms must be separated from highly pathogenic avian influenza A viruses. The level of virulence of isolated NDV strains is determined from experiments *in vivo*. The indicator is the middle day of death of infected 10-day-old chicken embryos, the intracerebral pathogenecity index (ICPI) for 1-day-old chickens, and the intravenous pathogenecity index (IPVI) for 6-week-old chickens.

Control and prophylaxis. A system of international control of NDV has been developed[23] and has been implemented by European Union countries.[32,33] The main element of prophylaxis

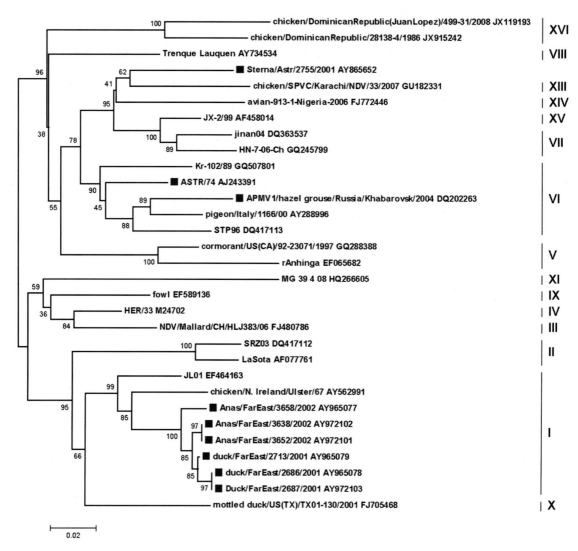

FIGURE 5.2 The phylogenetic tree constructed on the basis of aligned nucleotide sequences of the fusion protein (F) of different strains of NDV. The strains isolated in Russia from wild birds are designated with red squares. (For the interpretation of the references to color in the legend, the reader is referred to the online version of this book.)

actions is the wide use of anti-NDV vaccines based on the lentogenic (La Sota, Ulster, Bor-74 VGNKI, BI, F) or, rarely, mesogenic (H, GAM-61, Komarov, Roakin) strains. Chickens 7–14 days of age are vaccinated *per os*, intranasally, or through the conjunctiva of the eye. Depending on the concrete epizootological situation, the birds are vaccinated three to five times. Breeding birds and egg layers are vaccinated by inactivated vaccines, protecting both poultry and posterity. The egg layers are vaccinated before they lay. The level of anti-NDV immunity must be strictly controlled by serological tests (HIT, ELISA, NT). A decreasing level of specific antibodies is the reason for repeated vaccination.[32−36]

5.1.2 Genus *Morbillivirus*

The peculiarities of the *Morbillivirus* genus are its lack of neuraminidase activity (hemagglutinating activity is maintained) of the attachment protein (H) and its expression of an additional C-protein in the P locus (the P/C/V gene) by an RNA-editing process.[1,2] The size (15,690 nt for for canine distemper virus (CDV)) and structure of the genome of morbilliviruses are similar to those of members of the *Respiroviruses* genus, but the two viruses are not phylogenetically related. The morbilliviruses typically are formed of inclusion bodies (with a nucleocapsidlike structure) in the cytoplasm and nucleus of infected cells. Currently, six species in the genus *Morbillivirus* are known that can infect mammalians from different orders. The six are CDV (infecting animals of order Carnivora), measles virus (primates), peste-des-petits-ruminants virus (animals of order Artiodactyla, predominantly ruminants and swine), cetacean morbillivirus virus (CeMV, infecting animals of order Cetacea), phocine distemper virus (PDV, infecting seals), and rinderpest virus (infecting dolphins). Phylogenetically, PDV is closely related to CDV. In contrast, other morbilliviruses from marine mammals are more closely related to the rinderpest and peste-des-petits-ruminants viruses.[3−5]

5.1.2.1 Phocine Distemper Virus

History. PDV (order Mononegavirales, family Paramyxoviridae, subfamily Paramyxovirinae, genus *Morbillivirus*) was described in the former USSR in 1987, when the virus turned to be the cause of mass deaths of seals in Lake Baikal.[6] A year later, another mass death of seals occurred, this time in northwestern Europe,[7] and serological analysis revealed that the etiological agent was PDV.[8] In 1997, an outbreak among more than 6,000 seals took place on the Absheron peninsula on the western coast of the Caspian Sea; then, in the spring of 2000, PDV infection with high mortality began all over the northern Caspian region.[9,10] A second PDV epizootic in seals in northwestern European waters occurred in 2002. Sequence analysis revealed a high level of similarity (98%) between the northwestern European strains (PDV-1) of 1988 and 2002, as well as with PDV strains isolated on the west coast of the United States in 2006.[11] In contrast, the strain PDV-2, isolated from Lake Baikal seals and originally referred to as just PDV, is more closely related to CDV and is not related to the PDV-1 strain circulating in northwestern European waters (Figure 5.3).[12] The nucleotide identity level of the G protein of PDV-2 with CDV (strain Onderstepoort) is 91% and does not exceed 63% for PDV-1. These data were confirmed by sequence analysis of the F-gene. It has been shown that the active virus population of CDV circulating among Lake Baikal seals consists of two or three phylogenetic lineages.[13] Sequence analysis of the morbilliviruses that caused epizooty in Caspian seals revealed that their closest relationship was to the Lake Baikal seal virus.[9]

Virion and genome. PDV has a virion envelope that is of pseudospherical form and about 150−250 nm in length; $M_r = 500$ MD; $\rho_{Sac} = 1.18-1.20$ g/sm^3; $s_{20,w} = 1,000$ S. PDV is sensitive to heating, fat-dissolving agents, formaldehydes, and oxidizers.[2,14] Its virion envelope contains two transmembrane proteins—H (hemagglutinin) and F (fusion)—forming peplomers 8−12 nm in length. H peplomers are tetramers, whereas F peplomers are trimers. The H protein is immersed in the lipid layer by its N-term and recognizes cell receptors (sialosides on the surface of target cells). The H protein has a high level of similarity to the HN protein of *Respirovirus* and *Rubulavirus*; nevertheless, it has no neuraminidase activity. The F protein consists of two subunits linked

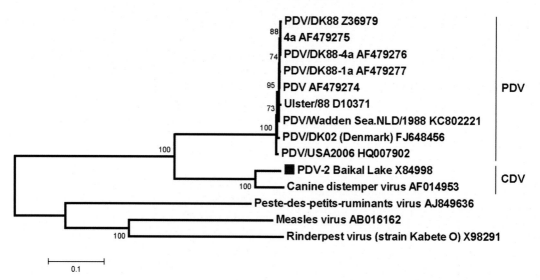

FIGURE 5.3 Phylogenetic analysis of the morbilliviruses (order Mononegavirales, family Paramyxoviridae, subfamily Paramyxovirinae, genus *Morbillivirus*), based on nucleotide sequences of the attachment protein (G) gene.

by a disulphide bond that is cleaved by cell protease into F1 and F2. (The last subunit is glycosylated.) The hydrophobic N-term of the F1 subunit initiates fusion of the virion envelope and cell membranes. The matrix M protein covers the internal lipoprotein surface of the virion. The nucleocapsid has a helical symmetry and contains the genome RNA and three proteins: the main nucleocapsid protein N (previously, this protein was named NP), a phosphoprotein P, and an RNA-dependent RNA polymerase (RdRp) L.[2,14,15]

The virus's genome is a single-stranded RNA ($-$) molecule 15,000–19,000 nt in length with 3′-leader (about 50 nt) and 5′-trailer (about 100 nt) sequences. The genome estructure is 3′-N-P-M-F-H-L-5′. The genes are separated from each other by intergene sequences.[2,14,15]

Epizootiology. The Baikal seal (*Pusa sibirica*), Caspian seal (*Phoca caspica*), and Greenland seal (*Phoca groenlandica*) are discussed in the literature as natural reservoirs of PDV.[6,7,9,10] Large outbreaks regularly take place in Lake Baikal, the Caspian Sea, the North Sea, and the Mediterranean Sea. Mortality during

PDV-linked outbreaks could reach ten thousand individuals, especially in naive populations. A trigger for PDV epizootics could be disturbances of ice conditions during the molting of seals, leading to crowding of the animals on the shoals.[10] When PDV penetrated into populations of northern fur seals (*Callorhinus ursinus*), it led to a tenfold decrease in the number of these animals.[16,17]

Pathogenesis and clinical features. In young seals, the disease develops as a general weakness, anorexia, dispepsia, conjunctivitis with formation of crusts, allocations from a nose, and pneumonia. Animals become exhausted. Petechia appears on mucous membranes. Necrotic lesions of the alveolar and bronchiolar epithelium often arise. Lymph nodes increase in size, and histological coloring may reveal necrotic lymphadenitis and the formation of polynuclear syncytia. The spleen is affected. In some cases, stomatitis with erosions and necrotic esophagitis appear.[16,18,19] Mortality is high.

Diagnostics. PDV is to be differentiated from the other members of the *Morbillivirus*

genus that cause similar distempers of seals (*Phocidae*), namely, CDV and CeMV. The most reliable approach to identifying PDV is isolating the virus with the help of the Vero cell line and then sequencing the viral genome.[9,20,21]

Control and prophylaxis. Symptomatic treatment of the disease is available in zoos. European centers of wild animal rescue tried to vaccinate sea animals with the vaccine against the closely related, but not identical, CDV. Vaccination did not prevent infection, but did prevent the development of clinical symptoms.[22]

5.2 FAMILY RHABDOVIRIDAE

The Rhabdoviriridae constitute a large and diverse family of enveloped, single-stranded, negative-sense RNA viruses that infect a wide range of vertebrates, invertebrates, and plants. Most members of the family Rhabdoviridae are grouped into one of six genera: *Vesiculovirus*, *Ephemerovirus*, *Lyssavirus*, and *Novirhabdovirus*, all of which infect vertebrates; and *Nucleorhabdovirus* and *Cytorhabdovirus*, which infects plants). Many of the rhabdoviruses can replicate in arthropods, which can serve as a vector for transmitting disease to animals or plants.[1–4]

The virion of the rhabdoviruses is rod shaped (or bullet shaped), 100–430 nm in length and 75 nm in diameter. The nucleocapsid comprises a closely packed ribonucleoprotein consisting of (1) genomic RNA and (2) three viral proteins: N (nucleoprotein, 47–62 kD), P (phosphoprotein, 35–50 kD), and L (RNA-dependent RNA polymerase, 150–190 kD). The nucleocapsid, containing RNA polymerase, has RNA replicative and transcriptional activity.[5,6] The lipid bilayer surrounding the nucleocapsid contains a glycoprotein (G) in the form of a trimer that protrudes on the surface of the virion.[1]

The length of genomic RNA varies from 11,000 to 15,000 nt for different genera. As with other members of the Mononegavirales order, the genomic RNA encodes at least five proteins in the direction 3′-N-P-M-G-L-5′; each of these proteins has its own transcription initiation and termination—polyadenylation signals. Transcription starts from the 3′-end of the genomic RNA—the end that has a binding site for RNA polymerase—and proceeds towards the 5′-terminus.[6] Each gene is transcribed as a separate mRNA with a cap(A) structure on the 5′-terminus and and a poly(A) structure on the 3′-terminus. Some members of the Rhabdaviridae family encode additional proteins.[7,8] Viruses of the *Vesiculovirus* genus have an additional ORF in the P gene that leads transcription of the small proteins C and C′.[9,10] Viruses of the *Ephemerovirus* genus encode up to five extra proteins, including the nonstructural glycoprotein G_{NS} and the small, accessory proteins $\alpha 1$, $\alpha 2$, β, and γ.[5,11,12]

5.2.1 Genus *Ephemerovirus*

The *Ephemerovirus* genus includes the rhabdoviruses, which are transmitted to vertebrate hosts by arthropod vectors (i.e., they are arboviruses). The genetic features that distinguish them from other genera of the Rhabdoviriridae family are the larger length of their genomic RNA (14,600–14,900 nt) and the presence of additional genes that encode the nonstructural glycoprotein G_{NS} and four or five accessory proteins: $\alpha 1$, $\alpha 2$, β, and γ (Figure 5.4).[1,2] This additional gene cluster is located between G and L. Of several known species of the *Ephemerovirus* genus, the bovine ephemeral fever virus (BEFV) and a number of related viruses are pathogenic for cattle and represent a serious threat to livestock.[3,4]

5.2.1.1 *Bovine Ephemeral Fever Virus*

History. BEFV (order Mononegavirales, family Rhabdoviridae, genus *Ephemerovirus*) originally was isolated in 1968 from the blood

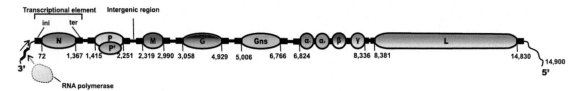

FIGURE 5.4 Scheme of the genome of the BEFV (genus *Ephemerovirus*, family Rhabdoviridae). Drawn by Tanya Vishnevskaya.

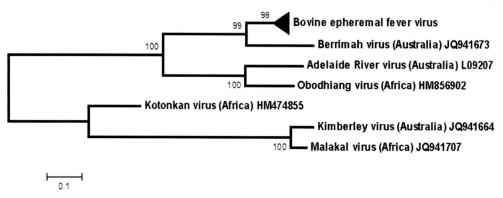

FIGURE 5.5 Phylogenetic structure of the ephemeroviruses (order Mononegavirales, family Rhabdoviridae, genus *Ephemerovirus*), constructed on the basis of aligned nucleotide sequences of the attachment protein (G) gene.

of a cow with fever in northern Queensland, Australia.[5,6] The first BEFV strain, which was deposited at the Russian State Collection of Viruses, was BEFV/JB76H, isolated from cattle during the 1976 epidemic in mainland China.[7] Epizootic outbreaks caused by BEFV occurred in 25 provinces of China at various times from 1952 to 1991.[7] Sporadic cases of BEFV infection broke out in central Asia on the territory of the former USSR from 1970 to 1990.[8–11]

Virion and genome. The genome of BEFV is ssRNA ($-$). Five structural proteins of BEFV have been described: nucleoprotein N, surface glycoprotein G, large RNA-dependent RNA polymerase (RdRp) L, polymerase associate protein P, and matrix protein M.[12,13] The virion envelope has a bulletlike taper form that is 60–80 nm (average, 70 nm) in diameter and 120–170 nm (average, 145 nm) in length, with spikes on the surface. Less frequently, giant particles about 100 nm in the diameter are observed.[14] Sodium hypochlorite and other disinfectants destroy BEFV. The strains of the virus that have been isolated in different regions have a low level of diversity. The identity of nucleotide sequences of the attachment protein (G) is in the range 89–99%, and that of putative amino acid sequences of the same attachment protein is in the range 94–100%. Nevertheless, phylogenetic analysis based on aligned nucleotide sequences of the G protein gene allows the sequences to be divided into three phylogenetic clusters (Figures 5.5–5.6). Cluster I contains the strains isolated from mainland China, Taiwan, and Japan. The Turkish and Israeli strains are grouped into

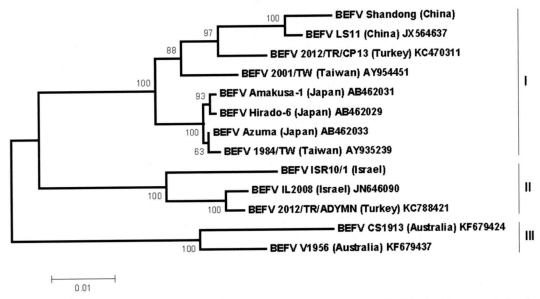

FIGURE 5.6 Rootless phylogenetic tree, constructed for selected representative BEFV (order Mononegavirales, family Rhabdoviridae, genus *Ephemerovirus*) strains isolated in different regions. The tree was constructed on the basis of aligned nucleotide sequences of the attachment protein (G) gene.

cluster II, and the Australian strains make up the independent cluster III (Figures 5.5–5.6).[7,14]

Epizootiology. BEFV is endemic in a belt of temperate, subtropical, and tropical regions in Africa, Australia, and Asia: Israel, Iraq, Iran, Syria, India, Pakistan, Bangladesh, the southern part of Mongolia, Korea, Japan, southern and central China, western and northern Australia, and all African countries.[7,11,14–19] Midges (family Ceratopogonidae) and sanguivorous mosquitoes (family Culicidae, e.g., *Culex annulirostris* and *Anopheles annulipes*) are vectors of BEFV.[13,17]

Sporadic cases of BEFV infection were detected in central Asia (Tajikistan and Uzbekistan) in the Amu Darya, Pyandzh, and Vahsh valleys.[11] Two BEFV strains (LEIV-7976Tur and LEIV-7938Tur) were isolated from midges *Culicoides puncticollis* collected on camels in September 1980 near Tedzhen (Turkmenistan)[10,20] within the framework of wide investigations in the area of biosafety and biodiversity in different ecosystems of Northern Eurasia as well as reinforcement of the database of the Russian State Collection of Viruses.[8,20,21] Natural foci in the Tedzhen oasis are secondary foci (so-called cattle natural foci).[22] The formation of such natural foc is usual for central Asia, where the biggest animal husbandry centers date from the Neolithic period. Domestic animals gradually superseded wild ones, becoming the components of local biocenoses and joining in virus circulation. Livestock growth and cattle populations increased the probability of the formation of secondary natural foci. In that manner, BEFV was introduced from adjacent countries and was fixed on the territory of Turkmenistan.[17,22]

Pathogenesis. BEFV leads to the appearance of a small amount of fibrin-rich fluid in the pleural, peritoneal, and pericardial cavities. Patchy edema may be found in the lungs. Petechial hemorrhages or edema may be found in the lymph nodes. Serofibrinous polysynovitis, polyarthritis, polytendovaginitis, and cellulitis are usually observed as well. Areas of focal necrosis in the major muscle groups is a common symptom.[11,16]

Clinical features. BEFV provokes economically important severe diseases of cows (*Bos taurus taurus*) and water buffalos (*Bubalus arnee*) with temperatures rising to 41−42°C and normalization in one to three days, emphysema, termination of rumination, decreasing yield of milk, lameness, and paralysis. Mortality is about 1−2%.[13,14,16]

Diagnostics. A virus neutralization test, the complement fixation test,[23] and ELISA are widely used serological methods to diagnose BEFV infection.[15,24] RT-PCR is used today as well.[25] BEFV infection is different from other viral (Rift Valley fever virus (family Bunyaviridae, genus *Phlebovirus*), foot-and-mouth disease virus (family Picornaviridae, genus *Aphthovirus*), and bluetongue virus (family Reoviridae, genus *Orbivirus*)) and bacterial (heartwater associated with *Ehrlichia ruminantium*; botulism with *Clostridium botulinum*; blackleg with *Clostridium chauvoei*; and babesiosis with *Babesia* sp.) infections.

Control and prophylaxis. The utilization of disinfection agents is relatively unimportant in preventing the spread of BEFV. This virus is not spread by casual contact. BEFV is rapidly inactivated in secretions and in the muscles of carcasses after death. Contact with potential insect vectors must be avoided. BEFV infection should be immediately reported to the proper authorities.[21] Several live attenuated, inactivated, and recombinant vaccines have been tested, with varying efficacy.[26,27]

5.2.2 Genus *Lyssavirus*

The genome of the lyssaviruses (about 12,000 nt in length) is of the same structure as that of the rhabdoviruses and consists of five genes (3′-N-P-M-G-L-5′), encoded structural proteins (N−G) and RNA-dependent RNA polymerase (L) (Figure 5.7).[1] The prototypical virus of the *Lyssavirus* genus is rabies virus (RABV), which has worldwide distribution and causes deadly acute encephalitis (rabies) in humans and animals. In addition to RABV, the *Lyssavirus* genus includes at least 11 other viruses, isolated mainly from bats in different regions of the world and antigenically related to RABV.[2−4] Phylogenetic analysis based on amino acid sequences of G-protein allows lyssaviruses to be divided into three phylogroups (Figure 5.8). RABV, Duvenhage virus (DUVV), European bat lyssavirus 1 (EBLV-1), European bat lyssavirus 2 (EBLV-2), Australian bat lyssavirus (ABLV), Aravan virus (ARAV), Khujand virus (KHUV), Irkut virus (IRKV), and Bokelho virus are assigned to phylogroup I.

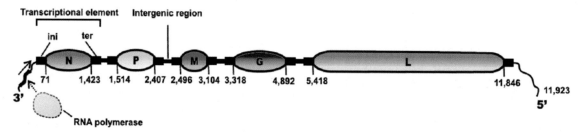

FIGURE 5.7 Scheme of the genome of the RABVs (family Rhabdoviridae, genus *Lyssavirus*). *Drawn by Tanya Vishnevskaya.*

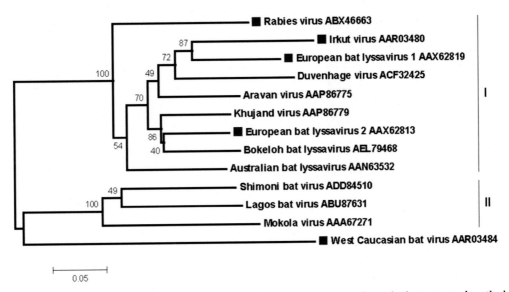

FIGURE 5.8 Phylogenetic tree of the lyssaviruses (family Rhabdoviridae, genus *Lyssavirus*), constructed on the basis of aligned amino acid sequences of the attachment protein (G). The viruses circulating in Northern Eurasia are designated with a red square. (For the interpretation of the references to color in the legend, the reader is referred to the online version of this book.)

Lagos bat virus (LBV), Shimoni bat virus (SHIBV), and Mokola virus (MOKV) form phylogroup II. West Caucasian bat virus (WCBV) represents a third distinct phylogroup.[2,5,6] The identity of the G-protein sequences between lyssaviruses from different phylogroups achieves 51–59%. The similarity of the G-protein sequence within the phylogroups is up to 70–75%.

5.2.2.1 Rabies Virus

History. RABV (order Mononegavirales, family Rhabdoviridae, genus *Lyssavirus*) is the etiological agent of rabies, a dangerous disease of animals and humans described as long as 3,000 years ago in the manuscripts of physicians of the ancient Orient.[7]

An experimental basis for the investigation of rabies was found by Louis Pasteur (Figure 5.9) and his followers in their experiments on RABV attenuation by subdural inoculation of rabbits: The crumbled spinal cord of an infected rabbit treated with an isotonic solution of sodium

FIGURE 5.9 Louis Pasteur (1822–1895).

chloride became the first vaccine against rabies. On July 6, 1885, Louis Pasteur used his vaccine on humans for the very first time, and as a result, Joseph Meister went down in history as the first person who made a complete recovery from previously incurable RABV infection.[8] Russia was the second—after France—country where antirabies stations were established—in Odessa,

FIGURE 5.10   Victor Babeş (1854–1926).

FIGURE 5.12   Maurice Nicolle (1862–1932).

FIGURE 5.11   Adelchi Negri (1876–1912).

FIGURE   5.13   Midat   Abdurakhmanovich   Selimov (1918–2001).

St. Petersburg, Moscow, Samara, and Warsaw in 1886 alone.

At the turn of the 20th century, Romanian bacteriologist Victor Babeş (Figure 5.10) and Italian pathologist Adelchi Negri (Figure 5.11) found a significant pathognomonic feature of rabies: inclusion bodies (2–10 mcm in diameter) in the cytoplasm of the infected nerve cells, especially in Ammon's horn of the hippocampus. In 1903, French microbiologist Maurice Nicolle (Figure 5.12) established that the etiological agent of rabies is filtrated and therefore belongs to the Virae kingdom. Great contributions to the study of rabies were made by Charles Calisher (see Figure 2.25), Midat Selimov (Figure 5.13), Sergey Gribencha (Figure 5.14), and other investigators.

Taxonomy of *Lyssavirus* genus. According to modern concepts, *Lyssavirus* genus comprises seven genotypes of prototype RABV as well as four other viruses: ARAV, IRKV (including the closely related Ozernoe virus, OZEV), KHUV, and WCBV.[9,10]

Genotype 1 of RABV comprises strains of "classical" street RABV (including arctic rabies and strains isolated in India)[11] isolated from ground mammalians and insectivorous, frugivorous, and bloodsucking bats, as well as vaccine strains. This genotype is widely distributed over Europe, Asia, Africa, and North and South America. Genotype 2, or LBV, comprises four subgenotypes (LBV-A, B, C, and D) and includes strains from dogs, cats, and

FIGURE 5.14   Sergey Vasilievich Gribencha (1937).

frugivorous bats in Central and South America. There is no modern evidence of pathogenecity for humans. Genotype 3, MOKV, was isolated from shrews, humans, dogs, and cats in Central and South Africa. Genotype 4, DUVV, was isolated from bats in South Africa and Zimbabwe, as well as from anyone bitted by the bats. Genotype 5, EBLV 1, circulates in Europe, including the European part of Russia. The strain belonging to this genotype was isolated from an 11-year-old girl bitten in the lip by a bat.[12] Genotype 6, EBLV 2, circulates in Europe, in Finland and Scotland.[5,9,13,14] Genotype 7, ABLV, circulates among bats in Australia and was isolated from two people as well.[15–17] Thus, six-sevenths (85.7%) of RABV genotypes are associated with bats. The main host of MOKV is unknown. Only genotype 1 is linked with *both* bats and carnivorous ground animals.[5] There is no modern evidence as to which variants, if any, were precursors of RABV, circulating in bats or in ground animals. Nevertheless, bats are frequently said to be the reservoirs for virus penetration into previously rabies-free territories.[2,18,19]

All viruses of the genus *Lyssavirus* except LBV can lead to lethal encephalitis in humans. All except MOKV are clinically indistinguishable. Sequencing of N- and G-genes of the members of *Lyssavirus* enables one to determine not only the virus's genotype, but also

different lineages within the genotype.[13,20–24] Global distribution and polyhostal ecology have led to the large number of RABV variants: fox (including the arctic fox), dog, skunk, coyote, raccoon, mongoose, and bat rabies viruses, as well as their geographical variants. Modern anti-RABV vaccines are produced on the basis of evidence that Genotype 1 is effective not only against all of its geographical lineages, but also against all RABV genotypes forming so-called phylogenetic supergroup I (comprising four to seven genotypes). Modern vaccines are not effective against LDV and MOKV, which together form phylogenetic supergroup II. G-protein distinctions between RABV and either LDV or MOKV are about 60%.

Epizootiology. From 1960 to 2011, cases of animals in Russia dying as the result of rabies fluctuated from 839 to 7,633 per year (foxes, 34.5%; dogs, 18%; cattle, 14%; cats, 11%; raccoons, 3.4%; other species, <1%).[25] The density of RABV infection among animals is extremely high in the southeastern, middle, and northern European part of Russia, south of western Siberia and low in the central north of western Siberia, in eastern Siberia, and in the Far East (Figures 5.15 and 5.16).[25–27]

Classical genotype 1 of RABV was found—besides in Carnivora—among bats (Chiroptera) in western (Novosibirsk and Omsk regions) and eastern (Yakut Republic) Siberia.[26,27] In 1985, EBLV 1 strain was isolated from a girl bitten on the lip by a bat in Belgorod, Russia, near the Ukranian border.[10] In 2002, IRKV was isolated in eastern Siberia and in China[5] and WCBV was isolated in the Caucasus region.[24,28,29] Like EBLV 1, OZEV was isolated from a human bitten by a bat, this time in the Far East in 2008.[30] The so-called steppe group of genotype 1 of RABV was found in the south and central parts of the European part of Russia, in the Ural district, and in western Siberia, with foxes, raccoon dogs (*Nyctereutes procyonoides*, introduced into Russia from Canada in 1930–1935), wolves, and other

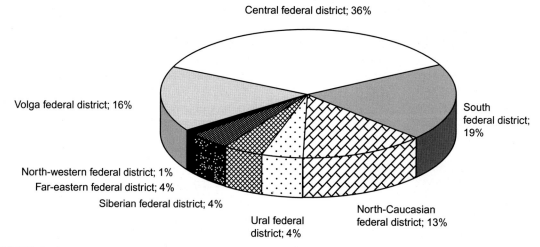

FIGURE 5.15   Rabies cases among humans in different regions of Russia.

Carnivora as hosts. Strains of what is called the "central group" circulate in the center of the European part of Russia, with raccoon dogs and foxes as hosts.[10]

Epidemiology. In Russia, 200,000—400,000 people annually suffer trauma from animals and need anti-rabies vaccine. The most unfavorable situation is observed on the territory around the central, southern, and middle parts of the VolgaRiver region and in the north Caucasus region (Figure 1.2). Six imported cases of rabies from Abkhazia, Ukraine, Moldova, Kirgizia, Azerbaijan, and India were revealed during 2007—2011, a period when 67 cases (about 13 cases per year, on average) were registered in Russia among humans.[25] In 2007 in the Primorsky Krai in the Far East, the bat was the source of infection for humans.[30] One more case of a bat infecting a human took place in 1985. On average, $58.2 \pm 6.0\%$ of human infection cases were connected with domestic Carnivora (dogs and cats), $41.6 \pm 6.0\%$ with wild Carnivora (foxes, raccoon dogs, wolfs). One of the rabies case was etiologically linked to a cow. For the last 50 years, out of 456 cases of rabies, 183 (40.1%) were the result of contact with dogs, 152 (33.3%) contact with foxes, 60 (13.2%) cats, 32 (7.0%) wolves, 25 (5.5%) raccoons, 2 (0.4%) arctic fox, and 2 (0.4%) cows.[25]

Pathogenesis. Rabies is discussed as a strict neuroinfection, with CNS being the main target organ. RABV inoculation almost always occurs (about 99.5—99.8% of cases) through being bitten by the sick animal. Nevertheless, inoculation may also occur as the result of saliva entering a scratch or wound on the skin. Aerosol inoculation is possible as well.[31] An intact layer of skin is impermeable to RABV.

RABV replicates in two stages. The first stage is in muscle cells within the wound following a bite, after which the virus infects nervous cells. In this case, RABV is detectable in the tissue at the location of inoculation for up to two months after infection. The ability of RABV to reproduce in muscle cells is explained by the presence of nicotinic acetylcholine receptors (nAChRs) on the surface of both muscle and nervous cells. These receptors interact with a receptor-binding site (174—203 aa) in viral protein G. The second way that RABV replicates starts in the wound itself, right after infection. RABV enters the wound

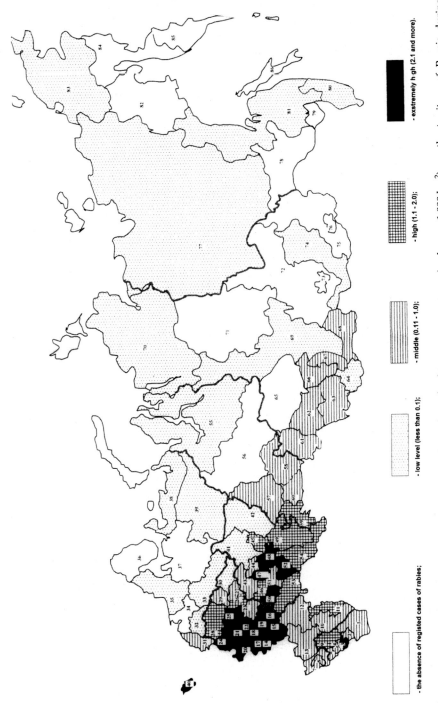

□ - the absence of registed cases of rabies;

□ - low level (less than 0.1);

▥ - middle (0.11 - 1.0);

▦ - high (1.1 - 2.0);

■ - exstremely h gh (2.1 and more).

FIGURE 5.16 Density of rabies (average annual number of rabies cases among animals per 1,000 km²) on the territory of Russia during 2007–2011.

and infects the nerve endings of motor or sensory neurons. Then the virus is transported to the CNS by the axoplasm of peripheral nerves.[32,33] Investigations into this phenomenon have enabled the World Health Organization (WHO) to establish a set of criteria for judging the likelihood of infection when one is bitten.[34] Thus, any case of a potentially rabid organism's saliva entering a scratch or wound on the skin must be considered a dangerous event falling into WHO's category III of severity. Injuries without blood appearing are considered an event falling into category II. Regardless of the length of the period that has passed after the bite, it is necessary to carry out a full regimen of anti-RABV vaccine in the case of a category II event and a full regimen of anti-RABV vaccine combined with inoculation of anti-RABV immunoglobulin around the wound in the case of a category III event.[34]

The receptor (nAChRs)-binding site of the G-protein is located in the range 174–203 aaand is homologous to the HIV-1 gp120 site (164–174 aa) and to neurotoxins of snake venom. Monoclonal antibodies against the nicotinic acetylcholine receptors (nAChRs)-binding site of the G-protein block both the interaction of gp120 with nAChRs and the interaction of snake venom with nAChRs.[35] It has been shown that sera of mice immunized against synthetic peptide that mimics 160–170 aa of gp120 cross-reacts with both RABV G-protein and snake venom. Thus, HIV-induced dementia could be provoked by interactions of gp120 with nAChRs on the surface of cells in the CNS.[31,35] Also, a high level of homology between RABV G-protein and HIV-1 gp120 could explain false-positive results in serological tests for HIV among individuals lately vaccinated against rabies.[36]

The ability to migrate into the CNS through the axoplasm of peripheral nerves is peculiar not only to RABV virions binding with neurotrophin receptor p75 with the help of G-protein, but also to nucleocapside binding with dynein molecules with the help of P-protein. Independently of clinical forms of rabies, RABV prefers to reproduce in the CNS (at first in the thalamus and basal ganglia and then in all sections of the brain). The mechanism of interneuronal transmission of RABV is not yet known in any detail. At first, it was thought that only nucleocapsides can penetrate synapses. However, the necessity for synaptic transmission in G-protein suggests that virions also can penetrate synapses. After it reaches the CNS, the virus begins to distribute itself in a peripheral direction and affect skeletal and heart muscles; the adrenal glands, kidneys, liver, retina, cornea, and pancreas; the walls of large blood vessels; and the salivary glands, which excrete RABV with saliva.

Several cases of rabies have arisen during the transplantation of organs from infected donors who were without clinical symptoms during the incubation period. Because the incubation period of rabies can be very long, physicians may not know whether the donor was bitten and infected earlier. What is needed is an algorithm that can be used to test organ donors for rabies before transplantation. To date, no such algorithm has been developed.[37]

Clinical features. The incubation period, from RABV inoculation till the first rabies symptoms appear, varies from 6–7 days to 6 years (1–3 months, on average).[33,38] Nevertheless, there are descriptions of incubation periods 10, 11, 13, and even 19.5 years. The duration depends on the number of wounds, their location and severity, the quantity of RABV inoculated into the wound, the biological properties of the given virus strain, and the quality of antirabies assistance.

The two main clinical forms, called impetuous and paralytic rabies, are characteristic of the disease. There are no correlations between either clinical form and the location of thebite.

The impetuous form of rabies is accompanied by anxiety attacks, psychical malaise, and excitation upon moving. Attacks of hydro-, aero (sensitiveness to drafts or fresh air)-, photo-, and acousticophobia, painful cramps, a diminished ability to swallow and to use the throat muscles, rich salivation, sweating, and fever are the most typical features. The temperature continues to increases, reaching a peak of 42–43°C just after death. The paralytic form of rabies begins with paralysis in the finiteness of the bite and then proceeds to para-, tri-, and quadriplegia. Paralytic rabies lasts longer (up to 30 days) than impetuous rabies (2–8 days, on average).[31,39] Clinical manifestations are not accompanied by any neuropathogenicity of neurons; however, neuron dysfunction does occur.[39] The particular mechanism determining which of the two clinical forms of rabies develops is at present unclear. Some authors maintain that the immune response of the host plays the key role in this process,[40] whereas others point to the genetic heterogeneity of RABV.[31,41] Wild RABV population could contain virus variants that provoke and the two clinical forms as well as chronic infection. The clinical form, then, seems to be determined by biological properties of the virus population.[41] This hypothesis of RABV population heterogeneity explains some previously unclear phenomena: (i) The ability of wild strains to provoke both the impetuous and the paralytic clinical forms is the result of the presence of different RABV variants in the limits of the strains; (ii) the transformation (carried out originally by Louis Pasteur Figure 5.9)) of the street RABV into the attenuated RABV is the result of positive selection of paralytic virus variants during successive passages through the brains of rabbits through faster reproduction of such variants; (iii) the high variability of some biological properties of RABV (e.g., the duration of incubation and of clinical periods, as well as the development of the particular clinical form) is the result of the mix of different viruses;[31] and (iv) there are no correlations between the N- and G-gene sequences of different isolates of RABV and the N- and G-gene sequences of clinical forms

of rabies:[42] A full genomic analysis is necessary for further progress in understanding a reverse genetics approach.

Rabies is considered an unavoidably lethal disease. Nevertheless, six cases of complete or partial recovery of sick humans are described in the modern scientific literature.[43] Five of the six (83.3%) patients who have recovered were vaccinated before RABV inoculation. Only one patient (a 15-year-old girl bitten by a bat) has recovered without an earlier vaccination: The patient achieved recovery by being placed into an artificial coma and being treated with antivirals.[43] This approach was ineffective with other patients. Although recovery from rabies continues to be a largely unsolved problem, some investigations of molecular mechanisms underlying the relation between immunity to RABVand abortion of the infection,[38] as well as a newfound ability to reproduce abortive RABV infections in different animals models[31] and with different forms of rabies,[41] instill optimism regarding the possibility of aborting a lethal infection.

Diagnostics. There are two types of rabies diagnostics: lifetime and postmortal. Lifetime diagnostics are based on the detection of protein N and virus RNA in a cryostatic slice of skin (e.g., a hair follicle); the detection of virus RNA in the urine, saliva, rheum, or cerebrospinal fluid; isolation of the virus in an *in vitro* or *in vivo* biological model; an indication of specific anti-RABV antibodies in the blood of unvaccinated persons; and an indication of specific anti-RABV antibodies in the cerebrospinal fluid, regardless of whether the person was vaccinatied. The reliability of lifetime diagnostics is insufficient and varies from 0% to 45%. Thus, the aforesaid methods are repeated until a diagnosis is obtained.

Postmortal diagnosis is highly reliable and is based on the detection of virus antigen and genome in the brain. The "golden standard" of rabies diagnostics consists of two methods: (i) isolating the virus with the use of *in vivo* or *in vitro* models; (ii) achieving an indication of nucleoprotein by RABV-specific fluorescent-labeled monoclonal antibodies. Molecular techniques—RT-PCR and sequencing of N- and G-proteins—have the highest sensitivity. Nevertheless, RT-PCR is not recommended for the routine diagnostics of rabies, because both false-positive and false-negative results are possible. Sequencing of N- and G-genes is used for the genotyping and identification of different RABV variants.[34]

Control and prophylaxis. Rabies is incurable and is always lethal if symptoms of the disease appear. According to WHO data, up to 100,000 people per year die as a result of rabies. However, the 100,000 figure seems to be underestimated. Prophylaxis is the only approach for the struggle against rabies. Every year, more than 12 million people are vaccinated against RABV.[34] In Russia, about 500,000 per year ask for antirabies treatment; every year, from 15 to 22 people die.[31]

Prophylaxis of rabies consists of two elements: prevention of the infection and prevention of the disease in the infected individual. Prevention of the infection among humans includes control of the number of potential RABV carriers and vaccination of animals that are around humans. Prevention of the disease after RABV inoculation includes antirabies treatment as rapidly as possible. From the very beginning, it is necessary to wash out a wound and receive anti-RABV vaccine or a combination of anti-RABV vaccine and specific anti-RABV immunoglobulin. Timely local washing of the wound is extremely significant and provides about 50% of the antirabies protection. The surface of the wound is washed carefully and with plenty of water and an alkaline soap or detergent for 15 min. Then an iodine solution or 70% ethyl alcohol solution is applied. If no chemicals are available, the surface of the wound is treated with plenty of pure water. The purpose of this entire procedure is maximal removal and chemical (alkaline) neutralization of RABV.

If possible, sutures on the wound are to be avoided. If the wound does need sutures, stitching is to be carried out after specific anti-RABV immunoglobulin inoculation and several hours of incubation so that antibodies can diffuse into the tissues.[31,34]

After the local treatment of the wound surface, it is necessary to start anti-RABV vaccination immediately. There are no contraindications for anti-RABV vaccination. Concentrated, refined, highly immunogenic ($\geq$2.5 ME/mL) culture vaccines, as well as human or horse anti-RABV immunoglobulin (in doses of 20 and 40 ME/kg, respectively) are used. Depending on the features of the animal bite, two treatment-and-prophylactic inoculation schemes are utilized. In the case of superficial bites and skin injuries falling into into category II, a sixfold vaccination regimen is used in Russia (fivefold in other developed countries): 1 mL at the 0th, 3rd, 7th, 14th, 30th, and 90th days. In the case of severe bites that could fall into category III, vaccine is combined with anti-RABV immunoglobulin,[31,34] inoculated simultaneously on the aforementioned days, regardless of the infection period. If anti-RABV is absent on the 0th day, one still may start vaccination and add immunoglobulin later (but not later than the 7th day after the sting). The entire volume of immunoglobulin is to be infiltrated into the tissues around and directly into the wound. In case anatomic peculiarities (e.g., if the wound is on the fingertips, in the ear, or in the nose) do not permit all of the recommended volume of immunoglobulin to infiltrate the wound, the remainder is to be inoculated intramusculary (but in an anatomical location different from the one that received vaccine inoculation). In the case of multiple bites that require a large volume of immunoglobulin for inoculation but the volume available is insufficient, the immunoglobulin may be diluted to one-third its concentration (but not more) with a sterile physiological solution. RABV-infected persons with compromised immune systems (those with AIDS, those requiring chemotherapy for transplanted organs, etc.) who have suffered bites falling into category II must be treated with both vaccine and immunoglobulin, like those who have suffered bites classified into the third category.[34]

Following are the main errors leading to misfortune in the treatment of rabies: (i) failure to use anti-RABV immunoglobulin for wounds of the third category; (ii) failure to inoculate the patient with anti-RABV immunoglobulin all around and directly into the wound. Instead, the immunoglobulin is inoculated as are antibiotics, into the buttock or in another anatomical location that is not the gate of infection (note that both experimental[31] and clinical[34] observations provide evidence of the insufficiency of anti-RABV immunoglobulin inoculation out of the infection gate); (iii) failure to use anti-RABV immunoglobulin at all, for any kind of bite; (iv) asking for antirabies treatment later than three days after the bite; (v) absence of local treatment of the wound surface.

Full and strict observance of the recommendations of the WHO Expert Committee on Rabies[34] guarantees the prevention of rabies in RABV-infected individuals.

5.2.3 Genus *Novirhabdovirus*

The genus *Novirhabdovirus* comprises a large group of viruses that infect a wide range of fish species. The genomic RNA of novirhabdoviruses has a size and structure common to the rhabdoviruses. In additional to the general proteins 3'-N-P-M-G-L-5', the novirhabdoviruses have an ORF for the nonstructural protein NV (12–14 kD) located between the G and L genes. The prototypical virus of the genus *Novirhabdovirus* is infectious hematopoietic necrosis virus (IHNV), whose genome is 11,131 nt in length. In the genus are included four recognized species (Hirame rhabdovirus, IHNV, snakehead rhabdovirus (SHRV), and Viral hemorrhagic septicemia virus (VHSV))

that have a common ecological niche and phylogenetic relationship.[1]

5.2.3.1 *Infectious Hematopoietic Necrosis Virus*

History. Infectious hematopoietic necrosis is an acute infection of young cultured salmonids and can occur in asymptomatic fish hosts. The host range of IHNV includes at least five species of Pacific salmon, several trout species, and the Atlantic salmon.[2] The first outbreaks of the disease, with isolation and identification of the etiologic agent (IHNV), were registered in sockeye salmon (*Oncorhynchus nerka*) fry at fish farm hatcheries in the United States during the 1950s.[3,4] Since then, IHNV has spread throughout the United States and to Canada and most Asian and European countries.[5,6] The primary route of transmission of the disease to Asian and European countries apparently was through the movement of infected eggs and fish.[7,8]

Taxonomy. Diversity among IHNV isolates from different geographical sources has been studied by various serological and genetic methods. Phylogenetic analyses based on G-protein and NV-protein gene sequences revealed that IHNV genetic types correlate with their geography.[9] On the basis of this finding, IHNV strains can be divided to three main genogroups corresponding to their place of their origin in North America.[6] The IHNV isolates from Europe and Asia have a high level of homology with American isolates, so it is presumed that IHNV was introduced from North American sources by shipments of infected eggs or fry.[2]

Distribution. IHNV is distributed over the Pacific coast of North America from California to Alaska, as well as over Asian countries (Republic of Korea, Japan, Taiwan, and other countries) and Europe (Austria, Croatia, Czech Republic, France, Germany, Italy, Netherlands, Poland, Slovenia, Spain, and other countries).[5]

Animal infection. The disease occurs in sockeye salmon (*Oncorhynchus nerka*), chinook salmon (*O. tshawytscha*), rainbow trout (*O. mykiss*), and other fish. Explosive outbreaks occur in young fish and losses may reach 100%. The onset of IHNV is abrupt. The fish may have superficial hemorrhagic lesions, exophthalmia, darkening of the body, and trailing fecal casts. The virus has been detected in the milt and ovarian fluid of spawning adults, and a fish may be a carrier its entire lifetime. The mean day-to-death time is related to the water temperature: The higher the temperature (up to 20°C), the more rapid is the clinical course to death. IHNV may be transmitted both vertically and horizontally through the water or by direct contact. Adult fish may be asymptomatic carriers of the virus, Which becomes activated during spawning.[3,7]

5.2.3.2 *Viral Hemorrhagic Septicemia Virus (VHSV)*

History. Viral hemorrhagic septicemia (also called "Egtved disease" because it first appeared in rainbow trout near the village of Egtved, Denmark) is the one of most important viral disease of cultured trout in Europe. VHSV was originally isolated from rainbow trout (*O. mykiss*) in Denmark in 1962; it was isolated in Denmark again in 1965.[10] Subsequently, it was found in many other European countries.[11] VHSV has also been isolated from many species of free-living marine fish in waters near Europe, Japan, and North America, indicating its widespread distribution.[12,13]

Taxonomy. Phylogenetic analysis of the different isolates of VHSV divided it to four genotypes, which appear to be distributed geographically, rather than by host or year of isolation. Genotypes I, II, and III are found mainly in Europe and Japan, while isolates of genotype IV have been found only in North America, Japan, and Korea.[14–19]

Distribution. The disease is prevalent in most of the countries of Europe. Also, VHSV was isolated from salmon and cod living in the Pacific coast waters of North America.[5,18]

Animal infection. External signs of infection include body darkening, pale gills, exophthalmia, hyperactivity, erratic swimming, and hemorrhages in the skin and gills. Internally, the kidneys and liver are swollen and discolored. Because, upon histological examination, the kidneys exhibit extensive necrosis (as does the liver), it is believed that they are a principal target of the virus. The disease results in high mortality and occurs at water temperatures of 15°C or lower. VHSV is able to infect rainbow trout during all life stages, and surviving fish, especially older fish, are likely to remain asymptomatic carriers. The rainbow trout is the species most affected by VHSV; however, brown trout, pike, grayling, whitefish, sea bass, and turbot also suffer natural infections. Natural transmission is through the water, probably via urine following the extensive renal necrosis. VHSV is virulent to marine fish species at 15°C or lower and has been found to kill 90% of sea bass and turbot under experimental conditions.

5.2.4 Genus *Vesiculovirus*

The genus *Vesiculovirus* comprises the antigenically related rhabdoviruses that can infect mammalians, fish, and insects.[1] The prototypical vesiculovirus is Vesicular stomatitis Indiana virus, whose genome structure is 3′-N-P-M-G-L-5′, a structure that is common to the rhabdoviruses. Ten species are recognized in the genus: Carajas virus (CJSV), Chandipura virus (CHPV), Cocal virus Indiana 2 (COCV-Ind2), Isfahan virus (ISFV), Maraba virus (MARAV), Piry virus (PIRYV), Spring viremia of carp virus (SVCV-VR1390), Vesicular stomatitis Alagoas virus Indiana 3 (VSAV), Vesicular stomatitis Indiana virus (VSIV-98COE), and Vesicular stomatitis New Jersey virus (NSNJV).[1]

5.2.4.1 *Spring Viremia of Carp Virus (SVCV)*

History. Spring viremia of carp (also called "infectious dropsy") is an acute hemorrhagic disease of the common carp, *Cyprinus carpio*. SVCV was originally isolated in a fish farm in Yugoslavia in 1971.[2] In Russia, heavy losses during the winter and high mortality after fish transplantation in the carp fishing ponds was first observed in 1960−1970. The clinical picture is coincident with the disease described in Yugoslavia. Subsequently, SVCV was detected in other European countries and in China, the United States, and Canada.[3−6]

Taxonomy. Phyllogenetic analysis based on a comparison of the G-protein gene of SVCV strains isolated in different regions of the world revealed four genotypes. Sequence identity between isolates within one genotype range from 97% to 100%, despite the different geography of isolation. Genotype 1a includes mostly the isolates from China, but also from the United States and the United Kingdom. Genotypes 1b and 1c comprise the isolates from Russia, Ukraine, and Moldova. Genotype 1d consists of isolates from the United Kingdom and Eastern Europe, including the reference strain Fijan from Yugoslavia (Figure 5.17).[5−8]

Distribution. The disease occurs throughout Europe. In 2002 it was found in the United States, in 2006 in Canada, and in 2004 in China.[6,9−11]

Animal infection. Infected fish show exophthalmia, abdominal swelling, long mucous fecal casts, and petechiation of the gills and skin. Internally, there are multiple focal hemorrhages with accompanying ascites and edema. Chronic infection is characterized by ulcers in the skin or more extensive edema. The organs affected include the kidney, liver, spleen, and swim bladder. Deaths cease when

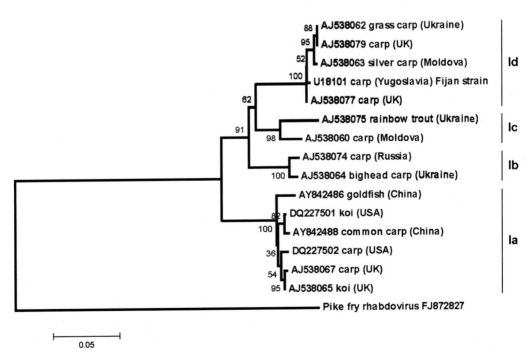

FIGURE 5.17 Phylogenetic tree based on nucleotide sequence analysis of the G gene of different isolates of SVCV, rooted to sequences of the Pike fry rhabdoviruses.

the water temperature reaches 25°C, and the disease does not occur if the temperature drops back below 15°C. The common carp is the natural host for SVCV, with mortality in both adult and young, and 1-year-old fish being the most susceptible. SVCV is spread from fish to fish through the water and also via parasites or mechanical transmission by leeches and fish lice. In addition to various carp (e.g., common, bighead, and grass carp), other species of fish (pike, guppies, tench, etc.) are susceptible to infection.[9,12]

References
5 Order Mononegavirales

1. Easton AJ, Pringle CR. Order Mononegavirales. In: King AMQ, Adams MJ, Carstens EB, Lefkowitz EJ, editors. *Virus taxonomy ninth report of the International Committee on Taxonomy of Viruses*. 1st ed. London: Elsevier;2012. p. 653–7.
2. Pringle CR, Easton AJ. Monopartite negative strand RNA genomes. *Semin Virol* 1997;(8):49–57.
3. Wertz GW, Whelan S, LeGrone A, et al. Extent of terminal complementarity modulates the balance between transcription and replication of vesicular stomatitis virus RNA. *PNAS* 1994;**91**(18):8587–91.
4. Barr JN, Whelan SP, Wertz GW. Transcriptional control of the RNA-dependent RNA polymerase of vesicular stomatitis virus. *Biochim Biophys Acta* 2002;**1577**(2):337–53.
5. Banerjee AK, Barik S, De BP. Gene expression of nonsegmented negative strand RNA viruses. *Pharmacol Therap* 1991;**51**(1):47–70.
6. Abraham G, Banerjee AK. Sequential transcription of the genes of vesicular stomatitis virus. *PNAS* 1976;**73**(5):1504–8.
7. Rose JK, Whitt MA. Rhabdoviridae: the viruses and their replication. In: Knipe DM, Howley PM, Griffin DE, et al., editors. *Fields virology*. 4th ed. Philadelphia, PA: Lippincott Williams and Wilkins;2001. p. 1221–44.
8. Whelan SP, Barr JN, Wertz GW. Transcription and replication of nonsegmented negative-strand RNA viruses. *Cur Top Microbiol Immunol* 2004;**283**:61–119.

5.1 Family Paramyxoviridae

1. Wang L-F, Collins PL, Fouchier RAM, et al. Family paramyxoviridae. In: King AM, Adams MJ, Carstens EB, Lefkowitz EJ, editors. *Virus taxonomy: classification and nomenclature of viruses: ninth report of the International Committee on Taxonomy of Viruses*. 1st ed. London: Academic Press;2012. p. 672–85.
2. Field HE, Mackenzie JS, Daszak P. Henipaviruses: emerging paramyxoviruses associated with fruit bats. *Cur Top Microbiol Immunol* 2007;**315**:133–59.
3. Vainionpaa R, Marusyk R, Salmi A. The paramyxoviridae: aspects of molecular structure, pathogenesis, and immunity. *Adv Virus Res* 1989;**37**:211–42.
4. Aguilar HC, Lee B. Emerging paramyxoviruses: molecular mechanisms and antiviral strategies. *Exp Rev Mol Med* 2011;**13**:e6.
5. Langedijk JP, Daus FJ, van Oirschot JT. Sequence and structure alignment of Paramyxoviridae attachment proteins and discovery of enzymatic activity for a morbillivirus hemagglutinin. *J Virol* 1997;**71**(8):6155–67.
6. Masse N, Ainouze M, Neel B, et al. Measles virus (MV) hemagglutinin: evidence that attachment sites for MV receptors SLAM and CD46 overlap on the globular head. *J Virol* 2004;**78**(17):9051–63.
7. Curran J, Kolakofsky D. Replication of paramyxoviruses. *Adv Virus Res* 1999;**54**:403–22.
8. Wang LF, Michalski WP, Yu M, et al. A novel P/V/C gene in a new member of the Paramyxoviridae family, which causes lethal infection in humans, horses, and other animals. *J Virol* 1998;**72**(2):1482–90.
9. Nagai Y, Kato A. Accessory genes of the paramyxoviridae, a large family of nonsegmented negative-strand RNA viruses, as a focus of active investigation by reverse genetics. *Cur Top Microbiol Immunol* 2004;**283**:197–248.
10. Lamb RA, Parks GD. Paramyxoviridae: the viruses and their replication. In: Knipe DM, Howley PM, editors. *Fields virology*. 5th ed. Philadelphia, PA: Lippincott Williams and Wilkins;2007. p. 1449–96.

5.1.1 Genus *Avulavirus*

1. Ganar K, Das M, Sinha S, Kumar S. Newcastle disease virus: current status and our understanding. *Virus Res* 2014;**184C**:71–81.
2. Miller PJ, Decanini EL, Afonso CL. Newcastle disease: evolution of genotypes and the related diagnostic challenges. *Infect Gen Evol J Mol Epidem Evol Gen Infect Dis* 2010;**10**(1):26–35.
3. Wang L-F, Collins PL, Fouchier RAM, et al. Family paramyxoviridae. In: King AM, Adams MJ, Carstens EB, Lefkowitz EJ, editors. *Virus taxonomy: classification and nomenclature of viruses: ninth report of the International Committee on Taxonomy of Viruses*. 1st ed. London: Academic Press;2012. p. 672–85.
4. Lamb RA, Parks GD. Paramyxoviridae: the viruses and their replication. In: Knipe DM, Howley PM, editors. *Fields virology*. 5th ed. Philadelphia, PA: Lippincott Williams and Wilkins;2007. p. 1449–96.

5. Kraneveld FC. Korte medeelingen uithet laboratorium voor veeurssennijkundig ondersock to Buitenzorg. *Nederlandsch-Indische Bladen voor Diergeneeskunde*, vol. 38; 1926. p. 448–51.

6. Doyle TM. A hitherto unrecorded disease of fowls due to a filter-passing virus. *Comp Pathol Therap* 1927;**40**: 144–69.

7. Syurin VN. *Pseudoplague of the birds (Newcastle disease)*. Moscow: Selkhozizdat;1963 [in Russian].

8. Alexander DJ. Gordon memorial lecture. Newcastle disease. *Br Poultry Sci* 2001;**42**(1):5–22.

9. Aldous EW, Mynn JK, Banks J, et al. A molecular epidemiological study of avian paramyxovirus type 1 (Newcastle disease virus) isolates by phylogenetic analysis of a partial nucleotide sequence of the fusion protein gene. *Avian Path J WVPA* 2003;**32**(3):239–56.

10. McMillan BC, Hanson RP. Differentiation of exotic strains of Newcastle disease virus by oligonucleotide fingerprinting. *Avian Dis* 1982;**26**(2):332–9.

11. Alexander DJ. Newcastle disease and other avian paramyxoviruses. *Rev Sci Tech* 2000;**19**(2):443–62.

12. Alexander DJ, Allan WH. Newcastle disease virus pathotypes. *Avian Path J WVPA* 1974;**3**(4):269–78.

13. Alexander DJ, Manvell RJ, Kemp PA, et al. Use of monoclonal antibodies in the characterisation of avian paramyxovirus type 1 (Newcastle disease virus) isolates submitted to an international reference laboratory. *Avian Path J WVPA* 1987;**16**(4):553–65.

14. Lancaster JE. Newcastle disease—à review of the geographical incidence and epizootology. *World Poultry Sci J* 1977;155–65.

15. Sazonov AA, Lvov DK, Zakstelskaya LY. Serological investigation of the sea complex birds in the north of the Far East. *Ecology of viruses*. Moscow; 1973. p. 17–23 [in Russian].

16. Shchelkanov M, Anan'ev V, Lvov DK, et al. Complex environmental and virological monitoring in the Primorye Territory in 2003–2006. *Vopr Virusol* 2007;**52** (5):37–48 [in Russian].

17. Shchelkanov M, Usachev EV, Fediakina IT, et al. Newcastle disease virus in the populations of wild birds in the south of the Primorye Territory in the period of autumn migrations in 2001–2004. *Vopr Virusol* 2006;**51**(4):37–41 [in Russian].

18. Silko N, Glushchenko AV, Shestopalova LV, et al. Biological properties of velogenic strains of the Newcastle disease virus isolated in the Northern Caucasian region. *Vopr Virusol* 2013;**58**(1):45–8 [in Russian].

19. Usachev EV, Fedyakina IT, Shchelkanov MY, et al. Molecular genetic characteristics of the Newcastle Sterna/Astrakhan/Z275/2001 virus isolated in Russia. *Molekul Genet Mikrobiol Virusol* 2006;(1):14–20 [in Russian].

20. Bogoyavlenskiy A, Berezin V, Prilipov A, et al. Characterization of pigeon paramyxoviruses (Newcastle disease virus) isolated in Kazakhstan in 2005. *Virol Sin* 2012;**27**(2):93–9.

21. Usachev EV, Shchelkanov M, Fedyakina IT, et al. Molecular virological monitoring of Newcastle disease virus strains (Paramyxoviridae, Avulavirus) in the populations of wild birds in the Volga estuary (the 2001 data). *Vopr Virusol* 2006;**51**(5):32–8 [in Russian].

22. Alexander DJ. Newcastle disease in the European Union 2000 to 2009. *Avian Path J WVPA* 2011;**40**(6):547–58.

23. Bennejean G. *Newcastle disease—control policies*. Boston: Academic Publishers;1988.

24. Choi KS, Lee EK, Jeon WJ, et al. Molecular epidemiologic investigation of lentogenic Newcastle disease virus from domestic birds at live bird markets in Korea. *Avian Dis* 2012;**56**(1):218–23.

25. Peoples ME. *Newcastle Disease virus replication. Newcastle Disease*. Boston: Kluwer Acad. Publ.;198845–78.

26. Beach JR. Avian pneumoencephalitis. Proc Annu Meet US Livestock Sanit Assoc; 1942. p. 203–23.

27. Beaudette FR. Newcastle disease in New Jersey. Proc Ann Meet US Livestock Sanit Assoc; 1946. p. 49–58.

28. Hitchner SB, Johnson EP. A virus of low virulence for immunizing fowls against Newcastle disease; avian pneumoencephalitis. *Vet Med* 1948;**43**(12):525–30.

29. McFerran JB, McCracken RM. *Newcastle disease. Newcastle Disease*. Boston: Kluwer Acad. Publ.;1988161–83.

30. Beard CW, Hanson RP. Newcastle disease. In: Hofstad MS, editor. *Diseases of poultry*. Ames: Iowa State University Press;1984. p. 452–70.

31. Russell PH, Samson ACR, Alexander DJ. Newcastle Disease virus variation. *Appl Virol Res* 1990;**II**:177–95.

32. Brandly C. *Recognition of Newcastle disease virus*. Madison: Univ. Wisconsin Press;1964.

33. Council of the European Communities. Council Directive 92/66/EEC of 14 July 1992 introducing Community measures for the control of Newcastle disease; 1992.

34. Cross JM. Newcastle disease—vaccine production. In: Alexander DJ, editor. *Newcastle disease*. Boston: Academic Publishers;1988. p. 333–46.

35. Goldhaft TM. Historical note on the origin of the LaSota strain of Newcastle disease virus. *Avian Dis* 1980;**24**(2):297–301.

36. Gough RE, Allan WH, Nedelciu D. Immune response to monovalent and bivalent Newcastle disease and infectious bronchitis inactivated vaccines. *Avian Path J WVPA* 1977;**6**(2):131–42.

5.1.2 Genus *Morbillivirus*

1. Lamb RA, Parks GD. Paramyxoviridae: the viruses and their replication. In: Knipe DM, Howley PM, editors. *Fields virology*. 5th ed. Philadelphia, PA: Lippincott Williams and Wilkins;2007. p. 1449–96.

2. Wang L-F, Collins PL, Fouchier RAM, et al. Family para-myxoviridae. In: King AM, Adams MJ, Carstens EB, et al., editors. *Virus taxonomy: classification and nomenclature of viruses: ninth report of the International Committee on Taxonomy of Viruses*. 1st ed. London: Academic Press;2012. p. 672−85.

3. Di Guardo G, Marruchella G, Agrimi U, et al. Morbillivirus infections in aquatic mammals: a brief overview. *J Vet Med A Physiol Pathol Clin Med* 2005;**52**(2):88−93.

4. Martella V, Elia G, Buonavoglia C. Canine distemper virus. *Vet Clin North Am Small Anim Pract* 2008;**38**(4):787−97.

5. Van Bressem MF, Raga JA, Di Guardo G, et al. Emerg Infect Dis in cetaceans worldwide and the possible role of environmental stressors. *Dis Aquatic Organ* 2009;**86**(2):143−57.

6. Likhoshway YeV, Grachev MA, Kumarev VP, et al. Baikal seal virus. *Nature* 1989;**339**(6222):266.

7. Visser IK, Kumarev VP, Orvell C, et al. Comparison of two morbilliviruses isolated from seals during out-breaks of distemper in north west Europe and Siberia. *Arch Virol* 1990;**111**(3−4):149−64.

8. Visser IK, van Bressem MF, van de Bildt MW, et al. Prevalence of morbilliviruses among pinniped and cetacean species. *Rev Sci Tech* 1993;**12**(1):197−202.

9. Butina TV, Denikina NN, Kondratov IG, et al. Molecular-genetic comparison of morbilliviruses which caused epizooty in Baikal (*Phoca siberica*) and Caspian (*Phoca caspica*) seals. *Molekul Genet Mikrobiol Virusol* 2003;(4):27−32 [in Russian].

10. Shestopalov AM, Beklemishev AB, Khuraskin LS, et al. Para- and Orthomyxoviridae of Caspian seals. *Proceedings of second international conference "Sea animals of Holarctic"* (Irkutsk, Russia; September, 10−15, 2002). Irkutsk; 2002. 234p. [in Russian].

11. Muller G, Kaim U, Haas L, et al. Phocine distemper virus: characterization of the morbillivirus causing the seal epizootic in northwestern Europe in 2002. *Arch Virol* 2008;**153**(5):951−6.

12. Mamaev LV, Denikina NN, Belikov SI, et al. Characterisation of morbilliviruses isolated from Lake Baikal seals (*Phoca sibirica*). *Vet Microbiol* 1995;**44**(2−4):251−9.

13. Butina TV, Denikina NN, Belikov SI. Canine distemper virus diversity in Lake Baikal seal (*Phoca sibirica*) popu-lation. *Vet Microbiol* 2010;**144**(1−2):192−7.

14. Kaverin NV, Lvov DK, Shchelkanov MY. Paramyxoviridae. In: Lvov DK, editor. *Handbook of virology viruses and viral infection of human and animals*. Moscow: MIA;2013. p. 192−7 [in Russian].

15. Varich NL. Structural organization of the genome of paramyxoviruses. *Molekul Genet Mikrobiol Virusol* 1988;(4):3−14 [in Russian].

16. Fernandez A, Esperon F, Herraez P, et al. Morbillivirus and pilot whale deaths, Mediterranean Sea. *Emerg Infect Dis* 2008;**14**(5):792−4.

17. Visser IK, Van Bressem MF, de Swart RL, et al. Characterization of morbilliviruses isolated from dolphins and porpoises in Europe. *J Gen Virol* 1993;**74**(4):631−41.

18. Belliere EN, Esperon F, Fernandez A, et al. Phylogenetic analysis of a new Cetacean morbillivirus from a short-finned pilot whale stranded in the Canary Islands. *Res Vet Sci* 2011;**90**(2):324−8.

19. Van Bressem M, Waerebeek KV, Jepson PD, et al. An insight into the epidemiology of dolphin morbillivirus worldwide. *Vet Microbiol* 2001;**81**(4):287−304.

20. Barrett T, Visser IK, Mamaev L, et al. Dolphin and por-poise morbilliviruses are genetically distinct from pho-cine distemper virus. *Virology* 1993;**193**(2):1010−12.

21. Duignan PJ, House C, Geraci JR, et al. Morbillivirus infection in cetaceans of the western Atlantic. *Vet Microbiol* 1995;**44**(2−4):241−9.

22. Vaughan K, Del Crew J, Hermanson G, et al. A DNA vaccine against dolphin morbillivirus is immunogenic in bottlenose dolphins. *Vet Immunol Immunopath* 2007;**120**(3−4):260−6.

5.2 Family Rhabdoviridae

1. Dietzgen RG, Calisher CH, Kurath G, et al. Family rhab-doviridae. In: King AM, Adams MJ, Carstens EB, et al., editors. *Virus taxonomy: classification and nomenclature of viruses: ninth report of the International Committee on Taxonomy of Viruses*. 1st ed. London: Elsevier;2012. p. 686−713.

2. Bourhy H, Cowley JA, Larrous F, et al. Phylogenetic relationships among rhabdoviruses inferred using the L polymerase gene. *J Gen Virol* 2005;**86**(10):2849−58.

3. Tesh RB, Travassos Da Rosa AP, Travassos Da Rosa JS. Antigenic relationship among rhabdoviruses infecting terrestrial vertebrates. *J Gen Virol* 1983;**64**(1):169−76.

4. Calisher CH, Karabatsos N, Zeller H, et al. Antigenic relationships among rhabdoviruses from vertebrates and hematophagous arthropods. *Intervirology* 1989;**30**(5):241−57.

5. Assenberg R, Delmas O, Morin B, et al. Genomics and structure/function studies of Rhabdoviridae proteins involved in replication and transcription. *Antiviral Res* 2010;**87**(2):149−61.

6. Rose JK, Whitt MA. Rhabdoviridae: the viruses and their replication. In: Knipe DM, Howley PM, Griffin DE, et al., editors. *Fields virology*. 4th ed. Philadelphia, PA: Lippincott Williams and Wilkins;2001. p. 1221−44.

7. Allison AB, Palacios G, Travassos da Rosa A, et al. Characterization of Durham virus, a novel rhabdovirus that encodes both a C and SH protein. *Virus Res* 2011;**155**(1):112−22.

8. Gubala AJ, Proll DF, Barnard RT, et al. Genomic characterisation of Wongabel virus reveals novel genes within the Rhabdoviridae. *Virology* 2008;**376**(1):13–23.

9. Spiropoulou CF, Nichol ST. A small highly basic protein is encoded in overlapping frame within the P gene of vesicular stomatitis virus *J Virol* 1993;**67**(6):3103–10.

10. Springfeld C, Darai G, Cattaneo R. Characterization of the Tupaia rhabdovirus genome reveals a long open reading frame overlapping with P and a novel gene encoding a small hydrophobic protein. *J Virol* 2005;**79**(11):6781–90.

11. McWilliam SM, Kongsuwan K, Cowley JA, et al. Genome organization and transcription strategy in the complex GNS-L intergenic region of bovine ephemeral fever rhabdovirus. *J Gen Virol* 1997;**78**(6):1309–17.

12. Walker PJ, Byrne KA, Riding GA, et al. The genome of bovine ephemeral fever rhabdovirus contains two related glycoprotein genes. *Virology* 1992;**191**(1):49–61.

5.2.1 Genus *Ephemerovirus*

1. McWilliam SM, Kongsuwan K, Cowley JA, et al. Genome organization and transcription strategy in the complex GNS-L intergenic region of bovine ephemeral fever rhabdovirus. *J Gen Virol* 1997;**78**(6):1309–17.

2. Walker PJ, Byrne KA, Riding GA, et al. The genome of bovine ephemeral fever rhabdovirus contains two related glycoprotein genes. *Virology* 1992;**191**(1):49–61.

3. Nandi S, Negi BS. Bovine ephemeral fever: a review. *Comp Immunol Microbiol Infect Dis* 1999;**22**(2):81–91.

4. St George TD. Bovine ephemeral fever: a review. *Trop Animal Health Product* 1988;**20**(4):194–202.

5. Ergunay K, Ismayilova V, Colpak IA, et al. A case of central nervous system infection due to a novel Sandfly Fever Virus (SFV) variant: Sandfly Fever Turkey Virus (SFTV). *J Clin Virol* 2012;**54**(1):79–82.

6. Doherty RL, Carley JG, Standfast HA, et al. Virus strains isolated from arthropods during an epizootic of bovine ephemeral fever in Queensland. *Austr Vet J* 1972;**48**(3):81–6.

7. Zheng F, Qiu C. Phylogenetic relationships of the glycoprotein gene of bovine ephemeral fever virus isolated from mainland China, Taiwan, Japan, Turkey, Israel and Australia. *Virol J* 2012;**9**:268.

8. Lvov DK. *Ecological soundings of the former USSR territory for natural foci of arboviruses. Sov Med Rev Virol.* UK: Harwood Academ. Pub. GmbH;19931–47.

9. Lvov DK. Ecology of viruses. In: Lvov DK, editor. *Handbook of virology: viruses and viral infection of human and animals.* Moscow: MIA;2013. p. 66–86 [in Russian].

10. Skvortsova TM, Gromashevsky VL, Sidorova GA, et al. Results of virological testing arthropod vectors in Turkmenia. In: Lvov DK, editor. *Ecology of viruses.* Moscow: USSR Academy of Medical Sciences;1982. p. 139–44 [in Russian].

11. Syurin VM, Samuilenko AY, Soloviev BV, et al. *Viral diseases of animals.* Moscow: Institute of biological industry;1998 [in Russian].

12. Dhillon J, Cowley JA, Wang Y, et al. RNA polymerase (L) gene and genome terminal sequences of ephemeroviruses bovine ephemeral fever virus and Adelaide River virus indicate a close relationship to vesiculoviruses. *Virus Res* 2000;**70**(1–2):87–95.

13. Walker PJ. Bovine ephemeral fever virus. In: Mahy BWJ, Regenmortel MHV, editors. *Encyclopedia of Virology.* Oxford: Elsevier Acad. Press;2008. p. 354–66.

14. Walker PJ. Bovine ephemeral fever in Australia and the world. *Cur Top Microbiol Immunol* 2005;**292**:57–80.

15. Zheng FY, Lin GZ, Qiu CQ, et al. Development and application of G1-ELISA for detection of antibodies against bovine ephemeral fever virus. *Res Vet Sci* 2009;**87**(2):211–12.

16. Lvov DK, Aliper TI. Ephemeral (3-day long) fever of cattle. In: Lvov DK, editor. *Handbook of virology: viruses and viral infection of human and animals.* Moscow: MIA;2013. p. 902–3 [in Russian].

17. Lvov DK, Klimenko SM, Gaidamovich SY. *Arboviruses and arbovirus infections.* Moscow: Meditsina;1989 [in Russian].

18. Nagano H, Hayashi K, Kubo M, et al. An outbreak of bovine ephemeral fever in Nagasaki Prefecture in 1988. *Nihon Juigaku Zasshi* 1990;**52**(2):307–14 [in Japanese].

19. Zheng FY, Lin GZ, Qiu CQ, et al. Serological detection of bovine ephemeral fever virus using an indirect ELISA based on antigenic site G1 expressed in *Pichia pastoris. Vet J* 2010;**185**(2):211–15.

20. Lvov DK. Arboviral zoonoses of Northern Eurasia (Eastern Europe and the commonwealth of independent states). In: Beran GW, editor. *Handbook of zoonoses Section B: Viral zoonoses.* London–Tokyo: CRC Press;1994. p. 237–60.

21. Organization of ecological-epidemiological monitoring in Russian Federation for anti-epidemic defense of the civilians and army. Moscow: Russian Ministry of Public Health, Federal Office of Biomedical and Extreme Problems, D.I. Ivanovsky Institute of Virology; 1993 [in Russian].

22. Sidorova GA. Secondary natural focies of arboviruses associated with ticks in the plain and foothill regions of Middle Asia. In: Gaidamovich SY, editor. *Arboviruses.* Moscow: D.I. Ivanosky institute of Virology of USSR Academy of Medical Sciences;1978. p. 87–94 [in Russian].

23. Bai WB, Tian FL, Wang C, et al. Preliminary studies of the complement fixation test to confirm the diagnosis of bovine ephemeral fever. *Austr J Biol Sci* 1987;**40**(2):137–41.

24. Zakrzewski H, Cybinski DH, Walker PJ. A blocking ELISA for the detection of specific antibodies to bovine

ephemeral fever virus. *J Immunol Meth* 1992;**151** (1–2):289–97.

25. Finlaison DS, Read AJ, Zhang J, et al. Application of a real-time polymerase chain reaction assay to the diagnosis of bovine ephemeral fever during an outbreak in New South Wales and northern Victoria in 2009–10. *Austr Vet J* 2014;**92**(1–2):24–7.

26. Aziz-Boaron O, Leibovitz K, Gelman B, et al. Safety, immunogenicity and duration of immunity elicited by an inactivated bovine ephemeral fever vaccine. *PLoS One* 2013;**8**(12):e82217.

27. Uren MF, Walker PJ, Zakrzewski H, et al. Effective vaccination of cattle using the virion G protein of bovine ephemeral fever virus as an antigen. *Vaccine* 1994;**12**(9):845–50.

5.2.2 Genus *Lyssavirus*

1. Tordo N, Poch O, Ermine A, et al. Walking along the rabies genome: is the large G-L intergenic region a remnant gene?. *PNAS* 1986;**83**(11):3914–18.

2. Calisher CH, Ellison JA. The other rabies viruses: the emergence and importance of lyssaviruses from bats and other vertebrates. *Travel Med Infect Dis* 2012;**10** (2):69–79.

3. Rupprecht CE, Turmelle A, Kuzmin IV. A perspective on lyssavirus emergence and perpetuation. *Cur Opin Virol* 2011;**1**(6):662–70.

4. Banyard AC, Hayman D, Johnson N, et al. Bats and lyssaviruses. *Adv Virus Res* 2011;**79**:239–89.

5. Botvinkin AD, Poleschuk EM, Kuzmin IV, et al. Novel lyssaviruses isolated from bats in Russia. *Emerg Infect Dis* 2003;**9**(12):1623–5.

6. Le Mercier P, Jacob Y, Tordo N. The complete Mokola virus genome sequence: structure of the RNA-dependent RNA polymerase. *J Gen Virol* 1997;**78** (7):1571–6.

7. Savateev AI. *Rabies*. Leningrad–Moscow: Gosizdat;1927 [in Russian].

8. Selimov MA. *Rabies*. Moscow: Meditsina;1978 [in Russian].

9. Dietzgen RG, Calisher CH, Kurath G, et al. Family rhabdoviridae. In: King AM, Adams MJ, Carstens EB, et al., editors. *Virus taxonomy: classification and nomenclature of viruses. Ninth report of the International Committee on Taxonomy of Viruses*. 1st ed. London: Elsevier;2012. p. 686–713.

10. Gribencha SV, Lvov DK, Shchelkanov MY. Rhabdoviridae. In: Lvov DK, editor. *Handbook of virology: viruses and viral infection of human and animals*. Moscow: MIA;2013. p. 197–202 [in Russian].

11. Nadin-Davis SA, Turner G, Paul JP, et al. Emergence of Arctic-like rabies lineage in India. *Emerg Infect Dis* 2007;**13**(1):111–16.

12. Selimov MA, Tatarov AG, Botvinkin AD, Klueva EV, Kulikova LG, Khismatullina NA. Rabies-related Yuli virus; identification with a panel of monoclonal antibodies. *Acta Virol* 1989;**33**(6):542–6.

13. Kuzmin IV, Botvinkin AD, McElhinney LM, et al. Molecular epidemiology of terrestrial rabies in the former Soviet Union. *J Wildlife Dis* 2004;**40**(4):617–31.

14. Metlin AE, Rybakov S, Gruzdev K, et al. Genetic heterogeneity of Russian, Estonian and Finnish field rabies viruses. *Arch Virol* 2007;**152**(9):1645–54.

15. St George TD. Australian bat lyssavirus. *Austr Vet J* 1997;**75**(5):367.

16. Guyatt KJ, Twin J, Davis P, et al. A molecular epidemiological study of Australian bat lyssavirus. *J Gen Virol* 2003;**84**(Pt 2):485–96.

17. Samaratunga H, Searle JW, Hudson N. Non-rabies Lyssavirus human encephalitis from fruit bats: Australian bat Lyssavirus (pteropid Lyssavirus) infection. *Neuropath Appl Neurobiol* 1998;**24**(4):331–5.

18. Kuzmin IV, Shi M, Orciari LA, et al. Molecular inferences suggest multiple host shifts of rabies viruses from bats to mesocarnivores in Arizona during 2001–2009. *PLoS Pathogens* 2012;**8**(6):e1002786.

19. Calisher CH, Childs JE, Field HE, et al. Bats: important reservoir hosts of emerging viruses. *Clin Microbiol Rev* 2006;**19**(3):531–45.

20. Bourhy H, Kissi B, Tordo N. Molecular diversity of the Lyssavirus genus. *Virology* 1993;**194**(1):70–81.

21. Kuzmin IV, Hughes GJ, Botvinkin AD, et al. Arctic and Arctic-like rabies viruses:distribution, phylogeny and evolutionary history. *Epidemiol Infect* 2008;**136**(4):509–19.

22. Franka R, Constantine DG, Kuzmin I, et al. A new phylogenetic lineage of rabies virus associated with western pipistrelle bats (*Pipistrellus hesperus*). *J Gen Virol* 2006;**87**(8):2309–21.

23. Shankar V, Orciari LA, De Mattos C, et al. Genetic divergence of rabies viruses from bat species of Colorado, USA. *Vector Borne Zoonot Dis* 2005;**5**(4):330–41.

24. Kuzmin IV, Hughes GJ, Botvinkin AD, et al. Phylogenetic relationships of Irkut and West Caucasian bat viruses within the Lyssavirus genus and suggested quantitative criteria based on the N gene sequence for lyssavirus genotype definition. *Virus Res* 2005;**111**(1):28–43.

25. Rabies in Russian Federation. *Informational bulletin of the Institute of natural foci infections*. Omsk: Omsk Institute of pedagogic;2013 [in Russian].

26. Poleshchuk EM, Sidorov GN, Gribencha SV. A summary of the data about antigenic and genetic diversity of rabies virus circulating in the terrestrial mammals in Russia. *Vopr Virusol* 2013;**58**(3):9–16 [in Russian].

27. Poleshchuk EM, Botvinkin AD, Tkachev SE, et al. Rabies of wild animals on the south of Eastern Siberia in the beginning of XXI century. *J Infektsionnoi Patologii* 2010;**17**(3):112–14 [in Russian].

28. Kuzmin IV, Wu X, Tordo N, et al. Complete genomes of Aravan, Khujand, Irkut and West Caucasian bat viruses, with special attention to the polymerase gene and non-coding regions. *Virus Res* 2008;**136**(1−2):81−90.

29. Poleshchuk EM, Botvinkin AD, Tkachev SE, et al. West Caucasian lyssavirus of Chiroptera: absence of vaccin protection. *Plecotus et al* 2003;(6):67−71 [in Russian].

30. Leonova GN, Belikov SI, Kondratov IG, et al. A fatal case of bat lyssavirus in Primorye territory of the Russian Far East. *Rabies Bull Eur* 2009;**33**:5−8.

31. Gribencha SV, Lvov DK. Rabies. In: Lvov DK, editor. *Handbook of virology: Viruses and viral infection of human and animals*. Moscow: MIA;2013. p. 811−18 [in Russian].

32. Finke S, Conzelmann KK. Replication strategies of rabies virus. *Virus Res* 2005;**111**(2):120−31.

33. Jackson AC. Rabies pathogenesis. *J Neurovirol* 2002;**8**(4):267−9.

34. WHO expert consultation on rabies: first report; 2004.

35. Neri P, Bracci L, Rustici M, et al. Sequence homology between HIV gp120, rabies virus glycoprotein, and snake venom neurotoxins. Is the nicotinic acetylcholine receptor an HIV receptor? *Arch Virol* 1990;**114**(3−4):265−9.

36. Henderson S, Leibnitz G, Turnbull M, et al. False-positive human immunodeficiency virus seroconversion is not common following rabies vaccination. *Clin Diagnost Labor Immunol* 2002;**9**(4):942−3.

37. Srinivasan A, Burton EC, Kuehnert MJ, et al. Transmission of rabies virus from an organ donor to four transplant recipients. *N Engl J Med* 2005;**352**(11):1103−11.

38. Jackson AC. Rabies: new insights into pathogenesis and treatment. *Cur Opin Neurol* 2006;**19**(3):267−70.

39. Hemachudha T, Wacharapluesadee S, Mitrabhakdi E, et al. Pathophysiology of human paralytic rabies. *J Neurovirol* 2005;**11**(1):93−100.

40. Hooper DC. The role of immune responses in the pathogenesis of rabies. *J Neurovirol* 2005;**11**(1):88−92.

41. Gribencha SV, Gribanova L, Malkov GB, et al. Population structure of some street rabies virus strains. *Arch Virol* 1989;**104**(3−4):347−50.

42. Franka R, Wu X, Jackson FR, et al. Rabies virus pathogenesis in relationship to intervention with inactivated and attenuated rabies vaccines. *Vaccine* 2009;**27**(51):7149−55.

43. Willoughby Jr. RE, Tieves KS, Hoffman GM, et al. Survival after treatment of rabies with induction of coma. *N Engl J Med* 2005;**352**(24):2508−14.

5.2.3 Genus *Novirhabdovirus*

1. Dietzgen RG, Calisher CH, Kurath G, et al. Family rhabdoviridae. In: King AM, Adams MJ, Carstens EB, et al., editors. *Virus taxonomy: classification and nomenclature of viruses: ninth report of the International Committee on Taxonomy of Viruses*. 1st ed. London: Elsevier;2012. p. 686−713.

2. Wolf K. *Fish viruses and fish viral diseases*. Ithaca, London: Cornell University Press;1988.

3. Guenther RW, Watson SW, Rucker RR. Etiology of sockeye salmon "virus" disease. *US Fish Wildl Serv Spec Sci Rep Fish* 1959;**296**:1−10.

4. Rucker RR, Whipple WJ, Parvin JR, et al. A contagious disease of salmon possibly of virus origin. *US Fish Wildl Serv Fish Bull* 1953;**54**:35−46.

5. OIE. *Manual of Diagnostic Tests for Aquatic Animals*. 6th ed. Paris, France: World Organisation for Animal Health;2009.

6. Kurath G, Garver KA, Troyer RM, et al. Phylogeography of infectious haematopoietic necrosis virus in North America. *J Gen Virol* 2003;**84**(4):803−14.

7. Enzmann PJ, Castric J, Bovo G, et al. Evolution of infectious hematopoietic necrosis virus (IHNV), a fish rhabdovirus, in Europe over 20 years: implications for control. *Dis Aquat Organ* 2010;**89**(1):9−15.

8. Crane M, Hyatt A. Viruses of fish: an overview of significant pathogens. *Viruses* 2011;**3**(11):2025−46.

9. Nichol ST, Rowe JE, Winton JR. Molecular epizootiology and evolution of the glycoprotein and non-virion protein genes of infectious hematopoietic necrosis virus, a fish rhabdovirus. *Virus Res* 1995;**38**(2−3):159−73.

10. Jensen MH. Research on the virus of Egtved disease. *Ann NY Acad Sci* 1965;**126**(1):422−6.

11. Olesen NJ. Sanitation of viral haemorrhagic septicaemia. *J Appl Ichthyol* 1998;**14**:173−7.

12. Hedrick RP, Batts WN, Yun S, et al. Host and geographic range extensions of the North American strain of viral hemorrhagic septicemia virus. *Dis Aquat Organ* 2003;**55**(3):211−20.

13. Mortensen HF, Heuer OE, Lorenzen N, et al. Isolation of viral haemorrhagic septicaemia virus (VHSV) from wild marine fish species in the Baltic Sea, Kattegat, Skagerrak and the North Sea. *Virus Res* 1999;**63**(1−2):95−106.

14. Lopez-Vazquez C, Raynard RS, Bain N, et al. Genotyping of marine viral haemorrhagic septicaemia virus isolated from the Flemish Cap by nucleotide sequence analysis and restriction fragment length polymorphism patterns. *Dis Aquat Organ* 2006;**73**(1):23−31.

15. Snow M, Cunningham CO, Melvin WT, et al. Analysis of the nucleoprotein gene identifies distinct lineages of viral haemorrhagic septicaemia virus within the European marine environment. *Virus Res* 1999;**63**(1−2):35−44.

16. Kahns S, Skall HF, Kaas RS, et al. European freshwater VHSV genotype Ia isolates divide into two distinct subpopulations. *Dis Aquat Organ* 2012;**99**(1):23−35.

17. Knusel R, Bergmann SM, Einer-Jensen K, et al. Virus isolation vs RT-PCR: which method is more successful in detecting VHSV and IHNV in fish tissue sampled under field conditions? *J Fish Dis* 2007;**30**(9):559−68.

18. Elsayed E, Faisal M, Thomas M, et al. Isolation of viral haemorrhagic septicaemia virus from muskellunge, Esox masquinongy (Mitchill), in Lake St Clair, Michigan, USA reveals a new sublineage of the North American genotype. *J Fish Dis* 2006;**29**(10):611–19.

19. Einer-Jensen K, Ahrens P, Forsberg R, Lorenzen N. Evolution of the fish rhabdovirus viral haemorrhagic septicaemia virus. *J Gen Virol* 2004;**85**(5):1167–79.

5.2.4 Genus *Vesiculovirus*

1. Dietzgen RG, Calisher CH, Kurath G, et al. Family rhabdoviridae. In: King AM, Adams MJ, Carstens EB, Lefkowitz EJ, editors. *Virus taxonomy: classification and nomenclature of viruses: ninth report of the International Committee on Taxonomy of Viruses*. 1st ed. London: Elsevier;2012. p. 686–713.

2. Fijan N, Petrinec Z, Sulimanivic D, et al. Isolation of the causative agent from the acute form of infectious dropsy of carp. *Vet Arch Zagreb* 1971;**41**:125–38.

3. Miller O, Fuller FJ, Gebreyes WA, et al. Phylogenetic analysis of spring virema of carp virus reveals distinct subgroups with common origins for recent isolates in North America and the UK. *Dis Aquat Organ* 2007;**76**(3):193–204.

4. Stone DM, Ahne W, Denham KL, et al. Nucleotide sequence analysis of the glycoprotein gene of putative spring viraemia of carp virus and pike fry rhabdovirus isolates reveals four genogroups. *Dis Aquat Organ* 2003;**53**(3):203–10.

5. Bucke D, Finlay J. Identification of spring viraemia in carp (*Cyprinus carpio* L.) in Great Britain. *Vet Rec* 1979;**104**(4):69–71.

6. Garver KA, Dwilow AG, Richard J, et al. First detection and confirmation of spring viraemia of carp virus in common carp, *Cyprinus carpio* L., from Hamilton Harbour, Lake Ontario, Canada. *J Fish Dis* 2007;**30**(11):665–71.

7. Fijan N. *Spring viraemia of carp and other viral diseases and agents of warm-water fish. Fish Diseases and DisordersViral, Bacterial, and Fungal Infections*. London: CAB International;1999.

8. Sheppard AM, Le Deuff RM, Martin PD, et al. Genotyping spring viraemia of carp virus and other piscine vesiculo-like viruses using reverse hybridisation. *Dis Aquat Organ* 2007;**76**(2):163–8.

9. Walker PJ, Winton JR. Emerging viral diseases of fish and shrimp. *Vet Res* 2010;**41**(6):51.

10. Warg JV, Dikkeboom AL, Goodwin AE, et al. Comparison of multiple genes of spring viremia of carp viruses isolated in the United States. *Virus Genes* 2007;**35**(1):87–95.

11. Teng Y, Liu H, Lv JQ, et al. Characterization of complete genome sequence of the spring viremia of carp virus isolated from common carp (*Cyprinus carpio*) in China. *Arch Virol* 2007;**152**(8):1457–65.

12. Ahne W, Bjorkina HV, Essbauer S, et al. E. Spring viraemia of carp. *Dis Aquat Org* 2002;**52**:261–72.

6

Order Picornavirales

The Picornavirales order unites small, nonenveloped viruses that infect eukaryotes and that possess a number of prominent features: (1) The genome is represented by single-stranded (or bipartite), positive-sense RNA; (2) the icosahedral virion (approximately 30 nm in size) has a pseudo $T = 3$ architecture; (3) the genome translates into autoproteolytically processed polyprotein; (4) a three-domain replicate module Hel-Pro-Pol (containing a superfamily III helicase, a (cysteine) proteinase, and an RNA-dependent RNA polymerase) is present in the virus.[1,2] Currently, the order comprises five different families (Picornaviridae, Iflaviridae, Dicistroviridae, Secoviridae, and Marnaviridae) that infect a wide range of hosts, including mammalians, birds, reptiles, fish, arthropods, and plants.[3]

The length of genomic RNA varies from 7,000 to 12,500 nt. The RNA is covalently bonded with small viral (VP4) protein on the 5′ terminus and has a poly-(A) tail on the 3′ terminus. Some taxa of Picornavirales have a bipartite genome with RNA1 (length 5,800–8,400 nt) and RNA2 (length 3,200–7,300 nt). In these taxa, both the RNA1 and the RNA2 are covalently bonded with VP4 on the 5′ end. Proteins are generally sunthesized in the form of an extended polyprotein precursor (P0), which then undergoes proteolytic processing involving cellular and viral proteases.[1,2]

6.1 FAMILY PICORNAVIRIDAE

The Picornaviridae family consists of 12 genera (*Aphthovirus*, *Avihepatovirus*, *Cardiovirus*, *Enterovirus*, *Erbovirus*, *Hepatovirus*, *Kobuvirus*, *Parechovirus*, *Sapelovirus*, *Senecavirus*, *Teschovirus*, and *Tremovirus*) that can infect a wide variety of animals, including birds and humans.[3] The genome of the picornaviruses is single-stranded RNA 7,000–8,900 nt in length. The RNA is infectious, and after infection of the cell, it serves as mRNA to synthesize viral proteins in the cytoplasm. The 5′-NTR contains the hairpin structure necessary for cap-independent ribosome binding. The length of the 5′-NTR varies from 740 to 1,300 nt. The 3′-NTR is contain *cis*-acting elements required for genome replication. All viral proteins are synthesized as a polyprotein precursor (P0), the ORF of which extends over the entire RNA, excluding 5′- and 3′-nontranslated regions (NTR). The P0 protein precursor can be divided into three parts: P1, P2, and P3. The P1 gene region encodes the viral structural proteins (VP4 (1A), VP2 (1B), VP3 (1C), and VP1 (1D)), whereas the P2 and P3 gene regions encode nonstructural proteins involved in protein processing and RNA replication (2A–2C) and (3A–D). Processing of the P0 precursor is coupled with autocatalytic cleavage followed by enzymatic activity of the viral

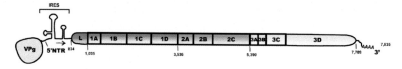

FIGURE 6.1 Genome structure of the encephalomyocarditis virus (EMCV) (family Picornaviridae, genus *Cardiovirus*). *Drawn by Tanya Vishnevskaya.*

proteases (2A^{Pro}, 3C^{Pro}, and 3CD^{Pro}). Genome replication involves the nonstructural proteins 2B, 2C, 3AB, 3B^{VPg}, 3CD^{Pro}, and 3D^{Pol}.[4,5]

6.1.1 Genus *Cardiovirus*

Currently, the *Cardiovirus* genus is known to consist of two species: the encephalomyocarditis virus (EMCV) and *theilovirus* (ThV), which includes several distinct genotypes: Theiler's murine encephalomyelitis virus (TMEV), Theravirus (TRV, a Theiler's-like rat virus), Saffold virus (SAFV), and Vilyuisk human encephalomyelitis virus (VHEV).[6] In general, the cardioviruses infect rodents, but SAFV, VHEV, and a human Theiler's-like virus were isolated from humans.[7–10] The genome of the prototypical cardiovirus, EMCV, is about 7,835 nt in length and has a type II internal ribosome entry site in the 5′-NTR (Figure 6.1).[11–13]

6.1.1.1 Syr-Darya Valley Fever Virus

History. Syr-Darya valley fever virus (SDVFV) was originally isolated from the blood of a patient with fever in the Syr-Darya district of Kazakhstan in July 1973.[14–16] Subsequently, it was isolated from *Hyalomma as. asiaticum* Schulze et Schlottke, 1929, ticks (subfamily Hyalomminae) and from *Dermacentor daghestanicus* Olenev, 1929, ticks (subfamily Rhipicephalinae). The two species of ticks were collected in the floodplains of the Syr Darya River and Ili River, respectively. The SDVFV isolation rate from the ticks was

0.5%. Several strains of SDVFV also were isolated from *Ornithodoros capensis* Neumann, 1901, ticks (family Argasidae) collected in the nests of gulls (*Laridae* Vigors, 1825; and *Sternidae* Bonaparte, 1838) on the islands of the Kara-Bogaz-Gol Bay in the Caspian sea, in Turkmenia, in 1973.[17] Morphological studies by electron microscopy and an antigenic relationship (found with a complement fixation test) characterize SDVFV as a member of the *Cardiovirus* genus.[18,19]

Taxonomy. The nucleotide sequence of the SDVFV strain LEIV-Tur2833 (GenBank ID: KJ191558) showed a strong similarity to that of ThV. A full-length genome comparison of SDVF with other cardioviruses revealed the highest level (85% nt and 95% aa) of similarity with TMEV. By contrast, the similarity of SDVFV with EMCV does not exceed 55%. The level of similarity of SDVFV with VHEV and TMEV in the P1 region (which codes four structural proteins) is 75%–91% for the nucleotide sequence and 80%–93% for the predicted amino acid sequence. The similarity of SDVFV with TMEV and VHEV in the P2 and P3 regions reaches 96%–98%. Phylogenetic analysis carried out on the basis of the P1, P2, and P3 regions placed SDVFV within the lineage of TMEV-like viruses (Figure 6.2).

Human pathology. The onset of Syr-Darya Valley disease is acute, with fever up to 40°C, abundant polymorphic roseolar–petechial rashes (with localization in the extremities, chest, and abdomen), chills, weakness, and a favorable outcome in 10–14 days.[16] Typically,

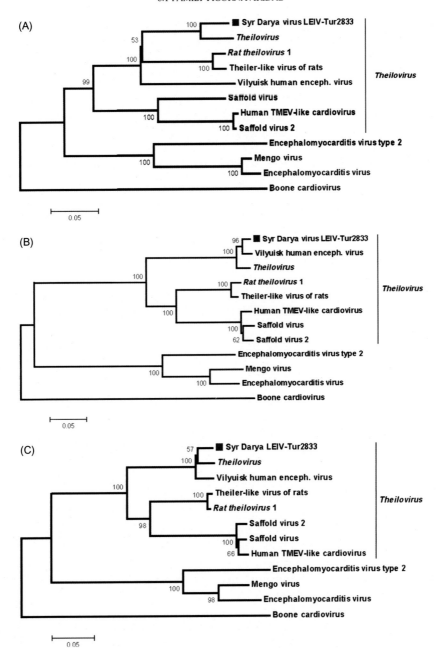

FIGURE 6.2 Phylogenetic analysis of predicted amino acid sequences of P1 (A), P2 (B), and P3 (C) regions of the cardio-viruses (family Picornaviridae, genus *Cardiovirus*). The place of SDVFV in the lineage of TMEV-like viruses is designated by a red square. (For the interpretation of the references to color in the legend, the reader is referred to the online version of this book.)

patients indicated sucking ticks 5 days before the disease. Similar fever, without laboratory confirmation, has been repeatedly registered in an endemic area. The population level of immunity to SDVFV in different regions of Kazakhstan varies from 0% to 3.5%, depending on the landscape. The SDVFV immune stratum of cattle reaches 11%−16% in floodplain landscapes of the Syr Darya River, Ili River, Emba River, and Talas River, whereas, in the steppe and mountain areas of Kazakhstan, there was only a single positive finding. These data indicate that natural foci in Kazakhstan are confined to floodplain-grazing biocenoses in desert zones.[15−18,20]

Animal pathology. ThV serotypes can be divided into two subgroups according to their pathogenesis in mice. Viruses of the TO subgroup (named for "Theiler's original strains") cause a biphasic infection in mice. The first (early) phase occurs with symptoms of poliomyelitis, with the virus replicating in the neurons of the brain. At the end of the first phase of infection, the virus enters the spinal cord and various types of cells, causing a persistent infection. Often, chronic infection results in chronic demyelination of neurons, a condition that can serve as a model for disseminated encephalomyelitis (multiple sclerosis) in humans.[21,22] The second subgroup includes the neurovirulent variants of ThV (currently, three different neurovirulent strains of ThV are known), which are characterized by the rapid development of fatal encephalitis in infected mice with active virus replication in neurons.[8,23,24] Like virulent isolates of ThV, SDVFV produces encephalitis in adult mice after intracerebral inoculation.[19] Subcutaneous infection of green monkeys resulted in clinically mild disease, but postmortem examination showed signs of meningoencephalitis with predominant involvement of the subcortical areas and cerebellum. Focal pneumonia, hepatitis, and spleen lesions have also been observed.[15]

References

1. Le Gall O, Christian P, Fauquet CM, et al. Picornavirales, a proposed order of positive-sense single-stranded RNA viruses with a pseudo-T = 3 virion architecture. *Arch Virol* 2008;**153**(4):715−27.
2. Putnak JR, Phillips BA. Picornaviral structure and assembly. *Microbiol Rev* 1981;**45**(2):287−315.
3. Sanfaçon H, Gorbalenya AE, Knowles NJ, Chen YP. Order Picornavirales. In: King AM, Adams MJ, Carstens EB, Lefkowitz EJ, editors. *Virus taxonomy: classification and nomenclature of viruses: ninth report of the International Committee on Taxonomy of Viruses.* 1st ed. London: Academic Press;2012. p. 835−9.
4. Racaniello VR. Picornaviridae: the viruses and their replication. In: Knipe DM, Howley PM, editors. *Fields virology.* 5th ed. Philadelphia, PA: Lippincott Williams & Wilkins;2007. p. 795−838.
5. Lin JY, Chen TC, Weng KF, et al. Viral and host proteins involved in picornavirus life cycle. *J Biomed Sci* 2009;**16**:103.
6. Knowles NJ, Hovi T, Hyypiä T, et al. Family Picornaviridae. In: King AM, Adams MJ, Carstens EB, Lefkowitz EJ, editors. *Virus taxonomy: classification and nomenclature of viruses: ninth report of the International Committee on Taxonomy of Viruses.* 1st ed. London: Academic Press;2012. p. 885−90.
7. Philipps A, Dauber M, Groth M, et al. Isolation and molecular characterization of a second serotype of the encephalomyocarditis virus. *Vet Microbiol* 2012;**161** (1−2):49−57.
8. Liang Z, Kumar AS, Jones MS, et al. Phylogenetic analysis of the species Theilovirus: emerging murine and human pathogens. *J Virol* 2008;**82**(23):11545−54.
9. Himeda T, Ohara Y. Saffold virus, a novel human Cardiovirus with unknown pathogenicity. *J Virol* 2012;**86**(3):1292−6.
10. Chiu CY, Greninger AL, Kanada K, et al. Identification of cardioviruses related to Theiler's murine encephalomyelitis virus in human infections. *PNAS* 2008;**105** (37):14124−9.
11. Kong WP, Roos RP. Alternative translation initiation site in the DA strain of Theiler's murine encephalomyelitis virus. *J Virol* 1991;**65**(6):3395−9.
12. Dvorak CM, Hall DJ, Hill M, et al. Leader protein of encephalomyocarditis virus binds zinc, is phosphorylated during viral infection, and affects the efficiency of genome translation. *Virology* 2001;**290** (2):261−71.
13. Bacot-Davis VR, Palmenberg AC. Encephalomyocarditis virus Leader protein hinge domain is responsible for interactions with Ran GTPase. *Virology* 2013;**443**(1): 177−85.

14. Lvov DK, Karimov SK, Kiriushchenko TV, et al. Isolation of the virus of Syr-Darya Valley fever. *Vopr Virusol* 1984;**29**(5):553–8 [in Russian].

15. Lvov DK. Syr-Darya fever. In: Lvov DK, editor. *Handbook of virology: viruses and viral infection of human and animals.* Moscow: MIA;2013. p. 409–11 [in Russian].

16. Lvov DK. Syr-Darya valley fever. In: Lvov DK, Klimenko SM, Gaidamovich SY, editors. *Arboviruses and arboviral infection.* Moscow: Meditsina;1989. p. 246–9 [in Russian].

17. Lvov DK. Natural foci of arboviruses in the USSR. In: Zhdanov VM, editor. *Soviet medical reviews virology.* UK: Harwood Academic Publishers GmbH;1987. p. 153–96.

18. Karimov SK, Lvov DK, Kiriushchenko TV. Syr-Darya Valley fever, a new virus disease in Kazakhstan. *Arch Virol* 1990;(Suppl. 1):345–8.

19. Karimov SK. *Arboviruses of the Kazakhstan regions* [Doctor of Medicine thesis]. 1983. p. 315 [in Russian].

20. Lvov DK. *Ecological soundings of the former USSR territory for natural foci of arboviruses. Soviet medical reviews virology.* UK: Harwood Academic Publishers GmbH;1993, p. 1–47.

21. Aubert C, Chamorro M, Brahic M. Identification of Theiler's virus infected cells in the central nervous system of the mouse during demyelinating disease. *Microb Pathogen* 1987;**3**(5):319–26.

22. Lipton HL, Twaddle G, Jelachich ML. The predominant virus antigen burden is present in macrophages in Theiler's murine encephalomyelitis virus-induced demyelinating disease. *J Virol* 1995;**69**(4): 2525–33.

23. Buckwalter MR, Nga PT, Gouilh MA, et al. Identification of a novel neuropathogenic Theiler's murine encephalomyelitis virus. *J Virol* 2011;**85**(14):6893–905.

24. Aubert C, Brahic M. Early infection of the central nervous system by the GDVII and DA strains of Theiler's virus. *J Virol* 1995;**69**(5):3197–200.

Double-Stranded RNA Viruses

7.1 FAMILY BIRNAVIRIDAE

The Birnaviridae family comprises four genera (*Aquabirnavirus*, *Avibirnavirus*, *Blosnavirus*, and *Entomobirnavirus*) of nonenveloped viruses whose genome consists of two segments (segments A and B) of double-stranded RNA. The sizes of the genomic dsRNAs are 3,100–3,600 nt (segment A) and 2,800–3,300 nt (segment B). The chain of each RNA segment is covalently linked to a VPg protein at the 5′ terminus. The 3′ terminus is not polyadenylated. The virion has icosahedral geometry with $T = 13$ symmetry. Birnaviruses are widely distributed and infect a range of hosts, including fish, mollusks (*Aquabirnavirus*, *Blosnavirus*), birds (*Avibirnavirus*), and insects (*Entomobirnavirus*).[1]

7.1.1 Genus *Aquabirnavirus*

The aquabirnaviruses have been isolated from many species of aquatic animals, including fish and mollusks. The genome of the prototypical infectious pancreatic necrosis virus (IPNV) consists of two segments (segment A, 3,097 nt; segment B, 2,784 nt) of dsRNA. Segment A encodes two viral structural proteins VP2 and VP3, as well as nonstructural proteins VP4 and VP5. Segment A contains a large open reading frame (ORF) that encodes a polyprotein precursor (106 kD) which is cleaved by VP4 to produce pre-VP2 (pVP2) and VP3. The outer capsid protein VP2 contains major neutralizing epitopes that are used for antigenic or genetic genotyping of IPNV. Segment B encodes VP1 protein (94 kD), which acts as the virion-associated RNA-dependent RNA polymerase (RdRp). The genus includes closely related species: IPNV, tellina virus (TV), yellowtail ascites virus (YTAV), marine birnavirus (MABV), and marine birnavirus H.[1]

7.1.1.1 Infectious Pancreatic Necrosis Virus

The development of aquaculture is complicated by spreading fish diseases, mostly of a virus etiology and leading to a loss of 15–20% of fish production annually. About 100 DNA (families Herpesviridae, Iridoviridae, and Adenoviridae) and RNA (families Birnaviridae, Rhabdoviridae, Paramyxoviridae, Coronaviridae, Picornaviridae, Retroviridae, and Reoviridae) viruses are associated with diseases of fish. A dramatic increase in the number of finfish viruses is a reflection of the rise in surveillance of both cultured and feral fish populations that has occurred because of the continued growth of the aquaculture industry in many countries. Many of the viruses that have been described have either little or no pathogenicity for their natural hosts.[2] Since 2000, IPNV (family

Zoonotic Viruses of Northern Eurasia.
DOI: http://dx.doi.org/10.1016/B978-0-12-801742-5.00007-6

TABLE 7.1   The Main Virus Infections of Fishes

Family	Genus	Virus	Distribution
Birnaviridae	*Aquabirnavirus*	IPNV	Northern and Southern America, Japan and other countries of Asia, Europe (including Baltic states), Russia (Murmansk region)
Rhabdoviridae	*Novirhabdovirus*	IHNV	Pacific coast of Northern America Europe, Russia (Far East, Karelia Republic)
		VHSV	Northern America, Europe including Germany, Finland, Norway, Sweden, Baltic states, Russia (Karelia Republic)
	Vesiculovirus	SVCV	Europe including Lithuania, Belorussia, Ukraine, Moldova, European part of Russia
Orthomyxoviridae	*Isavirus*	ISAV	Canada, Norway, Iceland, United Kingdom

Birnaviridae, genus *Aquabirnavirus*) and infectious hematopoietic necrosis virus (IHNV) (family Rhabdoviridae, genus *Novirhabdovirus*) were registered in Russia. In 2003—2006, several cases of fish death were associated with spring viremia of carp virus (SVCV) (family Rhabdoviridae, genus *Vesiculovirus*). The Viral hemorrhagic septicemia virus (VHSV) (family Rhabdoviridae, genus *Novirhabdovirus*) was isolated from wild population of salmon species. To date, the infectious salmon anemia virus (ISAV) (Table 7.1) has not been found in Russia.[3]

History. Infectious pancreatic necrosis disease has been observed in young hatchery-reared brook trout (*Salvelinus fontinalis*) and rainbow trout (*Oncorhynchus mykiss*) since 1940, when it was described as "acute catarrhal enteritis of salmonid fingerlings."[4—6] The virus etiology of the disease was identified in 1960, and the first strain (VR229) of IPNV was isolated on the fish cell line in the United States.[7] In Europe, the first strain of IPNV was isolated in 1971 and its antigenic differences from the strain VR229 have been demonstrated.[8] Since then, IPNV has been isolated in many different regions of the world. The initial hypothesis that IPNV infects only salmonids and is always pathogenic was revised when the virus was isolated from healthy white suckers (*Catostomus commersoni*).[9] Furthermore, numerous viruses

that reacted with anti-IPNV serum and, on the basis of that reaction, had been classified as IPNV were isolated from many species of aquatic hosts, including freshwater fish as well as marine mollusks and crustaceans.[10] Thus, IPNV-like birnaviruses, which are not pathogenic to salmonids, are named for their source of isolation (e.g., TV was isolated from *Tellina tenuis*, a bivalve mollusk), or they are simply called marine birnaviruses, isolated from different fish species.

Taxonomy. After numerous isolations of the IPNV strain, it was shown that these viruses have serological differences. Initially, two strains, Ab and Sp, which were different from each other and from strain VR229, were isolated from Danish rainbow trout.[11] These viruses formed the basis for dividing the strains into three main serogrups of IPNV. With the accumulation of more data, IPNV and IPNV-like isolates were classified into two serogroups: A and B. Serogroup A is divided to nine serotypes (A1—A9) that are all pathogenic to fish, whereas serogroup B comprises one avirulent serotype TV-1 (tellina virus).[12] Serotype A1 contains most of the U.S. isolates, and serotypes A6—A9 are found mainly in Canada. Four pathogenic serotypes of IPNV (A2—A5) are found in Europe. Phylogenetic analysis based on predicted amino acid sequences of segment A (or VP2 protein) revealed the existence of seven genotypes, with geographical and serological similarity. The sequence diversity between different genotypes is 18—30%, on average.[13,14]

Distribution. IPNV has worldwide distribution. A wide range of salmonid fish species are affected similarly. IPNV is found mainly in countries with a developed salmon culture industry: in North and South America, in several European countries, and in Japan and other Asian countries.[15] The disease was found on 174 fish farms in Norway in 2002. In the former Soviet Union, IPNV was registered in the Baltic states in 1986—1987. Epizooty occurred when the water temperature was 14°C; the outbreak led to the death of 80% of the fry.[3]

The disease often kills the fry of rainbow trout (*O. mykiss*), brook trout (*S. fontinalis*), brown trout (*Salmo trutta*), Atlantic salmon (*Salmo solar*), Japanese amberjack (*Seriola quinqueradiata*), the common dab (*Limanda limanda*), halibut (*Hippoglossus hippoglossus*), Atlantic cod (*Gadus morhua*), and other species. It has been shown that at least 40 species of hydrobionts, including freshwater fish and up to 70 species of sea invertebrates, can be asymptomatic carriers of IPNV.[16]

Animal infection. IPNV is an etiologic agent of epizootic disease in fish that causes necrotic lesions of the pancreas with high mortality (up to 90%) among salmon fry and fingerlings. Infection in older fish is usually asymptomatic. Affected fish swim in circles around their longitudinal axis. Other signs of the disease include darkening of the body, pale gills, and hemorrhage of the pyloric appendages. The most frequent histopathological finding is necrosis of the pancreas; necrosis also occurs in the kidney and intestinal mucosa. The mortality rate (10—90%) is typically dependent on a combination of various factors, such as the virulence of the strain, resistance of the host, and the nature of the environment. Mortality in fish can reach 45% of one-month-olds, 35% of two-month-

olds, and 7% in four-month-olds. Fish that recover or are asymptomatically infected often become lifelong carriers of the virus. IPNV is transferred by horizontal and vertical routes in the populations of river and sea fish and can be transmitted between hatcheries by personnel.[16,17] Restrictions on the movement of eggs and live fish have been imposed in an attempt to contain the spread of the virus.

Diagnostics. A diagnosis of IPN is based on histological examination and isolation of the virus in a cell culture, followed by serological identification through neutralization or enzyme-linked immunosorbent assay (ELISA) tests. RT-PCR tests have also been developed. In general, to detect carriers, at least 10% of a fish population should be tested.[18—22]

7.2 FAMILY REOVIRIDAE

The family Reoviridae comprises more than 500 viruses classified into 15 genera that can infect a wide range of hosts (fungi, plants, arthropods, and vertebrates). The main taxonomic characteristic of the reoviruses is a segmented (9—12 segments), double-stranded RNA genome and peculiar properties of the virion morphology.[1—3] The virion of the reoviruses has icosahedral symmetry and is formed by two or three protein layers. The inner capsid (approx. 80 nm in diameter), which generally has a $T = 2$ symmetry, contains genomic RNA, and at least three proteins possess enzymatic activity associated with RNA replication.[4] The shape of the outer layer of the virion varies with the different genera. The layer may be permeated with wide "towerlike" protein spikes in 12 icosahedral vertices of facets, or the virion has a spherical shape without ridges and spines. On the basis of this property, the reoviruses are divided into two subfamilies: the Spinareovirinae and the Sedoreovirinae.[1]

Members of the genera *Orthoreovirus* and *Rotavirus* can replicate only in the cells of vertebrates and are transmitted by a fecal–oral route or a respiratory route. Viruses of the genera *Orbivirus*, *Seadornavirus*, and *Coltivirus* infect vertebrates, but also can replicate in arthropods that serve as vectors (i.e., these viruses are arboviruses). Some species of the vertebrate reoviruses are important in human and animal pathology.

7.2.1 Genus *Orbivirus*

The genus *Orbivirus* includes nonenveloped viruses (diameter ≈90 nm), whose genome is represented by 10 segments of dsRNA (the length of the segments is from ≈800 to ≈4,000 nt, respectively; the total length of the genome of the prototype bluetongue virus (BTV) is 19,200 nt). Many orbiviruses are important human and animal pathogens: Peruvian horse sickness virus-1 (PHSV-1), epizootic hemorrhagic disease virus-1 (EHDV-1), Kemerovo virus (KEMV), Corriparta virus (CORV), Changuinola virus (CGLV), African horse sickness virus (AHSV), and BTV, among others. The orbiviruses belong to the ecological group of arboviruses (i.e., they are transmitted to vertebrate hosts through bloodsucking arthropods). Depending on the kind of arthropod vector and primary vertebrate host, there are three major environmental groups of the orbiviruses: *Culicoides* borne (transmitted by biting midges), mosquito borne (transmitted by mosquitoes), and tick borne (transmitted by ticks). Serological cross-reactions investigating intragenic relationships between orbiviruses identified four antigenic groups, designated A–D, which correspond, in general, to their ecological features, determined by the type of arthropod vectors. Group A includes viruses transmitted by biting midges (BTV, AHSV, etc.). Group B includes tick-borne viruses (KEMV, Great Island virus (GIV), Chenuda

virus (CNUV), Wad Medani virus (WMV), and Ieri virus (IERIV). Group C contains the mosquito-borne CORV and group D contains the Wongorr virus (WGRV). The separate species in the genus *Orbivirus* are considered to be a combination of different serotypes, which may have a significant divergence.[1] With the accumulation of additional genomic data, classification of the orbiviruses has been defined more precisely. Phylogenetic analysis, based on a comparison of various segments of the genome, allows the orbiviruses to be divided into three distinct lineages, which generally coincide with the antigenic and ecological groupings (Figure 7.1).[1,5–9]

7.2.1.1 Baku Virus

History. The Baku virus (BAKV) was originally isolated from ticks of the species *Ornithodoros capensis* (family Argasidae), collected in the nests of herring gulls (*Larus argentatus*) on the islands of the Baku archipelago in the Caspian Sea (2.5 km from the town of Älät, Azerbaijan) in the spring and summer of 1970.[1–3] Neutralized antibodies against BAKV have been found in the 15.6% of nesting birds on the islands. BAKV was also isolated during an epizootic outbreak in a pheasant farm in Barda district, Azerbaijan. During the outbreak, a high mortality was observed among young pheasants and BAKV was isolated from their internal organs. One strain of BAKV was isolated in the Ismail region of Azerbaijan from ticks of the species *Hyalomma marginatum*, collected from cows.[4] Subsequently, BAKV was repeatedly isolated from ticks of the species *Ornithodoros coniceps* (Canestrini, 1890), collected in the nests of gulls (*L. argentatus*) and terns (*Sterna hirundo*) on islands in Kara-Bogaz-Gol Bay in Turkmenia and in burrow nests of pigeons (*Columba livia neglecta*) in the foothills of the Chatkal Range of the West Tien-Shan Mountains in the Tashkent region of Uzbekistan.[5–8] In total, 69 strains of BAKV were isolated, as shown in Table 7.2. On the

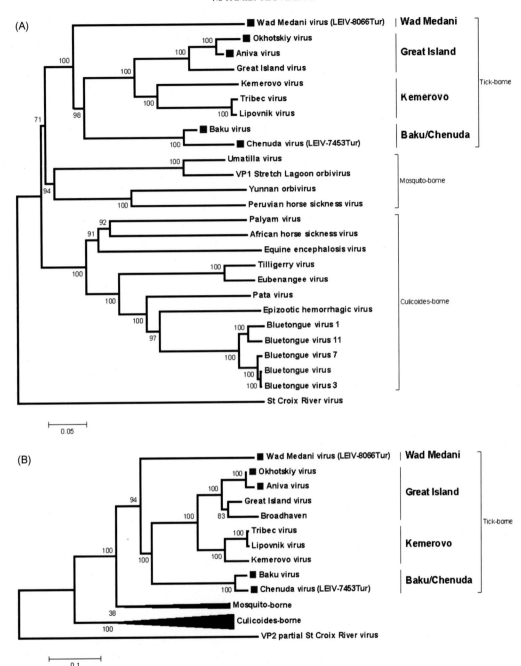

FIGURE 7.1 Phylogenetic analysis based on a comparison of amino acid sequences of the orbiviruses (family Reoviridae, genus *Orbivirus*): (A) VP1(Pol); (B) VP3 (T2) protein; (C) VP7 (T13) protein.

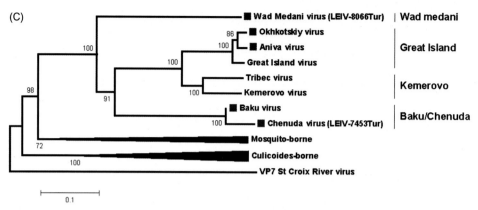

FIGURE 7.1 *Continued.*

TABLE 7.2   Isolation of BAKV from Argasid Ticks (taxon Acari, family Argasidae) in Nests of Colonial Birds

Place of isolation	Biotopes	Source of isolation[a]	Total strains (year)
Gil Island of Baku archipelago in Caspian Sea (Azerbaijan) 49°30′E, 39°50′N	Nests of herring gulls *L. argentatus* Dry subtropics, semi-arid landscape	Ticks *O. capensis* Neumann, 1901Rate of infection 1:120	14 (1970) 20 (1971) 25 (1973)
			Total—59
Islands in Kara-Bogaz-Gol bay in Caspian Sea (Turkmenistan) 53°E, 35°N	Nests of herring gulls *L. argentatus* and terns *S. hirundo* Dry subtropics, semi-arid landscape	Ticks *O. coniceps* Canestrini, 1980 Rate of infection 1:1500	1 (1972 г.) 2 (1973)
			Total—3
Near Parkent town (Tashkent region, Uzbekistan) Foothill of Chatkal Range of the West Tien-Shan (Tashkent region, Uzbekistan) 69°45′E, 40°40′N	Nests of gulls *Columba livia neglecta* Dry subtropics, semi-arid landscape	Ticks *O. coniceps* Canestrini, 1980 Rate of infection 1:130	1 (1972) 2 (1973) 1 (1974) 2 (1975)
			Total—6

[a]*One strain was isolated from ticks* H. marginatum, *collected on cows in Ismail region of Azerbaijan in 1973.*

basis of morphological and serological data (the structure of the virion, as seen through the electron microscope), BAKV was classified into the Kemerovo antigenic group of the genus *Orbivirus*.[3,9–12] According to the ninth report of the International Committee on Taxonomy of Viruses, BAKV was also classified into the Chenuda serocomplex.[13]

Taxonomy. The genome of BAKV was sequenced (KJ191548–57) and phylogenetic analysis was conducted. The tick-borne orbiviruses, which originally were consolidated into the Kemerovo antigenic group, can be divided into four subgroups: Kemerovo, Great Island, Chenuda, and Wad Medani, with subgroup named by the prototypical virus that was isolated (Figure 7.1). To date, genomic data are available only for a small number of tick-borne orbiviruses, belonging predominantly to the Great Island, Wad Medani, and Kemerovo subgroups.[14,15]

A comparison of amino acid sequences of the proteins of BAKV with those of other orbiviruses is shown in Table 7.3. The similarity of VP1 (Pol) protein between orbiviruses reaches 34.5–64.0%. VP1 (pol) of BAKV has, on average, 48% identity with VP1 (pol) of tick-borne orbiviruses and 41% with VP1 (pol) of mosquito-borne orbiviruses, whereas the similarity of VP1 (Pol) protein among viruses within the Great Island subgroup exceeds 90%. The similarity of VP1 (Pol) of viruses of the Kemerovo subgroup to that of viruses of the Great Island subgroup is in the range 72–73%.

The similarity of VP3 (T2) protein of BAKV to that of tick-borne orbiviruses reaches 64.8%, considerably more than the 42.4–47.3% and 34.0–37.7% for T2 protein of mosquito-borne and culicoides-borne orbiviruses, respectively. The lowest level of similarity that VP3 (T2) of BAKV has is to VP3 (T2) of St. Croix River virus (SCRV), which diverges most from the rest of the orbiviruses (Figure 7.1). Although SCRV was isolated from *Ixodes scapularis*, it has the lowest similarity (antigenically and genetically) to both the tick-borne and mosquito-borne orbiviruses (21–25%).

A phylogenetic analysis based on a comparison of proteins VP1 (Pol), VP3 (T2), and VP7 (T13) is presented in Figure 7.1. The topology of the tree shown indicates that BAKV and CNUV form a distinct phylogenetic lineage in the complex of tick-borne viruses of the Kemerovo antigenic group.

7.2.1.2 Chenuda Virus

History. CNUV has been isolated from *Argas reflexus hermannii* ticks (16 strains from 60 pools) collected in February 1954 in pigeons' nests in Chenuda village (30°N, 32°E), on the Nile River delta in Egypt.[1–3] CMUV has also been isolated in South Africa, where this virus is associated with *Argas perengueyi* in nests of cliff swallows.[4]

TABLE 7.3 Pairwise Comparison of Full-Length Amino Acid Sequences of the Proteins of BAKV with Those of Other Orbiviruses

Viruses	Proteins									
	VP1 (Pol)	T2	VP2	NS1	VP4	VP5	NS2	VP7 (T13)	VP6	NS3
Kemerovo virus	48.6	64.8	23.2	40	58	n.a	45	64	33	44
Tribec virus	48.6	64.3	26.1	40	52	n.a	48	64	32	43
Great Island virus	47.0	63.0	22.5	40	60	n.a	46	64	34	28
Yunnan virus	44.5	46.5	9.7	26	42	54	30	38	29	n.a.
Epizootic hemorrhagic disease virus	42.3	37.4	14.2	23	46	39	27	34	28	31
African horse sickness virus	42.1	36.8	8.8	23	41	34	30	36	32	29
Bluetongue virus	43.7	36.8	11.4	24	46	35	25	30	26	44

n.a.: not available.
Amino acid identity (%) are shown.

Genome and taxonomy. Like other orbiviruses, CNUV has a genome that consists of 10 segments of dsRNA with a length of approximately 900 to 3,900 nt.

The genome of CNUV (strain LEIV-7453Tur) was sequenced and phylogenetic analysis was conducted (Figure 7.1). The topology of CNUV on the trees shown in the figure was constructed on the basis of sequences of VP (Pol), VP3 (T2), or VP7 (T13) proteins and confirms that CNUV belongs to the lineage of tick-borne orbiviruses (Kemerovo antigenic group). The Chenuda serocomplex includes several orbiviruses (including BAKV) isolated mainly from *Ornithodoros* ticks.[5,6] Genomic data are currently available only for CNUV and BAKV.

Epizootiology. Complement-fixating antibodies against CNUV were found in Lower Egypt among camels (3.6%), buffalos (2.8%), pigs (7.0%), dogs (4.0%), donkeys (1.6%), and rodents (1.1%).[1]

In Northern Eurasia, CNUV was isolated (LEIV-7453Tur strain) in July 1979 from 15,400 mosquitoes from the species *Culex modestus*, *Aedes caspius*, and *Anopheles hyrcanus* collected on the coast of Mamed-Kul Lake within the lower part of the Atrek River in Turkmenistan.[7]

Epidemiology. Currently, there are no data on the epidemiological significance of CNUV.

7.2.1.3 Kemerovo Virus

History. An illness caused by this virus was first observed in secondary taiga landscape in the Kemerovo region in Western Siberia (50°40′N, 80°40′E) in May 1962. Twenty-two strains of an unknown virus subsequently named Kemerovo virus KEMV were isolated from the blood of patients in the acute phase during an outbreak. The strains of KEMV have also been isolated from healthy persons following tick bites and from *Ixodes persulcatus* ticks. Subsequently, the sera from 93 patients hospitalized with febrile illness and having no antibodies against tick-borne encephalitis virus (TBEV) were tested against KEMV in a neutralization test. Fifty-seven of the patients tested positive.

Twenty-one patients were tested for diagnostic increases in the neutralizing titers of acute and convalescent sera, and increases were found in 10 of these patients. Eighteen strains of KEMV have been isolated from 1,058 *I. persulcatus* ticks that were examined).[1,2] Later, KEMV was isolated from *Hyalomma anatolicum* ticks in Uzbekistan and from migrating *Phoenicurus phoenicurus* redstarts in Egypt.[3–5] The antigenically related Tribeč virus (TRBV) was isolated from *Ixodes ricinus* ticks in central Slovakia in 1963.[6–8] TRBV also had been isolated from the bank vole (*Clethrionomys glareolus*) and the pine vole (*Microtus pinetorum*). A subtype of TRBV—Lipovnic virus (strain LTP-91)—was also isolated from *I. ricinus* ticks in the eastern part of the former Czechoslovakia. Lipovnic virus is indistinguishable from TRBV.[7,8] Other subtypes or strains of TRBV include Koliba, Cvilin,[9] Brezová,[10] Mircha,[11] and Kharagysh.[12–14]

Epizootiology. KEMV causes a fatal illness in newborn, but not in adult, mice following intracerebral inoculation with a titer on the order of 10^{-6}; newborn white rats, cotton rats, hamsters and rhesus monkeys are also susceptible.[2,15] Neutralizing antibodies were found in the Kemerovo region in horses (16; 100% positive), cattle (151; 14.4%), the grey red-backed vole *Clethrionomys rufocanus* (35; 47.5%) the bank vole *C. glareolus* (17; 23.5%), and birds: the fieldfare *Turdus pilaris* (24; 21.7%) and the red-throated thrush *T. ruficollis* (6; 33.3%).[1,2]

Taxonomy. KEMV is a prototypical virus of the Kemerovo antigenic group, to which belong a large number of tick-borne orbiviruses. The group is divided into four subgroups: the Kemerovo subgroup, the Great Island subgroup, the Wad Medani subgroup, and the Chenuda subgroup, which have ecological, antigenic, and phylogenetic differences. The Kemerovo subgroup includes KEMV, TRBV, and their subtypes.[16–21] Gene pools of the Kemerovo group, like those of other orbiviruses, have a great reassortment potential because of the segmented dsRNA and resulting biological variability.[17,18,22]

Distribution. KEMV has been isolated from *I. persulcatus* ticks in western Siberia[1,2] and from migratory birds in Egypt.[4] Migratory birds have been implicated in the dispersal of viruses of the Kemerovo antigenic group over vast distances. TRBV has been isolated from *I. ricinus* ticks in the former Czechoslovakia[6−10] and in Moldova,[13,14,23] and from *Haemaphysalis punctata* in Romania.[24]

Ecology. The vertebrate hosts of TRBV are two species of rodents—the bank vole *C. glareolus* and the pine vole *M. pinetorum*—the hare *Lepus europeus*, goats,[6] and birds (the European starling *Sturnus vulgaris* and the common chaffinch *Fringilla coelebs*).[11,23] Antibodies were found in grazing ruminants (up to 45−88%).[10]

Clinical picture. The virus causes an outbreak of occasional febrile illness or aseptic meningitis in humans.[1,2,5,25,26] Antibodies occur in human populations at a higher frequency among patients with multiple sclerosis. The disease is probably underdiagnosed. There is a potential occupational risk for forestry workers and tourists. Additional studies are necessary to evaluate the importance of the virus for public health.

Diagnostic. Complement fixation tests and ELISA may be used in laboratory diagnosis, but not HI, because the virus does not produce hemagglutinin.

7.2.1.4 *Okhotskiy Virus and Aniva Virus*

History. Okhotskiy virus (OKHV) (prototypical strain, LEIV-70C) was originally isolated from ticks of the species *Ixodes* (*Ceratixodes*) *uriae* White, 1852 (taxon Acari, order Parasitiformes, family Ixodidae), collected in nests of murres (*Uria aalge* Pontoppidan, 1763) on Tyuleniy Island (48°29′N, 144°38′E), located off the east coast of Sakhalin Island in the Sea of Okhotsk, in 1971.[1−3] Later, several strains of OKHV were isolated in similar ecological conditions at high latitudes in other parts of the Sea of Okhotsk

as well as in the Bering Sea and the Barents Sea (Table 7.4).[4−8]

Another virus, named Aniva virus (ANIV) (prototypical strain, LEIV-1373C), was isolated from the same source (*I. uriae* ticks collected in Tyuleniy Island) and from ticks collected in Ariy Kamen Island (in the Bering Sea) (Table 7.4).[4−7]

OKHV and ANIV were classified into the genus *Orbivirus* through electron microscopy and into the Kemerovo antigenic group on the basis of the results of serological tests.[2,3] At least seven viruses isolated from *I. uriae* ticks were in the Kemerovo antigenic group in 1985 (Table 7.5). OKHV is probably identical to Cape Wrath virus (CWV), because it hybridizes well to all 10 genes.[9,10] CWV was isolated from *I. uriae* ticks in Scotland in 1973.[11,12] Antigenically and genetically, OKHV is closely related to GIV,[11−13] Bauline virus (BAUV),[11] Mykines virus (MYKV),[14,15] and Tindhólmur virus (TINDV)[16] (Table 7.5). Many orbiviruses that are similar or identical to OKHV, but are not registered as species, have been isolated (Table 7.6). BAUV was isolated in Canada in 1971[16] and in Norway in 1974.[10] Through RNA−RNA hybridization, the Norwegian strains have been found to be identical to the prototypical Canadian strain and indistinguishable from GIV.[9,10] But the two viruses can be differentiated by a virus neutralization test.[12] Some BAUV and GIV strains from Newfoundland have exhibited remarkable variation in all 10 genome segments.[13,17] All of the viruses described in this paragraph are classified as members of the Great Island subgroup of the Kemerovo antigenic group."[18,19] The status of some strains in the Great Island subgroup needs to be revised.[17,19−27]

Taxonomy. The genomes of OKHV and ANIV (KF981623-32; KJ191541-47) have been completely sequenced. Most OKHV and ANIV proteins have a high level of similarity (VP1 (Pol), 96% aa are identical; VP3 (T2), 99%; VP7 (T13), 98%; NS1, 94%; NS2, 98%; NS3, 72%;

TABLE 7.4   Isolation of Strains of OKHV and ANIV from *Ixodes* (*Ceratixodes*) *Uriae* White, 1852, in the Nests of Auks (*Alcidae* Leach, 1820) in the Basins of the Sea of Okhotsk, the Bering Sea, and the Barents Sea

Virus	Parameter	Far East of Russia				Europe
		Okhotsk Sea basin		Bering Sea basin		Barents Sea basin
		Sakhalin district		Kamchatka	Chukotka	Murman district
		Tyuleniy Island (48°29′N 144°38′E)	Iony Island (56°24′N, 143°23′E)	Arii Kamen Island (Komandorsky Islands) (55°13′N, 165°48′E)	Coast of Bering strait (64°50′N, 173°10′E)	Kharlov Island in Sirena bay of Barents Sea (68°49′ с.ш., 37°19′ в.д.)
OKHV	Number of isolated strains	5	2	3[a]	1	23[b]
	Infection rate of (%)	0.037	0.103	0.016	0.087	0.256
	Number of ticks examined	35,725				8,994
	Total isolated strains	11				23
Total	Infection rate (%)	0.031				0.256
ANIV	Number of isolated strains	1	0	1	0	0
	Infection rate (%)	0.007	0	0.005	0	0
	Number of ticks examined	35,725				8,994
	Number of strains examined	2				0
Total	Infection rate (%)	0.006				0

[a]One strain of OKHV was isolated from the tick Ixodes signatus *Birula, 1895.*
[b]Two strains were isolated from the black-legged kittiwake (R. tridactyla *Linnaeus, 1758).*

VP6, 93%). Although OKHV and ANIV are closely related, they are easily differentiated by serological methods because of the low similarity of their shell outer layer proteins (VP2, 62%; VP5, 68% aa). The VP2 and VP5 proteins carry the major species-specific antigenic determinants of the orbiviruses, and their lesser similarity explains the antigenic difference between OKHV and ANIV in their serologic reactions.

A comparison of the amino acid sequences of ANIV and OKHV with those of other orbiviruses revealed that the level of identity was greatest with those of GIV. The VP3 (T2) inner shell proteins of ANIV and OKHV are 92% identical to their GIV counterparts. The similarity level of VP7 (T13) among ANIV, OKHV, and GIV is 96%. VP7 (T13) is a protein of the outer layer and carries main-group–specific antigenic determinants. The high level of similarity of ANIV and OKHV to GIV is unambiguously attributed to the Great Island subgroup in the Kemerovo antigenic group. Major neutralizing determinants of the orbiviruses focused on two structural proteins of the outer layer: VP2 and VP5. The VP2

TABLE 7.5 The Orbiviruses Isolated from *Ixodes* (*Ceratixodes*) *Uriae* White, 1852 Ticks

Virus	Place of field material collection			Author
	Country, Region	Geographic coordinates	Biotope	
Aniva virus (ANIV)	Russia, Sakhalin district	Sakhalin Island, rocks in Aniva bay of Okhotsk Sea (46°30′N, 142°30′E)	Nests of guillemots	Lvov et al. (1993)
Bauline virus (BAUV)	Canada, Labrador province	Newfoundland Island, Avalon peninsula (52°46′N, 47°11′W)	Nests of guillemots and puffins	Main, et al. (1984)
Cape Wrath virus (CWV)	Great Britain, Scotland	Cape Wrath (58°36′N, 4°53′W)	Nests of guillemots	Main et al. (1984)
Great Island virus (GIV)	Canada, Labrador province	Newfoundland Island, Avalon peninsula (52°46′N, 47°11′W)	Nests of guillemots and puffins	Main et al. (1984)
Mykines virus (MYKV)	Danish	Faroe Islands (62°05′N, 7°30′E)	Nests of puffins	Main et al. (1984)
Nugget virus (NUGV)	Australia	Macquarie Island in Southern part of Pacific ocean between New Zealand and Antarctica (54°30′S, 159°00′E)	Nests of penguins	Doherty et al. (1984)
Okhotskiy virus (OKHV)	Russia, Sakhalin district	Tyuleniy Island in Okhotsk Sea (48°29′N, 144°38′E)	Nests of guillemots	Lvov et al. (1984)
Tindholmur virus (TDMV)	Danish	Faroe Islands (62°05′N, 7°30′E)	Nests of puffins	Main et al. (1984)
Yaquina Head virus (YHV)	USA, Oregon	Yakina Head rock in Yakina bay, Pacific coast (44°40′N, 124°05′W)	Nests of guillemots	Yunker et al. (1984)

These ticks are the obligate parasites of colonial seabirds.

proteins of OKHV and ANIV were, respectively, 67% and 69% similar to those of GIV. The similarity of the VP5 proteins reaches 72% and 79%, respectively (Table 7.7).

Phylogenetic analysis was carried out for conserved nonstructural proteins VP1 (Pol) and for structural proteins VP3 (T2) and VP7 (T13) by the neighbor-joining method (Figure 7.1). The topology of the resulting tree shows that OKHV and ANIV belong to the phylogenetic lineage of the Great Island subgroup of the Kemerovo antigenic group (Figure 7.1).

Common ecological features and a common distribution of OKHV and ANIV may explain the high level of identity of their proteins of the replicative complex and inner shell layer (93−99%). By contrast, the proteins of the outer shell layer (VP2 and VP5), which are associated with virulence, the production of neutralizing antibodies, and receptor recognition, have just 68−70% identity. Probably, this lower level of identity can be explained by a constant process of reassortment between OKHV and ANIV, which led to their convergence.

Ecology and distribution. Most of the ticks infected by OKHV (the rate of infection is 0.256%) were found in the northern part of the Kola Peninsula, in Europe, bordering the Barents Sea. By contrast, the proportion of OKHV-infected ticks in the Pacific region of their distribution is from 0.016% to 0.103%

TABLE 7.6   Other Viruses of the Great Island Subgroup of the Kemerovo Antigenic Group

Virus	Place of isolation
Above Maiden virus (ABMAV)	Scotland
Arbroath virus (ARTHV)	Scotland
Broadhaven virus (BHVNV)	Scotland
Colony virus (CLNV)	Scotland
Colony B North virus (CLNBV)	Scotland
Ellidaey virus (ELLDV)	Ireland
Foula virus (FOULV)	Scotland
Great Saltee Island virus (GSIV)	Ireland
Grimsey virus (GRIMV)	Iceland
Inner Farne virus (INFAV)	United Kingdom
Lundy virus (LUNV)	United Kingdom
Mill Door virus (MILDV)	Scotland
North Clett virus (NOCLV)	Scotland
North End virus (NENDV)	United Kingdom
Rost Island virus (ROISV)	Norway
Shiant Islands virus (SHISV)	Scotland
Thormodseyjarlettur virus (THORV)	Iceland
Vearoy virus (VEAV)	Norway
Wexford virus (WEXV)	South-East of Ireland

(i.e., eight times less in comparison with the European part of their distribution) (Table 7.4). In the Far Eastern part of the area over which *I. uriae* is distributed, the proportion of ANIV-infected ticks is 0.060%; in the European part, ANIV has not been isolated (Table 7.4).[2]

On the coast and the islands of the subarctic, arctic, and northern temperate zones, there is a vast nesting of colonial seabirds. Two species of ticks are obligate parasites of these seabirds: *I. uriae* and *I. signatus* Birula, 1895. *I. uriae* parasitizes at least 17 species of seabirds in the Northern hemisphere and 16 species in the Southern Hemisphere.[26–33] In the Northern Hemisphere, *I. uriae* has been found on auks (family Alcidae), among which are Brünnich's guillemot (*Uria lomvia*), the common guillemot (*U. aalge*), the razorbill (*Alca torda*), the crested puffin (*Fratercula cirrhata*), the deadlock (*F. arctica*), and the black guillemot (*Cepphus grylle*); on gulls (family Laridae), including the black-legged kittiwake (*Rissa tridactyla*), the herring gull (*Larus argentatus*), the great black-backed gull (*L. marinus*), and the burgomaster gull (*L. hyperboreus*); on the Arctic tern (*Sterna paradisaea*); on pelicans (family Pelecanidae); on cormorants (family Phalacrocoracidae), including Baird's

TABLE 7.7 Pairwise Comparison of Full-Length Amino Acid Sequences of ORFs of OKHV and ANIV with Other Orbiviruses

Virus	Proteins																			
	VP1 (Pol)		T2		VP2		NS1		VP4		VP5		NS2		VP7 (T13)		VP6		NS3	
	OKHV	ANIV	OKHV	ANIV	OKHV	ANIV	OKHV	ANIV	OKHV	ANIV	OKHV	ANIV	OKHV	ANIV	OKHV	ANIV	OKHV	ANIV	OKHV	ANIV
GIV	89	88	93	92	67	69	81	80	80	n.a.	72	79	60	88	96	96	65	66	78	74
Kemerovo	73	76	83	79	38	40	63	64	61	n.a.	64	63	81	65	80	79	48	49	57	51
Tribec	73	74	83	79	37	39	61	61	62	n.a.	63	64	57	62	76	76	48	49	57	46
Yunnan	48	49	44	43	10	9	28	29	39	n.a.	34	33	27	30	29	30	28	28	20	12
EHDV	45	44	38	35	11	15	20	22	39	n.a.	30	20	19	19	25	25	22	21	23	17
AHSV	46	48	36	34	10	12	21	21	38	n.a.	29	24	22	21	23	23	22	23	24	15
BTV	45	47	36	34	12	19	20	22	41	n.a.	31	33	17	18	21	22	20	20	18	11

n.a.: not available.
Amino acid (aa) identity is shown (%). The sequence of the VP4 gene of ANIV is not available. GenBank numbers: OKHV, KP981623-32; ANIV, KJ191541-47.

cormorant (*Phalacrocorax pelagicus*), the common shag (*Ph. aristotelis*), the large cormorant (*Ph. carbo*), and the red-faced cormorant (*Ph. urile*); on the northern gannet (*Morus bassanus*); and on the northern fulmar (*Fulmarus glacialis*). All these species, except the gannet, nest in Russia. In the Southern Hemisphere, *I. uriae* ticks have been found on penguins (family Spheniscidae), including the king penguin (*Aptenodytes patagonicus*), the Magellanic penguin (*Spheniscus magellanicus*), and the little penguin (*Eudyptula minor*); on five species of the order Procellariiformes: the Antarctic prion (*Pachyptila desolata*), Buller's albatross (*Thalassarche bulleri*), the black-browed albatross (*Th. melanophrys*), the light-mantled albatross (*Phoebetria palpebrata*), and the wandering albatross (*Diomedea exulans*); on cormorants, including the black-faced cormorant (*Ph. fuscescens*), the kerguelen shag (*Ph. verrucosus*), and the blue-eyed cormorant (*Ph. atriceps*); and on two petrels: the South Georgia diving petrel (*Pelecanoides georgicus*) and the smaller Kerguelen diving petrel (*Pelecanoides urinatrix exsul*). The main hosts of *I. uriae* are auks and penguins, in the Northern and Southern Hemispheres, respectively.[26−32,34]

In Russia, *I. uriae* is distributed over seabird colonies nesting on the Pacific shelf: on the Kuril Islands (in the Sea of Okhotsk), on the Commander Islands (in the Bering Sea), and on the southeast coast of Chukotka (off the Bering Sea). *I. uriae* is also found on the coast of the Kola peninsula (off the Barents Sea)[32] and on the south coast of Novaya Zemlya Island (between the Barents Sea and the Kara Sea).[34] Ticks are concentrated in the ground litter to a depth of 20 cm. Up to 7,000 ticks per square meter at all stages of metamorphosis can be found in the seabird colony. *I. uriae* ticks as well as *I. signatus* ticks can attack humans.[6]

The area over which *I. signatus* ticks are distributed is substantially smaller than that of *I. uriae* ticks and includes the islands of the Rimsky-Korsakov archipelago in the western

part of Peter the Great Bay (in the Sea of Japan), the Commander Islands (in the Bering Sea), Hokkaido (Japan's second-largest island) and the northern coast of Honshu (Japan's largest and most populous island) (both in the Japanese Archipelago), the Aleutian Islands (belonging to Alaska and between the Pacific Ocean and the Bering Sea), and the northern coast of California. The main host of *I. signatus* is Baird's cormorant (*Ph. pelagicus*).[35]

A serological survey of about 1,500 birds revealed the presence of antibodies against OKHV in 15% of northern fulmars (*F. glacialis*) on Iona Island in the Sea of Okhotsk and in 4−6% of common guillemots (*U. aalge*) on Tyuleniy Island and the Commander Islands. Also, anti-OKHV antibodies were found in 0.7% of Baird's cormorants (*Ph. pelagicus*) in the Commander Islands.[5,6] In addition, neutralizing antibodies to GIV and BAUV have been found in birds in places where these viruses have been isolated: in 37% of crested puffins (*F. cirrhata*), and 4−6% of streaked shearwaters (*Calonectris leucomelas*), in Canada.[36] Finally anti-OKHV antibodies have been found in 12% of samples from humans who lived in the Commander Islands a long time ago. A neutralization test confirmed the presence of these antibodies; however, no etiologically related human cases of OKHV have been registered.[5,6]

7.2.1.5 Wad Medani Virus

History. The Wad Medani virus (WMV) was originally isolated by R.M. Taylor and colleagues from *Rhipicephalus sanguineus* (family Ixodidae, subfamily Rhipicephalinae) ticks, collected from sheep near Wad Medani village (Sudan, Africa) in November 1952. On the basis of its serological properties, WMV was assigned to the Kemerovo antigenic group (family Reoviridae, genus *Orbivirus*).[1,2] Later, more than 20 strains of WMV were isolated from other species of ticks: *Hyalomma asiaticum* (Central Asia); *H. anatolicum* (Iran, Pakistan); *H. marginatum* (India); *Amblyomma cajennense* (Jamaica);

Rh. Microplus, formerly *Boophilus microplus*) (Malaysia); and *Rh. guilhoni* and *Rh. evertsi* (family *Ixodidae*, subfamily *Rhipicephalinae*) (both species found in Sudan, Egypt, Senegal, India, Pakistan, Iran, and Jamaica).[1,3]

Taxonomy. The genome of the orbiviruses has 10 segments of dsRNA, which encode seven structural (VP1−VP7) and four nonstructural (NS1−NS4) proteins.[4] Scientists have sequenced the genome of strain LEIV-8066Tur (GenBank ID, KJ425426−35), isolated from *H. asiaticum* ticks in Baharly district, Turkmenistan. The most conservative protein of the orbiviruses is an RdRp (Pol, VP1). A comparison of full-length VP1 amino acid sequences of WMV with those of other tick-borne orbiviruses revealed a 54% identity. A phylogenetic analysis based on a comparison of full-length VP1 (Pol), VP3 (T2), and VP7 (T13) amino acid sequences of the orbiviruses is presented in Figure 7.1. The major antigenic determinants of orbiviruses are located on three proteins of the outer and inner layers of the capsid. The most divergent are VP2 and VP5 proteins. The VP2 protein forms the outer layer of the capsid and carries the main neutralizing and receptor-binding sites. In addition, the VP2 protein is one of the virulence factors of the orbiviruses. The similarity of the VP2 amino acid sequences of WMV with those of other tick-borne arboviruses is 26−30%. The similarity of the VP5 protein of WMV with that of other tick-borne orbiviruses is 45%, on average. Among the structural proteins, the most conservative one is VP3 (T2), which forms the inner layer of the capsid. The VP7 protein (T13) is involved in the virus−cell interaction and, like the VP2 protein, is one of the virulence factors, defining, in particular, the infectivity of the virion. In an intact virion, the antigenic epitopes VP7 (T13) are hidden and therefore cannot be blocked by neutralizing antibodies. VP7 (T13) has group- and species-specific antigenic determinants. The similarity of the VP7 (T13) amino acid sequences of WMV to those of mosquito-borne orbiviruses is 46%, and to those of tick-borne viruses is 67%.

Genomic segment 9 of the orbiviruses encodes the viral enzyme VP6 (Hel), which possesses an RNA-binding and helicase activity. The similarity of VP6 (Hel) of WMV with that of other tick-borne orbiviruses is in the range 36−38%. It has been shown that segment 9 of GIV, BAKV, and BTV has an additional ORF that encodes a protein denoted VP6a or NS4 with an unknown function. Like other tick-borne orbiviruses, WMV also encodes VP6a ORF. Noted that, although the amino acid sequences VP6a of GIV, BAKV, and WMV have a low level of similarity (20−30%), they are the same size (195 aa) and have two closed start codons. The length of VP6a ORF in tick-borne arboviruses is almost two times larger than that of BTV.[5] The phylogenetic analysis shown in Figure 7.1, an analysis based on a comparison of structural and nonstructural proteins of the orbiviruses, reveals that the topology of WMV on the tree confirms the classification of the virus into antigenic group B of the genus *Orbivirus*.

Epizootiology. Antibodies against WMV have been found in camels and buffaloes.[1] Two strains of Seletar virus, which is closely related to WMV, were isolated by A. Rudnik from *B. microplus* ticks, collected in the Seletar area of Singapore in January 1961. Because *B. microplus* is a one-host tick, spending its entire life span on a single host, transovarial transmission would seem to be necessary for preservation of the viral population.[5,6] Fourteen strains of WMV were isolated during virological monitoring of different ecosystems in Northern Eurasia: 10 strains in Turkmenistan,[7−11] 2 in Kazakhstan,[12−16] 1 in Tajikistan, and 1 in Armenia.[17−19]

In Turkmenistan, all of the strains of WMV were isolated from *H. asiaticum* ticks, collected from sheep and camels in either Yölöten district of Mary province (in semiarid landscapes) or in Baharly district of Ahal province (in arid landscapes) in 1972, 1973, and 1981.[7−10] In

Kazakhstan, two strains were isolated from *H. asiaticum* in arid landscapes of Balkhash district (Almaty province) in 1977. The proportion of infected ixodids has been estimated as 0.094%.[12–16] In Tajikistan, one strain of WMV was isolated from *H. anatolicum* ticks. These ticks made up 76.5% of all ixodids in this territory, and the proportion of infected *H. anatolicum* ticks reached 0.002%.[17,19] In the arid climate of the southern part of Tajikistan, one or two generations of ixodid ticks can develop during the year. WMV was isolated from hungry, overwintered imagoes that exhibited transstadial transmission of the virus.[13,17] Experimental infection of calves has shown that the level of WMV viremia is sufficient to infect *H. anatolicum* larvae, which subsequently transferred the virus to imagoes that provides activity and stability for the natural foci of WMV.[20] Immunity to WMV among the human population in southern Tajikistan reaches 7.8–10.3%, while in the northern provinces it does not exceed 2.1%.[20,21] In Transcaucasia, WNV was isolated from *H. asiaticum* ticks collected near Nakhichevan, Azerbaijan, in 1985.[18] The results obtained from the isolation of WMV in Kazakhstan, Central Asia, and Transcaucasia indicate that *H. asiaticum* and *H. anatolicum* ticks play the main role in maintaining natural foci of the virus in pasture biocenoses and in arid and semiarid landscapes.

Epidemiology. There is no indication that WMV is pathogenic to humans.

References

7.1 Family Birnaviridae

1. Delmas B, Mundt E, Vakharia VN, et al. Family Birnaviridae. In: King AM, Adams MJ, Carstens EB, et al., editors. *Virus taxonomy: classification and nomenclature of viruses: ninth report of the International Committee on Taxonomy of Viruses.* 1st ed. London: Elsevier;2012. p. 496–507.

2. Hetrick FM. Viral deseases of fish and their relation to public health. In: Beran G, editor. *Handbook of Zoonoses.* 2nd ed. London, Tokyo: CRC Press;1994. p. 537–53.

3. Shchelkunov IS. Epizootic situation on virus diseases of cultivated fishes. *Veterinariya* 2006;(4):22–5 [in Russian].

4. McGonigle RH. Acute catarrhal enteritis of salmonid fingerlings. *Trans Am Fish Soc* 1940;**70**:297–303.

5. Roberts RJ, McKnight IJ. The pathology of infectious pancreatic necrosis. II. Stress-mediated recurrence. *Br Vet J* 1976;**132**(2):209–14.

6. McKnight IJ, Roberts RJ. The pathology of infectious pancreatic necrosis. I. The sequential histopathology of the naturally ocurring condition. *Br Vet J* 1976;**132**(1):76–85.

7. Wolf K, Snieszko SF, Dunbar CE, Pyle E. Virus nature of infectious pancreatic necrosis in trout. *Proc Soc Exp Biol Med* 1960;**104**:105–8.

8. Wolf K, Quimby MC. Salmonid viruses: infectious pancreatic necrosis virus. Morphology, pathology and serology of first European isolations. *Archiv fur die gesamte Virusforschung* 1971;**34**(2):144–56 [in German].

9. Sonstegard RA, McDermott LA, Sonstegard KS. Isolation of infectious pancreatic necrosis virus from white suckers (*Catastomus commersoni*). *Nature* 1972;**236** (5343):174–5.

10. Wolf K. *Fish viruses and fish viral diseases.* Ithaca, London: Cornell University Press;1988.

11. Jorgensen PEV, Kehlet NP. Infectious pancreatic necrosis (IPN) viruses in Danish rainbow trout. *Nord Vet Med* 1971;**23**:568–675.

12. Hill BJ, Way K. Serological classification of infectious pancreatic necrosis (IPN) virus and other aquatic birnaviruses. *Ann Rew Fish Dis* 1995;**5**:55–7.

13. Nishizawa T, Kinoshita S, Yoshimizu M. An approach for genogrouping of Japanese isolates of aquabirnaviruses in a new genogroup, VII, based on the VP2/NS junction region. *J Gen Virol* 2005;**86**(Pt 7):1973–8.

14. Blake S, Ma JY, Caporale DA, Jairath S, Nicholson BL. Phylogenetic relationships of aquatic birnaviruses based on deduced amino acid sequences of genome segment A cDNA. *Dis Aquat Organ* 2001;**45**(2):89–102.

15. Roberts RJ, Pearson MD. Infectious pancreatic necrosis in Atlantic salmon, *Salmo salar* L. *J Fish Dis* 2005;**28** (7):383–90.

16. Reno PW. Infectious pancreatic necrosis disease and its virulence. In: Woo PTK, Bruno DW, editors. *Fish diseases and disorders.* Wallingford: CABI Publishing;1999. p. 1–55.

17. Molloy SD, Pietrak MR, Bricknell I, Bouchard DA. Experimental transmission of infectious pancreatic necrosis virus from the blue mussel, *Mytilus edulis*, to cohabitating Atlantic Salmon (*Salmo salar*) smolts. *Appl Envir Microbiol* 2013;**79**(19):5882–90.

18. Munro ES, Gahlawat SK, Ellis AE. A sensitive non-destructive method for detecting IPNV carrier Atlantic salmon, *Salmo salar* L., by culture of virus from plastic adherent blood leucocytes. *J Fish Dis* 2004;27(3):129–34.

19. Lopez-Jimena B, Garcia-Rosado E, Infante C, et al. Detection of infectious pancreatic necrosis virus (IPNV) from asymptomatic redbanded seabream, *Pagrus auriga* Valenciennes, and common seabream, *Pagrus pagrus* (L.), using a non-destructive procedure. *J Fish Dis* 2010;33(4):311–19.

20. Gahlawat SK, Munro ES, Ellis AE. A non-destructive test for detection of IPNV-carriers in Atlantic halibut, *Hippoglossus hippoglossus* (L.). *J Fish Dis* 2004;27(4):233–9.

21. Milne SA, Gallacher S, Cash P, Porter AJ. A reliable RT-PCR-ELISA method for the detection of infectious pancreatic necrosis virus (IPNV) in farmed rainbow trout. *J Virol Meth* 2006;132(1–2):92–6.

22. Rodriguez Saint-Jean S, Borrego JJ, Perez-Prieto SI. Comparative evaluation of five serological methods and RT-PCR assay for the detection of IPNV in fish. *J Virol Meth* 2001;97(1–2):23–31.

7.2 Family Reoviridae

1. Attoui H, Mertens PPC, Becnel J, et al. Family Reoviridae. In: King AM, Adams MJ, Carstens EB, et al., editors. *Virus taxonomy: ninth report of the International Committee of Taxonomy of Viruses*. 1st ed. London: Elsevier;2012. p. 541–637.

2. Murphy FA, Borden EC, Shope RE, Harrison A. Physicochemical and morphological relationships of some arthropod-borne viruses to bluetongue virus—a new taxonomic group. Electron microscopic studies. *J Gen Virol* 1971;13(2):273–88.

3. Borden EC, Shope RE, Murphy FA. Physicochemical and morphological relationships of some arthropod-borne viruses to bluetongue virus—a new taxonomic group. Physiocochemical and serological studies. *J Gen Virol* 1971;13(2):261–71.

4. Mertens PP, Diprose J. The bluetongue virus core: a nano-scale transcription machine. *Virus Res* 2004;101(1):29–43.

5. Belaganahalli MN, Maan S, Maan NS, et al. Full genome sequencing and genetic characterization of Eubenangee viruses identify Pata virus as a distinct species within the genus Orbivirus. *PLoS One* 2012;7(3):e31911.

6. Belaganahalli MN, Maan S, Maan NS, et al. Full genome sequencing of Corriparta virus, identifies California mosquito pool virus as a member of the Corriparta virus species. *PLoS One* 2013;8(8):e70779.

7. Dilcher M, Hasib L, Lechner M, et al. Genetic characterization of Tribec virus and Kemerovo virus, two tick-transmitted human-pathogenic Orbiviruses. *Virology* 2012;423(1):68–76.

8. Belaganahalli MN, Maan S, Maan NS, et al. Umatilla virus genome sequencing and phylogenetic analysis: identification of stretch lagoon orbivirus as a new member of the Umatilla virus species. *PLoS One* 2011;6(8):e23605.

9. Belhouchet M, Mohd Jaafar F, Tesh R, et al. Complete sequence of Great Island virus and comparison with the T2 and outer-capsid proteins of Kemerovo, Lipovnik and Tribec viruses (genus *Orbivirus*, family *Reoviridae*). *J Gen Virol* 2010;91(12):2985–93.

7.2.1.1 Baku Virus

1. Baku virus. In: Karabatsos N, editor. *International catalogue of arboviruses including certain other viruses of vertebrates*. San Antonio, TX: American Society of Tropical Medicine and Hygiene;1985.

2. Gromashevsky VL, Lvov DK, Sidorova GA, et al. A complex natural focus of arboviruses on Glinyanyi Island, Baku Archipelago, Azerbaidzhan S.S.R. *Acta Virol* 1973;17(2):155–8.

3. Lvov DK, Timopheeva AA, Smirnov VA, et al. Ecology of tick-borne viruses in colonies of birds in the USSR. *Med Biol* 1975;53(5):325–30.

4. Mirzoeva NM. Results of study of arboviruses in Azerbaidzhan (1967–1976). *Arboviruses* 1978;3:27–31 [in Russian].

5. Andreev VP, Gromashevsky VL, Veselovskaya OV. Isolation of Baku virus in Western part of Turkmen SSR. In: *Ecology of Viruses*; 1973. p. 96–103 [in Russian].

6. Lvov DK, Gromashevsky VL, Sidorova GA, et al. Isolation of a new Baku arbovirus of the Kemerovo group from argasid *Ornithodoros coniceps* ticks in Azerbaijan. *Vopr Virusol* 1971;(4):434–7 [in Russian].

7. Sidorova GA, Andreev VP. Some properties of new arboviruses, isolated in Uzbekistan and Turmenistan. In: Lvov DK, editor. *Ecology of Viruses*. Moscow; 1980. p. 108–14 [in Russian].

8. Sidorova GA, Zhmaeva ZM, Gromashevsky VL. Isolation of Baku virus from ticks *Ornithodoros coniceps*, collected in nesting seats of pigeon in Tashkent region. In: Lvov DK, editor. *Ecology of Viruses*. Moscow; 1974. p. 102–05 [in Russian].

9. Lvov DK. Arboviruses in the USSR. In: Vasenjak-Hirjan J, Porterfield JS, editors. *Arboviruses in the Mediterranean countries*. 1st ed. NY: Gustav Fischer Verlag;1980. p. 35–48.

10. Lvov DK. Ecological soundings of the former USSR territory for natural foci of arboviruses. In: Zhdanov VM, editor. *Soviet Medical Review Virology*. UK: Harwood Academic Publishers GmbH;1993. p. 1–47.

11. Lvov DK. Natural foci of arboviruses in the USSR. In: Zhdanov VM, editor. *Sov. Med. Rev. Virol.* UK: Harwood Academic Publishers GmbH;1987. p. 153–96.

12. Sarkisyan BG, Novochatsky AC, Lvov DK. Orbiviruses. *Uspekhi Sovremennoi Biol* 1979;(5):210–25 [in Russian].

13. Attoui H, Mertens PC, Becnel J, et al. Family Reoviridae. In: King AM, Adams MJ, Carstens EB, et al., editors. *Virus taxonomy. Ninth Report of the International Committee on Taxonomy of Viruses.* London: Elsevier;2012. p. 541–637.

14. Belhouchet M, Mohd Jaafar F, Tesh R, et al. Complete sequence of Great Island virus and comparison with the T2 and outer-capsid proteins of Kemerovo, Lipovnik and Tribec viruses (genus *Orbivirus*, family *Reoviridae*). *J Gen Virol* 2010;**91**(12):2985–93.

15. Dilcher M, Hasib L, Lechner M, et al. Genetic characterization of Tribec virus and Kemerovo virus, two tick-transmitted human-pathogenic Orbiviruses. *Virology* 2012;**423**(1):68–76.

7.2.1.2 Chenuda Virus

1. Chenuda virus (CNUV). In: Karabatsos N, editor. *International catalogue of arboviruses including certain other viruses of vertebrates.* San Antonio, TX: American Society of Tropical Medicine and Hygiene;1985.

2. Taylor RM, Henderson JR, Thomas LA. Antigenic and other characteristics of Quaranfil, Chenuda, and Nyamanini arboviruses. *Am J Trop Med Hyg* 1966;**15**(1):87–90.

3. Taylor RM, Hurlbut HS, Work TH, et al. Arboviruses isolated from Argas ticks in Egypt: Quaranfil, Chenuda, and Nyamanini. *Am J Trop Med Hyg* 1966;**15**(1):76–86.

4. Jupp PG, McIntosh BM. Identity of Argasid ticks yielding isolations of Chenuda, Quaranfil and Nyamanini viruses in South Africa. *J Entomol Soc South Africa* 1986;**49**:392–401.

5. Brown SE, Morrison HG, Knudson DL. Genetic relatedness of the Kemerovo serogroup viruses: III. RNA–RNA blot hybridization and gene reassortment *in vitro* of the Chenuda serocomplex. *Acta Virol* 1989;**33**(3):221–34.

6. Chastel C, Main AJ, Bailly-Choumara H, et al. Essaouira and Kala iris: two new orbiviruses of the Kemerovo serogroup, Chenuda complex, isolated from *Ornithodoros (Alectorobius) maritimus* ticks in Morocco. *Acta Virol* 1993;**37**(6):484–92.

7. Skvortsova TM, Gromashevsky VL, Sidorova GA, et al. Results of virological testing arthropod vectors in. In: Lvov DK, editor. *Ecology of Viruses.* Moscow: USSR Academy of Medical Science;1982p. 139–44 [in Russian].

7.2.1.3 Kemerovo Virus

1. Chumakov MP. Report on the isolation from *Ixodes persulcatus* ticks and from patients in western Siberia of a virus differing from the agent of tick-borne encephalitis. *Acta Virol* 1963;**7**:82–3.

2. Chumakov MP, Sarmanova ES, Bychkova MV, et al. Identification of Kemerovo Tick-Borne Fever virus and its antigenic independence. *Fed Proc Transl Suppl* 1964;**23**:852–4.

3. Karabatsos N. *International catalogue of arboviruses including certain other viruses of vertebrates.* San Antonio, TX: American Society of Tropical Medicine and Hygiene;1985.

4. Schmidt JR, Shope RE. Kemerovo virus from a migrating common redstart of Eurasia. *Acta Virol* 1971;**15**(1):112.

5. Gresikova M. Kemerovo virus infection. In: Beran G, editor. *Hanbook series in Zoonoses. Section B. Viral.* Boca Raton, FL: CRC Press;1981.

6. Gresikova M, Nosek J, Kozuch O, et al. Study on the ecology of Tribec virus. *Acta Virol* 1965;**9**:83–8.

7. Libikova H, Rehacek J, Gresikova M, et al. Cytopathic viruses isolated from *Ixodes ricinus* ticks in Czechoslovakia. *Acta Virol* 1964;**8**:96.

8. Libikova H, Rehacek J, Somogyiova J. Viruses related to the Kemerovo virus in *Ixodes ricinus* ticks in Czechoslovakia. *Acta Virol* 1965;**9**:76–82.

9. Libikova H, Asmera J, Heinz F. Isolation of a Kemerovo complex orbivirus (strain Cvilin) from ticks in the North Moravian region. *Czecosl Epidemiol Mickrobiol Immunol* 1977;**26**:135–8 [in Czech].

10. Hubalek Z, Calisher CH, Mittermayer T. A new subtype ("Brezova") of Tribec orbivirus (Kemerovo group) isolated from *Ixodes ricinus* males in Czechoslovakia. *Acta Virol* 1987;**31**(1):91–2.

11. Vinograd IA, Vigovskii AI, Gaidamovich S, Obukhova VR. Characteristics of the biological properties of a Kemerovo group arbovirus isolated in Transcarpathia. *Vopr Virusol* 1977;**4**:456–9 [in Russian].

12. Duca M, Duca E, Buiuc D, Luca V. A seroepidemiologic and virological study of the presence of arboviruses in Moldavia in 1961–1982. *Rev Med Chir Soc Med Nat Iasi* 1989;**93**(4):719–33 [in Romanian].

13. Skofertsa PG, Gaidamovich S, Obukhova VR, et al. Isolation of Kemerovo group Kharagysh virus on the territory of the Moldavian SSR. *Vopr Virusol* 1972;**17**(6):709–11 [in Russian].

14. Skofertsa PG, Gaidamovich SY, Obukhova VR, et al. Isolation in the Moldavian S.S.R. of a Kemerovo group arbovirus from *Ixodes ricinus* ticks. *Acta Virol* 1972;**16**(4):362.

15. Gresikova M, Rajcani J, Hruzik J. Pathogenicity of Tribec virus for Macaca rhesus monkeys and white mice. *Acta Virol* 1966;**10**(5):420–4.

16. Attoui H, Mertens P, Becnel J, Belaganahalli S, et al. Family Reoviridae. In: King AMQ, Adams MJ, Carstens EB, et al., editors. *Virus taxonomy Ninth Report of the International Committee on Taxonomy of Viruses.* 1st ed. London: Elsevier;2012. p. 541–637.

17. Brown SE, Morrison HG, Buckley SM, et al. Genetic relatedness of the Kemerovo serogroup viruses: I. RNA–RNA blot hybridization and gene reassortment in vitro of the Kemerovo serocomplex. *Acta Virol* 1988;**32**(5):369–78.

18. Brown SE, Morrison HG, Knudson DL. Genetic relatedness of the Kemerovo serogroup viruses: II. RNA–RNA blot hybridization and gene reassortment *in vitro* of the Great Island serocomplex. *Acta Virol* 1989;**33**(3):206–20.

19. Jacobs SC, Carey D, Chastel C, et al. Characterization of orbiviruses of the Kemerovo serogroup: isolations and serological comparisons. *Arch Virol* 1986;**91**(1–2):107–16.

20. Libikova H, Buckley SM. Serological characterization of Eurasian Kemerovo group viruses. II. Cross plaque neutralization tests. *Acta Virol* 1971;**15**(1):79–86.

21. Libikova H, Casals J. Serological characterization of Eurasian Kemerovo group viruses. I. Cross complement fixation tests. *Acta Virol* 1971;**15**(1):65–78.

22. Gorman BM, Taylor J, Walker PJ, et al. The isolation of recombinants between related orbiviruses. *J Gen Virol* 1978;**41**(2):333–42.

23. Skofertsa PG, Yarovoi PI, Korchmar ND. Natural focies of Tribec arbovirus in Moldavian SSR. In: Akhundov VJ, editor. *Ecologia of Viruses.* Baku; 1976. p. 66–8.

24. Topciu V, Rosiu N, Georgescu L, Gherman D, Arcan P, Csaky N. Isolation of a cytopathic agent from the tick *Haemaphysalis punctata. Acta Virol* 1968;**12**(3):287.

25. Frankova V. Meningoencephalitis cased by Orbivirus infection in Czechoslovakia. *Sb Lek* 1981;**83**:234–5 [in Czech].

26. Libikova H, Heinz F, Ujhazyova D, Stunzner D. Orbiviruses of the Kemerovo complex and neurological diseases. *Med Microbiol Immunol* 1978;**166**(1–4):255–63.

7.2.1.4 Okhotskiy Virus and Aniva Virus

1. Lvov DK, Timopheeva AA, Gromashevski VL, et al. "Okhotskiy" virus, a new arbovirus of the Kemerovo group isolated from *Ixodes (Ceratixodes) putus* Pick.-Camb. 1878 in the Far East. *Archiv fur die gesamte Virusforschung* 1973;**41**(3):160–4.

2. Lvov DK. Arboviral zoonoses of Northern Eurasia (Eastern Europe and the Commonwealth of Independent States). In: Beran GW, editor. *Handbook of zoonoses. Section B: Viral zoonoses.* Boca Raton–London–Tokyo: CRC Press;1994. p. 237–60.

3. Okhotskiy. In: Karabatsos N, editor. *International catalogue of arboviruses including certain other viruses of vertebrates.* San Antonio, TX: American Society of Tropical Medicine and Hygiene;1985.

4. Lvov DK, Gromashevski VL, Skvortsova TM, et al. Arboviruses of high latitudes in the USSR. In: Kurstak E, editor. *Arctic and tropical arboviruses. Chapter 3.* NY–San-Francisco–London: Harcourt Brace Jovanovich Publ.;1979. p. 21–38.

5. Lvov SD. Arboviruses in high latitudes. In: Lvov DK, Klimenko SM, Gaidamovich SY, editors. *Arboviruses and arboviral infection.* Moscow: Meditcina;1989. p. 269–89 [in Russian].

6. Lvov SD. *Natural virus foci in high latitudes of Eurasia. Soviet Medical Review Section E: Virology Reviews.* Harwood, USA: Academic Publishers;1993. 137–185.

7. Lvov SD, Gromashevsky VL, Andreev VP. Natural foci of arboviruses in far northern latitudes of Eurasia. In: Calisher CH, editor. *Hemorrhagic fever with renal syndrome tick- and mosquito-borne viruses. Archives of Virology* 1991; Suppl. 1. p. 267–75.

8. Timofeeva AA, Pogrebenko AG, Gromashevskii VL, et al. Natural foci of infection on the Iona island in Okhotsk sea. *Zoologicheskii Zhurnal* 1974;**53**(6):906–11 [in Russian].

9. Brown SE, Morrison HG, Buckley SM, et al. Genetic relatedness of the Kemerovo serogroup viruses: I. RNA–RNA blot hybridization and gene reassortment *in vitro* of the Kemerovo serocomplex. *Acta Virol* 1988;**32**(5):369–78.

10. Brown SE, Morrison HG, Knudson DL. Genetic relatedness of the Kemerovo serogroup viruses: II. RNA–RNA blot hybridization and gene reassortment *in vitro* of the Great Island serocomplex. *Acta Virol* 1989;**33**(3):206–20.

11. Cape-Wrath virus. In: Karabatsos N, editor. *International catalogue of arboviruses including certain other viruses of vertebrates.* San Antonio, TX: American Society of Tropical Medicine and Hygiene;1985.

12. Main AJ, Shope RE, Wallis RC. Cape wrath: a new Kemerovo group orbivirus from *Ixodes uriae* (Acari: Ixodidae) in Scotland. *J Med Entomol* 1976;**13**(3):304–8.

13. Moss SR, Ayres CM, Nuttall PA. The Great Island subgroup of tick-borne orbiviruses represents a single gene pool. *J Gen Virol* 1988;**69**(Pt 11):2721–7.

14. Mykines virus. In: Karabatsos N, editor. *International catalogue of arboviruses including certain other viruses of vertebrates.* San Antonio, TX: American Society of Tropical Medicine and Hygiene;1985.

15. Main AJ. Tindholmur and Mykines: two a new Kemerovo group orbiviruses from the Faerel Islands. *J Med Entomol* 1978;**15**:11−14.

16. Tindholmur virus. In: Karabatsos N, editor. *International catalogue of arboviruses including certain other viruses of vertebrates.* San Antonio, TX: American Society of Tropical Medicine and Hygiene;1985.

17. Main AJ, Downs WG, Shope RE, Wallis RC. Great Island and Bauline: two new Kemerovo group orbiviruses from *Ixodes uriae* in eastern Canada. *J Med Entomol* 1973;**10**(3):229−35.

18. Attoui H, Mertens PPC, Becnel J, et al. Family Reoviridae. In: King AM, Adams MJ, Carstens EB, et al., editors. *Virus taxonomy: Ninth report of the International Committee on Taxonomy of Viruses.* 1st ed. London: Elsevier;2012. p. 541−637.

19. Belhouchet M, Mohd Jaafar F, Tesh R, et al. Complete sequence of Great Island virus and comparison with the T2 and outer-capsid proteins of Kemerovo, Lipovnik and Tribec viruses (genus Orbivirus, family Reoviridae). *J Gen Virol* 2010;**91**(12):2985−93.

20. Black F, Eley SM, Nuttall PA, et al. Characterisation of orbiviruses of the Kemerovo serogroup: comparison of protein and RNA profiles. *Acta Virol* 1986;**30**(4):320−4.

21. Carey D, Nuttall PA. Antigenic cross-reactions between tick-borne orbiviruses of the Kemerovo serogroup. *Acta Virol* 1989;**33**(1):15−23.

22. Gorman BM. On the evolution of orbiviruses. *Intervirology* 1983;**20**(2−3):169−80.

23. Nuttall PA, Booth TF, Carey D, et al. Biologocal and molecular characteristics of orbiviruses and orthomyxoviruses isolated from ticks. *Arch Virol* 1988;(Suppl. 1):219−25.

24. Nuttall PA, Carey D, Reid HW, Harrap KA. Orbiviruses and bunyaviruses from a seabird colony in Scotland. *J Gen Virol* 1981;**57**(1):127−37.

25. Nuttall PA, Moss SR. Genetic reassortment indicates a new grouping for tick-borne orbiviruses. *Virology* 1989;**171**(1):156−61.

26. Oprandy JJ, Schwan TG, Main AJ. Tick-borne Kemerovo group orbiviruses in a Newfoundland seabird colony. *Canad J Microbiol* 1988;**34**(6):782−6.

27. Yakina Head virus. In: Karabatsos N, editor. *International catalogue of arboviruses including certain other viruses of vertebrates.* San Antonio, TX: American Society of Tropical Medicine and Hygiene;1985.

28. Arthur DR. *British ticks.* London: Butterworths;1963.

29. Belopolskaya MM. Parasitofauna of the sea birds. *Scientific notes of Leningrad State University. Ch. "Biology"* 1952;**141**(28):127−80 [in Russian].

30. Bequaert JC. The ticks, or Ixodoidea, of the northeastern United States and eastern Canada. *Entom Am* 1946;**24**:73−120.

31. Clifford CM, Yunker CE, Easton ER, et al. Ectoparasites and other arthropods from coastal Oregon. *J Med Entomol* 1970;**7**(4):438−45.

32. Flint VE, Kostyrko IN. About biology of Ixodes putus Pick.-Camb ticks. *Zoologicheskii Zhurnal* 1967;**66**(8):1253−6.

33. Zumpt F. The ticks of seabirds. *Austr Nat Antarc Res Exped Rep* 1952;**1**(1):12−20.

34. Roberts FHS. *Australian ticks.* Melbourne (Australia): Commonwealth Sci. Industr. Res. Organ;1970.

35. Filippova NA. *Fauna of USSR. Arachnoidea. Ixodes ticks of subfamily Ixodinae.* Moscow, Leningrad: Nauka;1977 [in Russian].

36. Artsob H, Spence L. Arboviruses in Canada. In: Kurstak E, editor. *Arctic and tropical arboviruses.* NY−San-Francisco−London: Harcourt Brace Jovanovich Publishers;1979. p. 39−65.

7.2.1.5 Wad Medani Virus

1. Wad Medani virus. In: Karabatsos N, editor. *International catalogue of arboviruses including certain other viruses of vertebrates.* San Antonio, TX: American Society of Tropical Medicine and Hygiene;1985.

2. Taylor RM, Hoogstraal H, Hurlbut HS. Isolation of a virus (Wad Medani) from *Rhipicephalus sanguineus* collected in Sudan. *Am J Trop Med Hyg* 1966;**15**(1):1−75.

3. Volcit OV. Review on arboviruses isolated from Ixodidae ticks in Afganistan, Pakistan and India. In: Lvov DK, editor. *Ecology of viruses.* Moscow; 1982. p. 111−9 [in Russian].

4. Belhouchet M, Mohd Jaafar F, Firth AE, et al. Detection of a fourth orbivirus non-structural protein. *PLoS One* 2011;**6**(10):e25697.

5. Attoui H, Mertens PPC, Becnel J, et al. Family Reoviridae. In: King AM, Adams MJ, Carstens EB, Lefkowitz EJ, editors. *Virus taxonomy: Ninth report of the International Committee of Taxonomy of Viruses.* 1st ed. London: Elsevier;2012. p. 541−637.

6. Theiler M, Downs WG. *The arthropod-borne viruses of vertebrates.* London: Yale University Press;1973.

7. Lvov DK, Kurbanov MM, Neronov VM, et al. Isolation of Wad Medani virus from ticks *Hyalomma asiaticum* Sch. et Schl., 1929 in Turkmenskaya SSR. *Meditsinskaya Parasitologiya i Parasitarnye Bolezni* 1976;(4):452−5 [in Russian].

8. Sidorova GA, Andreev VL. Some features of novel arboviruses ecology isolated in Uzbekistan and Turkmenia. In: Lvov DK, editor. *Ecology of viruses.* Moscow: USSR Academy of Medical Sciences;1980. p. 108−14 [in Russian].

9. Sidorova GA. About relationship between some arboviruses and vectors. In: Lvov DK, editor. *Ecology of viruses.* Moscow: USSR Academy of Medical Sciences;1988. p. 15−22 [in Russian].

10. Sidorova GA. *Integrated natural foci of arboviruses in arid fields of the Central Asia [Doctor of Biological Sciences dissertation].* Moscow: D.I. Ivanosky institute of Virology of USSR Academy of Medical Sciences;1985 [in Russian].

11. Skvortsova TM, Gromashevsky VI, Sidorova GA, et al. Results of virological testing arthropod vectors in Turkmenia. In: Lvov DK, editor. *Ecology of viruses.* Moscow: USSR Academy of Medical Sciences;1982. p. 139–44 [in Russian].

12. Drobishchenko NI, Lvov DK, Rogovaya SG, et al. Reservoirs and vectors of arboviral infections in Ili-Karatal foci. In: Lvov DK, editor. *Ecology of viruses in Kazakhstan and Central Asia.* Alma-Ata; 1980. p. 64–74 [in Russian].

13. Karimov SK. Arboviruses in Kazakhstan region [Doctor Medical Sciences dissertation]. Alma-Ata, Moscow; 1983 [in Russian].

14. Karimov SK. Results and prospects of study arboviral infections in Kasakhstan. In: Lvov DK, editor. *Ecology of viruses in Kasakhstan and Central Asia.* Alma-Ata; 1980. p. 3–7 [in Russian].

15. Kiriushchenko TV, Karimov SK, Drobishchenko NI, et al. Identification of arboviruses, isolated in Kasakhstan. In: Lvov DK, editor. *Ecology of viruses in Kasakhstan and Central Asia.* Alma-Ata; 1980. p. 107–11 [in Russian].

16. Rogovaya SG, Karimov SK, Skvortsova TM, et al. Isolation of Wad Medani virus from ticks *Hyalomma asiaticum* in Alma-Ata region of Kazakhskaya SSR. In: Lvov DK, editor. *Ecology of viruses in Kasakhstan and Central Asia.* Alma-Ata; 1980. p. 93–4 [in Russian].

17. Kuyma AU, Daniyarov OA. The area structure of ticks Hyalomma in Tajikistan and their role in the ecology of arboviruses. In: Lvov DK, editor. *Ecology of viruses in Kasakhstan and Central Asia.* Alma-Ata; 1980. p. 85–9 [in Russian].

18. Zakaryan VA. Wad Medani F-57-Nakh. Depositor N 182 of Russian State Collection of viruses; 1985 [in Russian].

19. Pak TP, Lvov DK, Kostyukov MA, et al. The results of the virus scan in Tajikistan. In: Lvov DK, editor. *Ecology of viruses in Kasakhstan and Central Asia.* Alma-Ata; 1980. p. 7–10 [in Russian].

20. Pak TP. The problems of arboviruses conservation in the interepidemic period. In: Lvov DK, editor. *Ecology of viruses.* Moscow: USSR Academy of Medical Sciences;1980. p. 118–21 [in Russian].

21. Kostyukov MA, Gordeeva ZE, Rafiyev ChK, et al. Immunological structure of the population to Issyk-Kul virus and Wad Medani virus in Tajikistan. In: Lvov DK, editor. *Ecology of viruses.* Moscow: D.I. Ivanovsky Institute of Virology of USSR Academy of Medical Sciences;1976. p. 108–12 [in Russian].

Single-Stranded RNA Viruses

8.1 FAMILY BUNYAVIRIDAE

The Bunyaviridae family was named after the prototypical Bunyamwera virus (BUNV) isolated in 1943 from mosquitoes (*Aedes* spp.) in Bunyamwera, Uganda.[1] Currently, the Bunyaviridae family includes four genera of animal viruses (*Orthobunyavirus*, *Phlebovirus*, *Nairovirus*, and *Hantavirus*) and one genus (*Tospovirus*) of plant viruses.[2] Bunyavirus virions are spherical in shape (size, about 80–120 nm) and have an outer lipid bilayer with the viral envelope glycoproteins Gn and Gc exposed on the surface. The genome consists of three segments of single-stranded, negative-sense RNA with a total length from 11,000 to 19,000 nt. Depending on the size, the segments are designated L (large), M (medium), and S (small). The viral proteins are synthesized on the mRNA that is produced during replication and that is complementary to the genomic RNA. The length of segments varies for different genera, but in general, they have a common structure. The L-segment, whose length is from 6,400 nt (*Phlebovirus*) to 12,200 nt (*Nairovirus*), has a single open reading frame (ORF) encoding RNA-dependent RNA polymerase (RdRp). The M-segment of all of the genera also has a single ORF, which encodes a polyprotein precursor of envelope glycoproteins Gn and Gc. The length of the M-segment ranges from 3,288 nt for some of the phleboviruses to 4,900–5,366 nt for the nairoviruses. The mature glycoproteins Gn and Gc of the bunyaviruses are derived during complex endoproteolytic events leading to cleavage of the polyprotein precursor by cellular proteases. The S-segment of the bunyaviruses encodes a nucleocapsid protein. Additional nonstructural (NSs) protein is encoded by the S-segment of viruses of the *Phlebovirus*, *Tospovirus*, and *Orthobunyavirus* genera.[2,3]

The bunyaviruses are widely distributed in the world and are one of the most numerous known zoonotic viruses. Most of the zoonotic bunyaviruses are transmitted to animal or humans by bloodsucking arthropod vectors, usually mosquitoes or ticks. Viruses of the *Hantavirus* genus are the exception, being transmitted mainly by aerosol formed from virus-laden urine, feces, or saliva of infected rodents or insectivores that are their natural hosts.[4–6]

8.1.1 Genus *Hantavirus*

The genus *Hantavirus* consists of those bunyaviruses of vertebrates which do not have the ability to replicate in an arthropod's cell

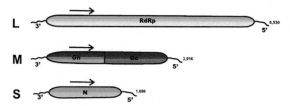

and which are transmitted by respiratory route through the formation of aerosols from urine or feces containing the virus.[1] The morphology of the virion and the genome structure of the hantaviruses are common to all bunyaviruses. The size of the negative-sense ssRNA genome of the prototypical Hantaan virus (HTNV) is 6,533 nt for the L-segment, 3,616 nt for the M-segment, and 1,696 nt for the S-segment (Figure 8.1).[1]

In nature, hantaviruses persist asymptomatically in rodents and insectivores, with each type of hantavirus associated predominantly with one host species. The phylogenetic relationships of hantaviruses enable virologists to divide them into three lineages, which correspond in general to their main hosts. In the S-segment of some hantaviruses carried by Arvicolinae and Sigmodontinae rodents, there is an additional ORF-encoded nonstructural protein NSs. But NSs is absent in the hantaviruses of the Murinae rodents.[2–4]

8.1.2 Hemorrhagic Fever with Renal Syndrome Virus and Related Viruses

History. Hemorrhagic fever with renal syndrome (HFRS) was originally described as a separate nosological category (called "endemic

(epidemic) hemorrhagic nephroso-nephritis" at that time) by Anatoly Smorodintsev (Figure 2.11) during 1935–1940 in the Far East. Later, Japanese scientists described HFRS in northeastern China as "Songo fever" and Swedish scientists as "epidemic nephropathy"; a similar disease was described in 1960 in China.[1] The abbreviation "HFRS" was suggested by Mikhail Chumakov (Figure 2.10) in 1954 and was recommended for adoption at a World Health Organization (WHO) Expert Meeting in 1982.

The viral nature of the HFRS etiological agent was established by Anatoly Smorodintsev (Figure 2.11) in 1940 during his experiments inoculating volunteers. The first historical and prototypical strain, Hantaan 76-118, was isolated by H.W. Lee in 1976 from a striped field mouse (*Apodemus agrarius coreae*) caught on the banks of the Hantaan River in South Korea.[2]

Hantaviruses. The hantaviruses are members of the *Hantavirus* genus of the Bunyaviridae family. The first serotype, —HTNV, included strains isolated from mouselike rodents (Muridae) in South Korea, China, and the southern part of the Russian Far East (Primorsky Krai).[2–4] The second serotype, Puumala virus (PUUV), was isolated from hamsterlike rodents (Cricetidae), mainly the bank vole (*Myodes glareolus*) in Finland and then in other European countries and the western part of Russia, as well we from Maximowicz's vole (*Microtus maximoviczii*) in the Far East.[5–8] The third serotype, Seoul virus (SEOV), was isolated from brown rats (*Rattus norvegicus*), black rats (*Rattus rattus*), and laboratory albino rats (*Rattus norvegicus f. domestica*) in South Korea and elsewhere, including the United States.[3,4] The fourth serotype, Dobrava–Belgrade virus (DOBV), was isolated from the striped field mouse (*Apodemus agrarius*) in Slovenia[9] and Yugoslavia.[10] The fifth serotype, Sin Nombre virus (SNV), literally "nameless virus" in Spanish, was isolated from the meadow vole (*Microtus pennsylvanicus*).[8]

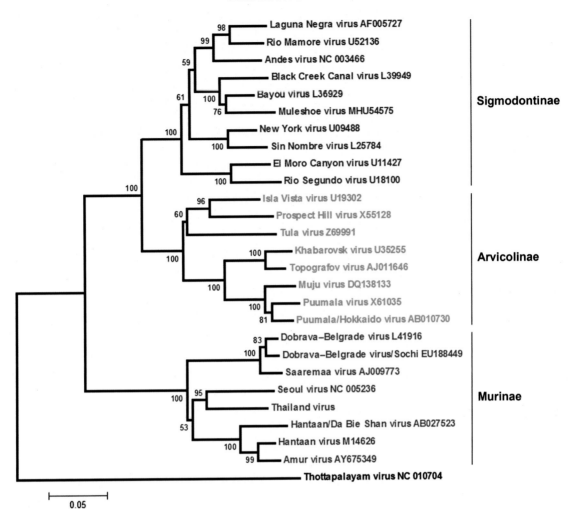

FIGURE 8.2 Phylogenetic analysis of S-segment sequences of certain hantaviruses (family Bunyaviridae, genus *Hantavirus*).

In addition to the 5 main serotypes, 15 other serotypes are known today, including 6 in Eurasia: Amur virus (AMRV), isolated from Asiatic forest mice (*Apodemus peninsulae*) in the Far East of Russia[11] and in China[12]; Tula virus (TULV), from common voles (*Microtus arvalis*) in central Russia[13,14]; Khabarovsk virus (KHAV), from from reed voles (*Microtus fortis*) and Siberian brown lemmings (*Lemmus sibiricus*) in the Far East[15]; Thottapalayam virus (TPMV),

from Asian musk shrews (*Suncus murinus*) in India[16]; Thailand virus (THAIV), from bandicoots (*Bandicota indica*) in Thailand[17]; and a new-found hantavirus, from Chinese mole shrews (*Anourosorex squamipes*) in Vietnam.[18]

Virion and Genome. The size of the negative-sense ssRNA genome of the prototypical HTNV is 6,533 nt for the L-segment, 3,616 nt for the M-segment, and 1,696 nt for the S-segment (Figures 8.1 and 8.2).

Epizootiology. Rodents (order Rodentia) are the main natural reservoir of hantaviruses. Nevertheless, strains have been isolated from birds in the Far East[19] and from bats in China.[20] Infection in rodents is asymptomatic, but the virus is expelled with saliva, urine, and excrement, most intensively during the first month after inoculation. (During this period, virus antigen can be detected in the lungs.)[4]

The evolution of hantaviruses is closely related to that of its rodent host (Figure 8.2).[4,6,21] At least 34 species of rodents (Rodentia), 2 species of lagomorphs (order Lagomorpha), 7 species of insectivores (order Insectivora), 1 species of predators (order Carnivora), and 1 species of artiodactyls (order Artiodactyla) are known to take part in hantavirus circulation on the territory of Northern Eurasia.[8,21,22] The main species of rodents, which are the hosts of hantaviruses in Russia, are presented in Table 8.1. The infection rate of mouselike rodents and insectivores lies within the limits $3.3 \pm 0.5\%$.[23] Hantavirus antigens have been detected in birds as well: the Oriental turtle dove (*Streptopelia orientalis*), coal tit (*Parus ater*), marsh tit (*Parus palustris*), Daurian redstart (*Phoenicurus auroreus*), nuthatch (*Sitta europaea*), black-faced bunting (*Emberiza spodocephala elegans*), Eurasian jay (*Garrulus glandarius*), hazel grouse (*Tetrastes bonasia*), pheasant (*Phasianus colchicus*), Ural owl (*Strix uralensis*), green-backed heron (*Butorides striatus*), and grey heron (*Ardea cinerea*).[19] Hantavirus (Magboi virus, or MGBV) was isolated in 2012 from the hairy slit-faced bat (*Nycteris hispida*) in Africa (Sierra Leone),[24] but the role of bats in the circulation of hemorrhagic fever with renal syndrome virus (HFRSV) is yet to be investigated in detail.

In western Siberia, the main natural reservoir of HFRSV is rodents of the hamsterlike (Cricetidae) family—in particular, bank voles

TABLE 8.1   Hantaviruses: Etiological Agents of Hemorrhagic Fever with Renal Syndrome in Russia

Virus	Rodents—Natural reservoirs	Distribution
Amur virus (AMRV)	Asiatic forest mouse (*Apodemus peninsulae*)	Far East of Russia, China
Dobrava–Belgrade virus (DOBV)	Yellow-necked field mouse (*Apodemus flavicollis*)	Balkans, central part of Russia, Slovenia, Yugoslavia
	Field mouse (*Apodemus agrarius*), small forest mouse (*A. uralensis*)	Central sector of European part of Russia
	Black Sea field mouse (*Apodemus ponticus*)	South of European part of Russia
	Unidentified	Western Siberia
Hantaan virus (HTNV)	Field mouse (*Apodemus agrarius*)	Far East of Russia, China, Japan, North and South Korea
Khabarovsk virus (KHAV)	Reed vole (*Microtus fortis*)	Far East of Russia, China
Puumala virus (PUUV)	Bank vole (*Myodes glareolus*)	Northern and central Europe, Russia, Balkans
Seoul virus (SEOV)	Brown rat (*Rattus norvegicus*), black rat (*Rattus rattus*)	Far East of Russia, China, Japan, North and South Korea
Tula virus (TULV)	Common vole (*Microtus arvalis*), Russian common vole (*Microtus rossiaemeridionalis*)	European part of Russia, Czech Republic, Slovenia

(*Myodes glareolus*), with a susceptibility up to 70%; red-backed voles (*Myodes rutilus*), susceptibility 9%; and, in the north, Siberian brown lemmings (*Lemmus sibiricus*), 14%. The infection rate of other rodents and insectivores is about 0.4—3.0%.[8,22]

In eastern Siberia, the maximum susceptibility is demonstrated in grey red-backed voles (*Myodes rufocanus*), 70%; house mice (*Mus musculus*), 15%; water voles (*Arvicola terrestris*), 8%; and tundra voles (*Microtus oeconomus*), 8%.[8]

In the Far East, HFRSV was revealed to circulate among field mice (*Apodemus agrarius*) with a susceptibility of about 35%; Asiatic forest mice (*A. peninsulae*), susceptibility 30%; reed voles (*Microtus fortis*), 4—18%; grey red-backed voles (*Myodes rufocanus*), 12%; and other rodents (Rodentia), 0.7—4.3%.[21,22,25]

Epidemiology. HFRSV infection starts by aerogenic penetration of the virus during the inhalation of waste products (saliva, urine, excrement) of latently infected animals. An alimentary pathway (with contaminated food and water) of the infection is also possible.[4,8,22,26,27]

HFRS is distributed over Eurasia (Russia, Belarus, Ukraine, Moldova, the Baltic countries, the Czech Republic, Slovakia, Bulgaria, Romania, Serbia, Slovenia, England, France, Germany, Belgium, Hungary, Denmark, Fennoscandia, Kazakhstan, Georgia, Azerbaijan, China, North and South Korea, Japan), as well as American and African countries.[7,28,29]

During 2000—2009, in 58 of 83 regions in Russia, 74,890 cases of HFRS were registered (Table 8.2).[8] Annual morbidity of HFRS in Russia is in the range from 2,700 to 11,400 cases (1.3—7.8%) and is decreasing. About 95% of cases take place in European forest landscapes. PUUV associated with the bank vole (*Myodes glareolus*) provokes about 90% of HFRS cases in Russia (especially in Bashkortostan, Udmurtia, Mari El, Tatarstan, the Chuvash Republic, Orenburg, Ulyanovsk, and the Penza region).[8,30] Morbidity in the urban population is higher

TABLE 8.2 Hemorrhagic Fever with Renal Syndrome in Russia (2013)

Federal District	Number of cases (%)	Infection rate per 100,000 of population
Central	624 (14.5)	1.62
Northwestern	134 (3.1)	0.98
Southern	1 (0.02)	0.04
North Caucasian	0	0
Lower Volga	3,378 (78.6)	11.32
Ural	49 (1.2)	0.40
Siberian	1 (0.02)	0.01
Far Eastern	110 (2.6)	1.75
Total	4,297 (100)	

(65%) than in the rural one. The peak of the disease occurs during July—October in forests and in gardens and kitchens closely situated to the forests.[4,31—33] DOBV associated mainly with field mice (*Apodemus agrarius*) and small forest mice (*A. uralensis*) is of leading epidemiological significance in the central and southwestern sectors of the European part of Russia (the Voronezh, Lipetsk, Orel, and Belgorod regions), as well as in Georgia.[8,31,34,35] PUUV and TULV are associated with the common vole (*Microtus arvalis*) and the bank vole (*Myodes glareolus*) and are also distributed over this territory.[4,8,36] A similar situation is observed in other regions of the Central Federal District: in the Moscow, Yaroslavl, Ryazan, Tver, Kaluga, Vladimir, Ivanov, Kostrom, Smolensk regions. HFRS morbidity in the Moscow region is associated with PUUV,[31] the infection rate of which is 12—57% among bank voles (*Myodes glareolus*), 10—20% in the common vole (*Microtus arvalis*), 11% in Major's pine vole (*Microtus majori*), and in 4—6% other rodent species.[1] In Krasnodar Krai, the Black Sea field mouse (*Apodemus ponticus*) and Major's pine vole (*Microtus majori*) play the main role in human morbidity.[31,37]

Human morbidity in the European part of Russia is registered beginning at a relatively low level in March–April, decreasing to yet a lower level in May–August, increasing in September–November, and then increasing again during December–January.[1] The hyper-endemic territory is the southwestern Ural region (especially the Bashkortostan Republic and the Chelyabinsk and Orenburg regions), the Volga-Vyatka economic region (especially the Udmurt Republic), the Chuvash Republic, and the Tatarstan, Mari El, Samara, Penza, Saratov, and Ulyanovsk regions.[4,8]

The main human morbidity occurs among those 20–40 years old (chiefly men). In Russia, HFRS represents a significant part of all natural-foci zoonotic diseases. The immune layer to HFRSV in the European part of Russia is a mean 4.7%; in the Bashkortostan Republic, it reaches up to 40% (mean, 17%).[4]

The immune layer among the populations of western and eastern Siberia is about 2% for the entire region, 0.2% in Krasnoyarsk Krai, 1.1% in the Irkutsk region, 3.1% in the Omsk region, and 12.6% in the Tyumen region.[1,4]

The Far East provides about 2% of all HFRS cases in Russia.[23] The highest morbidity was revealed in Khabarovsk Krai, Primorsky Krai, and the Amur region.[1] In Khabarovsk Krai and Primorsky Krai, las in China and Japan, — HTNV is associated with grey red-backed voles (*Myodes rufocanus*).[2,3,21,37] The morbidity of SEOV (the third serotype) associated with the synanthropic brown rat (*Rattus norvegicus*) and black rat (*R. rattus*) was examed in both the Far East and the European part of Russia. The researchers found that SEOV provoked HFRS more often among the urban population, whereas HTNV did so more often among the rural population, of Primorsky Krai.[21] Morbidity in the Far East has a small uptick in May–July and reaches its main peak in November–December. The immune stratum in the Far East is about 1% (ranging from 0.3% in the Amur region to 1.5% in Primorsky Krai).[1,21]

Pathogenesis. Capillary damage is the basis of HFRS pathogenesis. In the first part of the disease, toxicoallergic phenomena predominate, caused by viral infection of the walls of vegetative centers, venules, and arterioles. Lesions on the sympathetic nodes of the neck are followed by hyperemia of the face and neck. Irritation of the vagus nerve leads to bradycardia and a fall in arterial pressure. Damage to the vascular permeability is accompanied by hemorrhages in mucous membranes and the skin. The cause of death is cardiovascular insufficiency, massive hemorrhages into the vital organs, plasmorrhea into the tissues, collapse, shock, swelled lungs, spontaneous rupture of the kidneys, a hypertrophied brain, and paralysis of the vegetative centers.[4,22]

Clinical Features. The incubation period is 4–30 days. HFRS starts with fever, headache, muscular pain, dizziness, nausea, vomiting, hyperemia of the face and neck, bradycardia, and a fall in arterial pressure. Abnormalities of the central nervous system (CNS) in the form of block, excitement, hallucinations, meningeal signs, and visual impairments often occur. Hemorrhagic syndrome becomes apparent as plasmorrhea into the tissues, together with microthrombosis; exanthema; petechial skin rash; nasal, pulmonary, and uterine bleeding: vomiting blood, hematuria, and visceral bleeding. In some cases, Pasternatsky syndrome, pain in the kidneys, oliguria, and albuminuria become morphologically apparent as interstitial and tubular nephritis. The duration of fever is 3–9 days. Two-wave temperature dynamics is possible.[22,38] Analyses of 5,282 cases of HFRS etiologically linked with PUUV in Sweden during 1997–2007 found 0.4% mortality in the first three months of the disease.[39,40] Defense immunity remains for at least 30 years.[8,22]

Diagnostics. Laboratory diagnostics are based on the fluorescent antibody method (FAM), enzyme-linked immunosorbent assay (ELISA), and reverse transcription polymerase chain reaction (RT-PCR) testing. The virus can be isolated with the use of Vero E6 (green monkey kidney cell line), 2Bs (diploid human embryo lung cell line), A-549 (human lung carcinoma cell line), or RLC (rat lung tissue primary cell culture).[8,22]

Control and Prophylaxis. Treatment of HFRS can be symptomatic, pathogenetic, or etiotropic (or any combination thereof). During the fever period, early hospitalization, disintoxical therapy, and strengthening of the walls of vessels are necessary. During the oliguria period, transfusion with desalinated human albumin, hemodes, a 5% glucose solution, and an isotonic NaCl solution (under the control of the emitted volume of urine) are given. In case of shock, antishock therapy is applied, and hemodialysis is prescribed for kidney insufficiency.[4,22]

Vaccination is the most effective approach to the prophylaxis of HFRS. The efficacy of vaccination was demonstrated in China and in North and South Korea. Nevertheless, it must be mentioned that vaccines in these countries are produced from HTNV and SEOV stains and do not defend against PUUV infection, which is the main etiological agent of HFRS in the European part of Russia (where 98% of all Russian morbidity occurs)[8].

For a long time, anti-HFRS vaccine was difficult to produce because there were no sensitive cell lines to accumulate hantavirus. However, the recent adaptation of PUUV and DOBV to the certified Vero E6 cell line affords an opportunity to produce candidate vaccines against HFRS. Experimental series of "Combi-HFRS-Vac" vaccine have passed compliance tests for medical immunoglobulin preparations for use in humans.[8,41,42]

8.1.3 Genus *Nairovirus*

The genus *Nairovirus* includes the tick-transmitted bunyaviruses, whose genome is the

FIGURE 8.3 Genome structure of nairoviruses (Family Bunyaviridae, genus *Nairovirus*). *Drawn by Tanya Vishnevskaya*

largest in the family Bunyaviridae. The size of L-segments of the Dugbe virus (DUGV), a prototypical species of the nairoviruses, is 12,255 nt. The M- and S-segments are 4,888 and 1,716 nt, respectively (Figure 8.3). As with other bunyaviruses, the L-segment of the nairoviruses encodes an RdRp, the M-segment encodes a polyprotein precursor of the envelope glycoproteins Gn and Gc, and the S-segment encodes the nucleocapsid (N) protein.[1,2]

The genus *Nairovirus* was established on the basis of antigenic relationships among viruses of the six antigenic groups of arthropod-borne viruses: the Crimean–Congo hemorrhagic fever (CCHF), Nairobi sheep disease (NSD), Qalyub (QYB), Sakhalin (SAK), Dera Ghazi Khan (DGK), and Hughes (HUGV) groups.[3–6] Subsequently, a seventh, Thiafora (TFA), group was assigned to the genus.[7,8] Currently, about 35 viruses are assigned to the genus *Nairovirus*, now united in the aforementioned seven groups.[1] Sequence analysis of previously unclassified bunyaviruses revealed that the nairoviruses actually number much more than 35. Three additional groups of nairoviruses— Issyk-Kul (ISK), Artashat (ARTSV), and Tamdy (TAM)—were established on the basis of phylogenetic analysis (Table 8.3).

8.1.3.1 Crimean–Congo Hemorrhagic Fever Virus

CCHFV belongs to the *Nairovirus* genus of the Bunyaviridae family and is the etiological agent of Crimean–Congo hemorrhagic fever (CCHF).

TABLE 8.3   Viruses of the *Nairovirus* Genus (Family Bunyaviridae) Isolated in Transcaucasia, Central Asia, Kazakhstan, and High Latitudes of Northern Eurasia

Serogroup	Virus	GenBank ID	Type of Biome	Source of isolation	Distribution
CCHFV	Crimean−Congo hemorrhagic fever (CCHFV)	NC005300−NC005302	Pasture	Ticks *Hyalomma* spp., predominantly *H. marginatum*; Hedgehogs; hares; rodents; farm animals; humans	Central Asia, Middle East, China, Kazakhstan, Transcaucasus, south of Europe, Africa
	Hazara (HAZV)	M86624, DQ076419		Ixodidae ticks	Central Asia
Qalyub	Chim (CHIMV)	KF801656	Burrows	Ticks *Ornithodoros tartakovskyi*, *O. papillipes*, *Rhipicephalus turanicus*, *Hyalomma asiaticum*; Great gerbil, *Rhombomys opimus*	Central Asia
	Geran (GRNV)	KF801649		*Ornithodoros verrucosus* ticks	Transcaucasus
Sakhalin	Sakhalin (SAKV)	KF801659	Seabird colony	*Ixodes uriae* ticks	High latitudes
	Paramushir (PMRV)	KF801657			
Tamdy	Tamdy (TAMV)	KF801653	Pasture	Ticks *Hyalomma asiaticum* and *H.* spp., *Rhipicephalus turanicus*, *Haemophysalis concinna*, gerbils, birds, human	Central Asia, Kazakhstan, Transcaucasus
	Burana (BURV)	KF801651		Ticks *Haemaphysalis punctate*, *Haem. concinna*	Central Asia
Hughes	Caspiy (CASV)	KF801659	Seabird colony	Ticks *Ornithodoros capensis*; seagull *Larus argentatus*	Eastern and western coasts of Caspian Sea
Issyk-Kul	Issyk-Kul (ISKV)	KF801652	Burrows	Bats (Vespertilionidae); ticks Argasidae; birds; human	Central Asia, Malaysia
	Uzun-Agach (UZAV)	KJ744032		Bat *Myotis blythii*	Central Asia
Artashat	Artashat virus (ARTSV)	KF801650	Burrows	Ticks *Ornithodoros alactagalis*, *O. verrucosus*	Transcaucasus

History. CCHF was first mentioned as "hunibini" and "hongirifta" by Tajik physician Abu-Ibrahim Djurdjani in the twelfth century. The viral nature of CCHF was originally established in 1945 during an expedition to Crimea headed by Mikhail Chumakov at the time of an outbreak.[1−3]

The modern history of CCHFV investigation starts in June 1944 with an epidemic of the disease in the northwestern steppe part of the Crimean Peninsula. More than 200 severe cases of the disease broke out, all exhibiting hemorrhagic syndrome, known in that time as "severe infectious capillary toxicosis." Mikhail Chumakov headed an expedition to the region, and much research revealed that the disease is transmitted by *Hyalomma plumbeum* (*marginatum*) ticks of the Ixodidae family. The disease was named Crimean hemorrhagic fever. A viral etiology was demonstrated by experimental infection of volunteers by an ultrafiltrated homogenate of *H. marginatum* nymphs collected in 1945 from local hares.[1] In 1963, the historical Hodzha strain was isolated from a patient with hemorrhagic fever in Uzbekistan, as was a set of strains from *H. marginatum* larvae and nymphs in the Astrakhan region, near the Caspian Sea.[2,3] In 1967, the similarity between the etiological agent of Crimean hemorrhagic fever and that of Congo virus, isolated from a patient in 1966 in Zaire (Congo), was demonstrated, so the virus was renamed CCHFV.[4,5]

Genome and Taxonomy. Like the genomes of all nairoviruses, that of CCHFV consists of three segments of negative ssRNA: a signed small (S) (1,672 nt) segment, a medium (M) (5,366 nt), and a large (L) (12,108 nt) segment. Each segment has a single ORF that encodes the nucleocapsid protein (N, 482 aa, S-segment), a polyprotein precursor of envelope glycoproteins Gn and Gc (1,684 aa, M-segment), and RdRp (3,945 aa, L-segment).

Genetic diversity among CCHFV strains may reach 31% nt and 27% aa differences for M-segment sequences, a reflection of pressure on the immune system and adaptation to various ecologic zones with different prevalences of *Hyalomma* tick species. The S- and L-segments are more conservative: The level of divergence of S-segment sequences is 20% nt and 8% aa, and that for L-segment sequences is 22% nt and 10% aa.

Phylogenetic analysis based on sequence data comparisons of S-segments shows that CCHFV isolates from different regions can be clustered into seven phylogeographic groups: West African isolates (group I), as well as isolates from Central Africa (Uganda and the Democratic Republic of the Congo) (group II); South Africa and West Africa (group III); the Middle East and Asia (group IV) (the Asian strain can be divided to two distinct subgroups: Asia 1 (IVa) and Asia 2 (IVb)); Europe and Turkey (group V); and Greece (group VI), a separate group detached from the rest of Europe (Figure 8.4).[6–8] In general, the genotypic structure defined on basis of the S-segment analysis is correlated strictly with geography. Cases of isolation of strains not typical for a given territory were attributed to possible transmission of the virus by infected ticks carried by migratory birds. The tree topology based on the L-segment comparison is, on the whole, similar to that generated on the basis of the S-segment. Exceptions are the viruses from Senegal, which represent a separate lineage in the S-segment analysis, and those clustered within group III in the L-segment analysis. Similarly, the division of group IV into group IVa (Asia 1) and IVb (Asia 2) is not clear (Figures 8.5 and 8.6).

In Russia, most of the strains of CCHFV that were isolated were isolated in the country's southern regions (Astrakhan, Volgograd, and Stavropol districts). Phylogenetic analysis showed that all of them are closely related to European and Turkish strains (group V).[9–12]

Epizootiology. Up to today, CCHFV has been found to circulate in 46 countries in Europe, Africa, and Asia.[4,13–15]

CCHFV was isolated from at least 27 species of mainly Ixodidae ticks, but their roles in maintaining virus circulation are different (Tables 8.4 and 8.5). The main significance for CCHFV reservation and transmission belongs to ticks of the *Hyalomma* genus: *H. marginatum* in the south of he European part of Russia, *H. anatolicum* and *H. detritum* in the Middle East and Asia, and *H. asiaticum* in Kazakhstan. According to our data, the viral load among imagoes of *H. marginatum* in the Astrakhan region in 2001–2005 was 1.33%; among nymphs, the load was 0.2%. The presence of transphase and transovarial transmission of CCHFV provides a reservation for viruses during the interepidemic period. Three hosts— for larvae (ground birds, mainly Corvidae;

FIGURE 8.4   Phylogenetic relationship between CCHFV strains isolated in different geographical regions, based on comparison of the full-length sequences of the S-segment. I—West Africa (Africa 1); II—Central Africa (Africa 2); III—Southern and Western Africa (Africa 3); IV—Middle East/Asia, divided into groups IVa and IVb, respectively, corresponding to groups Asia 1 and Asia 2; V—Europe/Turkey (Europe 1); VI—Greece (Europe 2).

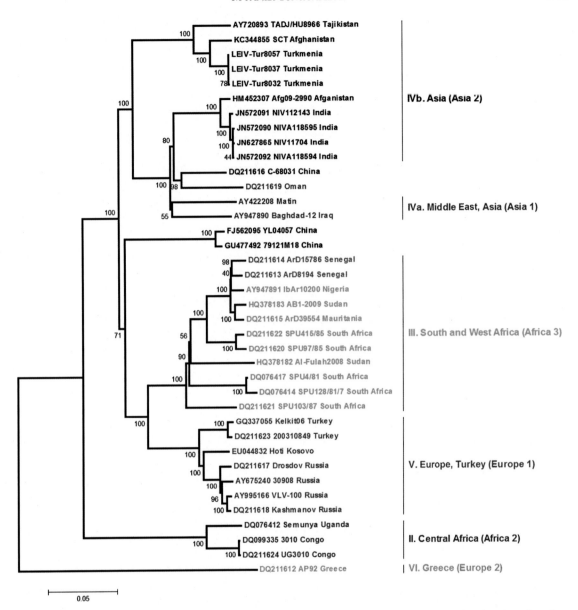

FIGURE 8.5 Phylogenetic relationship between CCHFV strains isolated in different geographical regions, based on comparison of the full-length sequences of the M-segment. I—West Africa (Africa 1); II—Central Africa (Africa 2); III—Southern and Western Africa (Africa 3); IV—Middle East/Asia, divided into groups IVa and IVb, respectively, corresponding to groups Asia 1 and Asia 2; V—Europe/Turkey (Europe 1); VI—Greece (Europe 2).

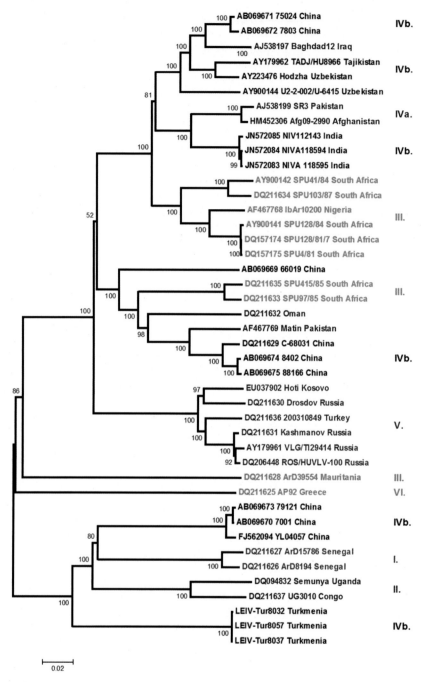

FIGURE 8.6 Phylogenetic relationship between CCHFV strains isolated in different geographical regions, based on comparison of the full-length sequences of the L-segment. I—West Africa (Africa 1); II—Central Africa (Africa 2); III—Southern and Western Africa (Africa 3); IV—Middle East/Asia, divided into groups IVa and IVb, respectively, corresponding to groups Asia 1 and Asia 2; V—Europe/Turkey (Europe 1); VI—Greece (Europe 2).

TABLE 8.4   Isolation of CCHFV from Ixodidae Ticks

Species of ticks	Place of collection
Alveonasus lahorensis	Armenia, Uzbekistan, Iran
Argas persicus	Uzbekistan
Hyalomma marginatum	Moldova, south of Ukraine, south of European part of Russia, central Asian countries, Kazakhstan
H. m. turanicum	Central Asian countries, Kazakhstan
H. m. rufipes	Africa (Nigeria, Senegal, Somali)
H. anatolicum	Central Asian countries, Kazakhstan, Pakistan, Africa (Nigeria)
H. an. excavatum	Africa (Nigeria)
H. asiaticum asiaticum	Central Asian countries, Kazakhstan, China
H. detritum	Europe (Spain)
H. impeltatum	Africa (Senegal, Nigeria, Ethiopia)
H. impressum	Africa (Senegal)
H. lusitanicum	Europe (Spain)
H. nitidum	Africa (Central African Republic)
H. punctatum	Europe (Spain)
H. truncatum	Africa (Senegal, Nigeria)
Amblyomma variegatum	Africa (Nigeria, Senegal, Uganda, Kenya)
B. annulatus (calcaratus)	Central Asian countries, Kazakhstan, Bulgaria, Africa (Senegal, Nigeria)
B. decoloratus	Pakistan
B. geigyi	Pakistan
B. microplus	Pakistan, Nigeria
Haemaphysalis punctata	Moldova, south of Ukraine
Dermacentor marginatus	South of Russia, Moldova, Uzbekistan
D. daghestanicus	Kazakhstan
Rhipicephalus sangnineus	Central Asian countries, south of Ukraine, Africa (Guinea)
Rh. turanicus	Central Asian countries
Rh. pumilio	Central Asian countries, Bulgaria
Rh. bursa	Central Asian countries, Azerbaijan, Greece
Rh. rossicus	South of Russia
Rh. pulchellus	Africa (Kenya)
I. ricinus	South of Russia, Moldova, Bulgaria, Hungary

TABLE 8.5   Infection Rate (%)[a] for CCHFV of Some Ixodidae Ticks[b]

	Ixodidae ticks										
	Hyalomma				Rhipicephalus		Dermacentor	Boophilus	Ixodes	Alveonasus	Argas
Region	Marginatum	Anatolicum	Asiaticum	Detritum	Turanicus	Bursa	Marginatus	Annulatus	Ricinus	Lahorensis	vespertilionis
South Europe	0.110	–	–	–	–	0.001	0.001	SI	–	0.108	–
Central Asia	0.100	0.011	0.013	SI	SI	–	–	–	SI	0.103	SI
Total	0.100	0.011	0.013	SI	SI	0.001	0.001	SI	SI	0.105	SI

[a] Abbreviations: SI, solitary isolations.
[b] According to data on monitoring conducted by the Center of Virus Ecology of the D.I. Ivanovsky Institute of Virology during 1972–1992.

mouselike rodents; and hares), nymphs (also ground birds, mouselike rodents, and hares), and imagoes (large mammals—mainly cattle, sheep, and camels)—provide a variety of ecological links of CCHFV to vertebrates.[1,16–19] In Nigeria, CCHFV was isolated from midges (*Culicoides* sp.)[4] The distribution of *H. marginatum* is limited by the isotherm of effective temperatures such that sum $(\Sigma_{T \geq 10°C}) = 3{,}000°C$, or 120 days with mean temperature $\geq 20°C$ per year.[20] So, the northern boundary of the distribution of CCHFV in the south of the European part of Russia lies in the dry steppe subzone.[1]

In Russia and South Africa, CCHFV is often isolated from hares.[1,21] CCHFV was isolated from hedgehogs (*Atelerix spiculus*) in Nigeria. Hares and mouselike rodents play the main role in CCHFV circulation.[1,21,22] Viremia in birds is not sufficient for vector transmission (although specific antibodies appear); nevertheless, ground birds are an important element of CCHFV transmission because they are the hosts for the preimaginal phases of *H. marginatum* development.[16,18,23] During field investigations of Chatkalsky Ridge in Kirgizia, nymphs and larvae of *H. marginatum* dominated among field-collected materials from birds. The highest number of ticks was found on rollers

(*Coracias garrulus*), crested larks (*Galerida cristata*), tree sparrows (*Passer montanus*), and black-billed magpies (*Pica pica*). In the Astrakhan region, rooks (*Corvus frugilegus*) are the main hosts for *H. marginatum* preimaginal phases.[16]

During migrations, birds can take part in dispersing preimago ticks that carry the virus. For example, in Spain in 2010, CCHFV of African origin (probably introduced by migrating birds) was isolated from *H. lusitanicum*.[24] European birds overwintering in Africa were also found to harbor ticks that carried the virus.[25]

CCHFV infection rates found as the result of an investigation of 40,711 domestic animal sera are presented in Table 8.6.[20] Domestic animals are one of the main reservoirs of CCHFV among vertebrates. Viremia (2.6–3.7 $(\log_{10}LD_{50})/20$ mcL) sufficient for the infection of ticks was detected 5–8 days after experimental inoculation of sheep. Viremia after up to 10 days post inoculation was detected in small gophers (*Citellus pygmaeus*), long-eared hedgehogs (*Hemiechinus auritus*), and wood mice (*Apodemus sylvaticus*). Experimental infection was revealed only in nymphs, and that is why hares and Corvidae birds—the main hosts for nymphs—play the chief role in CCHFV circulation.

TABLE 8.6 Detection of Specific Anti-CCHFV Antibodies Among Humans and Domestic Animals (1968—1992)

Region	Administrative unit, landscape	Number of tested sera: humans/ cattle—sheep—horses—camels—birds	Positive results, %
Crim peninsula	Crimea, steppe	0/442	ND/1.4[a]
North Caucasian	Rostov region	1,519/4,154—893—12—0—65	6.5/4.3—1.8—33.3—ND—0
	Krasnodar Krai, steppe	1,035/2,567	0/1.2
	Stavropol Krai	0/350—2,748	ND/0.3—0.5
Lower Volga	Kalmyk Republic, steppe	573/7,966—185	0/0.4—0
Kazakhstan	Southern deserts	0/1,688—2,782—136—181—0	ND/2.6—3.7—8.8—3.8—ND
Uzbekistan	Deserts	334/2,095—156—0—0—0	0.9/2.5—5.1—ND—ND—ND
Turkmenistan	Deserts	304/103—509—0—0—0	0/4.9—7.9—ND—ND—ND
Kirgizia	Mountain steppes, intermountain depression deserts	0/357—489—0—0—0	ND/12.6—7.4—ND—ND—ND
Tajikistan	Mountain steppes, intermountain depression deserts	7339/2,402—1,341—71—38 (donkeys)— 506 (other)	1.0/1.3—1.5—2.8—39.5 (donkeys)—0 (other)

[a]Abbreviation: ND, no data.

Epidemiology. CCHF distribution correlates with that of the main vector and natural reservoir of CCHFV—*H. marginatum* ticks—placing the virus within the limits of steppe, semidesert, and desert landscapes. CCHFV strains were have been isolated many times from patients in Russia,[1—3] Pakistan, Iraq, Iran, China, Greece, Bulgaria,[26] Turkey,[27,28] Romania, Yugoslavia, the United Arab Emirates, Senegal, Nigeria, the Republic of South Africa, Kenya,[29] Uganda, Tanzania, Ethiopia, Egypt, Burkina Faso, Mauritania, and Zimbabwe.[4] According to serological data, CCHFV is active in the south of France, Hungary, and India.[30] Figure 8.7 presents modern data about the known and predicted distribution of the virus.[4,31,32] CCHF outbreaks and sporadic cases of the disease were revealed on the territory of the former USSR in Ukraine, Moldova, Turkmenistan, Uzbekistan, Tajikistan, Kazakhstan,[20,33] Azerbaijan, the Rostov region, Krasnodar Krai, Stavropol Krai, the Kalmyk Republic, the Dagestan Republic, and the Ingush Republic.[1,34]

After 2006 until today, CCHFV is continuing to circulate in southeastern Europe (Greece and Bulgaria,[26]); in Asia (Turkey,[35,36] Iran,[37] Kazakhstan,[38] Pakistan, Afghanistan,[39] and China; and in all arid territories of Africa.[25,29,40,41] Cases of CCHF were revealed in Georgia.[42]

CCHFV infection rates found as the result of an investigation of 11,676 human sera are presented in Table 8.6.[20] CCHF mortality among humans was 12—16% in 1953—1967, 2.0% in 2006—2010, 3.6% in 2011—2013, and 5.1% in 2013. Starting in 1999, the epidemiological situation for CCHF worsened in the south of the European part of Russia, including Rostov, Astrakhan, the southern regions of Volgograd, the Kalmyk Republic, and the northern part of Dagestan, especially Stavropol Krai (Tables 8.7 and 8.8).[34,43] The deterioration may be explained by an increase in the population

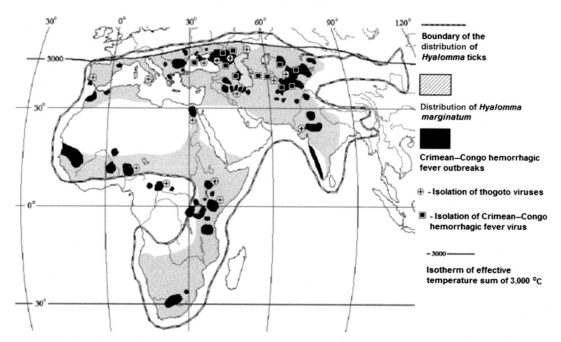

FIGURE 8.7   Distribution of *Hyalomma* ticks, Crimean–Congo hemorrhagic virus, and Thogotoviruses.

TABLE 8.7   Morbidity Due to CCHFV in Russia

Territory	Number of CCHF cases															Total
	1999	2000	2001	2002	2003	2004	2005	2006	2007	2008	2009	2010	2011	2012	2013	
Astrakhan	1	5	11	13	9	4	37	16	20	5	6	7	10	7	1	152
Volgograd	0	18	9	3	3	2	6	16	30	7	2	3	0	0	6	105
Dagestan	0	6	10	7	3	1	3	3	2	2	1	3	2	0	2	45
Ingush	0	0	0	0	0	4	0	0	1	0	0	0	0	0	0	5
Kalmyk	0	8	3	13	23	15	38	69	64	16	17	10	11	3	0	290
Karachaevo-Cherkessiya	0	0	0	0	0	0	0	0	1	0	0	0	0	0	0	1
Rostov	27	0	5	7	9	9	16	55	53	83	27	16	48	41	38	434
Stavropol	10	48	21	54	30	41	38	41	63	80	66	28	26	24	32	602
Krasnodar	0	0	0	0	0	0	1	0	0	0	0	0	0	0	0	1
Total	38	85	59	97	77	76	139	200	234	193	119	67	97	75	79	1,635

TABLE 8.8 CCHF in Russia in 2013[51]

Federal District	Number of cases (%)	Infection rate per 100,000 of Population
Central	1[a] (1.25)	0.01
Northwestern	0	0.00
Southern	45 (56.25)	0.32
North Caucasian	34 (42.50)	0.36
Lower Volga	0	0.00
Ural	0	0.00
Siberian	0	0.00
Far Eastern	0	0.00
Total	80 (100)	

[a]*Imported case.*

of Ixodidae (at first, *H. marginatum*) ticks in this region as the result of climatic changes.

During 1999−2010, 13,838 cases of CCHF[44] were recorded in Russia, including 520 in Stavropol Krai,[45] 307 in the Rostov region,[46] 276 in the Kalmyk Republic,[47] 134 in the Astrakhan region,[48] 99 in the Volgograd region,[49] 41 in the Dagestan Republic, 5 in the Ingush Republic, and 1 in the Karachaevo−Cherkesskaya Republic[50] (Table 8.7). In 2013, 80 cases of CCHF were recorded on the territory of the Southern Federal District and the North Caucasian Federal District (Table 8.8, Figure 8.8). The absence of any recorded cases of CCHF in Krasnodar Krai could be explained by a lack of attention to CCHF diagnostics.

A decrease in the proportion of severe clinical forms with hemorrhagic syndrome occurred after

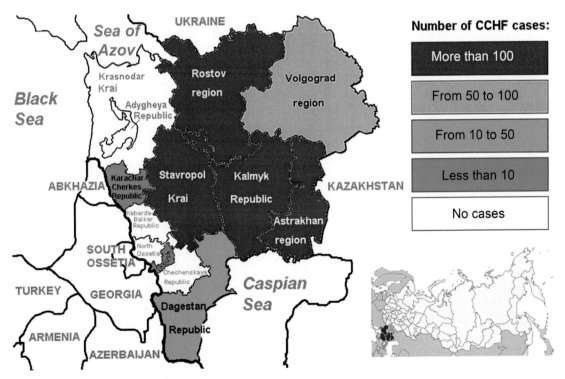

FIGURE 8.8   Distribution of CCHF in Russia (1999−2013).

2006. The drop could have been due to the introduction of high-grade express diagnostics methods into clinical practice and an intensification in seeking out and diagnosing those suspected of having CCHF. At the same time, the disease extended its incidence into the new territories of the Volgograd region, with nosocomial CCHF cases recorded there once again.[52] A warming of the climate promotes an expansion in the distribution of CCHFV to the north and the widening of endemic territory. Starting in 2002, CCHF was registered regularly in all of the units units of the Astrakhan region, accompanied by a mean morbidity of 3.7 per 100,000 population. More than 90% of CCHF cases occurred in the May–July period, and about 85% of the inhabitants of the rural area fell victim to the disease. Morbidity reached 26 units and three towns in the Rostov region. In the Volgograd region, CCHF was originally reported in 2000 in dry steppe on the boundary between that region and the Kalmyk Republic. Morbidity in the Kotelnichesky unit of the Volgograd region reached 37.5 per 100,000 people. Sporadic cases (with 16.7% lethality) were found in six units in the southwestern part of the Volgograd region, where the modern northern boundary of the distribution of CCHFV lies. During 2000–2006, 170 cases of CCHF were reported in the Kalmyk Republic. CCHF morbidity was found in 26 rural units of Stavropol Krai, mainly in dry steppes, where the highest infection rate from *H. marginatum* existed.

Pathogenesis. Pathogenesis is defined by lesions of the vascular and nervous systems.[17,51,53]

Clinical Features. The incubation period after transmissive CCHFV inoculation (as the result of a tick bite) is 2–7 days, whereas that after contact inoculation is 3–4 days. The difference is due to a much higher quantity of virus entering the system during contacts inoculation.[17,50,53] CCHF starts rapidly, with the temperature increasing to 39–40°C and the appearance of fever, skin hyperemia in the top half of the trunk, headache, lumbar pain, abdominal and epigastric pains, generalized

arthralgia, conjunctivitis, pharingitis, and diarrhea. About 50% of cases have two obvious waves of increasing temperature, with the temperature decreasing in 6–7 days after the end of the incubation period. Petechial rash appears in the majority of all CCHF patients in 3–4 days after the incubation period and is a marker of the second increasing-temperature wave. Hemorrhagic diathesis with nasal bleeding (in two-thirds of cases), bloody vomiting, blood in the sputum, and hematuria, all starting 3–5 days after the end of incubation period, are characteristic in 85% of cases. The duration of the hemorrhagic period is 8–9 days. Meningitis symptoms and signs of psychosis (depression, sleepiness, lassitude, photophobia) could develop as well. Lethality is 16–20% for transmissive inoculation and up to 50% for contact inoculation. Nevertheless, lethality is decreasing as the result of the introduction of modern testing systems and treatment with ribavirin. The convalescent period is about a month.[17,50,51,53]

E.V. Leshchinskaya has suggested the following clinical classification of CCHF: (i) severe form with hemorrhagic syndrome (1.a. without band bleeding; 1.b. with band bleeding); (ii) without hemorrhagic syndrome (2.a. medium-severe form; 2.b. light form).[50,53]

Diagnostics. Diagnosis is based on the detection of both specific antibodies via ELISA (IgM after 8 days post disease progression and IgG) and virus RNA via RT-PCR testing (earlier than 8 days post disease progression).[43,54] Both tests must be conducted for a definitive diagnosis of CCHF to be made. During the first week of infection with CCHF, positive results via RT-PCR are obtained in 93% of cases; during the second week, the percentage is 40%. During the second week of the disease, positive results in IgM via ELISA are obtained in 93% of cases; during the third week, the percentage of positive reults in IgG via ELISA is 80%.[55–58]

Control and Prophylaxis. Ribavirin is the most effective drug prescribed today.[53,59–61] Ribavirin is used for 5 days after symptoms first appear: 2,000 mg (10 capsules) or 30 mg/kg for the first time, then 600 mg × 2 times a day if the weight of the patient is more than 75 kg or 500 mg × 2 times a day if the weight of the patient is is less than 75 kg). The duration of treatment is 4—10 days. Ribavirin must not be used by pregnant women, except when the disease is considerd life threatening.

Vaccine development is currently just in the experimental stages,[62–64] so prophylaxis involves early detection of sick humans and the prevention of further contact infections. Nonspecific prophylaxis includes the eradication of Ixodidae ticks on livestock and acaricide treatment of locations inhabited by domestic animals. In pastures with large numbers of Ixodidae ticks, animals have to be led into box stalls and the humans leading them there must use special suits.

8.1.3.2 *Artashat Virus*

History. Artashat virus (ARTSV, strain LEIV-2236Ar) was originally isolated from *Ornithodoros alactagalis* ticks (family Argasidae) collected in the burrows of a small five-toed jerboa (*Allactaga elater*) near Arevashat village (40°02′N, 44°32′E; Artashat district, Ararat province), Armenia (Figure 8.9) in 1972 (authors: D.K. Lvov, V.A. Zakaryan, V.L. Gromashevsky, T.M. Skvortsova). A second strain of ARTSV was isolated at the same location and source in 1983. Later, in 1984–1985, 10 strains (topotypical strain, LEIV-9000Az) were isolated from *O. verrucosus* ticks, collected in the burrows of a Persian jird (*Meriones persicus*) in Azerbaijan (Figure 8.9, Table 8.9). On the basis of the morphology of the virion, ARTSV was classified as a member of the family Bunyaviridae, and because of the

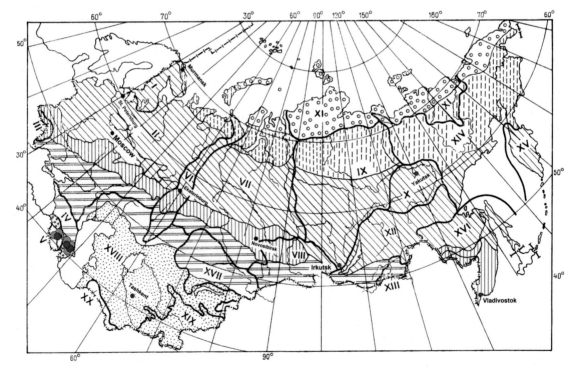

FIGURE 8.9 Places of isolation of ARTSV (family Bunyaviridae, genus *Nairovirus*) in Northern Eurasia. Red circle: strains of ARTSV with completely sequenced genome; Pink circles: strains of ARTSV identified by serological methods. (See other designations in Figure 1.1.)

TABLE 8.9   Isolation of ARTSV (Family Bunyaviridae, Genus *Nairovirus*)

Place of field material collection	Source	Date of isolation	Number of strains isolated
Armenia, Ararat province	*Ornithodoros alactagalis* ticks from burrows of small five-toed jerboa (*Allactaga elater*)	October 1972	1
Armenia, Ararat province	*Ornithodoros alactagalis* ticks from burrows of small five-toed jerboa (*Allactaga elater*)	July 1983	1
Azerbaijan, Ordubad district	*Ornithodoros verrucosus* ticks from a Persian jird (*Meriones persicus*)	September 1984	2
Azerbaijan, south part of Gobustan	*Ornithodoros verrucosus* ticks from a Persian jird (*Meriones persicus*)	September 1984	1
Azerbaijan, Goranboy District	*Ornithodoros verrucosus* ticks from a Persian jird (*Meriones persicus*)	September 1985	6
Azerbaijan, Yevlax District, near Mingachevir city	*Ornithodoros verrucosus* ticks from a Persian jird (*Meriones persicus*)	September 1985	1

absence of antigenic relationships with any known viruses, it was referred to as an "unclassified bunyavirus."[1–3]

Taxonomy. Three strains of ARTSV were sequenced.[4] A full-length genome comparison revealed that ARTSV has 42–60% nt similarity to other nairoviruses. Phylogenetic analysis revealed that the virus is a new species in the *Nairovirus* genus and forms a distinct genetic lineage on the nairovirus tree, which was constructed for all three segments of the genome (Figures 8.10–8.12).

The phylogeny of the nairoviruses is based mainly on analyses of the partial sequence of the conservative catalytic core domain of RdRp.[5,6] The similarity of this domain of ARTSV to other nairoviruses is 42–65% nt and 58–70% aa. The phylogenetic tree constructed by the maximum-likelihood method on the basis of the amino acid alignment of the RdRp catalytic core domain of nairoviruses confirms the topology of ARTSV on a newly formed genetic lineage (Figures 8.10–8.12). The nairoviruses on the tree can be divided into two main phylogenetic groups. The first group includes the nairoviruses, which are transmitted predominantly by

ixodids: the Crimean–Congo hemorrhagic fever group (*Hyalomma* and *Haemaphysalis*, as well as *Dermacentor*, *Rhipicephalus*, and *Ixodes*), the Dugbe group (mainly *Amblyomma*, but also *Hyalomma*, *Rhipicephalus*, and *Haemaphysalis*), the Sakhalin group (*Ixodes*), and the Tamdy group (*Hyalomma*). The first group also includes Erve virus (ERVEV), whose vectors are unknown.[7,8] The second phylogenetic lineage includes the nairoviruses from the Hughes, Issyk-Kul, Dera Ghazi Khan, and Qalyub groups, whose vectors are argasids: *Argas* and *Ornithodoros*. The tree topology of ARTSV shows that the virus is in the lineage of the nairoviruses transmitted predominantly by Ixodidae ticks, although all isolations of ARTSV were obtained from the Argasidae ticks *O. alactagalis* and *O. verrucosus* (Table 8.9). It can be assumed that the adaptation of ARTSV to argasids is the result of the the narrow ecologic niche occupied by those ticks, which are ticks of the subgenera *Theriodoros* and *Pavlovskyella*. Note that, although ERVEV, a European nairovirus, is phylogenetically close to the nairovirus transmitted by ixodids, the association of ERVEV with *Ixodes* spp. ticks has not been established in endemic areas (southern Europe).[8]

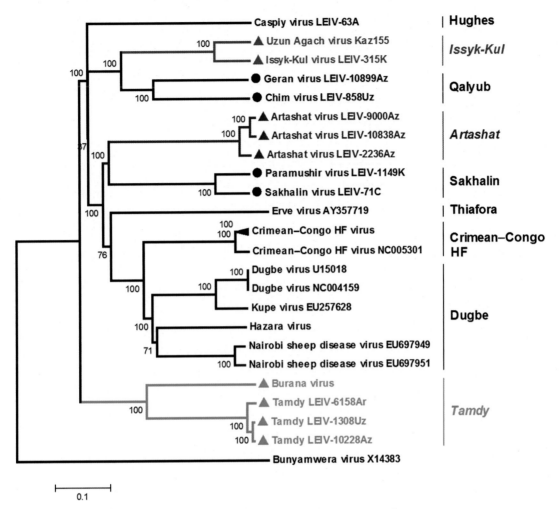

FIGURE 8.10   Phylogenetic structure of RdRp proteins of *Nairovirus* genus members.

ERVEV has been isolated from shrews (*Crocidura russula*).[9]

Arthropod vectors. The adaptation of viruses to Argasidae ticks facilitates the possibility of survival of viral populations in winter at low temperatures and in dry periods. The ability of argasids to fast (up to 9 years and more), the long life cycle of these ticks (up to 20—25 years), and their polyphagia and ecological plasticity determine the stability of the natural foci of arboviruses transmitted by argasids. These foci are confined mainly to the arid regions of the southern part of the temperate and subtropical zones.[1,2,10] The northern border of the range of argasids coincides with isolines denoting a frost-free period of 150—180 days per year and an average daily temperature above 20°C for no less than 90—100 days per year.[11]

Tick species from the subgenera *Theriodoros* (*Ornithodoros alactagalis, O. nereensis*) and *Pavlovskyella* (*O. papillipes, O. verrucosus,*

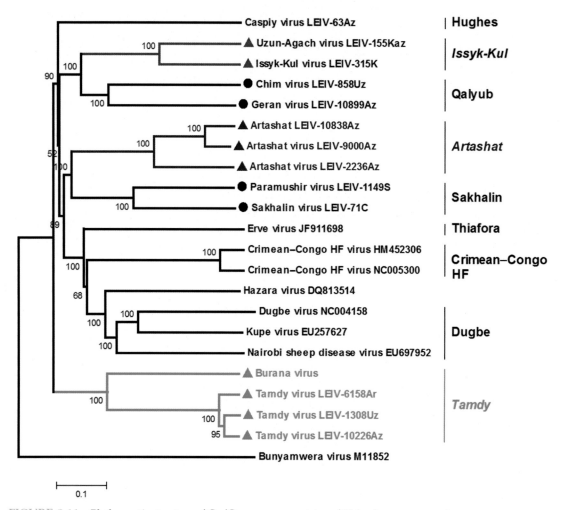

FIGURE 8.11　　Phylogenetic structure of Gn/Gc precursor proteins of *Nairovirus* genus members.

O. cholodkovskiy, O. tartakovskiy) are associated mainly with burrows of rodents.[11] This ecological peculiarity narrows the possibility of the spread of viruses that are adapted to ticks from the *Theriodoros* and *Pavlovskyella* subgenera.[2] It also applies to ARTSV associated with burrow—shelter biomes and found only in Transcaucasia.

8.1.3.3 Caspiy Virus

History. Caspiy virus (CASV, prototypical strain LEIV-63Az) was originally isolated from the blood of a sick herring gull (*Larus*

argentatus) caught on Gil Island in the Baku archipelago, off the western coast of Azerbaijan in the Caspian Sea (40°17′N, 49°55′E; Figure 8.13) in 1970.[1–4] On the basis of electron microscopy, CASV was classified as a member of the Bunyaviridae family, but antigenic relationships with known bunyaviruses have not been found. Thus, CASV was categorized into the unclassified bunyaviruses.[5,6–8] At the same time, and in the same place, three strains of CASV were isolated from *Ornithodoros capensis* (family Argasidae) ticks

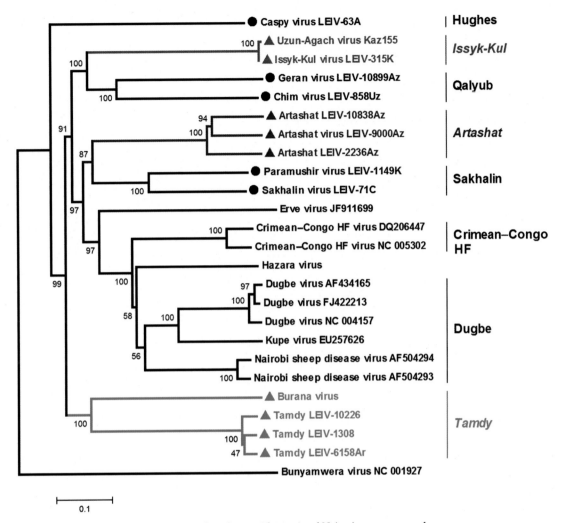

FIGURE 8.12 Phylogenetic structure of nucleocapsid protein of *Nairovirus* genus members.

(1,479 ticks were examined). Later, two more strains were isolated from *O. capensis* ticks (2,019 ticks were examined) collected in colonies of common terns (*Sterna hirundo*) nesting on islands in the Kara-Bogaz-Gol Bay in Turkmenistan, off the eastern coast of the Caspian Sea (41°02′N, 52°53′E; Figure 8.13) in 1974. In total, five strains of CASV were isolated from *O. capensis* ticks. (The rate of infected ticks was 0.21% in Azerbaijan and 0.1% in Turkmenistan.)[8–10]

Taxonomy. The genome of the prototypical strain LEIV-63Az of CASV was sequenced, and it has been shown that CASV is a member of the HUGV group of the *Nairovirus* genus.[11] The S-segment of CASV is about 1,594 nt in length and has a single ORF that encodes the nucleocapsid protein (N, 497 aa). The second start codon, in position 7, is located in the N-protein ORF of CASV. The identity of the amino acid sequence of the N-protein of CASV with those of other nairoviruses is only 28%, on average.

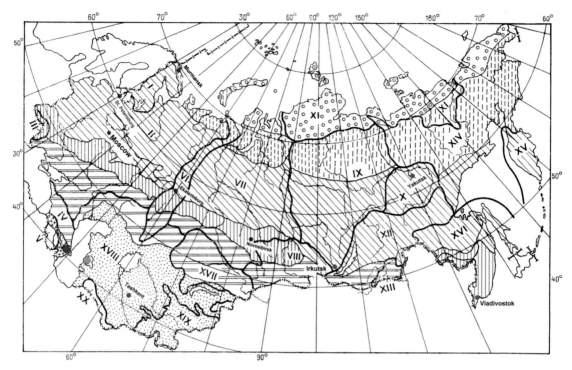

FIGURE 8.13   Places of isolation of CASV (family Bunyaviridae, genus *Nairovirus*) in Northern Eurasia. Red circle: prototype strains of CASV/LEIV-63Az with completely sequenced genome; Pink circle: two strains of CASV identified by serological methods (See other designations in Figure 1.1.)

The cleavage site for caspase-3 (D285EVD288) that has been found in the N-protein of CCHFV is absent in CASV. Cleavage of N by caspase-3 is required for effective replication of CCHFV.[12] Note that caspase cleavage sites in the nucleocapsid protein are also necessary for replication of human influenza A viruses.[13] The M-segment of CASV, like that of the other nairoviruses, has a single ORF-encoded polyprotein precursor of the envelope glycoproteins Gn and Gc. The length of the Gn/Gc precursor of CASV is 1,376 aa. According to the results of an analysis of polyprotein in the program SignalP server 4.1, the first 32 aa constitute the signal peptide that is cleaved on the SSA/SY site. The cleavage site between pre-Gn and pre-Gc is in position 699 (VSG/IK). These data are confirmed by the location of transmembrane domains in mature proteins Gn and Gc that was defined with the use of the program TMHMM server 2.0. Six potential sites of N-glycosylation are predicted in the mature Gn protein of CASV, only one in the Gc protein. In general, the level of identity of polyprotein in CASV is 25–27% aa with that of other members of the *Nairovirus* genus (Table 8.3). The L-segment of CASV has an ORF (4,001 aa) that encodes the viral enzyme RdRp, which is the most conservative viral protein. The similarity of the RdRp of CASV to that of other nairoviruses for which complete genome sequences were available is 38.8–43.0% aa.

Phylogenetic analysis based on the predicted full-length amino acid sequences revealed that CASV is equidistant from other nairoviruses, and forms a distinct branch, on the trees

(Figures 8.10–8.12). For many nairoviruses, only short sequences of the catalytic core domain of RdRp are available in GenBank. This domain of RdRp is very conservative and relevant to phylogenetic analysis.[1,14,15] The highest level of similarity (80% aa) that the RdRp core domain of CASV has is with the same sequences in viruses of HUG. On the dendrogram, constructed on the basis of a comparison of RdRp core domains, CASV is located on the branch of the HUG group (Figures 8.10–8.12). Note that viruses of this group (as well as CASV) have been isolated from *Ornithodoros* (*Carios*) ticks that are associated with seabirds on the coasts and islands of the world's oceans.[2,16] Thus, the phylogenetic relationship of CASV with HUG group viruses reflects the ecological features of those coasts and islands.

Arthropod Vectors. *Ornithodoros capensis* ticks inhabit the coasts and islands of the Atlantic, Indian, and Pacific Oceans from the southern part of the temperate zone to the equator, as well as some large inland ponds.[3,4] Ticks of the *O. capensis* group (*O. amblus*, *O. capensis*, *O. denmarki*, *O. maritimus*, *O. muesebecki*, *O. sawaii*) are obligate parasites of seabirds and replace *Ixodes* (*Ceratixodes*) *uriae* ticks in the south temperate, subtropical, tropical, subequatorial, and equatorial zones. *Ixodes* (*Ceratixodes*) *uriae* ticks remain common in the north temperate, subarctic, and subantarctic zones.[3,4,17] *O. capensis* ticks feed on many bird species, mainly those of the order Charadriiformes: gulls (family Laridae) and terns (Sturnidae), but also cormorants (Phalacrocoracidae) and pelicans (Pelecanidae).[4,17] These argasid ticks have a life cycle made up of six to eight stages: egg, larva, three to five stages of nymphs, and imago. According to laboratory study, the cycle is from 43 to 83 days and so can be completed during a single breeding season. These ecological peculiarities provide stability to the natural foci of the viruses, which are adapted to the *O. capensis* tick viruses and their transcontinental transfer by migrating birds.[5]

Vertebrate Host. In 1970, during the collection of field material on islands in the Baku archipelago, an epizootic among herring gulls was observed. The first strain of CASV was isolated from sick birds. Migrations in search of food, including migration between the western and eastern coasts of the Caspian Sea, result in a sharing of the argasids and viruses ranging over the area.

8.1.3.4 Chim Virus

History. The prototypical strain LEIV-858Uz of the Chim virus (CHIMV) was isolated from *Ornithodoros tartakovskyi* ticks collected in July 1971 in the burrows of great gerbils (*Rhombomys opimus*) in the vicinity of the town of Chim in the Kashkadarinsky region of Uzbekistan) (38°47′N, 66°18′E; Figure 8.14).[1–3] Isolation of CHIMV was carried out during monitoring of these arboviruses' foci on the territory of central Asia and Kazakhstan. CHIMV was investigated through serological testing with viruses from different families and with unclassified viruses isolated earlier in the USSR. Because no antigenic relationships of CHIMV were (and still have not been) found, CHIMV was assigned to the category of unclassified viruses.[3,4] Later, four strains of CHIMV were isolated from the ticks *O. tartakovskyi*, *O. papillipes*, and *Rhipicephalus turanicus* (*Rhipicephalinae*) respectively collected in the burrows of great gerbils in the Kashkadarya, Bukhara, and Syrdarya districts of Uzbekistan in 1972–1976.[5,6] Three strains of CHIMV also were isolated from *Hyalomma asiaticum* (*Hyalomminae*) ticks and from the livers of great gerbils, which were collected in the floodplains of the Or River and Karatal River (Dzheskazgan district, Kazakhstan) in April 1979 (Figure 8.14).[7,8]

Taxonomy. The genome of the prototypical strain LEIV-858Uz of CHIMV was sequenced, and, on the basis of sequence analysis, the virus was classified as a novel member of the *Nairovirus* genus.[9] Phylogenetic analysis based

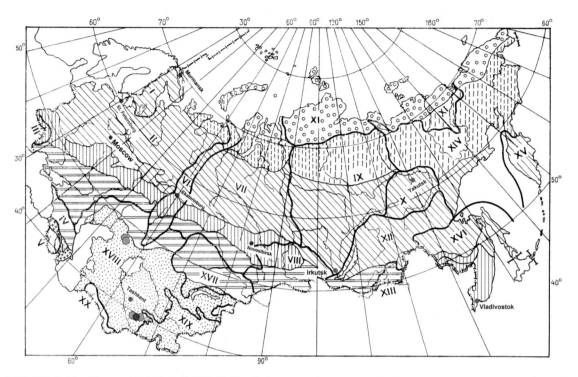

FIGURE 8.14   Places of isolation of CHIMV (family Bunyaviridae, genus *Nairovirus*) in Northern Eurasia. Red circle: prototype strains of CHIMV/LEIV-858 Uz with completely sequenced genome; Pink circles: strains of CHIMV identified by serological methods. (See other designations in Figure 1.1.)

on a partial sequence of a catalytic center of RdRp placed CHIMV on the genetic branch of the QYBV group.[9,10] The amino acid sequence of this domain of CHIMV has an 87% identity with QYBV, Geran virus (GERV), and Bandia virus (BDAV), the other members of the QYBV group.[11–14] All these data are consistent with the fact that viruses of the QYBV group, as well as CHIMV, have an environmental connection to ticks of the *Ornithodoros* genus and to the burrows of rodents. QYBV has repeatedly been isolated from *O. erraticus* ticks, collected in burrows of the African grass rat (*Arvicanthis niloticus*) in the Nile valley and the Nile delta in Egypt.[13] To date, only short sequences of the RdRp of QYBV are available in GenBank, but recently we gave a genetic characterization of GERV, isolated in Transcaucasia and, apparently,

closely related to QYBV.[11] The full-length amino acid comparison of CHIMV with GERV showed that their nucleocapsid proteins N (S-segment) have only a 55.6% identity. The similarity of complete amino acid sequences of RdRp (L-segment) is 74.8%. The similarity of the polyprotein precursor of Gn/Gc is 55.6%. The proteins of CHIMV have 30.3–42.4% aa (N-protein), 27.5–45.1% aa (Gn/Gc precursor), and 48.1–62.3% aa (RdRp) identities with their counterpart proteins in other nairoviruses. Among these nairoviruses, CHIMV has the highest level of similarity with ISKV, which is associated with bats in Central Asia (Figures 8.10–8.12).[15]

Arthropod Vectors. Most isolations of CHIMV were obtained from *Ornithodoros tartakovskyi* ticks. These ticks are common in the Irano-Turanian and mountain provinces of

Asia (Kazakhstan, the central Asian republics, northeastern Iran, and China (Xinjiang)). The western border of the area in question is the eastern shore of the Caspian Sea (53−54°E), the eastern border is in Xinjiang (87°E), and the northern border is 44−47°N. The typical biotopes that *O. tartakovskyi* ticks inhabit are the foothills of dry steppes with loess soils. The ticks also inhabit meadow steppes and deserts (floodplain terraces and canals). *O. tartakovskyi* ticks prefer burrows of small diameter (inhabited by rodents, including jerboas, ground squirrels, small predators, and hedgehogs, as well as by turtles and birds). Synanthropic biotopes are rarely inhabited.[16]

Vertebrate Hosts. The great gerbil (family Muridae, subfamily Gerbillinae, genus *Rhombomys*) is distributed from the shores of the Caspian Sea on the plains of central Asia and southern Kazakhstan, to the deserts of central Asia, Iran, and Afghanistan, and on eastward to northern China and Inner Mongolia. Great gerbils are typical inhabitants of sandy deserts and form a colony with complex multistory burrows that have a large number of entranceways and egresses (up to 200−500). These burrows are a specific biotope that exists for many decades, and they maintain natural foci (in particular, of plague) in arid areas.[6,8]

Animal Infection. The significance of CHIMV in the pathology of humans is unknown. Antibodies to CHIMV have been found in camels (9.5%) in the Kashkadarya region in Uzbekistan.[5] This finding shows the ability of CHIMV to infect camels, as does QYBV, but additional studies are necessary to clarify the pathogenicity of CHIMV in humans and cattle.[17]

8.1.3.5 Geran Virus

History. GRNV (strain LEIV-10899Az) was isolated from *Ornithodoros verrucosus* (family Argasidae, subfamily Ornithodorinae) ticks collected in a burrow of red-tailed gerbils (*Meriones* (*Cricetidae*) *erythrurus*) near Geran Station, Goranboy district, Azerbaijan; Figure 8.15). Serological methods have failed to identify GRNV, but the virus has been sequenced and classified into the *Nairovirus* genus (family Bunyaviridae).[1]

Taxonomy. The genome of GRNV was sequenced by a next-generation sequencing approach.[1] Full-length genome analysis revealed that the genetic similarity of GRNV to other known nairoviruses is, on average, 30−40% aa for the nucleocapsid protein (N, S-segment), 27−33% aa for the polyprotein precursor of the proteins Gn and Gc (M-segment), and 48.0−74.8% aa for RdRp (L-segment). The highest level of similarity all three proteins of GRNV have is to that of CHIMV (54.2−74.8% aa identity) and that of ISKV (42.4−62.3% aa identity).[2,3] Further analysis based on a comparison of partial sequences of the conservative core domain of RdRp of the nairoviruses showed that GRNV and CHIMV were most closely related to QYBV, which is the prototypical virus of the group of the same name.[4] The nucleotide sequence of the RdRp core domain of GRNV has 74.3% nt and 97.1% aa identities with the counterpart sequence of QYBV. The data obtained allow GRNV to be classified as a virus of the QYBV group (Figures 8.10−8.12). The phylogenetic relationship between GRNV and QYBV corresponds to their similar ecological characteristics. QYBV was first isolated in 1952 by R. Taylor and H. Dressler from argasid *Ornithodoros erraticus* ticks collected in a rodent burrow in the Nile River delta near Qalyub village, Egypt (30°N, 32°E).[5−7] Complement-binding antibodies to QYBV were found in humans (1.5%), camels, donkeys, pigs, buffalos, dogs, and rodents.[1,7] The antigenic group of Qalyub, a group that includes QYBV and antigenic-related BDAV, is one of the prototypical groups of the *Nairovirus* genus.[5,8] Previously, QYBV had been repeatedly isolated from *O. erraticum* collected in the burrows of rodents (*Arvicanthis*) in Africa. The second

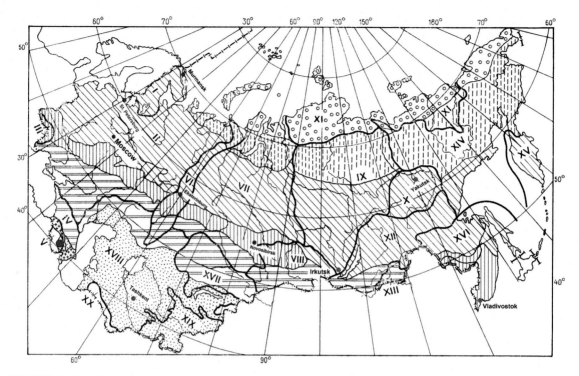

FIGURE 8.15 Place of isolation (red circle) of GRNV (family Bunyaviridae, genus *Nairovirus*) in Northern Eurasia. (See other designations in Figure 1.1.)

member of the QYBV group, BDAV, was isolated from *O. sonari* (a member of the *O. erraticus* group) collected in the burrows of rodents (mainly *Mastomys*) in Senegal.[9,10] The isolation of GERV, which is closely related to QYBV, is the first confirmation of the circulation of QYBV group viruses in Transcaucasia.

Arthropod Vectors. The area of distribution of *O. verrucosus* ticks covers the southern part of Moldova as well as Ukraine and the Caucasus region, and is limited by 47°30′N latitude. The area includes the southern part of Russia (the Krasnodar and Stavropol regions), the northern and eastern foothills of Dagestan, the foothills and lowland hills of Georgia, the valleys of the Hrazdan River in Armenia, the foothills of the Lesser Caucasus Mountains in Azerbaijan, and the Gobustan Plateau and the Absheron Peninsula, also in Azerbaijan.

O. verrucosus ticks inhabit shelter biotopes—in particular, the burrows of red-tailed gerbils (*Meriones* (*Cricetidae*) *erythrurus*), animals that are common in central Asia, southern Kazakhstan, and eastern Transcaucasia. Red-tailed gerbils tends to inhabit desert and semi-desert landscapes. Their burrows are deep and may have 5–10 entranceways and egresses.

8.1.3.6 Issyk-Kul Virus

History. ISKV (prototypic strain, LEIV-315K) was originally isolated from a pool of internal organs (liver, spleen, brain) of *Nyctalus noctula* bats, and their ticks (*Argas* (*Carios*) *vespertilionis*) were collected near Issyk-Kul Lake in Kyrgyzstan in 1970 (Figure 8.16).[1,2] Subsequently, ISKV was isolated from other bat species of the Vespertionidae family (*Vespertilio serotinus*, *Vespertilio pipistrellus*, *Myotis blythii*,

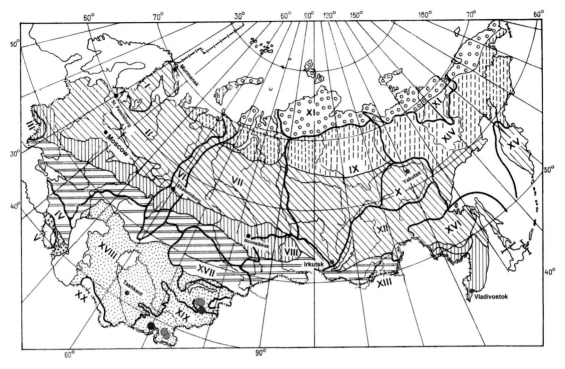

FIGURE 8.16 Places of isolation of ISKV and Garm virus (family Bunyaviridae, genus *Nairovirus*) in Northern Eurasia. Red circle: prototype strains of ISKV/LEIV-315K with completely sequenced genome; Pink circles: two strains of ISKV identified by serological methods; Dark-blue circle: strain LEIV-218 Taj of Garm virus. (See other designations in Figure 1.1.)

Rhinolophus ferrumequinum), and from birds, in different regions of Kyrgyzstan and Tajikistan.[3–11] Two strains were isolated from *Anopheles hyrcanus* mosquitoes and *Culicoides schultzei* biting midges, respectively (Figure 8.16, Table 8.10).[3,12,13] Complement-fixation testing showed that ISKV is closely related or identical to the Keterah virus, which was isolated from *Scotophilus temminckii* bats and *A. pusillus* ticks in Malaysia in 1960.[14,15] A strain that has a close, one-sided antigenic relationship to ISKV, LEIV-218Taj (named Garm virus), was isolated from a common redstart (*Phoenicurus phoenicurus*) caught in the village of Garm, Tajikistan, 39°10′N, 70°30′E (Figure 8.16) in the spring (migratory period) of 1976.

Morphological studies by electron microscopy characterized ISKV as a member of the Bunyaviridae family, and because no antigenic relation to any known viruses was found, it was assigned to the unclassified bunyaviruses.[16]

Taxonomy. The genome of the prototypical strain of ISKV, LEIV-315K, was sequenced, and, on the basis of sequence analysis, the virus was classified into the *Nairovirus* genus.[17] Like the genomes of other nairoviruses, that of ISKV consists of three segments of RNA (in negative polarity), each of which has a single ORF-encoded nucleocapsid protein (N, 485 aa, S-segment), a polyprotein precursor of the envelope glycoproteins Gn and Gc (1,631 aa, M-segment), and a RdRp (3,950 aa, L-segment). A pairwise comparison of the full-length nucleotide and deduced amino acid sequences of the ISKV ORFs with those of other nairoviruses revealed 48.2—51.1% nt

TABLE 8.10　Isolation of ISKV (Family Bunyaviridae, Genus *Nairovirus*)

Source of isolation		Place of isolation
Bats (Chiroptera)	*Nyctalus noctula*	Kirgizstan
	Myotis blythi	Kirgizstan
	Vespertilio serotinus	Kirgizstan
	V. murinus	Tajikistan
	V. pipistrellus	Tajikistan
	Rhinolopus ferrumequinum	Tajikistan
	Scotophilus temmencki	Malaysia
Birds (Aves)	Spanish sparrow (*Passer hispaniolensis*)	Kirgizstan
	White wagtail (*Motacilla alba*)	Kirgizstan
	Grey wagtail (*M. cinerea*)	Tajikistan
	Common redstart (*Phoenicurus phoenicurus*)	Tajikistan
	House swallow (*Hirundo rustica*)	Kirgizstan
	Wryneck (*Jynx torquilla*)	Tajikistan
	Common kingfisher (*Alcedo atthis*)	Kirgizstan
Ticks (Ixodidae)	*Argas (Carios) vespertilionis*	Kirgizstan, Tajikistan
	A. pusillus	Kirgizstan, Tajikistan
	Ixodes vespertilionis	Kirgizstan
Mosquitoes (Diptera: Culicidae)	*Aedes caspius*	Kirgizstan
	Anopheles hyrcanus	Kirgizstan
Horseflies (Diptera: Tabanidae)	*Tabanus agrestis*	Kazakhstan

(39.0–42.1% aa), 37.3–39.7% nt (23.2–26.5% aa), and 43.1–47.0% nt (31.9–34.5% aa) identity for RdRp, the precursor of Gn and Gc, and the N protein, respectively (Table 8.10).

Phylogenetic analysis carried out for the full-length amino acid sequences by the maximum-likelihood nearest-neighbor method showed that ISKV occupies a new and distinct branch on the phylogenetic trees relevant to all three nairovirus proteins (RdRp, Gn/Gc, and N) (Figures 8.10–8.12).

For the many known nairoviruses (i.e., QYBV, DGKV, and HUGV, as well as for a new nairovirus that was found in European bats by a metagenomics approach), there are only partial sequences of the conservative catalytic core domain of RdRp.[16,18,19] The level of identity for this domain of ISKV with other nairoviruses ranged from 59.6–66.1% for the nucleotide sequence and 64.8–75.2% for the amino acid sequence (Table 8.10). The ISKV RdRp core domain has the highest level of identity with QYBV (66.6% nt and 74.5% aa). The phylogenetic tree constructed on the basis of the amino acid alignment of the RdRp core domain of nairoviruses confirms the topology of ISKV on a new genetic branch of the nairoviruses (Figures 8.10–8.12).

Arthropod Vectors. Most isolates of ISKV were obtained from *Argas vespertilionis* ticks, and we can assume that these ticks are the main natural reservoir of the virus. The range of ticks of the *A. vespertilionis* group covers territory in central Asia, Africa, Oceania, and Australia (Figure 8.17).

Vertebrate Hosts. The natural vertebrate hosts of ISKV are apparently bats—specifically, the genera *Nyctalus*, *Vespertilio*, *Rhinolophus*, and *Myotis* (family Vespertilionidae). These bats are common in the temperate and subtropical zones of Europe, Asia, and North Africa, and widespread ISKV transmission and the appearance of an emergency are possible in all of their territories.

Human Pathology. The first case of Issyk-Kul fever was registered in Tajikistan in August 1975 when a staff member became ill after catching bats during surveillance for arbovirus. ISKV was isolated from his blood on the second

FIGURE 8.17 Geographical distribution of Argasidae ticks: vectors of ISKV (family Bunyaviridae, genus *Nairovirus*).

The highest percentage (9%) with antibodies to ISKV was found in the southeastern part of Turkmenistan.[12]

8.1.3.7 Uzun-Agach Virus

History. Uzun-Agach virus (UZAV), strain LEIV-Kaz155, was isolated from the liver of a *Myotis blythii oxygnathus* (order Chiroptera, family Vespertilionidae) bat caught in the vicinity of the village of Uzun-Agach, Alma-Ata district, Kazakhstan, during the virological sounding of territory in central Asia and Kazakhstan in 1977 (Figure 8.18).[1-3] On the basis of virion morphology, UZAV was classified into the Bunyaviridae family. No serological study of UZAV was ever conducted, but the place of UZAV isolation, Uzun-Agach, is close to where ISKV was originally isolated, namely, near Issyk-Kul Lake, and the source of both viruses is the same: bats.[4,5]

Taxonomy. The full-length genome of UZAV was sequenced, and, on the basis of phylogenetic analysis, the virus was classified into the *Nairovirus* genus.[6] The genome of UZAV, like those of other nairoviruses, consists of three segments of ssRNA with negative polarity. The L-segment encodes RdRp (3,988 aa), the M-segment encodes a polyprotein precursor of the envelope glycoprotein Gn and Gc (1,621 aa), and the S-segment encodes the nucleocapsid protein N (485 aa). A pairwise comparison of the sequence of the UZAV genome with those of other nairoviruses showed that the virus is related most closely to ISKV. Full-length sequences of the L- and M-segments of UZAV have, respectively, 69.3% nt and 64.1% nt identities with those of ISKV. Amino acid sequences of RdRp (S-segment) of UZAV and ISKV have 76.2% aa similarity. The similarity of the amino acid sequences of the precursor of Gn and Gc for UZAV and ISKV is 66.7% aa. A comparison of the S-segments of UZAV and ISKV revealed that they are almost identical (99.6%). Thus, we can conclude that UZAV is a reassortant virus that got an

day of the disease.[20] Later, one case of Issyk-Kul fever was registered in 1978 in Dushanbe, Tajikistan.[20] Local outbreaks of Issyk-Kul fever occurred in Tajikistan in 1982. That year, 22 patients with laboratory-confirmed Issyk-Kul fever were registered.[21] The disease occurs with fever (39–40°C), headache (94%), dizziness (50%), hyperemia of the throat (48%), cough (25%), and nausea (31%). The outcome is generally favorable, and no deaths have been registered.[18] Most of the cases were associated with the presence of bats in the attic of the residence. The primary route of human infection was apparently by argasid ticks, but respiratory or alimentary routes (via the feces and urine of bats) could not be excluded. Furthermore, a laboratory experiment showed that ISKV can be transmitted by *Aedes caspius* mosquitoes.[22] The percentage of the population immune to ISKV in the southern part of Tajikistan is 7.8%. In Kyrgyzstan, antibodies to ISKV have been found in 0.7–3.2% of the human population.

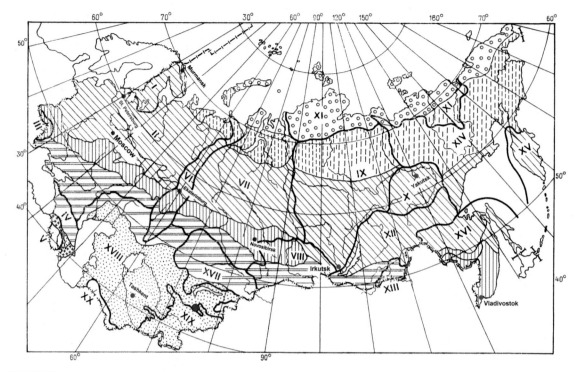

FIGURE 8.18   Place of isolation (red circle) of UZAV (family Bunyaviridae, genus *Nairovirus*) in Northern Eurasia. (See other designations in Figure 1.1.)

S-segment from ISKV. Phylogenetic analysis based on L- and M-segments placed UZAV in the lineage of ISKV (Figures 8.10–8.12).[6,7]

Vertebrate Hosts. The vertebrate host of UZAV is apparently bats, but because only a single isolation was obtained, this assertion is speculative. The finding that UZAV is a reassortant virus closely related to ISKV suggests that UZAV occupies the same ecological niche as ISKV and therefore is associated with bats and their argasid ticks. *Myotis blythii oxygnathus*, the bat from which UZAV was isolated, is common in the southern parts of the Russian Plain and in western Siberia, Caucasia, Kazakhstan, southern Europe, northern Africa, Middle and Central Asia, Iran, and Iraq. Bats are important natural reservoir of emerging viruses.[8–11] ISKV and UZAV are the first nairoviruses that appear to be associated with bats.

8.1.3.8 Sakhalin Virus and Paramushir Virus

Sakhalin virus (SAKV) has been isolated from *Ixodes (Ceratixodes) uriae* (family Ixodidae, subfamily Ixodinae) ticks, which are obligate parasites of auks (family Alcidae). The prototypical strain of SAKV (LEIV-71C) was isolated in 1969 from *I. uriae* ticks collected in a colony of the common murre (*Uria aalge*) on Tyuleniy Island near the southeastern coast of Sakhalin Island in the Sea of Okhotsk (48°29′N, 144°38′E; Figure 8.19).[1–4] Subsequently, 52 strains of SAKV were isolated from *I. uriae* ticks on Tyuleniy Island and Iona Island in the Sea of Okhotsk, the Commander Islands in the Barents Sea, and the southeastern coast of the Chukotka Peninsula in the Bering Strait (Table 8.11).[4–7] On the basis of virion morphology, SAKV has been classified into the Bunyaviridae family.

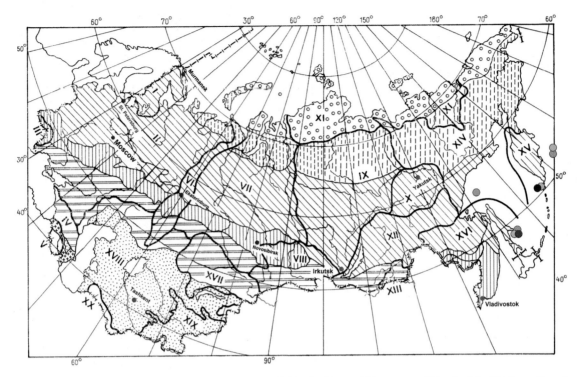

FIGURE 8.19 Places of isolation of SAKV and PMRV (family Bunyaviridae, genus *Nairovirus*) in Northern Eurasia. Red circle: strain of SAKV with completely sequenced genome; Pink circles: a number of strains of SAKV identified by serological methods; Dark-blue circle: strain of PRMV with completely sequenced genome; Light-blue circles: a number of strains of PRMV identified by serological methods. (See other designations in Figure 1.1.)

SAKV was the first of the eponymous viruses, which together have formed a basis for the *Nairovirus* genus.[8]

Paramushir virus (PMRV), prototypical strain, LEIV-2268, a virus of the SAKV group, was originally isolated from *Ixodes signatus* ticks collected in 1972 in a colony of cormorants (*Phalacrocorax pelagicus*) on Paramushir Island (in the Kuril Islands) (50°23′N, 155°41′E; Figure 8.19).[9,10] Later (in 1972–1987), 18 strains of PMRV were isolated from *I. uriae* ticks, collected in the nests of auks (family Alcidae) on Tyuleniy Island in the Sea of Okhotsk and on the Commander Islands in the Bering Sea (Table 8.11).[11–14]

At least five nairoviruses are included in the SAKV group.[3,10,15–17] Avalon virus (AVAV), which was isolated from engorged imagoes and nymphs of *I. uriae* collected in *L. argentatus*

nests on Great Island, Newfoundland, , in 1972, is apparently identical to PMRV.[15,18] Several strains of AVAV were isolated in 1979 in Cap Sizun, Brittany, France.[19] Clo Mor virus (CMV) was isolated in 1973 from nymphal *I. uriae* ticks collected in a *Uria aalge* colony of Clo Mor, Cape Wrath, Scotland.[20] CMV was found to be closely related to SAKV in a complement-fixation test. Two strains of CMV were isolated from *I. uriae* collected in seabird colonies on Lundy Island (England) and the Shiant Isles (Scotland) (Table 8.12).[18,20] Rukutama virus (RUKV) (strain LEIV-6269S), which previously had been included in the SAKV group, is now classified into the Uukuniemi virus (UUKV) group in the *Phlebovirus* genus.[9,21]

Taxonomy. Complete genomes of SAKV (strain LEIV-71C) and PMRV (LEIV-1149K)

TABLE 8.11   Isolation of SAKV and PRMV From *Ixodes* (*Ceratixodes*) *Uriae* Ticks (Obligate Parasites of Alcidae Birds) in the Basins of the Sea of Okhotsk and Bering Sea

Virus		Far East				European part
		Sakhalin District		Kamchatka	Chukotka	Murmansk District
		Tyuleniy Island (48°29′N, 144°38′E)	Iona Island (56°24′N, 143°23′E)	Ari Kamen Island (Commander Islands) (55°13′N, 165°48′E)	Bering Strait Coast (64°50′N, 173°10′E)	Kharlov Island near Kola Peninsula (68°49′N, 37°19′E)
SAKV	Number of strains	42	2	10	3	0
	% of infected ticks	0.307	0.103	0.033	0.26	—
Total	Number of strains	57				0
	Number of ticks examined	35,725				8,994
	% of infected ticks	0.160				—
PMRV	Number of strains	10	0	8	0	0
	% of infected ticks	0.073	—	0.042	—	—
Total	Number of strains	18[a]				0
	Number of ticks examined	35,725				8,994
	% of infected ticks	0.050				—

[a]*One strain was isolated from* I. signatus *ticks in a Bering cormorant colony on Paramushir Island in the Sea of Okhotsk (50°23′N, 155°41′E).*

were sequenced.[9] Also, partial sequences of RdRp of Tillamook virus (TILLV, identical to SAKV), isolated from *I. uriae* ticks on the Pacific coast (Oregon) of the United States, are available (Table 8.12).[18] A full-length genome comparison showed that SAKH and PMRV respectively share 75.6% nt and 88.0% aa identities in RdRp (L-segment), 59.7% nt and 57.9% aa in the precursor of Gn and Gc (M-segment), and 62.3% nt and 62.2% aa in the nucleocapsid protein (S-segment). SAKV N-protein ranges from 30% (CASV, HUGV) to 43% (CCHFV) similarity to other nairoviruses. The similarity of RdRp and the precursor of Gn and Gc proteins of SAKV to other nairoviruses ranges from 42.8% (CASV, HUGV) to 50.8%

TABLE 8.12   Viruses of the SAKV Group (Family Bunyaviridae, Genus *Nairovirus*) Isolated from *Ixodes* (*Ceratixodes*) *Uriae* Ticks and Penguins (*Spheniscidae*)

Virus	Place of material collection	Biome
Sakhalin (SAKV)	Sea of Okhotsk, Barents Sea (1969–1971)	Nests of common murre (*Uria aalge*), *Ixodes uriae* ticks
Paramushir (PMRV)	Sea of Okhotsk, Barents Sea (1969–1971)	Nests of common murre (*Uria aalge*) and pelagic cormorant (*Phalacrocorax pelagicus*), *Ixodes uriae* ticks
Taggert (TAGV)	Macquarie Island in southern part of Pacific Ocean (54°30′S, 159°00′E) (1972)	Nests of penguins (*Eudyptes schlegeli*), *Ixodes uriae* ticks
Clo Mor (CMV)	Cape Wrath (Scotland) (58°36′N, 04°53′W) (1973)	Nests of common murre (*Uria aalge*), *Ixodes uriae* ticks
Avalon (AVAV)	Avalon Peninsula, Newfoundland island, Labrador Province, Canada (52°46′N, 47°11′W) (1972)	Nests of common murre (*Uria aalge*), *Ixodes uriae* ticks
	France, Brittany	Nests of common murre (*Uria aalge*), *Ixodes uriae* ticks

(CCHFV), respectively, and from 25.9% (ERVEV, TFAV) to 28.9% (NSDV, DUGV), respectively.[9]

Arthropod Vectors. It has been shown that the infection rate of infected *Ixodes uriae* imagoes is 2 times higher than of the species' nymphs and 10 times higher than that of the larval stage. Transovarial transfer of SAKV has been found to be 10%. The infection rates of male and female ticks are approximately the same. The hypostome of male *I. uriae* ticks is vestigial; therefore, they cannot be infected by breeding on infected birds. The infection rate of *I. uriae* imagoes is at least 20 times higher than that of *I. signatus* imagoes.[4–6,22,23] Some other species of *Ixodes* ticks are parasites of seabirds and may be an additional reservoir of SAKV. *I. auritulus* and *I. zealandicus* ticks are distributed from Alaska to Cape Horn in South and North America.[24]

Laboratory experiments have demonstrated that *Aedes aegypti* and *Culex pipiens molestus* mosquitoes can be infected by SAKV as they suck blood. The virus was found in mosquitoes on 9, 14, and 19 days after infection in titers 1.0, 1.5, and 2.0 $\log_{10}(LD_{50})/10\,\mu L$, respectively. However, it was shown that infected mosquitoes could not transmit the virus to mice through a bite.[6,22]

Vertebrate Hosts. *Ixodes uriae* ticks and their host, the common murre (*Uria aalge*), are a natural reservoir of SAKV. Pelagic cormorants (*Phalacrocorax pelagicus*) and their obligate parasites (*I. signatus*) likely have only an additional influence. Antibodies to SAKV have been found in the common murre (*U. aalge*), pelagic cormorants (*P. pelagicus*), fulmars (*Fulmarus glacialis*), tufted puffins (*Lunda cirrhata*), and black-legged kittiwakes (*Rissa tridactyla*) in the Far East.[4–6,22] A serological examination of birds via an indirect complement-fixation test revealed that the northern boundary of the range of SAKV is the Commander Islands, where antibodies have been found in 2.2% of birds. The southernmost place where antibodies have been detected (1.1% birds) is Kunashir Island in the Kuril Islands. Antibodies were found most often (in 4.1–17.8% of birds) in the central part of the basin of the Sea of Okhotsk (on Sakhalin Island, Tyuleniy Island, and Iona Island). Antibodies were also found in the red-necked phalarope (*Phalaropus lobatus*), sanderling (*Calidris alba*), the long-toed stint (*C. subminuta*) (up to 8.4% of the population), fulmars (*F. glacialis*) (4.9%), Leach's petrels (*Oceanodroma leucorhoa*), tufted puffins (*L. cirrhata*) (4.6%), the common murre (*U. aalge*) (3.8%), Japanese cormorants (*Phalacrocorax filamentosus*) (1.0%), and black-legged kittiwakes (*R. tridactyla*) (0.6%). No antibodies were detected in other species of birds in the Alcidae family or in geese, ducks, or poultry. In the Arctic zone (Novaya Zemlya and Wrangel Island in the Arctic Ocean) and in the northern part of the subarctic (the Pacific coasts of Chukotka and the Kamchatka Peninsula), as well as on Moneron Island and Furugelm Island in the Sea of Japan, no antibodies to SAKV have been found in birds.[5,6,22]

Neutralizing antibodies to AVAV, a virus closely related to PMRV, have been found in 27.6% of puffins (*Fratercula arctica*), petrels (*Calonectris leucomelas*), and herring gulls (*Larus argentatus*) in Canada.[24,25]

Findings of antibodies to SAKV in seabirds carrying out their annual seasonal migration to the Southern Hemisphere suggest the possibility of transcontinental transfer of the virus to the Southern Hemisphere. The closely related Taggert virus (TAGV) was isolated from *Ixodes uriae* ticks in penguin colonies on Macquarie Island, a phenomenon that may indicate a transfer of viruses by birds and their ticks between the Northern and Southern Hemispheres.

Human Infection. Three human cases of cervical adenopathy associated with AVAV were described in France.[25] Serological examination of farmers in Cap Sizun, Brittany, France, found only 1% of the population positive.[18]

8.1.3.9 Tamdy Virus

History. TAMV (prototypal strain, LEIV-1308Uz) was originally isolated from *Hyalomma asiaticum asiaticum* (family Ixodidae, subfamily Hyalomminae) ticks collected from sheep in the arid landscape near the town of Tamdybulak (41°36′N, 64°39′E; Figure 8.20) in the Tamdinsky

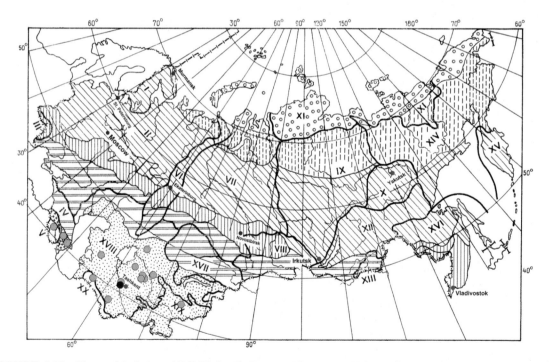

FIGURE 8.20 Places of isolation of TAMV (family Bunyaviridae, genus *Nairovirus*) in Northern Eurasia. Red circle: strain of TAMV with completely sequenced genome; Pink circles: strains of TAMV identified by serological methods. (See other designations in Figure 1.1.)

district of the Bukhara region of Uzbekistan in 1971.[1−3] Subsequently 52 strains of TAMV were isolated in Uzbekistan,[4−7] Turkmenistan,[8−11] Kyrgyzstan,[12,13] Kazakhstan,[11,14,15] Armenia,[6,16] and Azerbaijan[8,17−19] in 1971−1983 (Table 8.13). Most of the strains were obtained from *H. asiaticum* ticks, but several were isolated from birds, mammalians (including bats), and sick humans. On the basis of virion morphology, TAMV has been classified into the Bunyaviridae family. Serological studies by complement-fixation and neutralization tests revealed no antigenic relationships of TAMV with any known viruses.[2]

Taxonomy. Three strains of TAMV isolated in Uzbekistan (LEIV-1308Uz), Armenia (LEIV-6158Ar), and Azerbaijan (LEIV-10226Az) were completely sequenced.[20] Phylogenetic analysis of the full-length sequences showed that TAMV is a novel member of the *Nairovirus* genus, forming a distinct phylogenetic lineage (Figures 8.10−8.12). The similarity of the amino acid sequence of TAMV RdRp (L-segment) with those of other nairoviruses is 40% aa, on average. The similarity of the RdRp of TAMV with that of the nairoviruses associated predominantly with ixodid ticks (CCHFV, Hazara virus (HAZV), and DUGV) is higher (40% aa) than that with viruses associated with argasid ticks (ISKV and CASV) (38% aa). The similarity of the TAMV polyprotein precursor of Cn and Gc with that of other nairoviruses is less than 25% aa. The similarity of the amino acid sequence of the nucleocapsid protein (S-segment) of TAMV is 33% aa with ixodid nairoviruses and 28% aa with argasid nairoviruses. Phylogenetic analysis of the catalytic core domain of the RdRp of the nairoviruses confirms that TAMV forms a novel group in the *Nairovirus* genus (Figures 8.10−8.12).[20]

Genetic diversity among the three sequenced strains of TAMV is low. The prototypic strain LEIV-1308Uz, isolated in central Asia, has 99% nt identity in the L-segment with LEIV-10226Az from Transcaucasia. The L-segment of the strain LEIV-6158Ar has 94.2% nt and 96.3% aa identity with the L-segment of LEIV-1308Uz. The similarity of the M-segment of LEIV-1308Uz with those of LEIV-10226Az and LEIV-6158Ar is 93% nt and 89% aa, respectively. The similarity of the S-segment among the three strains is 93−95% nt.[20]

Arthropod Vectors. *H. asiaticum* ticks are apparently a main reservoir of TAMV. More than half (57%) of TAMV isolations were obtained from *H. asiaticum asiaticum* ticks, 6% from *H. asiaticum*, 8% from *H. anatolicum*, 6% from *H. marginatum*, 6% from *Rhipicephalus turanicus*, and 2% from *Haemaphysalis concinna*. The infection rates of male and female ticks in endemic territory were 1:210 and 1:200, respectively. The infection rate of *H. asiaticum* nymphs was 20 times lower.[7,10,14,16] Furthermore, TAMV was isolated from larvae of *H. asiaticum*, which were hatched from eggs in the laboratory, indicating transovarial transmission of the virus. *H. asiaticum asiaticum* ticks are the most xerophilous subspecies of the *Hyalomma* genus (Ixodinae subfamily),[21] a characteristic that allows TAMV to be distributed over the Karakum desert in Turkmenistan, the Moinkum desert in Kazakhstan, and the central part of the Kyzyl Kum desert in Kazakhstan and Uzbekistan.[7].

Animal Hosts. The larvae of *H. asiaticum* feed on ruminants, hoofed animals, small predators, hedgehogs, birds, and reptilians. One of the major hosts of *H. asiaticum* preimagoes is the great gerbil (*Rhombomys opimus*). Wild animals, as well as sheep and camels, are the hosts for *H. asiaticum* imagoes and may be involved in the circulation of TAMV (Table 8.13).

Human Pathology. Sporadic cases of the disease associated with TAMV was registered in Kyrgyzstan in October 1973, when TAMV was isolated from the blood of a patient with fever (39°C), headache, arthralgia, and weakness.[16] *H. asiaticum asiaticum* ticks rarely attack humans, and no outbreaks of TAMV fever have been registered; however, human infection by *H. asiaticum* ticks is still possible

TABLE 8.13   Isolations of TAMV (Family Bunyaviridae, Genus *Nairovirus*)

Region	Country	Location	Biotope	Source of isolation	Date of collection	Number of strains isolated
Central Asia	Uzbekistan	Buhara province, near Tamdy village	Sandy desert	*H. as. asiaticum* ticks	August 1971	3
					April 1972	6
					April 1973	1
					May 1974	1
					May 1983	1
					Total	**12**
	Turkmenistan	Near Karakum kanal (Sakar chaga village), Zahmet village, Sarygamysh lake)	Sandy desert	*H. as. asiaticum* ticks from camel	январь–May 1973	4
				H. marginatum ticks from sheep	June 1973	1
				H. as. asiaticum ticks from camel	June 1973	1
				H. as. asiaticum ticks from camel	Jule 1981	1
		Kyzyl Arvat village	Foothill desert	*H. as. asiaticum* ticks from sheep	April 1984	1
					Total	**8**
	Kyrgyzstan	Chu valley	Near desert	Human	May 1973	1
				Bat sp. (*Chiroptera*)	May 1973	1
				Pied wagtail (*Motacilla alba* Linnaeus, 1758)	May 1973	2
				European roller (*Coracias garrulus* Linnaeus, 1758)	May 1973	1
				Hoopoe (*Upupa epops* Linnaeus, 1758)	May 1973	1
				Starling (*Sturnus vulgaris* Linnaeus, 1758)	May 1973	1
				Red-tailed shrike (*Lanius meridionalis* Temminck, 1820)	May 1973	1
				Steppe polecat (*Mustela eversmanni* Lesson, 1827)	May 1973	1
				Rh. turanicus	May 1973	3
				Haem. concinna	May 1973	1
					Total	**13**

(Continued)

TABLE 8.13   (Continued)

Region	Country	Place of isolation Location	Biotope	Source of isolation	Date of collection	Number of strains isolated
Central Asia (*Continued*)	Kazakhstan	Suzak district	Near desert	*H. as. asiaticum* ticks from sheep	April 1979	1
		Kazaly district		*H. as. asiaticum* ticks from cows	April 1979	1
		Aral district		*H. as. asiaticum* ticks from camel	April 1979	1
		Kzyl-Orda district		*H. as. asiaticum* ticks from camel	May 1979	5
					Total	8
Transcaucasia	Armenia	Ashtrac district	Rocky desert	*H. as. caucasium* ticks from sheep	May 1976	1
					Total	1
	Azerbaijan	Qusar district	Near desert	*H. as. asiaticum* ticks from sheep	April 1985	1
		Apsheronsky district		*H. marginatum* ticks from sheep	May 1985	2
				H. as. caucasium ticks from sheep	May 1985	2
				H. anatolicum ticks from sheep	May 1985	4
				H. as. asiaticum ticks from sheep	May 1986	1
					Total	10

through contact with livestock necessitated by economic activities (e.g., sheep shearing).

8.1.3.10 Burana Virus (BURV)

History. The Prototypical strain LEIV-Krg760 of Burana virus (BURV) was originally isolated from *Haemaphysalis punctata* (family Ixodidae, subfamily Haemaphysalinae) ticks collected from cows in Tokmak Wildlife Sanctuary in the eastern part of the Chu valley (43°10′N, 74°40′E; Figure 8.21) in the foothills of the Kyrgyz Ala-Too Range near the village of Burana, Kyrgyzstan, in April 1971. Six strains of virus were isolated from 9,377 ticks of the species *Haem. punctata* and *Haem. concinna* (Ixodidae, Haemaphysalinae) during 1971–1975.[1,2] According to preliminary information, BURV is not able to agglutinate erythrocytes of birds and mammals and has no antigenic relationships with 59 arboviruses from different groups of the Togaviridae,

Flaviviridae, Bunyaviridae, Reoviridae, Orthomyxoviridae, and Arenaviridae families or with 35 unclassified viruses as well.

Taxonomy. The genome of BURV was sequenced, and the virus was classified into the *Nairovirus* genus, family Bunyaviridae. The genome consists of three segments: an L-segment (ORF, 11,919 nt; encodes RdRp); an M-segment (ORF, 4,035 nt; encodes a polyprotein precursor of the envelope proteins Gn and Gc); and an S-segment (ORF, 1,482 nt; encodes the nucleocapsid protein N).[3,4]

A comparison of RdRp sequences of BURV with those of other nairoviruses demonstrated that the virus is distantly related to TAMV (59% aa similarity). The similarity of the RdRp catalytic core domain of BURV to that of TAMV is 82% aa, compared with about 60% aa for viruses in other phylogenetic groups.

The level of similarity for the nucleotides sequences of this part of the RdRp of BURV is 68% nt with those of TAMV and 45−50% nt with those of other viruses (Figure 8.10).[3]

The M-segment of BURV has a long ORF and encodes a polyprotein precursor of the envelope glycoproteins Gn and Gc.[4] The size of the polyprotein precursor is 1,344 aa. The mature Gn and Gc proteins of nairoviruses are formed by complex processes involving cellular peptidases. By the NetNGlyc 1.0 server, 11 potential glycosylation sites were predicted, with only 5 within mature Gn or Gc proteins.[5,6] The level of similarity of the amino acid precursor of Gn and Gc in BURV is 45% with that of TAMV and no more than 27% with viruses of other phylogenetic groups. Phylogenetic analyses based on a comparison of the full-length polyprotein precursor

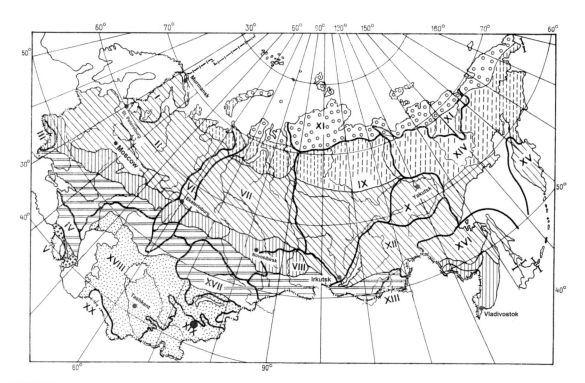

FIGURE 8.21   Place of isolation (red circle) of BURV (family Bunyaviridae, genus *Nairovirus*) in Northern Eurasia. (See other designations in Figure 1.1.)

demonstrated the position of BURV on the TAMV branch and was consistent with the RdRp data (Figure 8.11).[3]

The S-segment of nairoviruses encodes a nucleocapsid protein (N).[4,7] The size of the BURV nucleocapsid protein is 493 aa, corresponding to the average size of the N protein of other nairoviruses (480–500 aa). The level of similarity of the amino acid sequence of BURV N protein with that of TAMV is 44%, and that with the amino acid sequences of other nairoviruses is30–32%. Phylogenetic analyses of BURV N protein are represented in Figure 8.12. The phylogenetic position of BURV is on the TAMV branch, despite the virus's having the lowest level of similarity of the N protein compared with that of other virus proteins.

Arthropod Vectors. As mentioned earlier, six strains of BURV were isolated from the ticks *Haemaphysalis punctata* (five strains) and *Haem. concinna* (one strain) in 1971–1975. The rate of infected ticks was 2.2–2.6%. BURV is associated with *Haem. punctata* and *Haem. concinna* ticks in pasture biocenoses. The virus is phylogenetically close to TAMV, which is also associated with ixodes ticks in pasture and desert biocenoses.[8]

8.1.4 Genus *Orthobunyavirus*

The *Orthobunyavirus* genome consist of three segments of single-stranded negative-sense RNA designated as large (L), medium (M), and small (S) (Figure 8.22).[1] The L-segment of the prototypical BUNV (6,875 nt in length) encodes the viral RdRp.[2] The M-segment (4,458 nt) encodes two surface glycoproteins (Gn and Gc) and a nonstructural protein (NSm).[3,4] The S-segment (961 nt) encodes the nucleocapsid protein (N) and a nonstructural protein (NSs). The NSs protein is considered a pathogenic factor for vertebrates, because it may act as an antagonist of interferon, which is

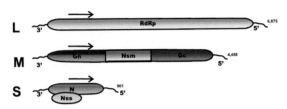

FIGURE 8.22 Scheme of genome organization of BUNV, a prototypical virus of the *Orthobunyavirus* genus. *Drawn by Tanya Vishnevskaya.*

involved in blocking the host's innate immune responses.[5–7] The genus *Orthobunyavirus* is subdivided into multiple serological groups.[8]

8.1.4.1 *Batai Virus and Anadyr Virus*

Batai virus (BATV) belongs to the BUNV group (family Bunyaviridae, genus *Orthobunyavirus*), which to date includes 22 viruses (Table 8.14). Among them the African Ilesha virus (ILEV), as well as the North American Northway virus (NORV), Cache Valley virus (CVV), and Tensaw virus (TENV), are the viruses most closely related to BATV.[1–3]

History. BATV was originally isolated by B. Elisberg (US Army Medical Research Unit) from *Culex gelidus* collected in 1988 on the outskirts of Kuala Lumpur, Malaysia.[4,5] BATV is identical to Chalovo virus, isolated in 1960 from *Anopheles maculipennis* (*An. messeae*) mosquitoes collected in Slovakia;[6,7] to Olyka virus, isolated in 1973 from *An. maculipennis* mosquitoes collected in western Ukraine;[8–11] and to Chittoor virus, isolated in 1957 from *An. barbirostris* mosquitoes collected in Brahmanpally, Chittoor district, Andhra Pradesh state, India.[12] The African Ngari virus (NRIV) is reassortant between BATV and BUNV.[12,13] In Russia, BATV was repeatedly isolated in different regions (Figure 8.23).

Anadyr virus (ANADV), strain LEIV-13395, was isolated by S.D. Lvov from a pool of *Aedes* mosquitoes collected in September 1986 in a swamp tundra landscape near the village of

TABLE 8.14 Viruses of the Bunyamwera Group (Family Bunyaviridae, Genus *Orthobunyavirus*)

| | Distribution | | | | | | Mosquito vector | Vertebrate hosts | | | | Disease of humans | |
| | America | | | | | Northern | | | Mammals | | | Clinical | |
Virus	Africa	North	South	Asia	Europe	Eurasia		Birds	Wild	Domestic	Humans	symptom	Epidemiology
Batai virus (BATV)	+	−	−	+	+	+	+	+	A	A	+	F	S, O
Birao virus (BIRV)	+	−	−	−	−	−	+	−	−	−	−	−	−
Bozo virus (BOZOV)	+	−	−	−	−	−	+	−	A	A	−	−	−
Bunyamwera virus (BUNV)	+	−	−	−	−	−	+	A	A	A	+	E	S, O
Fort Sherman virus (FSV)	−	+	−	−	−	−	+	−	−	−	+	F	S
Germiston virus (GERV)	+	−	−	−	−	−	+	−	+	A	+	E	S
Iaco virus (IACOV)	−	−	+	−	−	−	+	−	−	−	−	E	S
Ilesha virus (ILEV)	+	−	−	−	−	−	+	−	−	−	+	F	S
Cache Valley virus (CVV)	−	+	+	−	−	−	+[a]	−	+	+	A	F	S
Lokern virus (LOKV)	−	+	−	−	−	−	+[b]	+	+	A	A	F	S
Maguari virus (MAGV)	−	−	+	−	−	−	+	−	A	+	A	F	S
Mboke virus (MBOV)	+	−	−	−	−	−	+	−	−	−	−	−	−
Ngari virus (NRIV)	+	−	−	−	−	−	+	−	−	−	−	−	−
Northway virus (NORV)	−	+	−	−	−	−	+	−	−	−	A	F	S
Playas virus (PLAV)	−	−	+	−	−	−	+	−	−	−	−	−	−
Potosi virus (POTV)	−	+	−	−	−	−	+	−	−	−	−	−	−
Santa Rosa virus (SARV)	−	−	+	−	−	−	+	−	+	+	−	−	−
Tensaw virus (TENV)	−	+	−	−	−	−	+	A	+	+	+	F	S
Tlacotalpan virus (TLAV)	−	+	−	−	−	−	+	−	−	A	A	F	S
Tucunduba virus (TUCV)	−	−	+	−	−	−	+	−	−	−	−	F	S
Shokwe virus (SHOV)	+	−	−	−	−	−	+	−	+	+	−	F	S
Xingu virus (XINV)	−	−	+	−	−	−	+	−	−	−	+	E	S

[a]Designations: +, virus isolation; A, specific antibodies detection; F, fever; E, encephalitis, meningoencephalitis; S, sporadic cases; O, outbreak.
[b]Also isolated from midges.

Krasneno (64°37'N, 174°46'E) in the Anadyr District of the Chukotka Autonomous Okrug, Russia (Figure 8.23). On the basis of weak serological relationships, V.L. Gromashevsky classified the virus as a Batai-like virus. Sequence analysis then revealed that strain LEIV-13395 was a novel representative virus in the Bunyamwera group, and the virus was designated ANADV.[14]

Taxonomy. Phylogenetic analysis (Figures 8.24–8.26) revealed that the different strains of BATV can be divided into three groups according to their geographic spread: strains from China, Malaysia, Japan, and India form the Asian group; strains from Uganda and NRIV (to which BATV is considered to be a donor of the M-segment) belong to the African group; and strains isolated in Italy, Germany, the former Czechoslovakia, western

Ukraine, and Russia are members of the European group. Two strains of BATV—LEIV-Ast04-2-315 and LEIV-Ast04-2-336—isolated in Russia were completely sequenced and placed into the cluster of the European strains.[14] Within this group, they are phylogenetically close to strain 42, isolated in the Volgograd region in 2003 from *Anopheles messeae* (*maculipennis*) mosquitoes, for which the partial nucleotide sequences of the L- and M-segments are known. Between the strains LEIV-Ast04-2-315 and LEIV-Ast04-2-336, there is very high level of nucleotide and amino acid identity of three segments of the genome: 99.6/99.0% (L-segment/RdRp), 99.9/100.0% (M-segment/polyprotein predecessor), and 99.7/100.0% (S-segment/nucleocapsid). The levels of nucleotide identity of strain 42 with these strains on partial sequences of L- and M-segments are

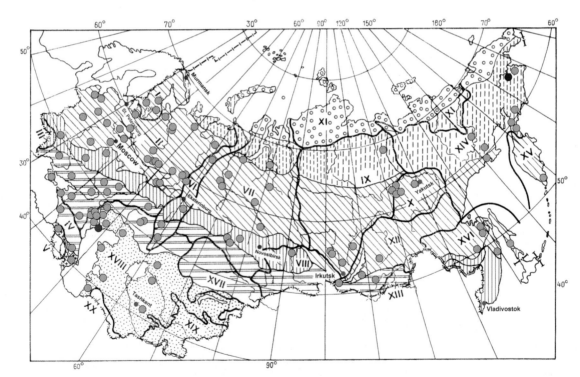

FIGURE 8.23 Places of isolation of viruses from the Bunyamwera group (family Bunyaviridae, genus *Orthobunyavirus*) in the former USSR. Red circle: strains of BATV with completely sequenced genome; Pink circles: strains of BATV identified by serological methods; Dark-blue circle: ANADV. (See other designations in Figure 1.1.)

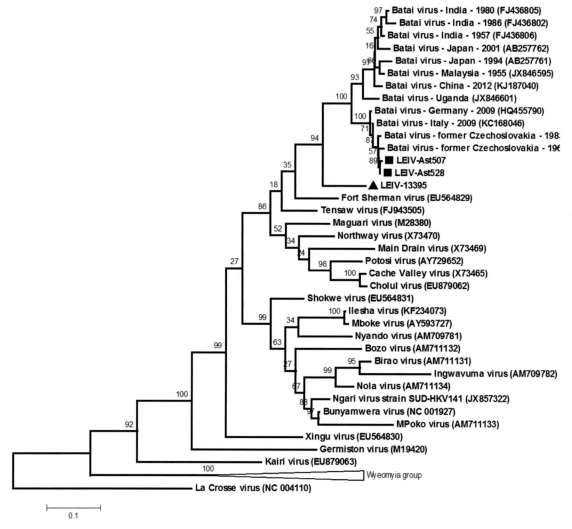

FIGURE 8.24 Phylogenetic analysis of S-segment of Bunyamwera group viruses (family Bunyaviridae, genus *Orthobunyavirus*). The trees were constructed by the maximum-likelihood method with thousandfold bootstrap analysis.

98.6/98.8% and 100/100%, respectively; that is, for the M-segment, all available nucleotide polymorphisms are synonymous. The lowest observed genetic differences and the temporal and geographical proximities of the various strains of these viruses suggest a common origin as different isolates of the same strain of BATV circulating in the southern part of Russia.

Phylogenetic analysis of ANADV (strain LEIV-13395) revealed its similarity to BATV. The L-segment of ANADV is from 76.5% to 79.7% identical with those of the different BATV strains (Figure 8.26, Table 8.15). The identity of the L-segment of ANADV with the L-segments of other viruses of the Bunyamwera group is 73.5% (BUNV), 74.1% (CVV), and 73.9% (TENV). The amino acid

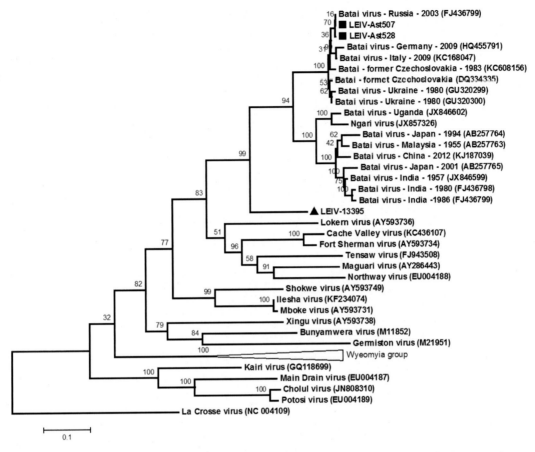

FIGURE 8.25 Phylogenetic analysis of M-segment of Bunyamwera group viruses (family Bunyaviridae, genus *Orthobunyavirus*). The trees were constructed by the maximum-likelihood method with thousandfold bootstrap analysis.

sequence of RdRp of ANADV is 91% similar to the various strains of BATV. The M-segment sequence of ANADV has the greatest similarity (76.9—80.8% nt and 83.1—84.4% aa) to the different strains of BATV. The identity of the S-segment sequence of ANADV with different strains of BATV ranges from 85.8% to 86.9% and is about 82.5% with TENV and CVV. The amino acid similarity of the nucleocapsid protein is 98.7% with that of BATV from Uganda.

Phylogenetic analysis of the nucleotide sequences of the S-, M-, and L-segments conducted with the use of a maximum-likelihood algorithm placed ANADV (LEIV-13395) on a distinct branch of the dendrogram that considers it a new representative of the Bunyamwera group.

Arthropod Vectors. BATV has been reported in Sudan, Africa.[15] The distribution of BATV in southeastern Asia includes Malaysia, India, Sri Lanka, Thailand, Cambodia, and Japan,[5,16] while in Europe BATV is distributed over Austria, Germany, Yugoslavia, Moldova, Ukraine, Belarus, and other countries.[2,17–19] In central Europe, BATV was isolated from *Anopheles claviger*, *An. maculipennis* (*An. messeae*), *Coquillettidia*

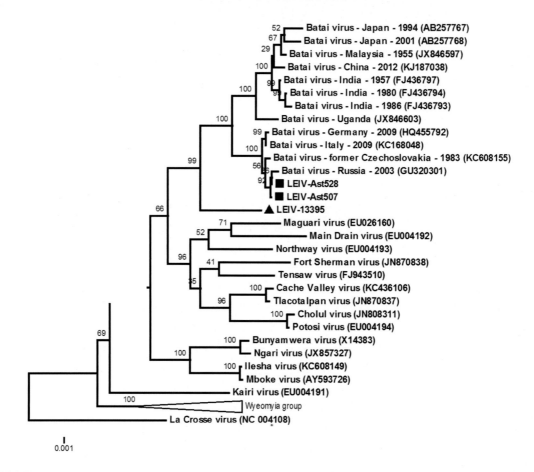

FIGURE 8.26 Phylogenetic analysis of L-segment of Bunyamwera group viruses (family Bunyaviridae, genus *Orthobunyavirus*). The trees were constructed by the maximum-likelihood method with thousandfold bootstrap analysis.

richiardii, Aedes (Ochlerotatus) punctor, and *Ae. communis.*[6,7,20] A wide distribution of BATV in different landscape belts of the European part of Russia, as well as in Siberia and the Far East, was demonstrated: In the temperate belt the main source of BATV isolation was the zoophilic *Anopheles* genus, whereas in high latitudes (tundra, northern taiga) it was the *Aedes* genus.[18,21–24]

In the European part of Russia, BATV has been isolated in the northern (Komi Republic), middle (Vologda region), and southern (Leningrad, Yaroslavl, and Vladimir regions;

Udmurt Republic); in the taiga and in deciduous forests (Kaliningrad, Nizhny Novgorod, Moscow, Smolensk, and Kaluga regions; Tatarstan Republic, Chuvash Republic, Mordovia Republic, Mari El Republic, and Bashkortostan Republic); and in forest—steppe and steppe landscape belts (practically everywhere; the Voronezh, Belgorod, Ulyanovsk, Kursk, Rostov, Volgograd, Saratov, Penza, and Lipetsk regions, as well as Stavropol Krai and Krasnodar Krai, are representative). BATV is widespread in western Siberia, from the northern taiga to the forest—steppe

TABLE 8.15 Pairwise Comparison of Full-Length Nucleotide and Amino Acid Sequences of the Viruses of the Bunyamwera Serogroup

L-SEGMENT (RDRP)[a]

Virus	LEIV-13395	BATV	BUNV	CVV	TENV	WYOV	LACV
LEIV-13395		90.6	81.4	83.1	83.0	67.8	55.3
BATV	79.5		81.6	83.1	83.2	68.1	56.1
BUNV	73.5	73.5		82.1	81.6	67.3	56.6
CVV	74.1	74.7	73.0		87.9	67.3	56.1
TENV	73.9	74.2	72.9	77.0		67.0	55.7
WYOV	67.5	67.3	66.9	66.7	66.4		56.1
LACV	60.9	59.3	59.3	58.7	59.1	60.3	

M-SEGMENT (CN/GC POLYPROTEIN PRECURSOR)[b]

LEIV-13395		84.2	64.0	75.2	74.0	51.2	41.3
BATV	77.5		64.3	75.5	74.5	51.4	42.0
BUNV	65.8	64.4		64.2	64.0	51.2	41.6
CVV	70.5	70.8	63.9		80.6	51.9	41.3
TENV	71.2	69.9	64.4	74.0		52.0	41.6
WYOV	58.8	58.0	58.0	57.3	58.7		40.7
LACV	52.9	53.0	52.1	53.0	53.4	52.9	

S-SEGMENT (NUCLEOCAPSID PROTEIN)[c]

LEIV-13395		97.8	93.1	94.0	95.3	62.7	42.9
BATV	86.3		92.6	93.9	94.3	62.6	43.5
BUNV	80.3	81.3		90.6	93.1	63.5	42.9
CVV	82.6	83.9	81.2		95.3	63.9	44.2
TENV	82.5	83.2	80.9	84.7		63.5	44.6
WYOV	56.8	57.2	58.5	58.4	58.3		48.5
LACV	51.3	51.2	49.0	50.3	49.9	51.8	

Nucleotide identities (%) are shown below diagonal. Amino acid similarity (%) is shown above diagonal.

[a]GenBank numbers of RdRp amino acid sequences: Batai virus BATV (AGM40000), Bunyamwera virus BUNV (P20470), Cache Valley virus CCV (KC436106), Tensaw virus TENV (FJ943510), Wyeomyia virus WYOV (JN801038), La Crosse virus LACV (NC_004108). GenBank numbers of nucleotide sequences of L segments: Batai virus BATV (KC168048), Bunyamwera virus BUNV (X14383), Cache Valley virus CCV (AGI03946), Tensaw virus TENV (ACV95629), Wyeomyia virus WYOV (AEZ35278), La Crosse virus LACV (NP_671968).

[b]GenBank numbers of Gn and Gc polyprotein amino acid sequences: Batai virus BATV (AGM39999), Bunyamwera virus BUNV (AAA42777), Cache Valley virus CCV (AAF33115), Tensaw virus TENV (ACV95627), Wyeomyia virus WYOV (AEZ35274), La Crosse virus LACV (NP_671969). GenBank numbers of nucleotide sequences of M segments: Batai virus BATV (KC168047), Bunyamwera virus BUNV (M11852), Cache Valley virus CCV (KC436107), Tensaw virus TENV (FJ943508), Wyeomyia virus WYOV (JN572081), La Crosse virus LACV (NC_004109).

[c]GenBank numbers of N protein amino acid sequences: Batai virus BATV (ADX97411), Bunyamwera virus BUNV (NP_047213), Cache Valley virus CCV (ADG62277), Tensaw virus TENV (ACV95624), Wyeomyia virus WYOV (AEZ35279), La Crosse virus LACV (NP_671970). GenBank numbers of nucleotide sequences of S segments: Batai virus BATV (KC168046), Bunyamwera virus BUNV (NC_001927), Cache Valley virus CCV (X73465), Tensaw virus TENV (FJ943505), Wyeomyia virus WYOV (FJ235921), La Crosse virus LACV (NC_004110).

(Tyumen region); in eastern Siberia, from the Arctic Ocean shoreline (Sakha–Yakutia Republic) to the middle and southern taiga (Irkutsk region, Sakha–Yakutia Republic, Republic of Buryatia); and in the Far East, in tundra (Magadan region, Chukotka Autonomous Okrug) and in northern and southern taiga and mixed forests (Khabarovsk Krai and Primorsky Krai).[18,21–24]

BATV is transmitted to vertebrates only by mosquitoes: in southeastern Asia (*Culex gelidus*, *Cx. bitaeniorhynchus*, *Anopheles subpictus*, *An. tessellatus*, *Aedes vexans*); in Czechoslovakia (*An. messeae*, *Ae. punctor*); and in Belarus, Ukraine, and other European countries (*An. Messeae*).[2,6,17,18,20] In the southern hyperendemic regions of Russia, the main vector of BATV is *An. messeae*. According to our data, the infection rate of *An. messeae* in the middle belt of the Volga delta (Astrakhan region) reaches 0.188% (approximately 1 infected mosquito out of 500). Because this species of mosquito attacks mainly domestic animals, it serves as a biological barrier, reducing risk of infection to humas. In the northern areas (the subarctic, the northern taiga), BATV circulation is due mainly to *Aedes* mosquitoes: *Ae. communis* complex and *Ae. punctor*. Under experimental conditions, BATV was isolated from hibernating females of *An. messeae*. Hibernation is one of the mechanisms by which BATV survives during the winter.[20,25]

Vertebrate Hosts. In anthropogenic biocenoses of the southern regions of Russia, domestic animals are the main vertebrate reservoir, because they (especially cattle) are the main hosts for *An. messeae*. BATV-neutralizing antibodies were found in India among rodents (*Mus cervicolor* (55.2%), *Rattus exulans* (36.4%), *Rattus rattus* (19.5%), *Bandicota indica* (15.5%)) and bats (*Cynopterus sphinx*) (2.6%).[2,5] This indicator is significantly higher in India among domestic animals: goats (41.8%), camels (100%), cows (60.9%), and buffalos (23.3%). In Finland, anti-BATV antibodies occasionally were found among cows (0.9%), but not

among reindeers.[19] The Chittoor strain is associated with mild illness, but is pathogenic to sheep and goats.[12] BATV was isolated from birds: crows (*Corvus corone*), coots (*Fulica atra*), and grey partridges (*Perdix perdix*).[9] Persistent avian infection was established experimentally with reactivation of viremia by cortisone six months after the acute infection period.[10]

An investigation of 5,000 sera of domestic animals in Russia during 1982–1992 revealed anti-BATV antibodies among these animals significantly more often than among people (Table 8.16). The largest immune layer was found in populations of horses (up to 80%), cattle (35–60%), sheep (up to 80%), and camels in forest–steppe, semidesert, and desert landscape belts. In contrast to the situation in Finland, antibodies were found in reindeer sera in a tundra landscape belt of the Chukotka Peninsula. No examinations of vertebrates in natural biocenoses were conducted.

Epidemiology. Epidemic outbreaks and sporadic cases caused by BATV, as well as outbreaks of hemorrhagic infection caused by Ngari virus, have been reported.[13,15,18,26,27] To date, no cases of laboratory infection are known. According to a serological examination of 10,000 people in the endemic regions of Russia, about 3–10% withstand BATV infection in an asymptomatic form. The highest infection rate was established in forest–steppe and steppe belts. (However, as a rule, the rate is higher for domestic animals than humans.) Some northern areas in Russia became hyperendemic for no apparent reason.[18,21,22]

Pathogenesis. No pathogenetic mechanism during BATV infection in humans has yet been described in detail. There are experimental data, however, on BATV infection in primates:[28] Green monkeys (*Chlorocebus sabaeus*) were found to be carriers of the virus 50 days after inoculation (the observation period); the virus was pantropic, destroying small vessels and producing vasculitis and perivascular focal lymphohistiocytic infiltrates.

TABLE 8.16   Detection of BATV-Neutralizing Antibodies Among Domestic Animals in Russia (1982−1992)[a]

| Territory | Federal subject | Number of sera tested | | |
		Total	Positive Number	%
Northern	Komi Republic, Karelia Republic, Vologda, Murmansk, Arkhangelsk regions	634/566	4/2	0.6/0.4
		2,171/180[b]	257/37	11.8/20.6
Northwestern	Leningrad, Novgorod, Pskov regions	206/0	1/−	0.4/−
Central	Ivanovo, Kostroma, Ryazan regions	285/0	2/−	0.3/−
	Vladimir, Smolensk, Tverskaya, Tula, Kaluga regions	357/0	5/−	1.4/−
		274/0	8/−	2.2/−
	Moscow, Bryansk, Oryol regions	228/0	10/−	4.4/−
		59/0	6/−	10.2/−
Central Chernozemny	Tambov, Kursk, Voronezh, Belgorod, Lipetsk regions	350/0	5/−	1.4/−
		296/32[b]	15/15	5.1/40.9
North Caucasian	Krasnodar Krai, Kabardino-Balkaria Republic, Chechen Republic	212/0	4/−	1.9/−
Volga	Penza, Ulyanovsk, Samara, Saratov, Astrakhan, Volgograd regions, Tatarstan Republic	466/70	2/13	0.4/18.6
		370/159[b]−24[c]−30[d]	9/36−13[c]−12[d]	2.4/22.6−54.2[c]−40.0[d]
Volga-Vyatka	Nizhny Novgorod, Kirov regions, Chuvashia Republic, Mordovia Republic, Udmurtia Republic	488/162[b]	8/57	1.6/35.2
		85/174[b]	50/56	58.8/32.2
Ural	Bashkortostan Republic, Perm, Orenburg regions	286/256[b]	3/9	1.0/3.5
		116/56[b]	12/6	10.3/10.7
Western Siberian	Altai Krai	125/0	0/−	0/−
Eastern Siberian	Irkutsk, Chita regions, Krasnoyarsk Krai, Sakha-Yakutia Republic, Buryatia Republic	1,760/1,360[b]	15/19	0.8/1.4
		40/1,845[b]	3/63	1.5/3.4
Far Eastern	Magadan, Sakhalin regions, Khabarovsk Krai, Kamchatka Krai	486/133[b]−62[e]	7/0−5[e]	1.4/0−8.1[e]
		460/0	3/−	0.7/−

[a]Human/domestic animal.
[b]Cattle.
[c]Sheep.
[d]Camels.
[e]Reindeers.

Clinical Features. The disease etiologically linked with BATV proceeds mainly as influenzalike disease complicated by meningitis, malaise, myalgia, and anorexia.[13,15,18,26,27] At the same time, Ngari virus (reassortant between BATV and BUNV) infection in east Africa appears as outbreaks of hemorrhagic fever.[13] Diseases associated withtheclosely related ILEV in Africa and Madagascar also proceed with hemorrhagic phenomena and with lethal outcomes.[29,30]

Diagnostics. A highly specific test based on RT-PCR has been developed, as have ELISA tests for the detection of specific anti-BATV IgM and IgG.[24,31]

8.1.4.2 California Encephalitis Complex Viruses: Inkoo Virus, Khatanga Virus, Tahyna Virus

California encephalitis (CE) and related diseases are etiologically linked with 13 currently known viruses of the CE serocomplex (family Bunyaviridae, genus *Orthobunyavirus*). Eight viruses from the CE serocomplex are distributed in North America, three in South America, three in Eurasia, and one in Africa (Table 8.17).[1,2]

History. The prototypical member of the CE serocomplex, California encephalitis virus (CEV), was originally isolated in 1943 by W.M. Hammon from mosquitoes of the genus *Aedes*. Tahyna virus (TAHV) was isolated in 1958 by V. Bardos and V. Danielova from *Ae. vexans* and *Ae. Ochlerotatus* mosquitoes collected in the vicinity of the village of Tahyna in eastern Slovakia. (TAHV is identical to Lumbo virus,[3] isolated in 1960 in Mozambique in the southeastern part of Africa.) Snowshoe hare virus (SSHV) was isolated in 1959 by W. Burgdorfer from the snowshoe hare (*Lepus americanus*). La Crosse virus (LACV) was isolated in 1960 by W.H. Thompson from the brain of a patient who died as the result of meningoencephalitis. Jamestown Canyon virus (JCV) was isolated in 1961 by L.C. La Motte et al. from the *Culiseta*

inornata mosquito. Inkoo virus (INKV) was isolated in 1964 by M. Brummer-Korvenkontio from *Ae. communis* and *Ae. punctor* mosquitoes in the south of Finland. Khatanga virus (KHTV) was isolated in 1982 from *Aedes* mosquitoes by D.K. Lvov.[2,4−10]

Genome and Taxonomy. The genome of the CE group of viruses consists of three segments of ssRNA with negative polarity. The L-segment of LACV, a prototypical virus of the group, is 6,980 nt in length, the M- and S-segments 4,527 and 984 nt, respectively. As in other bunyaviruses, the L-segment encodes RdRp, the M-segment a polyprotein precursor of the envelope glycoproteins Gn and Gc, and the S-segment nucleocapsid protein (N). Two nonstructural proteins are found in infected cells: NSs, which encodes by adding an ORF in the S-segment; and NSm, which forms during the maturation of the Gn and Gc proteins from the precursor.[11] Phylogenetic relationships of viruses of the CE serogroup are presented in Figures 8.27−8.29.

Arthropod Vectors. On the territories located to the West of Russia, arthropod vectors of CE serocomplex viruses are mosquitoes (subfamily Culicinae): *Aedes vexans, Ae. trivittatus, Ae. triseriatus, Ae. dorsalis, Ae. caspius, Ae. cantans, Ae. punctor, Ae. communis, Ae. flavescens, Ae. excrucians, Culiseta annulata, Culex modestus, Cx. pipiens, Anopheles hyrcanus, An. crucians,* and *An. punctipennis.*[4,12−24] *Ae. triseriatus* is the main vector in North America,[25,26] but viruses of the CE serocomplex were isolated from *Ae. albopictus* (a known vector for at least 22 arboviruses), which was imported from southeastern Asia and spread into 30 states of the United States.[27,28] Transovarial transmission was established in *Ae. vexans*[29] and *Cs. annulata.*[16] Overwintering of TAHV was documented in *Cx. modestus* and *Cs. Annulata* females.[16]

Mosquito species have been defined and classified only partially in connection with the huge volume of this laborious work.

TABLE 8.17 Viruses of the CE Serocomplex[a]

| Virus | Distribution | | | | | | Mosquito vector | Vertebrate hosts | | | | Disease of humans | |
| | Africa | America | | Asia | Europe | Northern Eurasia | | Birds | Mammals | | | Clinical symptoms | Epidemiology |
		North	South						Wild	Domestic	Humans		
California encephalitis virus (CEV)	–	+	–	–	–	–	+	–	+	+	+	F, E	S, O
Guaroa virus (GROV)	–	+	+	–	–	–	+	–	+	+	+	F	S
Inkoo virus (INKV)	–	–	–	–	+	+	+	–	+	–	–	F, E	S
Jamestown Canyon virus (JCV)	–	+	–	–	–	–	+	–	+	+	A	F, E	S
Keystone virus (KEYV)	–	+	–	–	–	–	+	–	+	–	–	–	–
Khatanga virus (KHTV)	–	–	–	+	+	+	+	–	A	A	A	F, E	S, O
La Crosse virus (LACV)	–	+	–	–	–	–	+	–	+	+	+	F, E	S, O
Melao virus (MELV)	–	–	+	–	–	–	+	–	–	–	–	–	–
San Angelo virus	–	+	–	–	–	–	+	–	A	A	A	–	–
Snowshoe hare virus (SSHV)	–	+	–	–	–	–	+	–	A	+	+	F, E	S
Serra do Navio virus (SDNV)	–	–	+	–	–	–	+	–	A	–	A	–	–
Tahyna virus (TAHV)	+	–	–	+	+	+	+	A	+	–	+	F, E	S, O
Trivittatus virus (TVTV)	–	+	–	–	–	–	+	–	A	–	A	–	–

[a]Designations: +, virus isolation; A, specific antibodies detection; F, fever; E, encephalitis, meningoencephalitis; S, sporadic cases; O, outbreak.

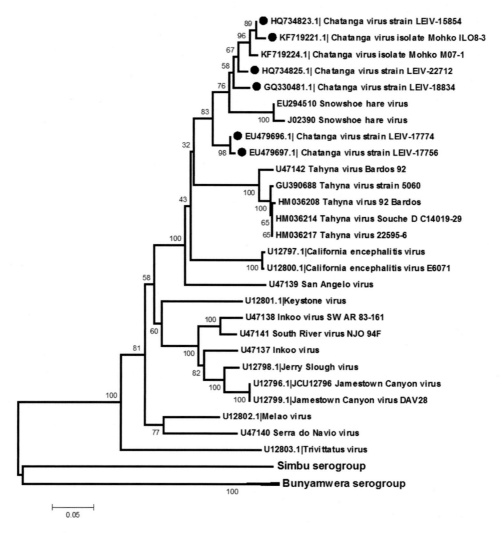

FIGURE 8.27 Phylogenetic analysis of full-length amino acid sequences of nucleocapsid (S-segment) of the California group viruses.

The majority of strains were isolated from pools of mosquitoes belonging to different species. Of 250 strains that were isolated (1 strain was isolated from a wild population of the common house mouse, *Mus musculus*), only 112 were isolated from strictly defined species (Table 8.18). The other 138 strains were isolated from *Aedes* mosquitoes of unidentified species: 34% of strains were from *Ae. communis*, 18% from the mixed pools, in which *Ae. communis* prevailed. Strains were isolated from other species significantly less often. Only one strain was isolated from *Anopheles maculipennis* (*An. messeae*) and *Culiseta alaskaensis*.[30]

The dynamics of the seasonal infection rate of mosquitoes was investigated for two years on the model of the northern part of the

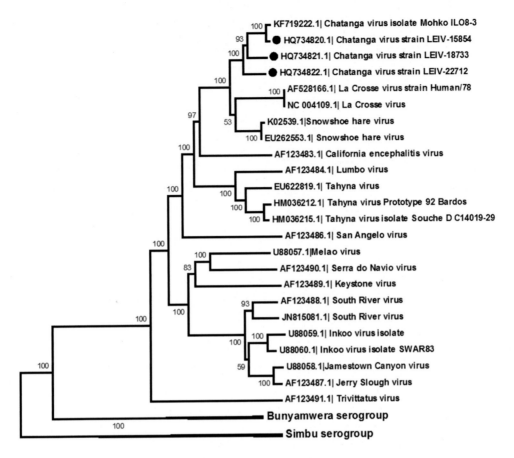

FIGURE 8.28 Phylogenetic analysis of full-length amino acid sequences of Gn/Gc precursor (M-segment) of the California group viruses.

Russian Plain and the eastern part of Fennoscandia. In tundra, the epizootic period begins with the second decade of July and proceeds to the beginning of August, when the activity of mosquitoes comes to an end. In forest tundra, the epizootic period begins with the first decade of July and proceeds for 1.5 months; in the northern taiga, this period lasts at least 2 months (July−August); in the middle and southern taiga, the first strains began to be isolated in the second decade of June. The mosquito infection rate increases significantly in the third decade of July and reaches a maximum in the middle to end of August, when the total number of mosquitoes decreases.[30,31]

The data collected testify to an almost universal distribution of CE serocomplex viruses in all landscape belts, except the Arctic, in all six physicogeographical lands examined in the north of Russia,[32] located on a territory of more than 10 million km^2.

The infection rate of mosquitoes increases ($p < 0.01$) in moving from the subarctic (tundra) ($0.0090 \pm 0.0018\%$) to the landscape belt of the middle taiga ($0.0196 \pm 0.0020\%$).

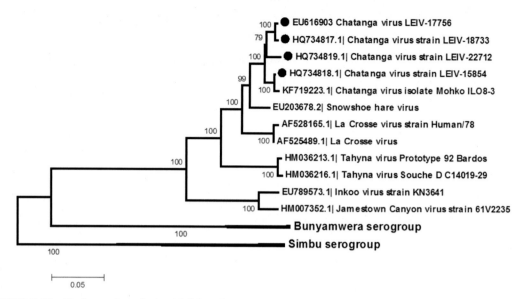

FIGURE 8.29   Phylogenetic analysis of full-length amino acid sequences of RdRp (L-segment) of the California group viruses.

This indicator in tundra and in the forest tundra is close to that in the southern taiga of the Russian Plain (0.0122%), in North America (0.01%), and in the forest steppes of the Russian Plain (0.0100−0.0017%). In the steppe belt of the Russian Plain, the infection rate of mosquitoes appeared to be the smallest (0.001%). In the leaf forests of the Russian Plain (0.0148%) and of the former Czechoslovakia (0.0210%), the infection rate of mosquitoes is comparable to that for landscape belts of the northern and middle taiga.

To date, at least s63 CE serocomplex virus strains were isolated from mosquitoes in the central and southern parts of the Russian Plain. Among them, 4 strains were isolated from the blood and spinal fluid of patients, and 3 strains from the internal parts of rodents (2 from the bank vole, *Myodes glareolus*; and 1 from the wood mouse, *Apodemus sylvaticus*). The infection rate of mosquitoes depends on the landscape belt and the particular season in which field material was collected. The rate decreases, as a rule, from the north to the south. Data indicating an absence of viruses in semideserts can be explained by an insufficient quantity of mosquitoes collected, but in wet subtropical zones in Azerbaijan CE serocomplex viruses were isolated from *Anopheles hyrcanus*.[33]

In the southern taiga belt and mixed forests, the infection rate of mosquitoes was defined to be from the third week of May to the second week of August and two peaks were noted: at the end of June (the emergence of the first generation of *Aedes* mosquitoes) and at the end of July to the beginning of August (the emergence of the second generation of *Aedes* mosquitoes). In the majority of the southern belts, the infection rate was registered from the second week of June until the end of August with a small peak in the first week of August caused by the emergence of the second generation of *Aedes* mosquitoes and by the peak of activity of *Culex*, *Coquillettidia*, and *Anopheles* mosquitoes.[17]

TABLE 8.18   Isolation of CE Serocomplex Viruses From *Aedes* Mosquitoes in High Latitudes of Russia

Physicogeographical land	Landscape belt	Collected		Strains isolated[b]					Susceptibility, %
		Places	Mosquitoes, thousands	TAHV	INKV	KHTV	CE complex	Total[a]	
Fennoscandia	Tundra	3	10.6	0	0	0	0	0	0
	Forest–tundra	2	5.7	0	0	0	0	0	0
	Northern taiga	19	25.8	0	2	2	2	6	0.023
	Middle taiga	8	99.4	0	0	0	18	18	0.018
	Total	32	141.5	0	2	2	20	24	0.017
Russian Plain	Tundra	11	26.3	0	1	1	0	2	0.0076
	Forest–tundra	5	23.8	0	0	0	0	1	0.0042
	Northern taiga	11	63.6	0	6	1	3	10	0.0167
	Middle taiga	14	70.2	0	2	2	0	3	0.0114
	South taiga	1	5.3	0	1	7[c]	0	9[c]	0.0589
	Total	42	189.2	0	10	11[c]	4	25[c]	0.013
Western Siberia	Tundra	19	81.0	0	9	1	1	11	0.0148
	Forest–tundra	12	40.1	0	13	1	0	14	0.0349
	Northern taiga	13	52.8	1	1	2	1	5	0.0095
	Middle taiga	32	91.4	0	4	6	17	27	0.0031
	South taiga	24	53.5	0	2	2	4	8	0.0156
	Forest–steppe	13	43.3	1	1	2	7	11	0.0255
	Steppe	12	41.6	0	2	11	29	42	0.0966
	Total	125	390.6	3	32	25	59	118	0.03
Central Siberia	Tundra	9	56.1	0	0	1	4	5	0.0089
	Forest–tundra	7	57.4	0	0	0	4	4	0.008
	Northern taiga	4	14.8	0	0	0	0	0	0.0068
	Middle taiga	14	53.1	0	0	3	5	8	0.0151
	South taiga	2	6.2	0	0	0	0	0	0.016
	Total	36	187.6	0	0	4	13	17	0.009

(Continued)

TABLE 8.18   (Continued)

Physicogeographical land	Landscape belt	Collected		Strains isolated[b]					Susceptibility, %
		Places	Mosquitoes, thousands	TAHV	INKV	KHTV	CE complex	Total[a]	
Northeastern Siberia	Tundra	10	46.1	0	0	3	2	5	0.0108
	Forest–tundra	8	31.2	0	0	2	0	2	0.0064
	Northern taiga	26	85.1	0	0	12	9	21	0.0247
	Middle taiga	36	130.9	0	0	19	10	29	0.0222
	Total	80	293.3	0	0	36	21	57	0.01
North Pacific land	Tundra	14	76.4	0	0	2	2	4	0.0052
	Forest–tundra	10	30.0	0	0	1	1	2	0.0067
	Northern taiga	11	30.7	0	0	2	2	4	0.013
	Middle taiga	5	19.6	0	0	0	0	0	0.0051
	Total	40	156.7	0	0	5	5	10	0.0064
All lands	Tundra	66	296.5	0	10	8	9	27	0.0091
	Forest–tundra	44	188.2	0	13	4	6	23	0.0122
	Northern taiga	84	272.8	1	9	19	17	46	0.0169
	Middle taiga	109	464.6	0	6	35[c]	50	91[c]	0.0196
	South taiga	27	65.0	0	3	4	4	11	0.0169
	Forest–steppe	13	43.2	1	1	2	7	11	0.0255
	Steppe	12	41.6	0	2	11	29	42	0.0966
	Total	355	1371.9	2	44	83[c]	122	251[c]	0.0183

[a]Identification of virus strains was carried out with a neutralization test.
[b]Abbreviations: TAHV, Tahyna virus; INKV, Inkoo virus; KHTV, Khatanga virus; CE, California encephalitis.
[c]Virus strain was isolated from a wild population of house mice (M. musculus).

In steppe and forest–steppe belts, CE serocomplex viruses were isolated from mosquitoes collected in the Rostov and Orenburg regions, as well as in the foothills of the Caucasus Mountains (Krasnodar Krai). Most of the strains were obtained from *Aedes* mosquitoes, which play the leading role in virus circulation. In these regions, *Anopheles* mosquitoes join the virus population maintenance (three strains were isolated), being ecologically connected with agricultural animals and, because of that connection, playing an important role as an indicator species in anthropogenic biocenoses. In the center and south of the Russian Plain, there is a mix of populations of INKV, TAHV, KHTV.[31,32]

Vertebrate Hosts. The principal vertebrate hosts of TAHV in Europe are Lagomorpha

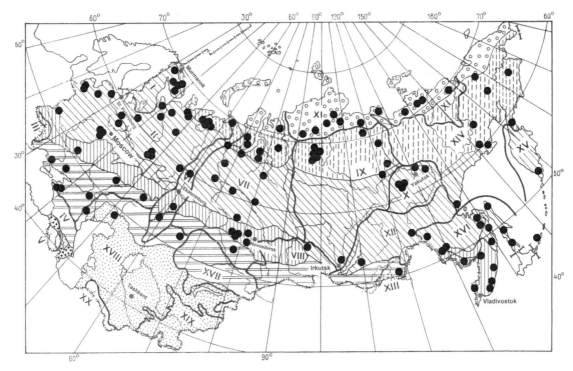

FIGURE 8.30 Places of isolation of CE complex viruses (●) in the former USSR. (See other designations in Figure 1.1.)

(hares (*Lepus europaeus*), rabbits (*Oryctolagus cuniculus*), hedgehogs (*Erinaceus roumanicus*), and rodents (Rodentia)). Experimental viremia has been established in lagomorphs, hedgehogs, ground squirrels (*Citellus citellus*), muskrats (*Ondatra zibethicus*), squirrels (*Sciurus vulgaris*), martens (*Martes foina*), polecats (*Putorius eversmanni*), foxes (*Vulpes vulpes*), badgers (*Meles meles*), bats (*Vespertilio murinus*), piglets, and puppies.[14,15,34,35]

In total, 251 strains of CE serocomplex viruses were isolated within all landscape belts of all physicogeographical lands (Figure 8.30, Table 8.19). According to our data, the susceptibility of mosquitoes increased from the tundra to the northern and middle taiga; however, the highest indicators were noted to be in the forest–steppe and the steppe of western Siberia (in Altai Krai). Identification of these strains revealed at least three viruses of the CE

complex: 2 strains of TAHV, 44 of INKV, and 183 strains of KHTV.[30]

In all landscape belts east of the Yenisei River (central and northeast Siberia and the physicogeographical lands bordering the North Pacific Ocean), only KHTV strains have been isolated. West of the Yenisei River, INKV strains predominated in the tundra and the forest–tundra of western Siberia, whereas KHTV prevailed in other landscapes located to the south. In the eastern part of Fennoscandia and in the north of the Russian Plain, INKV and KHTV strains were isolated in about equal proportions.[30]

The pattern of distribution of TAHV, INKV, and KHTV over Northern Eurasia suggests that the emergence of the ancestor of CE serocomplex viruses probably is connected to Oligocene Chinese–Manchurian fauna of the deciduous forests of eastern

TABLE 8.19   Isolation of CE Serocomplex Viruses from Different Species of *Aedes* Mosquitoes in Russia

Species of *Aedes* genus		Strains isolated	
Dominating	Attendant	Number	%
Communis		38	(33.9)
	Nigripes	7	
	Vexans	4	
	Cinereus	1	
	Cantans, excrucians, cinerius	4	
	Cantans, punctor	1	
	Excrucians, nigripes, ciprius	1	
	Intrudens, punctor	1	
	Riparius	1	
	Alaskaensis[a]	1	
	Total	58[a]	51.75
Excrucians		4	(3.5)
	Nigripes, cinerius, vexans	2	
	Mercurator, flavescens, cataphylla, nigripes, cinereus	1	
	Punctor, communis, nigripes, cataphylla	1	
	Cataphylla	1	
	Nigripes	1	
	Total	9	8.0
Cantans		7	6.25
Flavescens		4	(3.5)
	Cinereus, mercurator, hexodontus, nigripes	1	
	Excrucians, mercurator, nigripes, communis	1	
	Nigripes, mercurator	1	
	Total	7	6.25
Ciprius	*Excrucians, communis*	4	
	Cantans, caspius, cinereus	2	
	Nigripes, excrucians, flavescens	1	
	Total	7	6.25

(Continued)

TABLE 8.19   (Continued)

Species of *Aedes* genus		Strains isolated	
Dominating	Attendant	Number	%
Punctor		1	
	Intrudens, nigripes	4	
	Total	5	4.5
Vexans		4	
	Communis	1	
	Total	5	4.5
Cataphylla		3	
	Punctor, communis	1	
	Total	4	3.6
Nigripes	*Flavescens*	2	
	Mercurator	1	
	Impiger, communis, punctor	1	
	Total	4	3.6
Hexodontus		3	2.6
Eudes		1	0.9
Mercurator, nigripes, excrucians		1	0.9
Maculipennis[b]		1	0.9
	Total	112	100

[a]*One strain was isolated from the genus* Culiseta.
[b]*One strain was isolated from the genus* Anopheles.

Siberia evolving into Okhotsk fauna during the Upper Tertiary period. The Okhotsk fauna, in its turn, extended in early glacial times to the north, the west, and partially to the east in tundra through ancient Beringia and on into North America. The ancestral virus could then penetrate into North America together with this fauna and gradually extend in the southern direction, in the process laying the foundation for the appearance of some other viruses of the CE serocomplex now circulating mainly in North America.

The introduction of the virus population to the Western Hemisphere probably occurred through two pathways around the Central Siberian Plateau: (i) through the tundra lying to the north of the plateau and (ii) through southern taiga and forest–steppe territories. These pathways can explain the modern predominance of KHTV in the forest–steppe belt of Siberia and in a taiga belt west of the Yenisei River. In moving to other ecological systems further to the west, KHTV could have been transformed partially to INKV and TAHV. The INKV population penetrated into the western part of the Eurasian subarctic through the taiga belt and occupied that part of Eurasia, whereas TAHV proceeded into the deciduous forests of Europe, where it now prevails.[36]

Epidemiology. CEV is endemic in the United States in California, New Mexico, Texas, the southwestern part of Virginia, Tennessee, and Kentucky.[26,37] Sporadic morbidity with CNS lesions occurs in those states, but the main morbidity is linked to LACV, which is endemic in 20 states, predominantly the U.S. Census Bureau–defined East North Central states (Ohio, Wisconsin, Minnesota, Iowa, and Indiana), where morbidity reaches 0.1–0.4%.[26] Cases of LACV-associated encephalitis are within the distribution of the main vector—*Aedes triseriatus*—eastward from the Rocky Mountains.[38] During the last few decades, natural foci in West Virginia, North Carolina, and Tennessee, with sporadic cases occurring in Louisiana, Alabama, Georgia, and Florida, joined with previously known ones in Wisconsin, Illinois, Minnesota, Indiana, and Ohio. Thus, having traversed the distance from southeastern Asia to North America, *Ae. triseriatus* is now part of the North American virus circulation.[39] The clinical picture varies from an acute fever syndrome (in some cases with pharyngitis and other symptoms of acute respiratory disease) to encephalitis. Lethality is about 0.05%. From 40 to 100 cases occur

annually. Generally, the virus attacks children age 10 and under (60%), a phenomenon that may be explained by the existence of a layer of immunity in up to 40% of adults.[40] JCV (in the United States and Canada) and SSHV (in the northern part of the United States and in Canada) are associated with sporadic cases of fever and encephalitis.[26] Domestic dogs are susceptible to LACV, which provokes encephalitis.[21,34,41,42] The role of deer in virus circulation has been established as well. Horizontal and vertical transmission of viruses provides an active circulation of the virus, a high rate of infection in mosquitoes, and stability of natural foci under the relatively rough conditions of the central and northern parts of the temperate climatic belt.[43]

All three viruses (INKV, KHTV, and TAHV) of the CE serocomplex distributed in Eurasia have significance in human pathology.[43,44] These viruses were found in Czechoslovakia in 1959,[4,45] Austria in 1966,[13] Finland in 1969,[6,46] Romania in 1974,[12] Norway in 1978,[24] the former USSR(in Transcaucasia) in 1972,[47] and elsewhere in the European and Asian parts of Russia.[9,30,32,33,36,44,48–50] In Europe, human disease associated with TAHV presents as an influenzalike illness mainly in children with sudden-onset fever, headache, malaise, conjunctivitis, pharingitis, myalgia, nausea, gastrointestinal symptoms, anorexia, and (seldom) meningitis and other signs of CNS lesions.[16,42,51–56] The circulation of CE serocomplex viruses was established in China,[57] where they provoke human diseases with encephalitis[58] as well as acute respiratory disease, pneumonia, and acute arthritis.[59] In North America (the United States and Canada), LACV is the most important of these viruses,[60] but SSHV also is associated with human disease.[61] Between 1963 and 1981 in the United States, 1,348 cases of CE were reported.[60,62] So, CE serocomplex viruses have circumpolar distribution. In Russia, these viruses are found from subarctic to desert climes (Figure 8.30, Table 8.18).[32,44]

According to our summary data for 8,732 sera, the number of people with specific antibodies to CE serocomplex viruses in the tundra and forest—tundra belts (27.8%) is significantly lower than the number in the north and middle taiga belts (48% and 47%, respectively). These data correlate with the infection rate of mosquitoes in those landscape belts.[31,49]

Results obtained from serological investigation of the human population correlate with those obtained from virological investigation of the mosquitoes (Figure 8.31). The maximum immune layer of the healthy population is registered in the southern taiga. In the landscape and geographical zones located south of that landscape, a gradual decrease in this indicator takes place. Specific antibodies to INKV are seen everywhere that this virus circulates. In forest—steppes, specific antibodies to TAHV and INKV are marked out with an identical

frequency. In semideserts, anti-TAHV antibodies are found twice as often as anti-INKV ones. The small number of strains isolated in these natural zones precludes establishing a relationship between the circulation of viruses and an immune layer of the population. Active circulation of CE serocomplex viruses on the territory of Russia results in regular registration of the diseases caused by these viruses. More than 7% of all seasonal fevers are etiologically linked to such viruses, and in some natural zones (the southern taiga and the mixed forests), this indicator increases to 10—12%.

In mixed forests, the main etiological role most often belongs to INKV (50.4%), and in semideserts (Astrakhan region) to TAHV (76.5%). The diseases caused by CE serocomplex viruses in the center and south of the Russian Plain start appearing during the middle of May and reach a maximum in

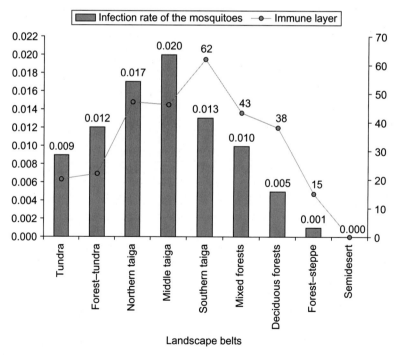

FIGURE 8.31   Immune layer among population and infection rate of mosquitoes for CE serocomplex viruses in different landscape belts in the north of Russia.

July—August. Cases are seen in September as well. The seasonal dynamics of the incidence of TAHV in various landscape zones correlate with the seasonal dynamics of the infection rate of mosquitoes. The etiological role of CE serocomplex members was confirmed in 183 (7.47%) of 2,451 patients with unstated fevers in the summertime . The etiological role of INKV was established in 64 cases (35%), TAHV in 40 (21.9%), and KHTV in 14 (7.6%). Almost equal titers of specific antibodies to more than one virus were revealed in 65 patients (35.5%) in a neutralization test.[31,43,49]

Diseases were registered from May to September: in May, 22 cases (12.02%); in June, 35 (19.13%); in July, 67 (36.61%); in August, 54 (29.51%); and in September, 5 (2.73%). The seasonal dynamics in all landscape zones were identical: The maximum number of diseases is noted in July—August. Diseases were registered everywhere in the form of sporadic cases and small outbreaks, but more often in the taiga and the deciduous forests of the European part of Russia and western Siberia. Most patients were 15—40 years old, with those up to 30 years making up 52.5% of all people infected.[9]

Pathogenesis. A systematic destruction of small vessels, together with the development of vasculitis and perivascular focal lymphohistiocytic infiltrates, underlies the pathogenesis of the diseases caused by CE serocomplex viruses. Lesions in the lungs, brain, liver, and kidneys are the most frequent complications.[31,49]

Clinical Features. The incubation period lasts from 7 to 14 days, but in some cases is only 3 days. Three main forms of disease linked with CE serocomplex viruses have been proposed: (i) influenzalike; (ii) with primary compromise of the bronchiopulmonary system; (iii) neuroinfectious, which proceeds with a syndrome of serous meningitis and encephalomeningitis.

Analysis of the clinical picture of cases examined showed that 79.8% of cases proceeded without signs of CNS lesion, 20.2% with a syndrome of acute neuroinfection, and 8.9% with radiologically uncovered signs of changes in the bronchi—lung system. A comparison of clinical forms and etiologic agents showed that INKV and TAHV often cause disease without CNS lesions (65.6% and 92.5%, respectively) and that INKV plays the leading role in acute neuroinfection (34.4%). The etiological role of KHTV was established in 14 cases without CNS symptoms of lesions.[63]

Eighty-three patients had an influenzalike form of the disease etiologically linked to CE serocomplex viruses. The incubation period was 7—14 days. The disease began abruptly, with a high temperature that reached a maximum of 39—40°C in 98.9% of patients on the first day. The duration of the fever was 4.48 ± 0.30 days. One of the main symptoms was an intensive headache (3.62 ± 0.26 days in duration) that developed in the first few hours and was often accompanied by dizziness, nausea (31.3%), and vomiting (21.7%).[43,63—66]

A survey of patients revealed infection of the sclera ($59.0 \pm 3.4\%$), hyperemia of the face and the neck ($10.8 \pm 3.4\%$), and, in some cases (3.6%), a spotty and papular rash on the skin of the trunk and the extremities. Violations of the upper respiratory airways were characterized by hyperemia of the mucous membranes of the fauces ($95.2 \pm 2.3\%$)and congestion of the nose and a dry, short cough ($13.2 \pm 3.7\%$). With regard to the lungs, $26.5 \pm 4.8\%$ of patients exhibited rigid breathing a dry, rattling cough during auscultation, and a strengthening of the bronchovascular picture on roentgenograms. Among CNS symptoms, the most common were a decrease in appetite, a stomachache without accurate localization and with liquid stool, and a small increase in the size of the liver with a short-term increase in aminotransferase activity in the blood. Inflammatory changes in the bronchi—lung system (bronchitis and pneumonia) occurred as well. In all cases in which it appeared, pneumonia had a focal character with full regression of inflammatory changes through 14.44 ± 1.36 days after the first symptoms appear. High fever (39.2 ± 0.21°C), nausea and vomiting (53.8%), and meningism phenomena (30.7%) also attracted attention.

The etiological role of different CE serocomplex viruses has been established in 8% of 463 cases with acute diseases of the nervous system (serous meningitis, encephalomeningitis, arachnoiditis, acute encephalomyelitis, and seronegative tick-borne encephalitis (TBE)): INKV (56.7 ± 8.1%), TAHV (8.1 ± 4.5%), and unidentified (35.1 ± 7.8%). The age of patients with CNS lesions was from 3 to 61 years, with the majority (51.5%) from age 21 to 30. Serous meningitis was observed in 29 patients who arrived at the hospital a mean 3.3 days after symptoms appeared. The disease began abruptly. The majority (58.6%) of patients complained of a high temperature that reached a maximum the first day, The duration of the fever was 4.54 ± 0.05 days, with a critical (37.9%) or steplike (62.1%) decrease. Headache was noted in 100% of patients and was accompanied by dizziness in 31%. Vomiting developed on the first (53.6%) or the second (46.4%) day and continued in 67.7% of patients. Meningeal signs appeared in 96.5% of patients but were weak and dissociated in most cases, with only 37.9% of patients exhibiting rigidity of the occipital muscles. The duration of the meningeal signs was 3.50 ± 0.4 days. The cells of the spinal fluid (investigated on the 4.57th ± 0.54 day of the disease) was lymphocytic, mostly reaching three digits and up to 500 cells (55.6%); the protein concentration was reduced (0.15 ± 0.02 g/L) in 41.4% of cases but was within the normal range (0.31 ± 0.01 g/L) in other cases. In 34.5% of patients exhibiting acute neuroinfection symptoms of bronchitis and focal pneumonia, their condition was confirmed radiologically.

Encephalomeningitis caused by INKV was characterized by an abrupt beginning and fast development of focal symptomatology (ataxy,

horizontal nystagmus, and discoordination) against a background of common infectious and meningeal syndromes, including inflammatory changes to the spinal fluid.[43,63−66]

The variability of the clinical picture of the diseases caused by CE serocomplex viruses and its similarity—especially at early stages—to that of other infections suggest the necessity of carrying out differential clinical diagnostics with a number of diseases. The influenzalike form needs to be differentiated, first of all, from influenza, especially in the presence of symptoms of neurotoxicity, as well as from other acute respiratory diseases (parainfluenza, adenoviral and respiratory−syncytial diseases), pneumonia (including a mycoplasma and chlamydia etiology), and enteroviral diseases. The main epidemiological features and clinical symptoms that lend themselves to carrying out differential clinical diagnostics for the influenzalike diseases described here are presented in Table 8.20. Note that considerable difficulties arise in implementing differential clinical diagnostics of the diseases that proceed with acute neuroinfection syndrome (serous meningitis, encephalomeningitis), especially when those diseases occur in the same season (Tables 8.20 and 8.21).[31,44,66]

The main criteria in differential clinical diagnostics of the disease etiologically linked with CE serocomplex viruses are as follows (see Tables 8.20 and 8.21): acute onset; high short-term fever (4−8 days, on average) reaching a maximum on the first day and decreasing critically at the end of the feverish period; and intensive headache, nausea, vomiting, and weakness. Also observed are insignificant catarrhal phenomena (nose congestion, rare dry cough) or their complete absence. A radiograph of the chest reveals signs of bronchitis and focal pneumonia with poor clinical symptomatology. An examination of the liver shows that its size, as well as its aminotransferase activity, has increased. Changes in urine, such as albuminuria and, in some cases, cylindruria, are frequently reported. Finally, symptoms relating to the vegetative nervous system (hyperemia of the face and the neck, subconjunctival hemorrhage, bradycardia, and persistent tachycardia) can be observed, as can both CNS lesions in the form of serous meningitis and encephalomeningitis in combination with compromise of the bronciopulmonary system, liver, and kidneys.

Diagnostics. Specific diagnostics of the diseases etiologically linked with CE serocomplex viruses could be based on virological testing (using sensitive biological models of newborn mice or cell lines to isolate the strains) or on serological testing.

In the presence of the sera taken from patients during the acute period of the disease (the first 5−7 days) and in 2−3 weeks, the best method of retrospective inspection is a neutralization test. A hemagglutination inhibition test is considerably less sensitive. Both complement-binding reactions and diffuse precipitation in agar have no diagnostic value today. For serological reactions, it is necessary to utilize HKTV, TAHV, and INKV antigens simultaneously. (In reference labs, SSHV antigen should be used as well.) A quadruple (or greater) increase in the titers of specific antibodies or the detection of specific antibodies in the second serological test in their absence in the first test are diagnostic criteria. ELISA for IgG indication and monoclonal antibody capture ELISA (MAC-ELISA) for IgM indication provide good diagnostic opportunities.

Control and Prophylaxis. Supervision of morbidity and of the activity of natural foci linked with CE serocomplex viruses offers the following instructions: (i) Monitor the patient clinically and the disease epidemiologically. (ii) Provide well-timed diagnostics and seroepidemiological investigations. (iii) Track the number and specific structure of mosquito vectors and possible vertebrate hosts.

8.1.4.3 Khurdun Virus

History. Khurdun virus (KHURV), strain LEIV-Ast01-5 (deposition certificate N 992, 04.11.2004, in the Russian State Collection of Viruses), was isolated from a pool of internal parts of the coot (*Fulica atra*; order Gruiformes, family Rallidae), collected August 3, 2001, in natural biomes in the western part of the Volga River delta, in Khurdun tract, Ikryaninsky District, Astrakhan region.[1] Later, nine more strains of KHURV were isolated from *F. atra* and the cormorant *Phalacrocorax pygmaeus*; order Pelecaniformes: family Phalacrocoracidae) in 2001−2004 (Figure 8.32).

At least six viruses associated with birds have been shown to circulate in the Volga River estuary.[2,3] KHURV has not been identified by any serological method,[1] including sera against viruses of the Flaviviridae, Togaviridae, Bunyaviridae, and Orthomyxoviridae families.[4]

Taxonomy. The genome of KHURV was sequenced, and phylogenetic analysis revealed that it is a new representative of the *Orthobunyavirus* genus (Figures 8.33−8.35).[5] The genome consists of three segments of ssRNA with negative polarity—an L-segment (6,604 nt), an M-segment (3,161 nt), and an S-segment (950 nt)—and has only 25−32% identity with those of other orthobunyaviruses. The terminal 3′- and 5′-sequences of KHURV genome segments, determined by rapid amplification of cDNA ends, are canonical for the orthobunyavirus (3′-UCAUCACAUG and CGTGTGATGA-5′).[6]

The L-segment of KHURV has a single ORF (6,526 nt) that encodes RdRp (2,174 aa). The similarity of KHURV RdRp with those of the orthobunyaviruses is 32%, on average. The similarity of the conservative polymerase domain III (A, B, C, D, and E motifs)[7] in RdRp reaches 62% (in BUNV).

The M-segment of KHURV is shorter than those of the orthobunyaviruses (3,161 nt vs. 4,451 nt for BUNV). The M-segment of KHURV

TABLE 8.20   Basic Differential Features of the Influenzalike Diseases Caused by CE Serocomplex Viruses, Enteroviruses, and Acute Respiratory Viruses

Symptom	Influenzalike form of the diseases caused by CE serogroup viruses	Diseases caused by enteroviruses	Influenza	Parainfluenza	Respiratory–syncytial disease
1	2	3	4	5	6
Epidemiological peculiarities	Sporadic morbidity during summer period (July–August)	Sporadic cases and outbreaks in organized collectives during summer–autumn period	Wide outbreaks and epidemics during autumn–winter period	Increased morbidity during autumn–spring period	Sporadic cases during autumn–spring period (often with worsening of chronic respiratory–syncytial disease against the background of an influenza epidemic)
Period of epidemiological activity[a]					
Onset of the disease	Acute	Acute	Acute, often sudden	Gradual, starting with catarrhal phenomena	Gradual, in some cases acute
Leading symptom	Toxicosis	Toxicosis	Toxicosis	Catarrh	Bronchitis, bronchiolitis
Level of toxicosis	Expressed	Expressed, sometimes moderate	Expressed, sometimes hypertoxicosis	Insignificant or absent	As a rule, poorly expressed
Temperature	High from the very beginning, but short term	High, often two waves	High, short term	Subfebrile, seldom high	More often subfebrile
Meningism	Often expressed from the very beginning	Possible	Could be expressed, as a rule, in severe cases	Absent	Absent
Cough	Dry, rare	As a rule, absent	Dry, painful, tracheitis	Dry, with laryngitis phenomena	Dry, paroxysmal, unproductive
Rhinitis	Nose congestion	Is not characteristic	Dryness of a mucous membrane, poorly serous separated	Swelling of a mucous membrane, plentifully separated	As a rule, absent
Scleritis	Moderate	Moderate	Moderate with cyanosis starting from second to third day of the disease	Moderate	Absent
		Hyperemia			Moderate hyperemia

(Continued)

TABLE 8.20   (Continued)

Symptom	Influenzalike form of the diseases caused by CE serogroup viruses	Diseases caused by enteroviruses	Influenza	Parainfluenza	Respiratory—syncytial disease
1	2	3	4	5	6
Change of a mucous membrane of fauces	Hyperemia with pharyngitis phenomena		Hyperemia with cyanosis, discrete hemorrhages	Diffuse hyperemia	
Bronchitis	Often expressed	Absent	Seldom expressed at the early stage	Often expressed	Leading symptom
Pneumonia	Lobular pneumonia at the early stage	Not described	Complications	Complications	Complications
Radiological data from researching lungs	Strengthening of vascular and bronchi—lung outlines	Not described	Strengthening of vascular outlines	As a rule, absent	Strengthening of bronchi—lung outlines with cellular structure
Stomachache	Moderate, seldom	For the majority of patients	Absent	Absent	Absent
Liquid stool	Possible, not often	Possible, not often	Absent	Absent	Absent
Increase in size of liver	Minor, seldom	Minor, seldom	Absent	Absent	Absent
Increase in size of spleen	Absent	Minor	Absent	Absent	Absent
Peripheral blood	Normocytosis, rapid erythrocyte sedimentation speed	Normocytosis, turn to leukopenia, rapid erythrocyte sedimentation speed	Normocytosis, turn to leukopenia	Normocytosis	Normocytosis, moderate leukocytosis
Changes in the urine	Moderate albuminuria, cylinderuria, changed erythrocytes in the beginning of the disease	Seldom in case of severe disease	Albuminuria, cylinderuria in case of severe disease	Absent	Absent

^aDesignations:

 Winter;   Spring;    Summer;    Autumn.

TABLE 8.21   Basic Differential Features of Serous Meningitis Caused by CE Serocomplex Viruses and Other Viruses

Symptom	Serous meningitis caused by CE serocomplex viruses	Serous meningitis caused by enteroviruses	Meningitis form of TBE	Meningitis caused by parotitis	Lymphocytic choriomeningitis
1	2	3	4	5	6
Epidemiological peculiarities	Sporadic morbidity during summer, with July–August peak	Sporadic outbreaks (more often among children) during summer–autumn period	Sporadic cases and outbreaks on natural-foci territories during spring–summer	Sporadic morbidity during winter–spring period	Sporadic cases (outbreaks are possible) during winter–spring period
Period of epidemiological activity[a]					
Transmission	Mosquito borne	Airborne and fecal–oral	Tick borne, alimentary	Airborne	Airborne and fecal–oral
Leading syndrome	Intracranial hypertension	Intracranial hypertension	Meningitis	Intracranial hypertension	Meningitis, intracranial hypertension
Temperature	High, starting from the first day of the disease	High	High, often in two waves	High	High, wavelike
Headache	Expressed	Expressed	Sharp	Expressed	Excruciating
Meningeal signs	As a rule, moderate (often dissociated)	Moderately expressed	Clearly expressed	Expressed	Expressed
Emesis	Repeated, but not frequent	Frequent	Frequent	Frequent	Frequent
Bronchitis	Often expressed	Absent	Seldom expressed	Absent	Possible
Pneumonia	Lobular pneumonia at the early stage is possible	Absent	Lobular pneumonia during the acute period, hypostatic pneumonia during the late period	Absent	Possible
Increase in size of liver	Often	In some cases	Often	Seldom	In some cases
Increase in size of spleen	Absent	Seldom	Absent	Absent	Seldom
Rash	Papular, seldom	Polymorphic, spotty–papular (10–25% of patients), often rich	Not typical	Not described	Not described

(Continued)

TABLE 8.21   (Continued)

Symptom	Serous meningitis caused by CE serocomplex viruses	Serous meningitis caused by enteroviruses	Meningitis form of TBE	Meningitis caused by parotitis	Lymphocytic choriomeningitis
1	2	3	4	5	6
Stomachache	Seldom	Rarely	Rarely	Rarely	In a few cases
Spinal liquid	Normal or decreased protein concentration, lymphocytosis (often in three-digit range, up to 500 cells)	Normal or decreased protein concentration, lymphocytic pleocytosis during the early stage, mixed and not high	Increased protein concentration, mixed cytosis	Increased protein concentration, high lymphocytic cytosis	Increased protein concentration, high lymphocytic cytosis
Changes in conjunctiva of eye	Plethora of expanded veins during acute period (about 30% of cases)	More than seldom	Typical	More than seldom	Hypostasis of optic nerve (about 50% of cases)
Peripheral blood	Normocytosis, rapid erythrocyte sedimentation speed	Normocytosis, shift to the left	Leukocytosis; more rarely, leukopenia	Normocytosis turning to leukopenia, lymphocytosis, rapid erythrocyte sedimentation speed	Lymphocytosis, rapid erythrocyte sedimentation speed

[a]Designations:

 Winter;   Spring;    Summer;    Autumn.

has a single ORF (2,997 nt), which encodes a polyprotein precursor (998 aa) of the envelope glycoproteins Gn and Gc. Apparently, the M-segment of KHURV does not contain a nonstructural protein NSm, which is common in most of the orthobunyaviruses.[8,9] The putative cleavage site between Gn and Gc of KHURV was found in position 319/320 aa (ASA/EN). This site corresponds to the cleavage site between NSm/Gc of the orthobunyaviruses and the conservative amino acid A/E (VAA/EE in BUNV). The size of the Gn protein of KHURV is the same as that of the other orthobunyaviruses, 320 aa. The similarity of KHURV Gn is 23–29% aa, on average, to that of the other orthobunyaviruses

(28.5% aa to BUNV). The size of the Gc protein of KHURV, 679 aa, is shorter than that of the other orthobunyaviruses (cf. 950 aa for the Gc protein of BUNV). The C-part (approx. 500 aa) of the Gc protein, which includes the conservative domain G1 (pfam03557), has about 30% aa similarity to the C-part in the other orthobunyaviruses, whereas the N-part (approximately 170 aa) has no similarity to that of any proteins in the Genbank database.

The S-segment of KHURV is 950 nt in length and encodes a nucleocapsid protein (227 aa). The similarity of the N protein to that of the orthobunyaviruses is 22–26 aa%. Most orthobunyaviruses have an additional ORF that encodes

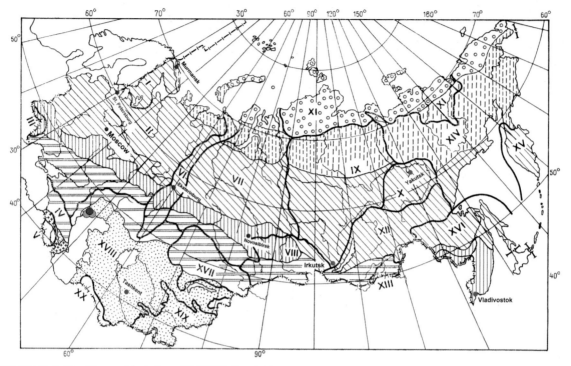

FIGURE 8.32   Places of isolation of KHURV (family Bunyaviridae, genus *Orthobunyavirus*) in Northern Eurasia. Red circle: strains LEIV-Ast01-5 of KHURV with completely sequenced genome; Pink circles: strains of KHURV identified by serological methods. (See other designations in Figure 1.1.)

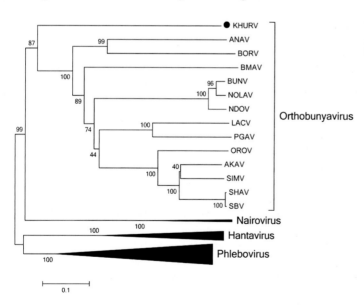

FIGURE 8.33   Phylogenetic tree for nucleo-capsid protein (S-segment) of the viruses belonging to the *Orthobunyavirus* genus.

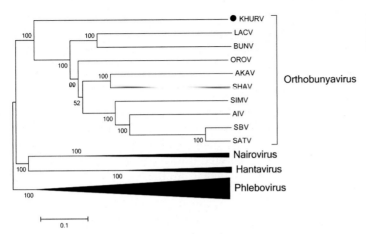

FIGURE 8.34 Phylogenetic tree for Gn/Gc precursor protein (M-segment) of the viruses belonging to the *Orthobunyavirus* genus.

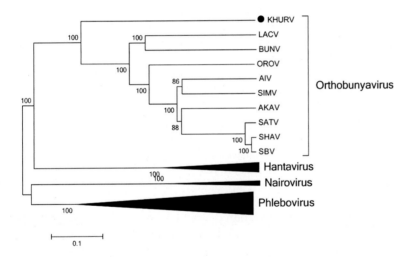

FIGURE 8.35 Phylogenetic tree for RdRp-protein (L-segment) of the viruses belonging to the *Orthobunyavirus* genus.

the nonstructural protein NSs. This protein is considered to be a factor in the pathogenicity of the orthibunyaviruses to vertebrates, because it has the ability to block the action of the interferon pathway. Some orthobunyaviruses (Anopheles A, Anopheles B, Tete serogroups) do not encode any NSs protein.[9] Phylogenetic analysis of full-length sequences showed that KHURV forms a distinct external lineage of the *Orthobunyavirus* genus (Figures 8.33–8.35).

Arthropod Vectors. There are no known arthropod vectors of KHURV; the virus has been isolated only from birds. More than 20,000 *Aedes, Culex,* and *Anopheles* mosquitoes were examined during the survival period for arboviruses in this region, and no KHURV isolations were obtained. The family Ceratopogonidae of biting midges is a potential vector of KHURV, but these insects have not been surveyed.

Vertebrate Hosts. All isolations of KHURV were obtained from birds. Nine strains of the virus were isolated from coots (*Fulica atra*). (One hundred seventeen birds were examined and were found to have an infection rate of 8.5%.) One strain was isolated from the pygmy cormorant (*Phalacrocorax pygmaeus*). (Two hundred eighty-nine cormorants, mostly *Ph. carbo*, were examined and were found to have an infection rate of 0.3%.)

8.1.5 Genus *Phlebovirus*

The *Phlebovirus* genus comprises about 70 viruses that are divided into two main groups based on their ecological, antigenic, and genomic properties: mosquito-borne viruses and tick-borne viruses.[1,2]

The genome of the phleboviruses consists of three segments of ssRNA with negative polarity: L (about 6,500 nt), M (about 3,300—4,200 nt), and S (about 1,800 nt) (Figure 8.36). In general, the structure of the genome is the same for mosquito-borne and tick-borne phleboviruses, but the M-segment is shorter in tick-borne viruses and it does not encode the nonstructural protein NSm.[3] Phylogenetically, the phleboviruses can be divided into two branches in accordance with their ecological features. The tick-borne phleboviruses comprise viruses of the Uukuniemi group, the Bhanja group, and the two novel related viruses severe fever with thrombocytopenia syndrome virus (SFTSV) and Heartland virus (HRTV), which form separate clusters and are unassigned to any group

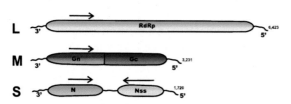

FIGURE 8.36   Structure of genome of UUKV (family Bunyaviridae, genus *Phlebovirus*). *Drawn by Tanya Vishnevskaya*

(Figures 8.37—8.39). The UUKV serogroup currently comprises 15 viruses, but the status of some of them may be revised with the accumulation of more genomic and serological data.

8.1.5.1 *Bhanja Virus and Razdan Virus (var. Bhanja virus)*

History. Bhanja virus (BHAV) was originally isolated from *Haemaphysalis intermedia* ticks that were collected from a paralyzed goat in the town of Bhanjanagar in the Ganjam district in the state of Odisha, India, in 1954 and was assigned to the unclassified bunyaviruses.[1] In Europe, the first isolation of BHAV was obtained from adult *Haem. punctata* ticks collected in Italy (1967) and then in Croatia and Bulgaria.[2,3,4] Palma virus (PALV), a virus closely related to BHAV, was isolated from *Haem. punctata* ticks in Portugal.[5] Two viruses—Kismayo virus (KISV) and Forécariah virus (FORV)—antigenically related to BHAV were isolated in Africa.[6,7] These viruses have been merged into the Bhanja group on the basis of their serological cross-reactions.[8,9] In Transcaucasia, BHAV (strain LEIV-1818Az) was isolated from Ixodidae ticks *Rhipicephalus bursa* collected from cows in Ismailli District, Azerbaijan, in 1972 (Figure 8.40). Closely related to BHAV, RAZV (strain LEIV-2741Arm) was isolated from ixodid ticks *Dermacentor marginatus* collected from sheep near the village of Solak in the Razdan district of Armenia (Figure 8.40).[10,11] Serological methods (detection of antibodies in animals and humans) have shown that BHAV circulates in many Mediterranean countries, the Middle East, Asia, and Africa.[12,13]

Taxonomy. Viruses of the BHAV group are not antigenically related to any of the other bunyaviruses, but they were assigned to the *Phlebovirus* genus on the basis of a genetic analysis of their full-length genome sequences.[14,15,16] Weak antigenic relationships were found between BHAV and SFTSV, a novel phlebovirus isolated in China.[16,17,18] SFTSV, in its turn, is antigenically related to viruses of the Uukuniemi group.[19] The genomes of certain viruses of the

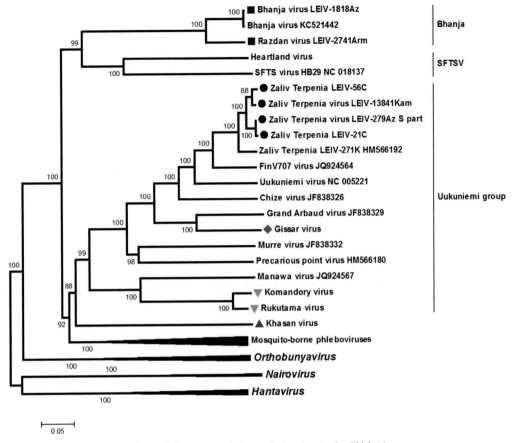

FIGURE 8.37   Phylogenetic analysis of S-segment of viruses belonging to the *Phlebovirus* genus.

Bhanja group were sequenced;[14,15,16] the size and structure of their genomes correspond to those of the tick-borne phleboviruses. The L-segment of BHAV (6,333 nt) encodes RdRp, which has a 34.8% aa identity with that of UUKV and 31.8% aa with that of Rift Valley fever virus (RVFV) (Table 8.22). The structure of the RdRp of the viruses with a negative genome includes three main domains. Domains I and II are formed around the conservative dipeptides HD and PD, respectively.[20] These dipeptides in the RdRp of BHAV are located in positions HD82 and PD113. (They were respectively found to be located in positions HD80 and PD111 in both UUKV and RVFV.)

The third, catalytic, domain, Domain III, consists of preA, A, B, C, D, and E motifs. The preA motif is formed around the conservative position K924 in BHAV (K919 and K922 in RVFV and UUKV, respectively) and includes the site of binding of genomic RNA (R941MIQFSIELLAR in BHAV; conservative positions R941, Q944, E948, and R952 are bolded). The A motif includes the dipeptide KW, which is conservative for viruses with a negative RNA genome.[20] Viruses of the Bhanja group in this position contain a substitution (TW1000) (Figure 8.41). Substitutions in this site are also found in Gouleako virus, an unclassified bunyavirus.[21]

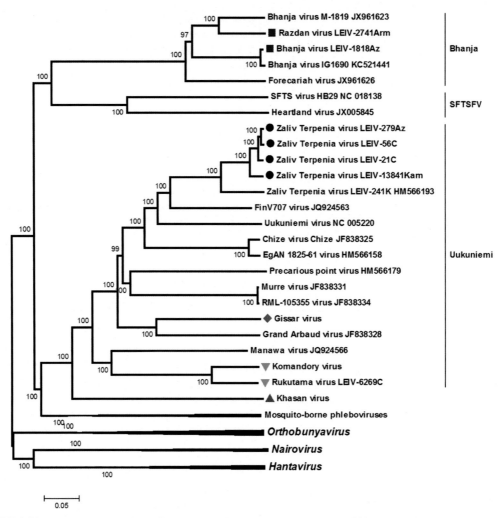

FIGURE 8.38   Phylogenetic analysis of M-segment of viruses belonging to the *Phlebovirus* genus.

The M-segment of BHAV (3,307 nt) encodes a polyprotein precursor (1,069 aa) of the envelope glycoproteins Gn and Gc. Like the M-segments of other tick-borne phleboviruses, that of BHAV has no NSm proteins that are common to mosquitoes-borne phleboviruses. The predicted cleavage site between Gn and Gc proteins has been found by Signal IP software (http://www.cbs.dtu.dk/services) to be in position 559/560 of the polyprotein precursor (motif MHMALC/CDESRL). A dipeptide

CD in the cleavage site is also typical for SFTSV and HRTV, which were associated with human disease in China and the United States, respectively.[17,18,22] Other phleboviruses, including UUKV and RVFV, contain a dipeptide CS in this position.

The S-segment (1,871 nt) of BHAV has two ORFs (N and NSs proteins) disposed in opposite orientations (an ambisense expression strategy) and separated by an intergenic spacer (139 nt). The similarity of the nucleocapsid

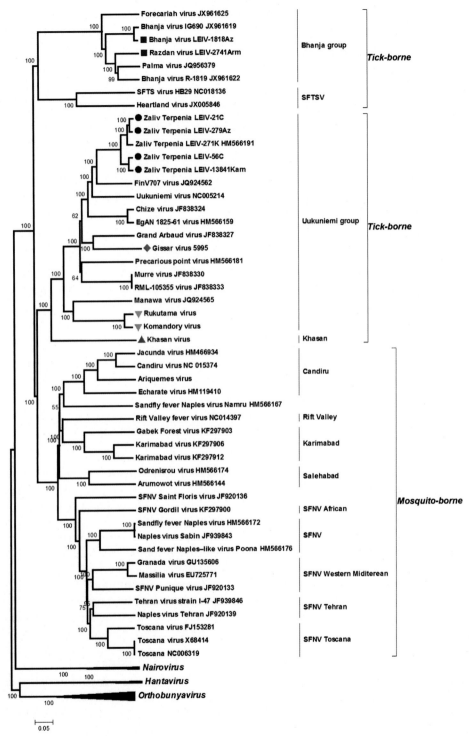

FIGURE 8.39   Phylogenetic analysis of L-segment of viruses belonging to the *Phlebovirus* genus.

II. ZOONOTIC VIRUSES OF NORTHERN EURASIA: TAXONOMY AND ECOLOGY

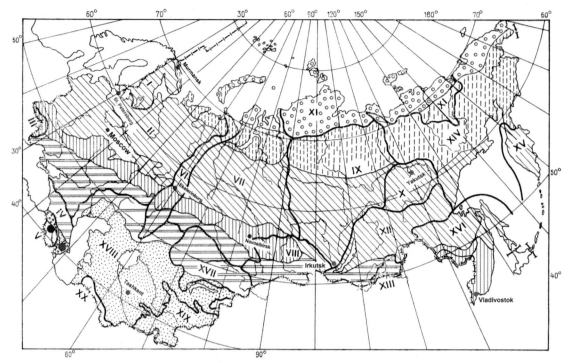

FIGURE 8.40 Places of isolation of BHAV and RAZV (family Bunyaviridae, genus *Phlebovirus*) in Northern Eurasia identified by complete genome sequencing. Red circle: strain LEIV-1818Az of BHAV; Dark-blue circle: strain LEIV-2741Arm of RAZV. (See other designations in Figure 1.1.)

protein (N-protein) of BHAV to that of UUKV is 29.3% and to that of RVFV is 37.4% (Table 8.22).

Phylogenetic analysis has shown that the Bhanja group forms a new separate lineage in the *Phlebovirus* genus (Figures 8.37–8.39).

Arthropod vectors. The main vector of BHAV is Ixodidae ticks. In Europe, BHAV was repeatedly isolated from *Haemaphysalis punctata*, *Haem. sulcata*, and *Dermacentor marginatus*; elsewhere, it has been isolated from *Haem. intermedia*, *Boophilus decoloratus*, *B. annulatus*, *B. geigy*, *Amblyomma variegatum*, *Hyalomma marginatus*, *H. detritum*, *H. dromedarii*, *H. truncatum*, *Rhipicephalus bursa*, and *Rh. appendiculatus*.[12,23]

Vertebrate Hosts. The ungulates, including domestic cows, sheep, and goats, are apparently involved in the circulation of BHAV.[24] Usually, BHAV infection in adult animals is

asymptomatic, but it is pathogenic to young ones (lamb, calf, suckling mouse), causing fever and meningoencephalitis.[13,25–27] Experimental infection of rhesus monkeys by BHAV induced encephalitis.[28] Several strains of BHAV were isolated from the four-toed hedgehog (*Atelerix albiventris*) and the striped ground squirrel (*Xerus erythropus*) in Africa. Antibodies have been detected in dogs, roe deer (*Capreolus capreolus*), and wild boars (*Sus scrofa*).[12]

Human Pathology. BHAV infection in human is mainly asymptomatic, but several cases of fever and meningoencephalitis caused by BHAV have been described.[29–31]

8.1.5.2 Gissar Virus

History. Gissar virus (GSRV) was isolated from *Argas reflexus* ticks collected in a dovecote in the town of of Gissar in Tajikistan (38°40′N,

TABLE 8.22 Similarity of Amino Acid Sequences of RAZV and BHAV (Strain LEIV-1818Az) to Those of Certain Phleboviruses (%)

	RdRp (L-segment)		G (M-segment)		N (S-segment)		NSs (S-segment)	
Virus	**RAZV**	**BHAV**	**RAZV**	**BHAV**	**RAZV**	**BHAV**	**RAZV**	**BHAV**
RVFV	32.1	31.8	18.4	19.4	34.3	37.4	13.2	13.2
TOSV	34.4	34.1	16.5	16.8	32.3	35.9	13.9	11.9
AGUV	34.8	34.8	17.3	17.9	33.5	39.8	10.7	11.1
UUKV	34.8	34.8	19.0	19.0	26.3	29.3	13.6	14.0
SFTSV	40.4	40.7	25.3	25.7	40.5	43.2	20.0	19.3
BHAV_IG690	95.9	99.3	89.0	99.2	92.5	99.0	88.8	99.0
BHAV_ibAr2709	94.7	93.7	84.5	86.2	93.2	94.9	87.2	84.0
BHAV_M3811	98.3	96.3	94.9	90.7	94.7	97.1	92.3	86.6
BHAV_R1819	98.2	96.0	93.7	90.1	93.6	96.3	91.7	86.9
PALV	98.1	96.5	92.5	90.5	92.5	96.2	92.7	86.3
FORV	93.9	93.6	84.9	86.2	92.8	95.6	87.5	83.7
BHAV_LEIV-Azn1818	95.8	–	90.3	–	92.5	–	87.9	–
RAZV LEIV-Arm2741	–	95.8	–	90.3	–	92.5	–	87.9

Abbreviations: VFV, Rift Valley Fever virus; TOSV, Toscana virus; UUKV, Uukuniemi virus; AGUV, Aguacate virus; SFTSV, Severe fever with thrombocytopenia syndrome virus; BHAV, Bhanja virus; PALV, Palma virus; FORV, Forécariah virus.

	983		
RVFV	V W T C A T S D D A R	**K W**	N Q G H F
TOSV	- - - - - - - - - - -	- -	- - - - Y
AGUV	- - - - - - - - - - K	- -	- - - - -
UUKV	H E - V - - - - - - A	- -	- - C - H
PPV	H E - V G - - - - - A	- -	- - C - H
SFTSV	S I N I N S - N - - K	- -	- - - - Y
HEAV	T I N V N S - N - - K	- -	S - - - Y
FORV	M V - L C S - N - - -	**T** -	- - - - Y
BHAV (M3811)	M I - L C S - N - - -	**T** -	- - - - Y
BHAV (ibAr2709)	M V - L C S - N - - -	**T** -	- - - - Y
BHAV (IG690)	M I - L C S - N - - -	**T** -	- - - - Y
BHAV (R1819)	M I - L C S - N - - -	**T** -	- - - - Y
PALV (M3443)	M I - L C S - N - - -	**T** -	- - - - Y
BHAV (LEIV-Az1818)	M I - L C S - N - - -	**T** -	- - - - Y
RAZV (LEIV-Arm2741)	M I - L C S - N - - -	**T** -	- - - - Y
GOUV	K S N V S S - N - - K	**V** -	- - - - H

FIGURE 8.41 Alignment of the partial aa sequence of motif A of RdRp of certain phleboviruses. Conservative dipeptide KW is in frame. (RVFV, Rift Valley fever virus; TOSV, Toscana virus; UUKV, Uukuniemi virus; AGUV, Aguacate virus; PPV, Precarious Point virus; SFTSV, Severe fever with thrombocytopenia syndrome virus; HRTV, Heartland virus; BHAV, Bhanja virus; PALV, Palma virus; FORV, Forécariah virus; GOUV, Gouleako virus.)

68°40′E; Figure 8.42). With the use of electronic microscopy, GSRV has been classified into the Bunyaviridae family.[1–3]

Taxonomy. The genome of GSRV (strain LEIV-5995Taj) has been sequenced.[4] Phylogenetic analysis shows that GSRV is a member of the *Phlebovirus* genus of the Uukuniemi group (Figures 8.37–8.39). GSRV is closely related to Grand Arbaud virus (GAV), which was isolated from a pool of *Argas reflexus* ticks collected in a dovecote near Gageron in Arles in the Rhône River delta in the Camargue region of France in 1966.[5] GAV is classified as virus belonging to the Uukuniemi group.[6] The identity of the nucleotide and amino acid sequences of GSRV and GAV is 76% nt for the S-segment (94% aa for the nucleocapsid protein), 73% nt for the M-segment (82% aa for the polyprotein

precursor of Gn/Gc), and 76% nt for the L-segment (87.5% aa for RdRp).

Arthropod Vectors. Regardless of their geographical distribution, GSRV and GAV occupy a narrow ecological niche associated with ticks (*Argas reflexus*) and birds (most likely, pigeons (Columbidae)). In laboratory experiments, GSRV reproduced in *A. reflexus* ticks in 30 days with titers up to 2.0 $\log_{10}(LD_{50})/20$ mcL.[7]

The distribution of *Argas reflexus* ticks is limited between 51°N on the north and 31°N on the south. The *A. reflexus* metamorphosis cycle is about three years. The ticks inhabit pigeons' habitats, which are also used by other birds, such as swallows and swifts. *A. reflexus* larvae were found in Europe on a rock swallow (*Ptyonoprogne rupestris*), in Egypt on a little owl (*Athene noctua*), in Israel on a rock dove (*Columba livia*) and a fan-tailed raven

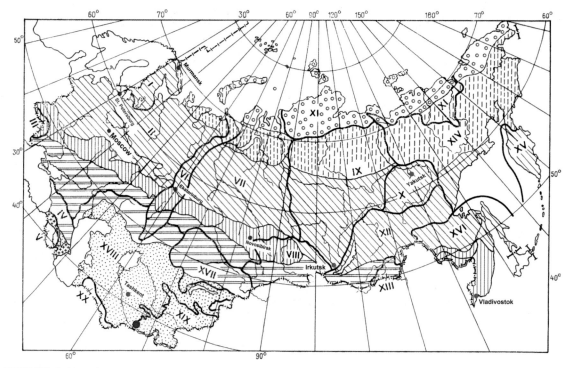

FIGURE 8.42 Place of isolation of GSRV (family Bunyaviridae, genus *Phlebovirus*) in Northern Eurasia. Red circle: prototype strain of GSRV/LEIV-5295-Taj with completely sequenced genome. (See other designations in Figure 1.1.)

(*Corvus rhipidurus*), and in Crimea on the western jackdaw (*Corvus monedula*). The mass reproduction of mites in a dovecote has a negative impact on pigeons' bereeding behavior. Worse, at night the ticks can go down to the living space and bite people if the dovecote is built into a house.[8]

Vertebrate Hosts. The main vertebrates involved in the circulation of GSRV are apparently birds, particularly the Columbidae. In laboratory experiments, GSRV was isolated from the blood of small doves (*Streptopelia senegalensis*) 5, 9, 22, and 30 days after infection. The virus titer in the blood was 1.5–2.5 $\log_{10}(LD_{50})/20$ mcL, on average. Serological examination of birds in Tajikistan found antibodies to GSRV 2% of doves (*Columba livia*).[7]

8.1.5.3 Khasan Virus

History. Khasan virus (KHAV) was isolated from *Haemaphysalis longicornis* ticks collected from spotted deer (*Cervus nippon*) in 1971 in the forest in Khasan District in the south of Primorsky Krai, Russia (Figure 8.43).[1] Morphologic studies showed that KHAV belongs to the Bunyaviridae family. The virion of KHAV has structural elements (filaments up to 10 nm) that are typical for UUKV, but no antigenic relationships between KHAV and UUKV (as well as Zaliv Terpeniya virus, ZTV) have been found.[1,2] In a complement-fixation test, KHAV did not react with serum used in the identification of certain bunyaviruses, so it was categorized in with the unclassified bunyaviruses.[3]

Taxonomy. The genome of KHAV (strain LEIV-Prm776) was sequenced, and the virus

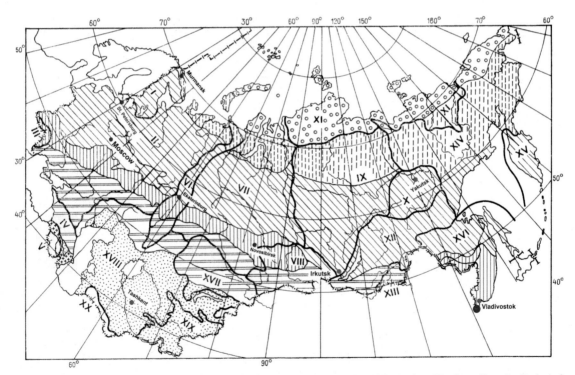

FIGURE 8.43 Place of isolation of KHAV (family Bunyaviridae, genus *Phlebovirus*) in Northern Eurasia. Red circle: prototype strain KHAV/LEIV-Prim776 with completely sequenced genome. (See other designations in Figure 1.1.)

was classified into the *Phlebovirus* genus of the Bunyaviridae family.[4] The genome of KHAV consists of three segments of ssRNA whose size and ORF structure correspond to the size and ORF structure of the other tick-borne phleboviruses. A full-length pairwise comparison of L-segments revealed a 53.1% nt identity between KHAV and UUKV and 45.3% between KHAV and RVFV. The predicted amino acid sequence of RdRp of KHAV has 48.6% and 35.3% aa identities with UUKV and RVFV, respectively. As in other tick-borne phleboviruses, the M-segment of KHAV does not contain any NSm protein. The similarity between the M-segments of KHAV and UUKV is 45.6% nt, and that between the polyprotein precursors of KHAV and UUKV is 35.9% aa. The S-segment of KHAV has 35% nt (25.5% aa for the N-protein) identity, on average, with that of the Uukuniemi group viruses and 35% nt (27.8% aa), on average, with the mosquito-borne phleboviruses.

On phylogenetic trees constructed on the basis of the alignment of full-length genome segments, KHAV forms a distinct branch external to the Uukuniemi group viruses (Figures 8.37−8.39). At least 14 viruses with unsettled taxonomy are included in the Uukuniemi group.[5] Some of them can be considered variants of the species UUKV, Manawa virus (MWAV), Precarious Point virus (PPV), and GAV. Two tick-borne phleboviruses, SFTSV and HRTV, are more closely related to the Bhanja group than the Uukuniemi group.[6,7]

Arthropod Vectors. Only a single isolation of KHAV was ever obtained, and the ecology of the virus has not been studied. *Haemaphysalis longicornis* ticks, from which KHAV was isolated, are distributed in the Far East of Russia, the northeastern part of China, the northern islands of Japan, Korea, Fiji, New Zealand, and Australia.[8] *Haem. longicornis* ticks also are the main vector of SFTSV (oterwise called Huaiyangshan virus, HYSV), which

caused a large outbreak of febrile illness with a high mortality rate (30%) in 2009 in China.[9]

Vertebrate Hosts. The principal vertebrate host of KHAV is unknown. KHAV was isolated from ticks collected on deer.[1] *Haemaphysalis longicornis* ticks are repeatedly found on cows, goats, horses, sheep, badgers, and dogs.[8]

8.1.5.4 Sandfly Fever Naples Virus and Sandfly Fever Sicilian Virus

History. The sandfly fever virus group includes Naples and Sicilian subtypes.[1] Epidemics of the comparatively mild acute febrile disease of short duration brought on by these viruses in countries bordering the Mediterranean have been known since the Napoleonic Wars.[2] The same disease was common among newly arrived Austrian soldiers on the Dalmatian coast each summer.[3] Experiments conducted by an Austrian military commission proved that the disease was caused by a filterable agent in the blood of patients and that the sandfly *Phlebotomus papatasi* can serve as a vector to transmit the disease.[4] During World War II, epidemics occurred among troops in the Mediterranean and two antigenically distinct strains were isolated from the blood of patients in 1943 in Sicily and Naples. These strains have been designated the sandfly fever Sicilian virus (SFSV) and sandfly fever Naples virus (SFNV), with prototype virus TOSV.[5,6] Dr. A. Sabin gave a clinical description of the disease and demonstrated that immunity developed to one type of virus does not protect from infection caused by the other type. Later, several viruses related to SFNV (Anhanga (ANHV), Bujaru (BUJV), Candiru (CDUV), Chagres (CHGV), Icoaraci (ICOV), Itaporanga (ITPV), and Punta Toro (PTV)) were isolated from humans and rodents in South America.[2,3,7] To date, viruses related to TOSV have been found in all regions of the world, including the Palearctic, Neotropical, Ethiopian, and Oriental zoogeographical regions.[2] The prototype strain of TOSV was isolated from *Phlebotomus papatasi* sandflies in 1971 in Monte Argentario in central Italy.[8] Two viruses antigenically related to TOSV—Karimabad virus (KARV) and Salehabad virus (SALV) were isolated from *Phlebotomus* flies collected in 1959 near Karimabad village and Salehabad village, respectively, in Iran.[9,10] Several related viruses were isolated in the Mediterranean: sandfly fever Cyprus virus (SFCV;[11] Adria virus (ADRV, Salehabad-like), isolated in Saloniki (alternatively, Thessaloniki), Greece;[12] and Massilia virus, isolated near Marseilles, France.[13] Epidemic outbreaks of sandfly fever whose agents could not be typified occurred in some central Asian countries and in Crimea during and after World War II and in Turkmenistan after the devastating earthquake of 1948. Antibodies to SFSV, SFNV, and KARV were found in the blood of humans in Tajikistan, Azerbaijan, and Moldova.[14] Antibodies were also found in wild animals in Turkmenistan: the great gerbil (*Rhombomys opimus*), the long-clawed ground squirrel (*Spermophilopsis leptodactylus*), and the hedgehog (*Erinaceus auritus*). Three strains of SFNV and two strains of SFSV were isolated in 1986–1987 from the blood of patients in Afghanistan.[14,15]

Taxonomy. The genome of TOSV consists of three segments of negative-polarity ssRNA: L-segment (6,404 nt in length), M-segment (4,214 nt) and S-segment (1,869 nt). Phylogenetic analysis revealed that viruses of the SFNV complex are divided into five genetic clades that differ in their geographical distribution: (i) from Africa (Saint Floris virus and Gordil virus (GORV)); (ii) from the western Mediterranean (Punique virus (PUNV), Granada virus (GRV), and Massilia virus); (iii) TOSV; (iv) viruses from Italy, Cyprus, Egypt, and India; (v) strains from Serbia and Tehran virus.[16]

Distribution. SFNV and SFSV are distributed over those areas of the southern parts of Europe and Asia, and over those areas of Africa, which are within the range of the vector.[15,17–22] TOSV is distributed over Italy;

Spain; Portugal; the south of France; Slovenia; Greece, including the Ionian Islands: Cyprus; Sicily; and Turkey.[13,17,23–32] Both the Naples and Sicilian strains were isolated from the blood of patients with febrile illness in the vicinity of Aurangabad, Maharashtra state, in northern India. Sandfly virus fever also circulates in western India, as well as in Pakistan.[33] The cocirculation of two TOSV genotypes was uncovered in the southeast of France.[13,15,33] A case of disease associated with TOSV befell a tourist returning from Elba to Switzerland in 2009, and another struck an American tourist returning from Sicily the same year.[27] TOSV from France is genetically different from that in Spain.[3,13,33,34] Periodic outbreaks of sandfly fever occurred in the first half of the twentieth century in some central Asian republics, Transcaucasia, Moldova, and Ukraine.

Arthropod Vectors. The primary vector of SFNV and SFSV is *Phlebotomus Papatasi*; for TOSV, the primary vectors are *Ph. perniciosus* and *Ph. perfiliewi*. The viruses can be transmitted by the transovarial route and therefore may not require amplification in wild vertebrate hosts.[35] The infection rate of sandflies can reach 1:220.[36] The active period of *Phlebotomus* in the southern part of Europe lasts from May to September. Sandflies are peridomestic; the immature stages feed on organic matter in soil and do not require water, but are sensitive to desiccation and therefore are often found in association with humid rodent burrows.

Vertebrate Hosts. The main vertebrates involved in the circulation of SFNV are rodents, particularly the great gerbil (*Rhombomys opimus*) and the long-clawed ground squirrel (*Spermophilopsis leptodactylus*), as well as a hedgehog (*Erinaceus auritus*). The great gerbil is distributed over areas ranging from near the Caspian Sea to the arid plains and deserts of central Asia. The northern border of the animal's distribution is from the mouth of the Ural River on northward to the Aral Karakum and Betpak-Dala deserts, to the southern coast of Lake Balkhash, and thence to northern China and Inner Mongolia. The habitats of *Rh. opimus* are sandy and clayey deserts. TOSV was isolated from the brain of the bat *Pipistrellus kuhlii*.[8]

Animal and Human Pathology. Sandfly virus fever does not cause disease in domestic or wild animals. The hosts of *Phlebotomus* sandflies are usually rodents, which may develop antibodies. Over 100 human experimental volunteers were infected at the time of World War II.[36,37] The incubation period is between 2 and 6 days, and the onset of fever and headache in those patients was sudden. Nausea, anorexia, vomiting, photophobia, pain in the eyes, and backache were common and were followed by a period of convalescence with weakness, sometimes diarrhea, and usually leucopenia. Viremia was present 24 h before and 24 h after the onset of fever.[37] TOSV was established as the cause of one-third of previously undiagnosed human aseptic meningitis and encephalitis cases examined in central Italy. SFCV was associated with a large outbreak in the Ionian Islands of Greece.[28] ADRV is associated with serious illness with tonic muscle spasms, convulsions, difficulty urinating, and temporary loss of sight. Human disease frequently goes unrecognized by local health-care workers. Studies of antibodies in people indicate that the most infections occur in children. When large numbers of unimmunized adults are introduced into an endemic area, the incidence of disease can be high. Human exposure to sandflies can be reduced by repellents, air-conditioning, and screens on windows. Because sandflies have a flight range of not more than 200 m, human habitats can be constructed at a distance from potential domestic sandflies' breeding places, such as chicken houses and quarters for other farm animals.[19]

8.1.5.5 *Uukuniemi virus and Zaliv Terpeniya virus*

History. UUKV was originally isolated from *Ixodes ricinus* ticks collected in 1959 from cows in southeastern Finland.[1,2] Antigenically similar isolates (strains LEIV-540Az and LEIV-810Az) have been obtained from blackbirds (*Turdus merula*) and *I. ricinus* ticks collected in the foothills of the Talysh Mountains in the southeast of Azerbaijan in 1968 and 1969, respectively.[3–5] UUKV is distributed in the mid- and southern boreal zones of Fennoscandia and adjacent areas of the Russian Plain. Twelve strains of UUKV were isolated from *I. ricinus* ticks (the infection rate was 0.5%), and one strain from *Aedes communis* mosquitoes, in landscapes in the mideastern region of Fennoscandia.[6,7] Three strains were isolated from *I. persulcatus* ticks collected in Belozersky District, Vologda Region, Russia, in 1979.[8,9] UUKV was also isolated from the mosquitoes *Ae. flavescens* and *Ae. punctor* in the west of Ukraine, as well as at the border between Poland and Belarus.[10,11] Twenty-eight strains of UUKV were isolated from *I. ricinus* ticks collected in Lithuania and Estonia in 1970–1971.[6,7,12–14] UUKV was isolated as well from birds and *I. ricinus* ticks in western Ukraine and Belarus.[11,15,16] In central Europe, UUKV was found in the Czech Republic, Slovakia, and Poland.[17–20]

The prototypical strain LEIV-21C of ZTV was isolated from *Ixodes uriae* ticks collected in 1969 in a colony of common murres (*Uria aalge*) in Tyuleniy Island in Zaliv Terpeniya Bay in the Sea of Okhotsk).[21,22] In accordance with the results of electron microscopy, ZTV was assigned to the Bunyaviridae family. Complement-fixation testing revealed that ZTV is most closely related to UUKV, but the two viruses are easily distinguishable in a neutralization test.[21,22] More than 60 strains of ZTV were isolated from *I. uriae* ticks collected in colonies of seabirds on the shelf and islands of the Sea of Okhotsk, the Bering Sea, and the Barents Sea (Table 8.23, Figure 8.44).[9,21,23,24] Two strains of ZTV were isolated from *I. signatus* ticks collected on Ariy Kamen Island in the Commander Islands, but their infection rate was less than 1:10,000 (<0.01%).[9] A similar virus was found in Norway.[25] One strain of ZTV (LEIV-279Az) was isolated from the mosquito *Culex modestus* collected in 1969 in a colony of herons (genus *Ardea*) in the district of Kyzylagach in the southeastern part of Azerbaijan (Figure 8.44).[3] Natural foci of ZTV and UUKV associated with bloodsucking mosquitoes (subfamily Culicinae) are found in continental areas in the European part of Russia, particularly Murmansk region.[7]

Taxonomy. The viruses of the *Phlebovirus* genus can be divided into two main ecological groups: those transmitted by bloodsucking mosquitoes (subfamily Culicinae) and midges (subfamily Phlebotominae), together called mosquito borne, and those transmitted by ticks (tick borne). UUKV is a prototypical virus of the Uukuniemi antigenic group, which includes at least 15 related tick-borne phleboviruses (Figures 8.37–8.39).[26] The genome of UUKV consists of three segments of ssRNA: an L-segment 6,423 nt long, an M-segment 3,229 nt long, and an S-segment 1,720 nt long. The M-segment of UUKV, and indeed, that of all tick-borne phleboviruses, is shorter than the M-segment of mosquito-borne phleboviruses, owing to the absence of the nonstructural protein NSm, which is common in the mosquito-borne phleboviruses. Originally, ZTV was described as a virus closely related to UUKV. A full-length sequence comparison showed that the similarity of ZTV to UUKV is 77.3% nt identity of the L-segment (90.9% aa of RdRp) and 70.9% nt identity of the M-segment (81.5% aa).

Arthropod Vectors. Most isolations of UUKV and ZTV were obtained from *Ixodes*

TABLE 8.23 Isolation of Zaliv Terpeniya Virus (ZTV) (Family Bunyaviridae, Genus *Phlebovirus*) from *Ixodes* (*Ceratixodes*) *Uriae* Ticks, Obligate Parasites of Alcidae Birds in the Basins of the Sea of Okhotsk, the Bering Sea, and the Barents Sea

	Far East				European part
	Basin of Sea of Okhotsk		Basin of the Bering Sea		Basin of Barents Sea
	Sakhalin District		Kamchatka	Chukotka	Murmansk District
Results of examination	Tyuleniy Island (48°29′N, 144°38′E)	Iona Island (56°24′N, 143°23′E)	Ari Kamen Island (Commander Islands) (55°13′N, 165°48′E)	Bering Strait Coast (64°50′N, 173°10′E)	Kharlov Isl., Near Kola Peninsula (68°49′N, 37°19′E)
Number of strains isolated	3	2	20	0	41
Infecion rate (%)	0.022	0.103	0.096	<0.087	0.456
Total Number of strains	25				41
Number of ticks tested	35,725				8,994
Infection rate total (%)	0.070				0.456

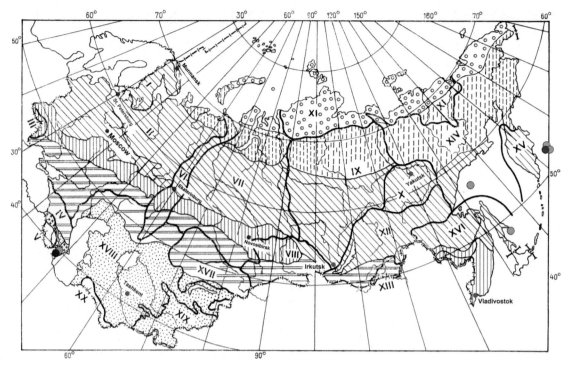

FIGURE 8.44   Places of isolation of ZTV (family Bunyaviridae, genus *Phlebovirus*) in Northern Eurasia. Red circle: strain of ZTV/LEIV-13841Kam with completely sequenced genome; Pink circles: strains of ZTV/LEIV-279Az identified by complete genome sequencing; Dark-blue circle: strains of ZTV identified by serological methods. (See other designations in Figure 1.1.)

ricinus and *I. uriae* ticks, respectively. The infection rates of nymphs and larvae of *I. uriae* are 5 and 13 times lower, respectively, than that of the imago. These rates indicate a high frequency (8–10%) of transovarial transmission of ZTV.[7,9] Probably, ZTV has a more pronounced ability to replicate in mosquitoes that are active in the subarctic climate zone (tundra landscapes) in July through the first half of August at temperatures sufficient for the accumulation of virus in the salivary glands.[7,9] During this period, circumpolar species predominate: *Aedes communis*, *Ae. punctor*, *Ae. hexodontus*, *Ae. impiger*, and *Ae. nigripes*; *Culiseta alaskaensis* is rarer. *I. ricinus*, together with *I. persulcatus*, *I. pavlovskyi*, *I. nipponensis*, *I. kashmiricus*, *I. kazakstani*, *I. granulatus*,

I. nuttallianus, and *I. hyatti*, forms a single phylogenetic branch within the genus *Ixodes*.[7,9] The origin of this group of species is said to be southeast Asia. Currently, *I. ricinus* inhabits mixed and deciduous forests of the eastern coast from the Atlantic Ocean to the Middle Volga, but zoogeographical data show that the tick emerged in the Paleocene epoch in a belt of mesophilic forests.[27] Comparative ontogenetic analysis revealed a relation of *I. ricinus* to the southern group of *I. persulcatus* (*I. kazakstani*, *I. kashmiricus*, *I. nipponensis*, and *I. hyatti*) and, through that group, with *I. pavlovskyi*.

Vertebrate Hosts. Complement-binding antibodies to ZTV were found in 1% of common murres (*Uria aalge*) on the Commander

Islands. In the Murmansk Region, which lies to the north of the European part of Russia, antibodies were found in 6% of common murres (*U. aalge*), 4% of black-legged kittiwakes (*Rissa tridactyla*), and 1% of voles (*Microtus oeconomus*).[7,9] Apparently, ruminants could be infected by mosquitoes or by eating fallen birds. On the north coast of the Kola Peninsula, antibodies were found in 6% of thick-billed murres (*U. lomvia*), in 7% of black-legged kittiwakes, and in 1% of voles.[7,9]

In central and eastern Europe, a number of vertebrate hosts are involved in the circulation of UUKV: forest rodents (*Myodes glareolus*, *Apodemus flavicollis*) and terrestrial passerine birds—the blackbird (*Turdus merula*), pale trush (*T. pallidus*), ring ouzel (*T. torquatus*), European robin (*Erithacus rubecula*), hedge sparrow (*Prunella modularis*), wheatear (*Oenanthe oenanthe*), European starling (*Sturnus vulgaris*), carrion crow (*Corvus corone*), magpie (*Pica pica*), brambling (*Fringilla montifringilla*), hawfinch (*Coccothraustes coccothraustes*), yellow bunting (*Emberiza sulphurata*), turtle dove (*Streptopelia turtur*), and ring-necked pheasant (*Phasianus colchicus*).[20,28−32] Viremia and long-term persistence of the virus were demonstrated in experimentally infected birds of many species. Specific antibodies were detected in cows and reptiles.

Human Pathology. An association was revealed between UUKV and different forms of disease, including neuropathy.[33,34] A serological survey of 1,004 people in Lithuania concluded that antibodies existed in 1.8−20.9% of the population. Human antibodies to UUKV were detected in less than 5% of the human population in central Europe[33−35] and 13−14% in Belarus.[16] The people living in the tundra landscape had antibodies to ZTV in 3.3% of cases, while in the forest no such antibodies were detected (via a neutralization reaction).[7]

8.1.5.6 Komandory Virus and Rukutama Virus (var. Komandory virus)

History. Komandory virus (KOMV), strain LEIV-13856, was isolated in 1986 from *Ixodes (Ceratixodes) uriae* ticks in Ari Kamen Island (54°30′N, 166°20′E; Figure 8.45), a part of the Commander Islands archipelago in the Bering Sea.[1] RUKV, strain LEIV-6269, was isolated from *I. uriae* ticks collected in September 1976 in a colony of common murres (*Uria aalge*) on Tyuleniy Island.[1] RUKV was named for the Rukutama River, which flows into Zaliv Terpeniya Bay, off Sakhalin Island, near the place of first isolation of the virus (Figure 8.45). Later, two additional strains of RUKV were isolated from the same source and place (Table 8.24). In previous studies, RUKV was mistakenly included in the Sakhalin serogroup in the *Nairovirus* genus.[1]

Taxonomy. The genome of KOMV (strain LEIV-13856) and RUKV (strain LEIV-6269) were completely sequenced, and the two viruses were classified into the *Phlebovirus* genus.[2,3] A full-length comparison showed that the genetic similarity between KOMV and RUKV is 93.0−95.5% nt. Among other tick-borne phleboviruses, KOMV and RUKV are most closely related to MWAV, which was isolated from *Argas abdussalami* ticks in 1964 in Pakistan.[4] The similarities of the genomes of KOMV and RUKV to that of MWAV are 67.1% nt for the L-segment (73.0% aa for RdRp), 59.6% nt of the M-segment (58% aa for the polyprotein precursor), and 66.8% nt for the S-segment (58.4% aa for the N-protein). In phylogenetic trees, KOMV and RUKV were placed into the Uukuniemi group (Figures 8.37−8.39).[5]

The ecology and area of distribution of KOMV and RUKV are the same as those of ZTV, which is closely related to UUKV. Several strains of ZTV isolated on the Commander Islands were sequenced, and no reassortants of ZTV with KOMV were found.[6,7]

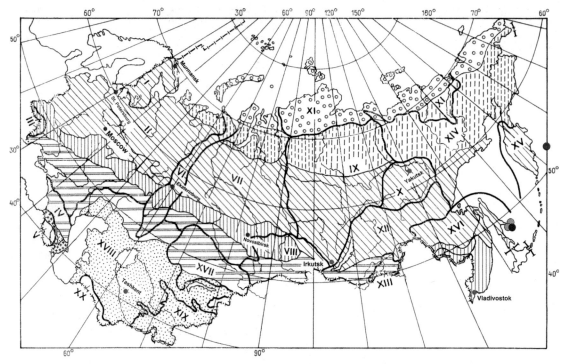

FIGURE 8.45   Places of isolation of KOMV and its variant RUKV (family Bunyaviridae, genus *Phlebovirus*) in Northern Eurasia. Red circle: strain of KOMV/LEIV-13856 identified by complete genome sequencing; Pink circles: strains of KOMV identified by serological methods; Dark-blue circle: strain of KOMV/LEIV-6269 (RUKV) identified by complete genome sequencing. (See other designations in Figure 1.1.)

Arthropod vectors. All isolations of KOMV and RUKV were obtained from *Ixodes uriae* ticks, the obligate parasite of Alcidae birds. The Commander Islands are located on the border of the temperate and subarctic climatic zones, and many different viruses belonging to the Bunyaviridae (ZTV, SAKV, PMRV), Flaviviridae (Tyuleniy virus, TYUV), and Reoviridae (OKHV) families have been isolated from *I. uriae* ticks collected from birds living in colonies there.[8–11] Note that the KOMV infection rate of the *I. uriae* ticks in the Commander Islands is 10 times less than the ZTV (1:900) and TYUV (family Flaviviridae, genus *Flavivirus*) infection rates of the same ticks.

Vertebrate Hosts. The main vertebrate host of KOMV and RUKV is apparently Alcidae birds, especially the common murre (*Uria*

aalge), but their involvement in the circulation of KOMV and RUKV has not been studied sufficiently.

Human Pathology. UUKV group viruses, in general, do not play a role in human infectious pathology, although serological studies have detected antibodies to various viruses of this group in people.

8.2 FAMILY FLAVIVIRIDAE

The Flaviviridae family (from the Latin *flavus,* "yellow," as well as from yellow fever virus (YFV)) includes three genera: *Flavivirus, Pestivirus,* and *Hepacivirus.*[1] The Flaviviridae are small (40–60 nm) enveloped viruses. The genome is represented by ssRNA

TABLE 8.24 Isolation of KOMV, or Synonymous RUKV, from *Ixodes Uriae* Ticks in Colonies of Alcidae Birds in the Basin of the Sea of Okhotsk.

Virus		Far East	
		Sakhalin district	Kamchatka
		Tyuleniy Island (48°29'N, 144°38'E)	Ari Kamen Island (55°13'N, 165°48'E)
Commander Islands	Number of strains	3	1
	% of infected ticks	0.022	0.005
Total	Number of strains	4	
	Number of ticks examined	35,725	
	% of infected ticks	0.011	

FIGURE 8.46 Genome organization of YFV (family Flaviviridae, genus *Flavivirus*). *Drawn by Tanya Vishnevskaya.*

long ORF of a polyprotein precursor flanked by 5' and 3' untranslated regions. Mature viral proteins are produced during a complex process of proteolytic cleavage of the polyprotein precursor by cellular and viral proteases. Structural proteins (core, M, and E) occupy one-third of the RNA (the N part of the polyprotein) on the 5' part of the genome, followed by nonstructural proteins (NS1-NS5b) (Figure 8.46).[2,4]

Most of the flaviviruses are arboviruses; that is, they can be transmitted to vertebrate hosts by bloodsucking arthropod vectors (Figure 8.47). Approximately 50% of known flaviviruses are transmitted by mosquitoes, about 30% by ticks. The arthropod vectors of some flaviviruses are unknown. There is also a group of flaviviruses that infect only insects and not vertebrates. Some flaviviruses (e.g., West Nile virus, WNV) have ecological plasticity and can be transmitted either by mosquitoes or by ticks. Flaviviruses are distributed over all continents, with mosquito-borne viruses found mainly in regions with an equatorial and tropical climate And tick-borne viruses found mostly in regions with a temperate climate zone. Many flaviviruses are associated with birds, which can transfer them during the birds' seasonal migration. Flaviviruses belongs to natural foci zoonoses. Certain flaviviruses, such as YFV, dengue virus (DENV), and West Nile virus (WNV), pose a serious threat to humans.[5-7]

(9,100–12,300 nt) with positive polarity. The genomic RNA is infectious and acts as mRNA in the cytoplasm of infected cells.[2] Members of the the *Flavivirus* genus are zoonotic viruses infecting a wide range of vertebrate and arthropod hosts.

8.2.1 *Flavivirus* Genus

The *Flavivirus* genus includes more than 70 viruses classified into 15 antigenic groups.[1,3] The *Flavivirus* virion is spherical (50 nm) and consists of a nucleocapsid (30 nm) and a lipid bilayer envelope covering it. The lipid envelope contains two transmembrane glycoproteins: M (matrix protein, 8 kD) and E (envelope protein, 50 kD). The genome of the flaviviruses is a single molecule of RNA about 11,000 nt in length and capped on the 5' terminus. The genomic RNA encodes a

8.2.1.1 *Omsk Hemorrhagic Fever Virus*

History. The first hint that Omsk hemorrhagic fever (OHF) was etiologically linked

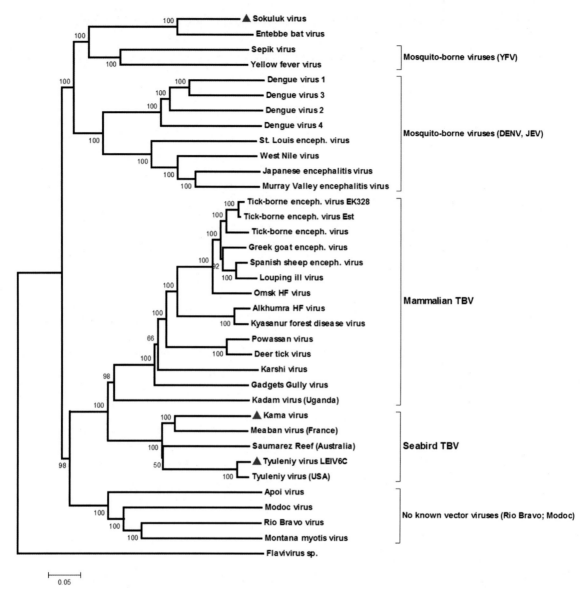

FIGURE 8.47   Phylogenetic analysis of viruses belonging to the *Flavivirus* genus, based on full-length genome comparison.

with Omsk hemorrhagic fever virus (OHFV) (family Flaviviridae, genus *Flavivirus*, antigenic complex of TBE) came in 1940–1945.[1,2] The disease arose in the spring–summer in the northern forest–steppe landscape zone in the region around Omsk, western Siberia; Figure 8.48) an area with a wide network of lakes. About 200 cases with two lethal outcomes ("atypical tularemia" and "atypical leptospirosis") were investigated (without

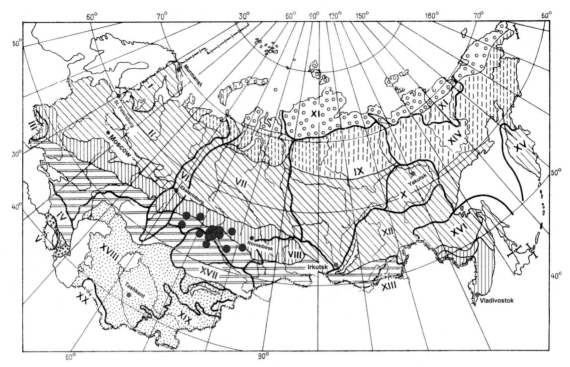

FIGURE 8.48   Places of isolation of OHFV (family Flaviviridae, genus *Flavivirus*) strains (●) in the former USSR. (See other designations in Figure 1.1.)

laboratory confirmation) from the end of April to the beginning of October in 1945. In the spring–summer of 1946, the endemic territory enlarged and the number of cases increased to 623, with 4 lethal outcomes.[3,4] In 1947, a large expedition (about 50 members) under the leadership of Mikhail Chumakov (Figure 2.10) began to work in the Omsk region; the virus etiology of OHF was proven, and the disease acquired its modern name: OHF.[3,5–7] The expedition produced prodigious results: The prototype strain OHFV/Kubrin was isolated from the blood of one patient; the mechanism of transmission of the virus by the Ixodidae tick *Dermacentor reticulatus* was established; the epidemiological and clinical features of OHF, as well as its pathogenesis and pathomorphology, were described; and inactivated vaccine

from mouse brain was developed and prepared for epidemiological trials.[5] Later, the role of another species of Ixodidae ticks (*D. marginatus*) as an OHFV vector was revealed.[8,9]

Taxonomy. OHFV belongs to the phylogenetic branch of the mammalian tick-borne virus group (Figure 8.47). The OHFV genome has a length of 10,787 nt, and its organization is common to the flaviviruses. Two genotypes of OHFV are known today: Prototypical strains for the first one are OHFV/Kubrin and OHFV/Bogolubovska, which have an extremely small genetic distance between them; the prototypical strain for the second genotype is OHFV/uve.[10–12] Only six nucleotide substitutions, which encode four amino acids, have been found in the entire genome.

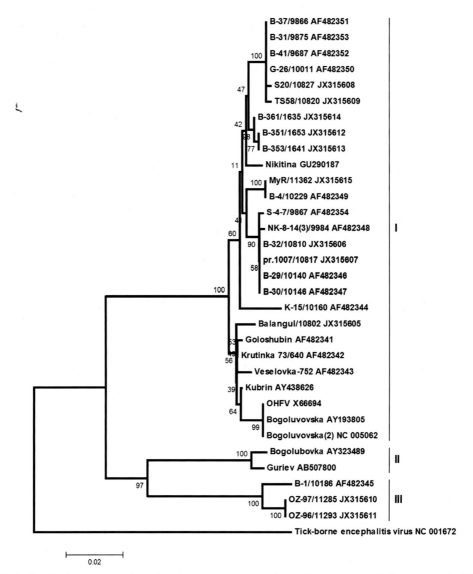

FIGURE 8.49 Phylogenetic analysis based on comparison of partial sequences of E protein of OHFV (family Flaviviridae, genus *Flavivirus*, TBEV group). Alignment of OHFV sequences available in GenBank were analyzed by neighbor-joining method and rooted on the corresponding sequence of TBEV.

Three of four amino acid changes were located in the envelope glycoprotein E.[11]

Phylogenetic analysis based on a comparison of partial sequences of the E gene available in GenBank showed that OHFV isolates can be divided to three genetic lineages (Figure 8.49).

The genetic diversity among strains of different lineage is up to 11.8%.

Arthropod Vectors. The natural foci of OHFV are found in the forest—steppe landscape zone of western Siberia, an area with numerous bogs and a wide network of lakes within the Omsk, Novosibirsk, Kurgan, and Tyumen regions (Figure 8.48). The natural foci border the area of distribution of TBEV, and the two virus's natural foci are intermingled.[13–15]

The principal Ixodidae tick vectors for OHFV are *Dermacentor pictus* (in the northern forest–steppe subzone) and *D. marginatus* (in the southern forest–steppe subzone).[3,8,9] The infection rate of *D. pictus* in epidemic years reaches 8%, in interepidemic years 0.1–0.9%. The main host for preimago phases of *D. pictus* is the narrow-headed vole (*Microtus gregalis*). This species of rodent is host to 70–90% of *D. pictus* nymphs and larvaein the northern forest–steppe subzone. In 1959–1962, when the number of *Microtus gregalis* voles fell significantly, there was a concomitant decrease in the number of *D. pictus* ticks in the center of an epidemic zone that was accompanied by a sharp decrease in the infection rate of ticks and an attenuation of the meadow natural foci of OHFV. In some of those years, however, a high number of *Ixodes apronophorus*, all phases of which feed on the water vole (*Arvicola terrestris*), become involved in the virus's circulation on a par with *D. pictus* ticks. *Ar. terrestris* makes fodder migrations in June–August from damp locales (where their infection takes place) to coastal meadows (where peak activity of the larvae and nymphs of *D. pictus* is observed during those months). Small animals living in those meadows become infected as they feed on the *D. pictus* larvae and nymphs. In damp locales, *I. apronophorus* could infect muskrats. Also, *D. marginatus*, whose optimum zone lies in a steppe landscape belt, plays some (though largely insignificant) role in the lake areas of the southern forest–steppe subzone.[16]

During epizootic and epidemic activity of OHF natural foci, Gamasidae ticks, as well as aquatic organisms belonging to the Hydracarinae, take part in OHFV circulation. Their involvement is confirmed by the identity of isolated strains with those isolated from muskrats and sick humans. Experiments with experimentally and spontaneously OHFV-infected Gamasidae ticks testify to the ability of longitudinal (more than six months) virus preservation.[17]

Vertebrate Hosts. The principal vertebrate host of OHFV, which is able to directly infect humans, is the muskrat (*Ondatra zibethicus*). This species was introduced into western Siberia from Canada in 1928. Their population density reached a modern-day high in the 1940s. Close interactions among *O. zibethicus* and local populations of *Arvicola terrestris* emerged. *Ar. terrestris* has periods of rapid population growth followed by epizootics of tularemia, leptospirosis, and OHFV. Muskrats suffered these epizootics together with other local species of rodents: *Microtus oeconomus*, *M. gregalis*, *Myodes rutilus*, *Apodemus agrarius*, and *Ar. terrestris*.[13] The OFV infection rate among muskrats is about 15% in both the autumn–winter and the spring–summer periods.[16] Latent infection was established in all rodents except the muskrat.[18]

OHFV was detected in birds and in mosquitoes, but the role of these two animals in virus circulation is not clear.[18–21]

Epidemiology. OHFV is transmitted both by Ixodidae tick bites and as the result of direct contact with infected muskrats, their flesh, and fresh fells.[1,5]

OHF morbidity during 1945–1949 reached 1.5–5.0%. Then there was a gradual decrease down to single cases. Most OHF cases (96.8%) were detected in the lake forest–steppe, in the south of the forest–steppe landscape zone, which occupies 14.5% of the territory where 15.3% of country people in the Omsk region live. The northern forest–steppe landscape zone is the youngest landscape of western Siberia, having evolved in place of the former southern taiga landscape zone.[22,23] In the south of western Siberia, the following territorial zones can be marked out: (i) the preferred territory of Tick-borne encephalitis virus (TBEV) (the southern taiga); (ii) intermediate territory (the boundary of the southern taiga with the northern forest—steppe); (iii) the preferred territory of OHFV (the northern and southern forest— steppe); and (iv) the territory of sporadic cases of OHF (part of the southern forest—steppe and steppe).[13,23] In the first zone, more than 90% of all cases of TBE in western Siberia are registered and only single OHF cases are found; in the second zone, 1% each of cases of TBE and OHF; in the third zone, 4% of TBE and 96% of OHF; and in the fourth zone, 4% of TBE and single cases of OHF.[13]

The seasonal incidence of OHF distinctly correlates with the activity of the principal Ixodidae tick vectors. Cases (a few) of OHF acquired by direct contact with muskrats occur mainly during the season in which the animals are hunted, in October—January. In the spring—summer season, OHF cases occur chiefly in rural areas. The age of patients ranges from 5 to 70 years, but cases occur mainly among middle-aged persons (40—50 years old). In the autumn—winter period, OHF occurs mainly among muskrats trappers (60%), adult members of their families (28%), and children (12%). It appears that all patients infected directly from muskrats develop symptomatic illness. Seroprevalence ranges from 0 to 32% in populations of endemic regions.[3,7,23] In the last decade of the twentieth century, an increase in OHF natural foci activity took place in the Tyumen (1987), Omsk (1988, 1999—2007), Novosibirsk (1989—2002; regular epidemic activity took place on the territory of only four administrative districts), and Kurgan (1992) regions. In the absolute majority of laboratory-confirmed cases, the nontransmissive pathway (direct contact with muskrats) of the infection dominated.[17]

Pathogenesis is determined first of all by the destruction of capillaries, the vegetative nervous system, and the adrenal glands.[16,24]

Clinical Features. The incubation period of OHFV is 2—4 days long. The disease begins abruptly, with fever, head and muscular pain, hyperemia, and injection in the sclera. The body temperature increases up to 39—40°C and stays that way for 3—4 days, then decreases a little and critically falls on the 7th to 10th day after symptoms appear. From the first days of the illness, there are diapedetic bleedings, especially in the nose. Recovery is usually complete, without any residual phenomena; lethal outcomes are possible, but are rare.[16,24—26]

Control and Prophylaxis. OHFV survives up to 20 days in lake water. Water can be contaminated by urine and feces of the infected muskrats or some other rodents. The water pathway in human infection has been discussed in the literature.[13,14]

Prevention of the infection depends on the use of protective respirators and rubber gloves in processing muskrat pelts and on personal protective measures against tick bites. TBE vaccine offers a high degree of protection against OHF.[10,23] Cases of laboratory-acquired OHF have been reported in unvaccinated persons, and TBE vaccine is recommended for laboratory personnel working with either virus.[23]

Interferon and its inductors have shown a high efficiency in preventing OHF in experiments using animal models.[27]

8.2.1.2 Powassan Virus

History. Powassan virus (POWV) (family Flaviviridae, genus *Flavivirus*, antigenic complex of TBE) was originally isolated by D.M. McLean and W.L. Donohue in September 1958 from the brain of a five-year-old child who was admitted to the hospital with blinking, tremors, and dizziness in the small town of Powassan in the north of Ontario, Canada.[1] The child later died of encephalitis. Subsequent virological and serological surveys carried out in the Powassan—North Bay and

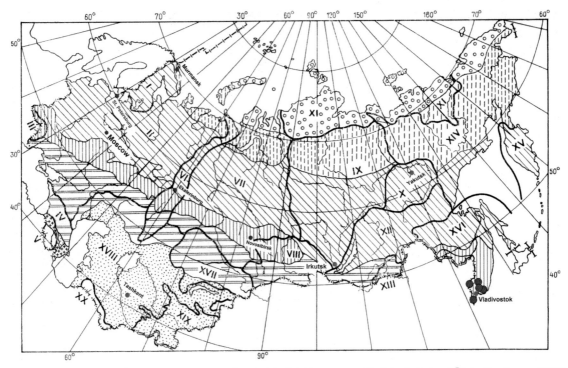

FIGURE 8.50 Places of isolation of POWV (family Flaviviridae, genus *Flavivirus*) strains (●) in the former USSR. (See other designations in Figure 1.1.)

Manitoulin Island areas of northern Ontario during 1959–1966 elucidated a summer transmission cycle involving Ixodidae ticks and small wild mammals.[2] Neutralizing antibodies were found in 1.1% of humans.[3] A virus strain isolated from a pool of *Dermacentor andersoni* collected in Colorado was identified as POWV.[4] In Russia, POWV was isolated in 1972 in Primorsky Krai from *Haemaphysalis neumanni* collected from spotted deer (*Cervus nippon*) (Figure 8.50).[5]

Taxonomy. POWV belongs to the tick-borne group of mammalian flaviviruses, together with TBEV, OHFV, Kyasanur Forest disease virus (KFDV), Alma-Arasan virus (AAV), Alkhurma fever virus (AHFV), Langat virus (LGTV), Gadgets Gully virus (GGYV), Louping ill virus (LIV), and Royal Farm virus (RFV).[6,7]

The genome of POWV is a about 10,835 nt in length. The virus comprises two genetic lineages, formed by POWV (lineage I) and the closely related deer tick virus (DTV, lineage II) (Figure 8.51).[8] Phylogenetic analysis based on partial sequences of the E gene showed that the population of POWV in Russia has a low genetic diversity.[9] The strains of POWV isolated in Russia have a high genetic similarity to the strains of lineage I isolated in North America. A full-length genome comparison revealed that Far Eastern isolates (LEIV-3070Prm, Spassk-9, and Nadezdinsk-1991) have a 99.5% identity with strain POWV/LB from Canada (Figure 8.51).

Arthropod Vectors. POWV was isolated from Ixodidae ticks collected in the Russian Far East and in the U.S. states of California,

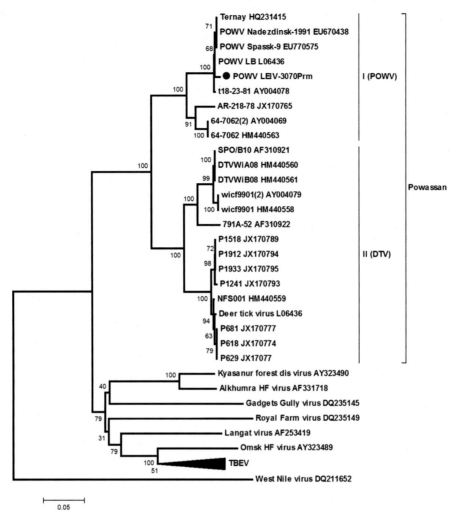

FIGURE 8.51   Phylogenetic analysis of the Eurasian and North American isolates of POWV (family Flaviviridae, genus *Flavivirus*) based on the partial sequences of E gene.

Colorado, Connecticut, Massachusetts, South Dakota, and West Virginia. Serological investigations of wild mammals indicate that POWV also circulates in the Canadian provinces of Alberta, British Columbia, and Nova Scotia.[3,10–12]

In North American natural foci, POWV was isolated from *Ixodes cookei, I. spinipalpus, I. marxi*, and *Dermacentor andersoni* ticks.[3,10,11] In the Far East, known vectors of POWV are *Haemaphysalis*

longicornis, Haem. concinna, Haem. japonica, D. silvarum, and *I. persulcatus* ticks.[5,9,13,14] Transphase and transovarial transmission of POWV in Ixodidae ticks has been established.

Vertebrate Hosts. In North America, POWV was isolated from wild mammals: the woodchuck (*Mormota monax*, the main reservoir), American red squirrel (*Tamiasciurus hudsonicus*), deer mouse (*Peromiscus maniculatus*), red fox (*Vulpes fulva*), eastern gray squirrel

(*Sciurus carolinensis*), North American porcupine (*Erethizon dorsatum*), striped skunk (*Mephitis mephitis*), raccoon (*Procyon lotor*), long-tailed weasel (*Mustela frenata*), and gray fox (*Urocyon cinereoargenteus*).[2,4,15] Infection of wild vertebrates most often is inapparent.[2,10] In the south of the Russian Far East (in Primorsky Krai), POWV was isolated from aquatic birds: the common teal (*Anas crecca*) and the mallard (*Anas platyrhynchos*).[9,13,14,16]

Epidemiology. Human infections of POWV were reported in Canada (Ontario and Quebec), the United States (New York and Pennsylvania),[2] and Russia (Primorsky Krai).[14,17,18] Nevertheless, human infection rarely develops.

Clinical Features. The clinical picture of developing meningitis and encephalomeningitis includes high temperature, dryness in the gullet, drowsiness, headache, disorientation, convulsions, vomiting, difficulty breathing, coma, and paralysis, with 11% lethality in the severe phase of the disease. Autopsy has revealed widespread perivascular and focal parenchymatous infiltration. In 50% of recoveries, consequent damage to the CNS develops, which could lead to death in 1−3 years.[2,18]

Control and Prophylaxis. The vaccine against TBEV is not effective against POWV.[2,17,19]

8.2.1.3 Tick-Borne Encephalitis Virus and Alma-Arasan Virus (var. Tick-borne encephalitis virus)

TBEV (family Flaviviridae, genus *Flavivirus*, TBEV antigenic group) is the natural foci for neuroinfection transmitted by Ixodidae ticks.

History. In 1931−1934, the Russian military medical doctor−neuropathologist A.G. Panov, together with his colleagues A.N. Shapoval and D.A. Krasnov, described a neuroinfection with a high level of mortality in the Far East. This neuroinfection later was called "spring−summer encephalitis."[1,2] During field expeditions in 1937−1940, the historical strain TBEV/Sofjin was isolated from the brain of a patient with encephalitis who died in Khabarovsk

Krai (Figure 8.52). In that period, the main vector of TBEV—*Ixodes persulcatus* ticks— was established, epidemiological peculiarities of TBE were studied, and the first anti-TBEV vaccine was developed on the basis of intracerebrally infected mouse brain and was successfully used in medical practice.[2] Complex expeditions were undertaken by a number of prominent virologists (L.A. Zilber (Figure 2.9), M.P. Chumakov (Figure 2.10), A.A. Smorodintsev (Figure 2.11), E.N. Levkovich (Figure 2.14), A.D. Sheboldaeva, and A.K. Shubladze (Figure 2.15)), bacteriologists (V.D. Soloviev (Figure 2.13) and N.V. Ryzhkov), parasitologists (Ye. N. Pavlovsky (Figure 2.12), A.V. Gutsevich, B.I. Pomerantsev, A.S. Monchadsky, and A.N. Skrynnik), and clinicians (A.G. Panov, A.N. Shapoval, and Z.I. Finkel).

Taxonomy. The antigenic complex of TBEV includes TBEV proper (with three genotypes; see next paragraph); GGYV (Australia, Oceania); KFDV (Hindustan) and the closely related AHFV (Arabian Peninsula); LGTV (Malay Peninsula); LIV (Europe), with British (LIV-Brit), Irish (LIV-I), Spanish (LIV-Spain), Turkish (Turkish sheep encephalitis virus (TSEV)), and Greek (Greek goat encephalitis virus (GGEV)) subtypes; OHFV; POWV and the closely related DTV; RFV; and Karshi virus (KSIV).[3−10]

The three genotypes of TBEV were established by antigenic or phylogenetic analysis: Far Eastern (TBEV-FE) (prototypical strain, TBEV/Sofjin; KC806252), Siberian (TBE-Sib) (TBEV/Aina; JN003206; TBEV/Vasilchenko; L40361), and European (TBE-Eur) (TBEV/Neudoerfl; U27495) (Figure 8.53). Genetic diversity among strains of different genotypes is about 32−33% nt and for the polyprotein precursor is 25−26% aa.

Strain TBEV/LEIV-1380Kaz (the former AAV) was isolated from *Ixodes persulcatus* in the low-mountain part of southeastern Kazakhstan (Alma-Ata Region) in 1977.[11]

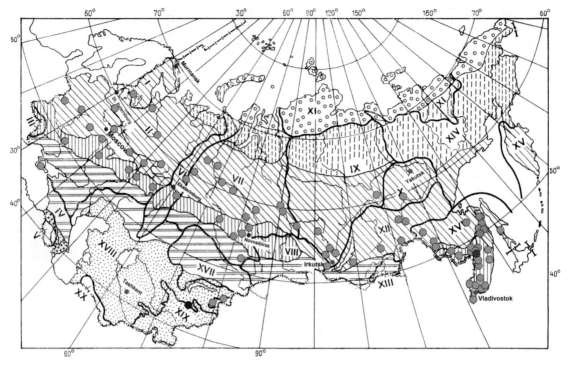

FIGURE 8.52   Places of isolation of TBEV (family Flaviviridae, genus *Flavivirus*) in the former USSR. (See other designations in Figure 1.1.)

Preliminary investigation revealed a one-sided antigenic relation between AAV and POWV.[12] AAV was associated with human cases of meningitis. Specific antibodies to AAV were found among ground squirrels (*Citellus fulvus*), agricultural animals, and humans. Later, the AAV genome was sequenced (GenBank ID: KJ 744033).[13] A full-length genome comparison showed that AAV has the highest similarity (94.6% nt and 98.3% aa identities) to the TBEV/Chita-653, TBEV/Irkutsk-12, TBEV/Aino, and TBEV/Vasilchenko strains belonging to the Siberian genotype (Figure 8.53).

Recent genetic studies of TBEV revealed two additional genotypes of this virus on the territory of eastern Siberia (Irkutsk Region): for the first one, only a single strain is known today; for the latter, there are five strains in Mongolia.[14] Thus, TBEV has a high level of genetic diversity in Northern Eurasia. TBEV-Sib genotype dominates in Europe, western Siberia, and eastern Siberia, TBEV-FE in the Far East.[15,16] The TBEV-FE genotype, which was widely distributed in Siberia and northeastern Europe, is now being displaced by TBEV-Sib. TBEV-FE strains are often pathogenic to laboratory mice, whereas TBEV-Sib frequently provokes severe and lethal disease.[15] Local populations of all genotypes of TBEV could be stable for a long time.[16]

Distribution. TBEV is distributed within the areas of distribution of its main vectors: *Ixodes persulcatus* and *I. ricinus* ticks (Figure 8.54—see details in the detailed work of E.I. Korenberg[17]). In Russia, those areas are the Far East, Siberia, the Ural region, and the European part of Russia[18–24] (Tables 8.25 and 8.26); in Fennoscandia, Finland,[20,25,26] Sweden,[19,27,28] and

FIGURE 8.53 Phylogenetic analysis of the complete genome nucleotide sequences of certain strains of TBEV (family Flaviviridae, genus *Flavivirus*).

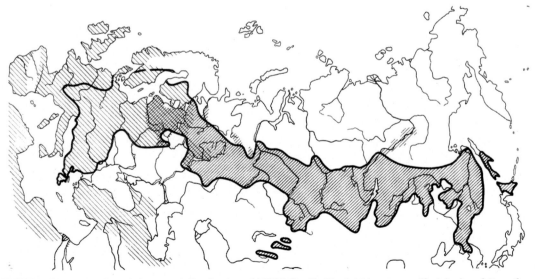

FIGURE 8.54 Coincidence of areas of distribution of TBEV (family Flaviviridae, genus *Flavivirus*) and its arthropod vectors *Ixodes persulcatus* and *I. ricinus* in Northern Eurasia. (See other designations in Figure 1.1.)

TABLE 8.25   Incidence of TBE in Europe

Year	Austria	Hungary	Germany	Denmark	Italy	Latvia	Lithuania	Norway	Poland	Slovakia	Slovenia	Finland	France	Croatia	Czech Republic	Sweden	Switzerland	Estonia
1996	128	224	114	0	8	309	716	0	257	101	406	10	1	57	571	44	62	177
1997	99	99	211	0	8	874	645	0	201	76	274	19	1	25	415	76	123	404
1998	62	84	148	1	11	1,029	548	1	209	54	136	17	2	24	422	64	68	387
1999	41	51	115	4	5	350	171	1	101	57	150	12	5	26	490	53	112	185
2000	60	45	133	3	15	544	419	2	170	92	190	41	0	18	719	133	91	272
2001	54	76	253	1	19	303	298	1	205	76	260	33	0	27	411	128	107	215
2002	60	80	226	1	6	153	168	2	126	62	262	38	2	30	647	105	53	90
2003	82	114	278	4	14	365	763	1	339	74	275	16	6	36	606	105	116	237
2004	54	89	274	8	23	251	425	3	262	70	204	31	7	38	500	160	138	182
2005	100	52	431	4	22	142	242	3	174	28	297	17	0	28	642	130	206	164
2006	84	56	546	0	14	170	462	3	316	91	373	18	6	20	1029	163	259	171
2007	45	62	238	2	4	171	234	13	233	46	199	20	7	12	542	190	113	140
2008	86	70	285	1	34	181	220	9	202	77	246	23	10	20	630	224	127	90
2009	79	64	313	1	32	328	617	8	335	71	307	26	0	0	816	211	118	179

TABLE 8.26   TBE Cases in Russia in 2013

Federal District	Number of TBE cases/portion (%)	Infection rate per 100,000 population
Central	40/1.8	0.10
Northwestern	288/12.8	2.11
Southern	0	0.00
North Caucasian	0	0.00
Volga	272/12.0	0.91
Ural	315/13.9	2.60
Siberian	1,307/58.0	6.79
Far Eastern	33/1.5	0.53
Total	2,255/100.0	

Norway;[29–31] in the rest of Europe, the Czech Republic,[8,32] Slovakia,[6,33,34] Bulgaria,[35] Hungary,[36,37] Poland,[38,39] Croatia,[40] Latvia,[41] Lithuania,[42] Estonia,[43,44] Denmark,[31] Germany,[45–48] Austria,[49] Slovenia,[50] France,[51] Italy,[52,53] and Spain[54] (Table 8.25); And in Asia, the Russian Far East and Siberia,[1,16,55] Japan (Hokkaido),[55] North and South Korea,[56,57] China,[58] Mongolia,[59] Kazakhstan,[13] and Kyrgyzstan.[60]

Arthropod Vectors. Natural TBEV infection has been observed in 16 species of Ixodidae ticks. The principal arthropod vectors for TBEV in Russia are the Ixodidae ticks *Ixodes persulcatus* (in the Far East, Siberia, and the north of the European part of the country) and *I. ricinus* (in the south of the European part) (Figure 8.54).

The infection rate in *I. ricinus* is about 0.2% in the northwestern part of the Russian Plain, 0.4% in Lithuania, and 3.8% in Crimea. The infection rate in *I. persulcatus* is, as a rule, higher, ranging from 0.6% to 4.8%. Occasional vectors are *I. hexagonus*,[61] *I. pavlovskii*,[17,62] *Haemaphysalis inermis*, *Haem. concinna*, *Haem. punctata*, *Haem. longicornis*, *Haem. japonica*, and *Dermacentor reticulatus*.[35,63–66] The main TBEV vector in Japan (Hokkaido) is *I. ovatus*.[67,68] In the Korea Peninsula, where both *I. ricinus* and *I. persulcatus* are absent, the main TBEV vector is *Haem. longicornis* and *Haem. flava*.[69]

The northern boundary of *I. persulcatus* and *I. ricinus* lies within the limits of an effective temperature sum isoline of about 1,000–1,300°C (the middle taiga landscape belt). The most suitable climatic conditions for these ticks are within the south taiga. Imago tick activity begins in the third d decade of April and reaches a maximum in the second and third decades of May or in the first and second decades of June, with activity beginning to decrease in the third decade of June. This time frame correlates with morbidity dynamics having an 8- to 10-day lag (Figure 8.55).[70]

The ecological links of TBEV during its circulation in natural foci are extremely diverse as the result of wide distribution of this virus (Figures 8.52 and 8.54). Ixodidae ticks, mainly *I. persulcatus*, are the natural reservoir of TBEV and the core of natural foci.[12,62,71,72]

From the very beginning of the tick's larval stage, a suctional, tarlike liquid appears around the hypostome and becomes rosin.[62,73] The quantity of virus in this rosin plug is comparable to that in the tick's body (10^3–10^4 PFU/mcL).[74] The place of suction on the body of the host is significant for the development of infection; for example, suction in the axillary hollow results in the highest lethality (16.1%, 1.5 times more in comparison to suction in the neck and in the head.[73]

Ticks become infected as they suck blood from a vertebrate host with a level of viremia that is equal to or higher than the threshold required for infection. Ticks can also become infected from an *uninfected* vertebrate host as they suck blood together with infected ticks.[73,75] Transovarial and transphase transmission of TBEV has been described in the literature; nevertheless, the effectiveness of vertical transmission of TBEV is low. (About 1% of progeny turn out to be infected).[52,76] The sexual pathway of TBEV transmission from male to female is quite effective (about 50%).[77–79] The aggressiveness and activity of TBEV-infected Ixodidae ticks increases with the TBEV titer in their bodies.[62,75] Infected ticks have been found on the clothing of

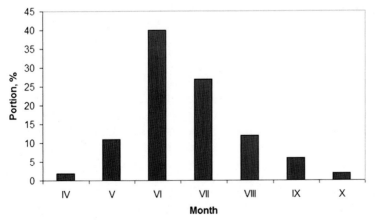

FIGURE 8.55 Trends in the incidence of TBE in Russia, by month (as a percentage of the amount of disease for the year, according to long-term data).

humans at a fequency 5–20 times higher than uninfected ticks have been found.[48,62,75]

TBEV has been isolated from the mosquitoes *Anopheles hyrcanus* in Kyrgyzstan[80] and *Aedes* sp. in western Siberia.[81] The strain TBEV/Malyshevo was isolated from *Aedes vexans nipponii* mosquitoes collected in 1978 on the coast of Petropavlovskoe Lake in Khabarovsk Krai in the Russian Far East (48°40′N, 135°41′E).[82–84] A preliminary investigation[82] concluded that this strain belonged to Negishi (NEGV) virus,[85] and later the possibility was discussed that the strain belonged to a separate, Malyshevo virus. Then, phylogenetic analysis using a next-generation sequencing approach revealed that Malyshevo is a strain of TBEV and is closely related to TBEV strains isolated in the Far East: TBEV/1230, TBEV/Spassk-72, TBEV/Primorye-89.[13]

TBEV has been isolated many times from-ticks and fleas of the superfamily Gamasoidea living in nests of rodents and birds (Table 8.27), even during the winter period.[2,47,86–89]

Vertebrate Hosts. Hosts for the preimago stage of Ixodidae ticks—Asian chipmunks (*Tamias sibiricus*), shrews (members of the Soricidae family), bank voles (*Myodes glareolus*), field voles (*Microtus agrestis*), mountain hares (*Lepus timidus*), and 74 species of birds (Table 8.28)—have great significance in TBEV circulation.[10,12,62,64,71,72,90,91] Persistent TBEV infection in bank voles and field voles has been found during the winter period.[26]

Infection among vertebrates occurs mainly by tick bites. In rare instances, alimentary transmission of TBEV through milk containing viruses is possible.[34,92]

Epidemiology. There are two basic modes of human infection by TBEV: (i) as the result of being bitten by infected Ixodidae ticks (the main mode); and (ii) as the result of consuming infected raw goat, sheep, and cow meat, milk, or dairy products (mainly in natural foci linked to *Ixodes ricinus*).[23,32,93] The latter pathway of TBEV distribution often involves whole families. As much as 70% of cases in Belarus have been alimentary.[70] TBEV can persist in milk at 60°C for more than 10 min, and some of the viruses can remain viable even after pasteurization at 62°C for 20 min. Nor is TBEV inactivated after 24 h at 4°C and pH 2.8. Many laboratory infection cases (usually by aerosol) have been described.

Several hundred cases are recorded in Europe (Table 8.25) and in Russia (Table 8.26) each year, with considerable interannual variation.[17,70,94–96] The highest level of TBE morbidity is registered in the Baltic states (Latvia, 6.2–10.8 per 100,000 population); Lithuania, 6.5–13.5; and Estonia, 10.4–13.5) and in Slovenia (10.2–18.6) and the Czech Republic (5.0–10.0). In neighboring Austria, where the vaccination rate is higher, the index is lower (0.6–1.2).[97] Seasonal TBE morbidity in Russia is connected with seasonal activity of the Ixodidae tick vectors (Figure 8.55).

The risk of infection depends upon the frequency of tick bites, which is different for populations living in the different landscape belts. Results of an investigation of almost 200,000 people demonstrate that the highest risk is for the population living in the southern taiga belt, where about 20% of adults were found to have tick bites during one epidemic season (Table 8.29).

In rural localities of the southern taiga belt, about half of schoolchildren and about 80% of adults have antibodies to TBEV. For comparison, only 14–20% of adult citizens of Kemerovo, a city of about half a million in western Siberia, and 2–3% of citizens of Moscow have antibodies specific to TBEV (Table 8.29).[98]

A mathematical model for evaluating the infection rate and the probability of developing the disease as a function of the density of the tick population, its infection rate and biting activity, and the level of the immune human layer was developed by D.K. Lvov and coauthors.[98–102] The same approach, which is also suitable for other arboviral infections, was

TABLE 8.27 TBEV Isolation from Birds and their Ectoparasites in Northern Eurasia

	Taxonomy of birds			Place of collection	
Order	Family	Genus	Species	Physicogeographical land	Region of Russia
1	2	3	4	5	6
Anseriformes	Anatidae	Clangula	Long-tailed duck (C. hyemalis)	Russian Plain	European part
		Melanitta	White-winged scoter (M. fusca)	Russian Plain	European part
		Anas	Garganey (An. querquedula)	Russian Plain	European part
Falconiformes	Accipitridae	Accipiter	Goshawk (A. gentilis)	Eastern Siberia	Irkutsk region
		Butastur	Buzzard eagle (B. indicus)	Southern part of the Far East	Primorsky Krai
Galliformes	Phasianidae	Tetrastes	Grouse (T. bonasia)	Western Siberia, southern part of the Far East	Tomsk region, Khabarovsk Krai
			Black grouse (T. tetrix)	Eastern Siberia, southern part of the Far East	Irkutsk region, Primorsky Krai
Ralliformes	Rallidae	Crex	Corn crake (C. crex)	Eastern Siberia	Irkutsk region
Charadriiformes	Charadriidae	Charadrius	Little ringed plover (C. dubius)	Eastern Siberia	Irkutsk region
		Scolopax	Woodcock (S. rusticola)	Eastern Siberia	Irkutsk region
Columbiformes	Columbidae	Streptopelia	Oriental turtle dove (S. orientalis)	Southern part of the Far East	Primorsky Krai
		Columba	Pigeon (C. livia)	Eastern Siberia	Irkutsk region
Mesostigmata	Dermanyssidae	Dermanyssus	Mites D. gallinae from pigeons (C. livia)	Eastern Siberia	Irkutsk region
			Mites D. hirundinis from common starling (S. vulgaris)	Eastern Siberia	Irkutsk region
Cuculiformes	Cuculidae	Cuculus	Himalayan cuckoo (C. saturatus)	Eastern Siberia, southern part of the Far East	Irkutsk region, Primorsky Krai
			Lesser cuckoo (C. poliocephalus)	Eastern Siberia, southern part of the Far East	Irkutsk region, Primorsky Krai
Piciformes	Picidae	Jynx	Eurasian wryneck (J. torquilla)	Eastern Siberia	Irkutsk region
		Picoides	Three-toed woodpecker (P. tridactylus)	Eastern Siberia	Buryatia Republic
		Dendrocopos	Great spotted woodpecker (D. major)	Ural	Perm region
			Lesser spotted woodpecker (D. minor)	Eastern Siberia, southern part of the Far East	Buryatia Republic, Primorsky Krai
			White-backed woodpecker (D. leucotos)	Southern part of the Far East	Primorsky Krai

(Continued)

TABLE 8.27 (Continued)

	Taxonomy of birds			Place of collection	
Order	Family	Genus	Species	Physicogeographical land	Region of Russia
1	2	3	4	5	6
Coraciiformes	Meropidae	Merops	European Bee-eater (M. apiaster)	Central Asia	Kyrgyzstan
Ixodida	Ixodidae	Ixodes	Ticks I. persulcatus from the nests of sand martins (R. riparia)	Northern part of the Far East	Yakut Republic
Siphonaptera	Ceratophyllidae	Ceratophyllus	Fleas C. maculates from the nests of house martins (Delichon urbicum)	Ural	Perm region
Passeriformes	Alaudidae	Alauda	Eurasian skylark (A. arvensis)	Eastern Siberia, central Asia	Buryatia Republic, Kyrgyzstan
	Hirundinidae	Riparia	Sand martin (R. riparia)	Eastern Siberia	Irkutsk region
	Motacillidae	Anthus	Tree pipit (A. trivalis)	Western Siberia, eastern Siberia	Tomsk region, Buryatia Republic
			Pechora pipit (A. gustavi)	Southern part of the Far East	Khabarovsk Krai
		Motacilla	Pied wagtail (M. alba)	Eastern Siberia	Irkutsk region
			Yellow wagtail (M. flava)	Southern part of the Far East	Khabarovsk Krai
	Campephagidae	Pericrocotus	Rosy minivet (P. roseus)	Southern part of the Far East	Primorsky Krai
	Laniidae	Lanius	Brown shrike (L. cristatus)	Southern part of the Far East	Primorsky Krai
			Chinese grey shrike (L. sphenocercus)	Eastern Siberia	Irkutsk region
	Troglodytidae	Troglodytes	Northern wren (T. troglodytes)	Southern part of the Far East	Primorsky Krai
	Turdidae	Turdus	Siberian thrush (T. sibirica)	Eastern Siberia	Irkutsk region
			Song thrush (T. philomelos)	Eastern Siberia, western Siberia	Tomsk region, Buryatia Republic
			Naumann's thrush (T. naumanni)	Southern part of the Far East	Primorsky Krai
			Red-throated thrush (T. ruficollis)	Western Siberia	Tomsk region
			Grey-backed thrush (T. hortulorum)	Southern part of the Far East	Khabarovsk Krai
			Pale thrush (T. pallidus)	Southern part of the Far East	Primorsky Krai
			Common blackbird (T. merula)	Western European part	Kaliningrad region
			Eyebrowed thrush (T. obscurus)	Southern part of the Far East	Khabarovsk Krai

Family	Genus	Common name (species)	Region	Location
Muscicapidae	*Monticola*	White-throated rock thrush (*M. gularis*)	Southern part of the Far East	Primorsky Krai
	Tarsiger	Red-flanked bluetail (*T. cyanurus*)	Southern part of the Far East	Primorsky Krai
	Luscinia	Siberian blue robin (*L. cyane*)	Southern part of the Far East	Khabarovsk krai
Bombycillidae	*Bombycilla*	Bohemian waxwing (*B. garrulus*)	Eastern Siberia	Buryatia Republic
Sylviidae	*Sylvia*	Lesser whitethroat (*S. curruca*)	Eastern Siberia	Buryatia Republic
Muscicapidae	*Ficedula*	Mugimaki flycatcher (*F. mugimaki*)	Eastern Siberia	Buryatia Republic
	Muscicapa	Brown flycatcher (*M. latirostris*)	Southern part of the Far East	Primorsky Krai
Acrocephalidae	*Acrocephalus*	Blyth's reed warbler (*A. dumetorum*)	Ural	Perm region
Paridae	*Parus*	Great tit (*P. major*)	Western European part of Russia, Eastern Siberia	Kaliningrad region, Buryatia Republic
	Periparus	Coal tit (*P. ater*)	Eastern Siberia	Buryatia Republic
Remizidae	*Remiz*	Black-headed penduline tit (*R. macronyx*)	Eastern Siberia	Buryatia Republic
Aegithalidae	*Aegithalos*	Long-tailed tit (*Ae. caudatus*)	Eastern Siberia	Buryatia Republic
Sittidae	*Sitta*	Eurasian nuthatch (*S. europaea*)	Southern part of the Far East	Khabarovsk Krai
Emberizidae	*Emberiza*	Yellowhammer (*E. citrinella*)	Western Siberia	Tomsk region
		Chestnut-eared bunting (*E. fucata*)	Southern part of the Far East	Khabarovsk Krai, Primorsky Krai
		Yellow-breasted bunting (*E. aureola*)	Eastern Siberia	Irkutsk region
		Meadow bunting (*E. cioides*)	Eastern Siberia, Southern part of the Far East	Buryatia Republic, Primorsky Krai
		Little bunting (*E. pusilla*)	Southern part of the Far East	Primorsky Krai
		Yellow-throated bunting (*E. elegans*)	Southern part of the Far East	Primorsky Krai
		Tristram's bunting (*E. tristrami*)	Southern part of the Far East	Primorsky Krai
		Black-faced bunting (*E. spodocephala*)	Southern part of the Far East	Primorsky Krai

(Continued)

TABLE 8.27 (Continued)

	Taxonomy of birds			Place of collection	
Order	Family	Genus	Species	Physicogeographical land	Region of Russia
1	2	3	4	5	6
Passeriformes (Continued)	Fringillidae	Acanthis	Common redpoll (A. flammea)	Eastern Siberia	Irkutsk region
		Eophona	Chinese grosbeak (E. migratoria)	Southern part of the Far East	Primorsky Krai
			Japanese grosbeak (E. personata)	Southern part of the Far East	Primorsky Krai
		Pyrrhula	Eurasian bullfinch (P. pyrrhula)	Eastern Siberia	Buryatia Republic
		Uragus	Long-tailed rosefinch (Ul. sibiricus)	Southern part of the Far East	Primorsky Krai
		Pinicola	Pine grosbeak (P. enucleator)	Eastern Siberia	Buryatia Republic
		Loxia	Common crossbill (L. curvirostra)	Eastern Siberia	Buryatia Republic
		Fringilla	Brambling (F. montifringilla)	Western Siberia	Tomsk region
	Passeridae	Passer	Tree sparrow (P. montanus)	Western European part of Russia	Ukraine
	Sturnidae	Sturnus	Common starling (S. vulgaris)	Eastern Siberia	Irkutsk region
	Corvidae	Nucifraga	Spotted nutcracker (N. caryocatactes)	Eastern Siberia	Irkutsk region
		Garrulus	Eurasian jay (G. glandarius)	Eastern Siberia	Irkutsk region
		Perisoreus	Siberian jay (P. infaustus)	Eastern Siberia	Buryatia Republic
	13	33	53	76	**Within the boundaries of distribution of I. persulcatus and I. ricinus ticks**

TABLE 8.28 Number (Thousands) of Humans Investigated /Frequencies (%) of Ixodidae Tick Bites in the Various Western Siberian Landscape Belts (One Epidemic Season)

Landscape belt	Age groups (years)				
	<3	3−7	8−17	≥18	Sum
Southern taiga	2.6/5.0	6.2/12.9	13.4/19.9	32.2/18.2	54.4
Mountain taiga	0.7/0.9	2.1/1.4	5.6/3.2	9.9/3.9	18.3
Forest−steppe	7.0/0.6	15.1/4.5	31.2/7.9	67.7/8.0	121.0
Total	10.3	23.4	50.2	109.8	193.7

TABLE 8.29 Specific Antibodies ((Number of Samples Investigated)/(Portion (%) of Positive Results) to TBEV Among Populations Living in the Various Landscape Belts of Western Siberia and in Moscow (As An Outside Point)

Landscape belt	Setting	Serological method[a]	Age groups (years)		
			≤7	8−17	≥18
Southern taiga	Rural	HIT	25/16	100/53	274/78
		NT	90/12	146/60	376/88
	Urban	HIT	−	−	568/59
		NT	−	−	476/60
Mountain Taiga	Rural	HIT	81/10	191/26	422/64
		NT	76/17	64/20	121/44
	Urban	HIT	−	−	80/20
		NT	−	−	92/20
Forest−steppe	Rural	HIT	64/11	138/17	746/47
		NT	−	−	15/47
	Urban	HIT	−	−	296/21
		NT	−	−	103/32
Steppe	Rural	HIT	34/12	49/12	111/21
Kemerovo	Urban	HIT	−	−	454/14
		NT	−	−	54/9
Moscow	Urban	HIT	−	−	266/3
		NT	−	−	49/2
Total		HIT	204/11	478/28	3,217/42
		NT	166/14	210/48	1,286/58
Sum		HIT	3,899/38		
		NT	1,662/52		

[a]Abbreviations: HIT, hemagglutination inhibition test; NT, neutralization test; −, no data.

used for landscape-epidemiological zoning of TBEV natural foci in Altai Krai in the southern part of western Siberia: More than 10,000 residents living in the different landscape belts on a territory about 250,000 km^2 were tested by serological methods (Figure 8.56). The tests produced a good fit between calculated and registered morbidity data (Table 8.30).

Pathogenesis. TBE can be realized in several pathogenetic variants. An inapparent clinical form is characterized by short-term localization of TBEV in lymph nodes and immune cells, as well as by extranervous reproduction without viremia. Infection is terminated by the development of stable immunity. About 95% of cases of infection are inapparent.[102] Clinical fever is expressed as a common infectious process, but both the central and the peripheral nervous system are involved in the pathology.[103] Neuroinfection is characterized by lesion of the envelope and substance of the spinal cord and CNS.

Clinical Features. The incubation period ranges from 1 to 30 days, but usually is 7–12 days. The onset of illness in typical cases is abrupt and with a headache. The temperature

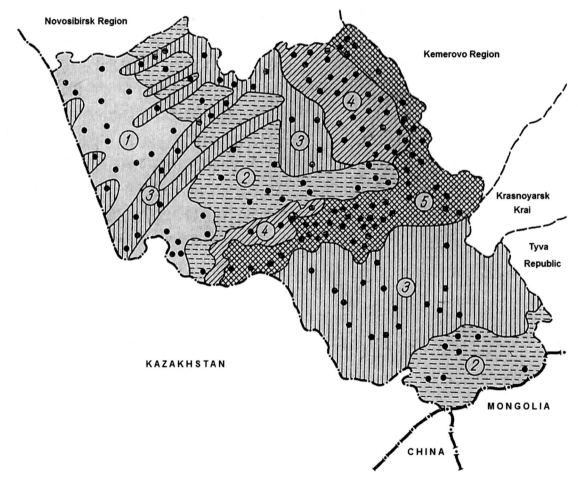

FIGURE 8.56 Landscape-epidemiological zoning of the territory of Altai Krai for TBE. Designations for epidemiological regions in respect to the level of immune layer: 1. 0-5%; 2. 6-10 %; 3. 11-20%; 4. 21-40%; 5. 41-60%.

TABLE 8.30 Calculated and Registered Tick-Borne Encephalitis Morbidity Per 100,000 in a Rural Population of Western Siberia

| | Age groups (years) | | | | | |
| | 8−12 | | 13−22 | | 23−57 | |
Landscape belt	Registered	Calculated	Registered	Calculated	Registered	Calculated
Southern taiga	99	107	84	93	53	48
Forest-steppe	41	47	42	38	39	45

rises to 38−40°C. Malaise, vomiting, general hyperesthesia, and photophobia develop.

Clinical symptoms of TBE, as well as the severity of the disease, are at least partially determined by biological properties of the virus.[104] There are two main clinical forms of TBE: the Far Eastern variety, associated with Far Eastern and Siberian strains of the virus, and the European variety (also known as Western biphasic meningoencephalitis or biphasic milk fever), associated chiefly with European strains. Human disease of the first type is usually clinically more severe in the acute phase, but is associated with a lower rate of chronic CNS sequelae.

The first phase starts with sudden fever, flulike symptoms (pronounced headache, weakness, nausea, myalgia, arthralgia), and conjunctivitis. The second phase appears after 4−7 days of apparent recovery, but then the CNS is affected (meningoencephalitis appears), accompanied with fever, retrobulbar pain, photophobia, stiff neck, sleeping disorders, excessive sweating, drowsiness, tremors, nystagmus, meningeal signs, ataxia, pareses of the extremities, dizziness, confusion, psychic instability, excitability, anxiety, disorientation, and/or memory loss. TBEV produces diffuse degenerative changes in neurons, perivascular lymphocytic infiltration, and damage to Purkinje cells in the CNS. Mortality ranges from 1% (TBEV-Eur), to 8% (TBEV-Sib), to 20−40% (TBEV-FE). Convalescence is prolonged, and neurological and psychotic sequelae often include paresis and atrophic paralysis of the neck and shoulders.[27,45,104] A chronic form of the disease occasionally combines with a progressive course (called Kozhevnikov's epilepsy), in which progressive neuritis of the shoulder plexus, multiple sclerosis, and progressive muscle atrophy often develop.[105,106] The chronic form is registered in 1−2% of all TBE cases and is said to be the result of virus−immunity interactions.[19]

Many authors have noted a decreasing number of severe TBE cases.[103]

Diagnostics. Laboratory diagnosis of TBE involves both serological (ELISA, hemagglutination inhibition test (HIT), neutralization testing) and virological methods (virus isolation using a biological model of intracerebrally inoculated newborn mice, 5−6 g mice, cell culture), as well as highly sensitive RT-PCR and real-time RT-PCR.

Control and Prophylaxis. Specific and nonspecific prophylaxis tools are highly efficient if they are utilized correctly. Personal safety includes protection against ticks. Vaccination against TBEV has a long history of success. Mass vaccination of populations in the endemic territory is necessary. A full course of vaccination provides 98% safety.[102] All vaccines produced in Russia are effective in the entire area of distribution of TBEV, independently of the strain used to prepare the vaccine. Vaccination has reduced TBE morbidity down to single cases in Austria, the Czech Republic, and Slovakia.[107]

Single cases of TBEV among vaccinated persons need to be investigated because possible causes are personal peculiarities of the immune system and errors in the control of vaccine production.[108] The presence of brain tissue in vaccines produced on the basis of intracerebrally inoculated newborn mice was a source of danger for a long time: Demyelinating encephalitis could develop. This danger was eliminated after vaccines were developed which used TBEV strains that reproduced in cell cultures. In the 1960s, cell culture vaccines against TBEV were developed by E.N. Levkovich and G.D. Zasukhina, and their high efficiency was demonstrated during 1961−1964 in controlled epidemiological trials carried out by D.K. Lvov in Kemerovo Region, western Siberia (Figure 2.36). The total number of people tested was 1,779,000.[101] Wide use of vaccination is the most effective instrument for reducing TBE morbidity.[109−111]

Etiotropic treatment includes three groups of antivirals: (i) serological (specific anti-TBEV immunoglobulin, immune blood plasma); (ii) enzymes (ribonuclease); and (iii) interferon.[27,103,104]

8.2.1.4 Japanese Encephalitis Virus

History. Japanese encephalitis virus (JEV) was originally isolated by H. Hayashi in 1933 from a patient who died with encephalitis and then, again, in 1935 from a patient who died with a fever in Tokyo.[1,2] Before that, however, Japanese encephalitis (JE) epidemics was documented in Japan in 1903 and onward as "Ioshiwara cold." In the south of the Russian Far East, strains of JEV were known since the end of the 1930s (Figure 8.57).

JE played a role in the historical events of World War II. American military personnel massed on Okinawa and preparing to invade

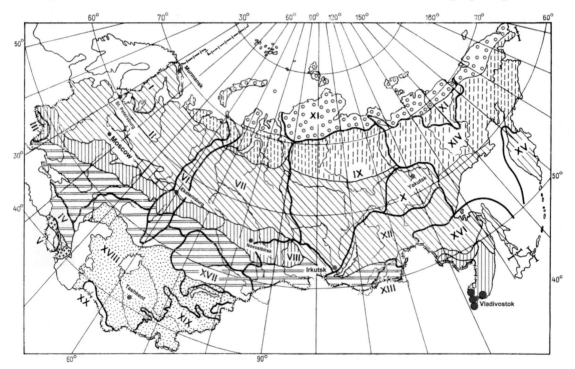

FIGURE 8.57   Places of isolation of Japanese encephalitis virus (JEV) (family Flaviviridae, genus *Flavivirus*) strains (●) in the former USSR. (See other designations in Figure 1.1.)

Japan were demoralized by an outbreak of encephalitis among the indigenous people. A fictionalized account of the risk from JE for American soldiers during World War II underscores the military risk.[3]

Taxonomy. Phylogenetic studies indicated that JEV isolates be divided into five genotypes, the distributions of which overlapped (Figure 8.58). Genotypes I, II, and III are most prevalent and are spread throughout Asia (Japan, China, India, Korea, Malaysia, and Vietnam), the Far East of Russia, and northern Australia. Genotypes IV and V are rarer and were isolated in Indonesia and India, respectively. Genotypes I and III are found mostly in temperate zones, whereas genotypes II and IV predominate in tropical zones.[4–6] Genetic diversity between strains of the different genotypes ranges from 9.1% to 16.6%.

Arthropod Vectors. JEV circulation in the equatorial and subequatorial climatic zones is year-round and is seasonal in the tropical, subtropical, and temperate belts, with a peak at the end of summer and the beginning of fall. JEV is brought from the equatorial and tropical climatic belts to the subtropical and temperate belt during the spring migration of birds.

About 30 species of mosquitoes are able to transmit JEV; nevertheless, only some of them are effective vectors. The main vector in Japan, the Philippines, the Korean Peninsula, China, the Indochinese Peninsula (except Malaysia), Indonesia, Sri Lanka, India, and Nepal is

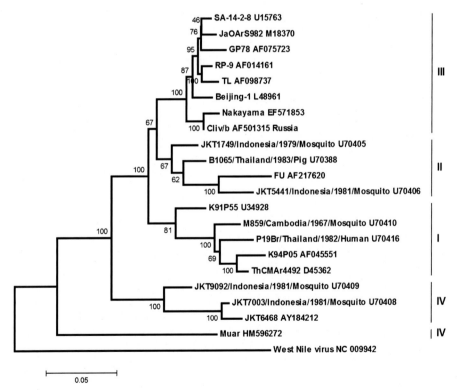

FIGURE 8.58 Phylogenetic analysis of E gene sequences of different isolates of JEV (family Flaviviridae, genus *Flavivirus*).

mosquitoes belonging to the *Culex vishnui* group (in particular, *Cx. tritaeniorhynchus*). Additional vectors are *Cx. vishnui* in India, Thailand, and southern China; *Cx. fuscocephala* in Malaysia, Thailand, and southern China; *Cx. gelidus* in Indonesia, Thailand, and Vietnam; and *Cx. annulus* —in southern China. The JEV contamination index reaches 1:200 among *Cx. tritaeniorhynchus* imagoes, 1:30,000 among *Cx. vishnui* larvae, and 1:550 among *Cx. gelidus* imagoes. Epidemics usually develop after plentiful precipitation and a long rise in environmental temperatures until they are no less than 25°C (but within the range 25−32°C).[7]

For a long time, the main vector for JEV in the south of Primorsky Krai in Russia was *Culex tritaeniorhynchus*. In the 1940s, as a result of both improvements in agriculture and meteorological changes, this species of mosquitoes consisted about 80% of all field collections. In subsequent years, however, their numbers abruptly declined, and by the 1960s the species represented only 0.15−0.75% of all mosquitoes collected. *Cx. pipiens* is an accessory vector, and *Aedes togoi* transmits JEV in seashore areas. JEV was also isolated in 1989 from *Ae. vexans*.[8,9]

Vertebrate Hosts. Aquatic and semiaquatic birds (especially herons) have the main significance in the natural cycle of JEV circulation. Regular transfer of JEV in migratory birds from endemic territories with year-round circulation of the virus to regions of the southern part of the temperate climatic belt (in particular, the southern part of Primorsky Krai, to the south from Lake Khanka[11]) is likely.[10] JEV transfer over hundreds of kilometers by infected mosquitoes is possible as well, especially in areas with a monsoonal climate (e.g., in Australia through the Torres Strait[12−14]). Birds transfer JEV from natural to synantropic biocenoses, where, thanks to *Culex tritaeniorhynchus* mosquitoes willingly attacking wild birds, pigs, persons, synantropic birds, and domestic animals (chiefly pigs), these all join into JEV circulation.[7] Infection in pigs could be inapparent, or it could be clinically expressed with encephalitis and a lethal outcome. The level of viremia in infected pigs is enough to infect mosquitoes. Such epizootics among pigs are, in effect, amplifiers for JEV, serving as prerequisites for the development of epidemics, first of all among people living in the countryside, but then among city dwellers as well.

Antibodies to JEV specifically were revealed among wild boars (83%), raccoons (59%),[14] and dogs (17%).[7] In the south of China, JEV was isolated from both Leschenault's Rousette (*Rousettus leschnaulti*), a species of fruit bat, and the little tube-nosed bat (*Murina aurata*),[15] and anti-JEV antibodies were identified in the blood of those animals.[16] JEV preservation in bats could be one of the mechanisms of the year-round circulation of the virus in its natural foci, with activation in the spring and subsequent replication and spreading in the summer and autumn.

In natural foci, birds are the principal vertebrate hosts contributing to transmission of the virus; in synantropic foci, pigs are the most important vertebrate hosts.[10,11] JEV has been isolated from the grey-headed bunting (*Emberiza fucata*), great cormorant (*Phalacrocorax carbo*), Japanese thrush (*Turdus cardis*), azure-winged magpie (*Cyanopica cyana*), Japanese wagtail (*Motacilla grandis*), barn swallow (*Hirundo rustica*), and night heron (*Nicticorax nicticorax*). Natural foci are situated in meadows. Of late, *Culex tritaeniorhynchus* has become more abundant in connection with intensive rice cultivation, portending the possibility of increased JEV circulation and epidemics.[17,18]

Epidemiology. All the territory of Japan, except for northern part of Hokkaido,[7] is endemic, but most diseases are registered near islands in a closed sea, as well as in Tokyo and adjacent prefectures.[3] Before 1966, outbreaks of JE emerged in Japan practically every year, with 1,200−2,700 patients seen.

Later, morbidity began to decrease to tens of cases per year. In the 1970a and 1980s, morbidity fell to the level of single cases per year. The main cause of the decrease was a significant drop in the population of the main JEV vector—*Culex tritaeniorhynchus* mosquitoes—as the result of a reduction in the acreage of rice fields as well as water pollution in places of mosquito habitation. In addition, the program of mass vaccination carried out annually among children of school age and a change in the structure of pork farms lessening the availability of pigs played a significant role in the falloff in the mosquito population.

JE is a serious problem in 20 countries of southeast Asia and Oceania.[19] During the last few years, more than 50,000 cases per year were registered, with about 20% lethality.[19] Morbidity increases annually in Bangladesh, Indonesia, Laos, Myanmar, North Korea, and Pakistan.[19,20] In addition, , the occurrence of an epidemic in southeastern Asian countries is becoming more and more likely because those countries are now seeking to increase their production of rice. The greatest risk of JE is said to be in China, Nepal, Sri Lanka, Thailand,[21] Laos, and Vietnam. JE is of the highest importance among all kinds of endemic encephalitis, potentially threatening nearly 50% of the population of our planet.[3] The disease especially affects military contingents, as it did the American army during the concentration of armies in Okinawa[3] and the Soviet army during the Battle of Lake Khasan (also called the Changkufeng Incident) in the south of Primorsky Krai.

Precursors of JEV circulated in Indonesia and then evolved into six genotypes.[22] Genotype III is widespread in a moderate climatic belt and often provokes epidemic outbreaks in eastern and southeastern Asia. Genotype I originated in Indonesia, circulated in Thailand and Cambodia in the 1970s and in South Korea and Japan in the 1990s, and has now completely replaced genotype

III.[23] Genotype I got into Japan in two ways: from southeastern Asia and from mainland China.[24,25] Two island territories—the Philippines and Taiwan, in both of which genotype III circulates—were free of genotype I—and the Philippines remains free—but the genotype appeared in Taiwan in 2008.[26] The evolution of JEV led to the emergence of two new subclusters in 2009−2010; the two together have replaced genotype III. Until recently, the Qinghai-Tibet Plateau, in China, was free of JEV, but in August 2009 the virus was isolated from *Culex tritaeniorhynchus* mosquitoes there.[27] During an epidemic in September−November 2009, genotype I circulated in Japan.[28] In Nepal, on the northern border of India, JE has been known since 1978, after which outbreaks were observed annually.[9] JEV circulates in the north of Australia as well.[12,21]

JE claimed morbidity in the south of the Russian Far East (in Primorsky Krai) in 1938 during an expedition headed by P.G. Sergiev and I.I. Rogosin. Epidemics of JEV broke out in the region in 1938, 1939, and 1943. More than 800 cases were recognized between 1938 and 1943, with 68% reported in the extreme south of Primorsky Krai. The northern extent of this area is limited by the southern part of the Ussuri Lowland (about 42−43°N, 130−133°E). Enzootic JEV circulation without human morbidity has been documented, with the seroprevalence of residents estimated at about 10−20%.[11,18,29,30] JE cases occur mainly in August−September (but also when heavy rains are combined with high temperatures from April to September: ≥21°C in April, ≥23°C in June, ≥25°C in August, and ≥21°C in September).

Clinical Features. The clinical picture of JE varies from asymptomatic and easy feverish forms to an encephalitis syndrome. The ratio of clinical to asymptomatic forms is from 1:300 to 1:1,000, although the ratio in India in the 1970s and 1980s was from 1:20 to 1:30.[31−33]

The start of the disease is sudden, with fever (80%), headache, vomiting (24%), and symptoms of CNS destruction (most often, hemiplegia and articulation lesions)—in 12% of cases, and at the height of the illness in 65% of cases. About one-third of patients with CNS lesions recover completely.[34] Lethal outcomes are preceded by unconsciousness and then coma (20−44% of the total number of patients). Death comes in two-thirds of cases during the first week, in one-fourth during the second week, and in the rest of the cases in one month, from the onset of symptoms. After the disease, residual phenomena in the form of paralysis and mental issues are quite often observed.[28,32]

Control and Prophylaxis. Inactivated vaccines are used to immunize people,[19,29,33,35−37] live vaccines to immunize pigs and horses.[31] Vaccination and protection of pigs from mosquito attack and protection of humans from mosquitoes (through the use of repellents, mosquito nets, bed curtains, etc.) are recommended during epidemics among people. Mass vaccination has been carried out successfully in Japan, South Korea, China, and India.[19,28,33,35−37] Live vaccine manufactured on the basis of the Chinese strain SA 14−22 is is given in China, South Korea, and other countries in government programs aimed at expanding immunization of children.[19,20,33,35−37]

8.2.1.5 Tyuleniy Virus and Kama Virus

History. The prototypical strain LEIV-6C (deposition certificate 526 in the Russian State Collection of Viruses; authors: D.K. Lvov, V.L. Gromashevsky) of TYUV was isolated from Ixodidae ticks *Ixodes* (*Ceratixodes*) *uriae* collected from nests of Alcidae birds in August 1969 in the territory of the Russian Far East on Tyuleniy Island in the Sea of Okhotsk (48°29′N, 144°38′E) (Figure 8.59).[1−5] On the basis of electron microscopy, TYUV was classified as a member of the Flaviviridae family.[6]

Subsequently, serological investigation of Tyuleniy antigenic complex revealed that TYUV belonged to the *Flavivirus* genus,[2,5] which includes (1) Meaban virus (MEAV), isolated from *Ornithodoros* (*Alectorobius*) *maritimus* ticks collected in July 1981 from the nests of herring gulls (*Larus argentatus*) in Meaban Bay in the French province of Brittany (47°31′N, 02°56′W);[7−9] (2) Saumarez Reef virus (SREV), isolated from *O. capensis* ticks collected in August 1974 from the nests of sooty terns (*Sterna fuscata*) on Saumarez Reef, a chain of coral islands in the Australian state of Queensland (22°00′S, 153°30′E)[10,11]; and (3) GGYV, isolated from *I. uriae* ticks collected in December 1976 in a nesting colony of penguins on Macquarie Island (a part of the Australian state of Tasmania) in the southern Pacific Ocean between New Zealand and Antarctica (54°30′S, 159°00′E).[12] Later, MEAV and SREV were categorized into the seabird tick-borne virus group whereas GGYV was classified into the mammalian tick-borne virus group.[13−15] Subsequently, TYUV was multiply isolated in the basins of the Sea of Okhotsk and the Bering and Barents Seas.[16−23]

Kama virus (KAMV) was isolated from *Ixodes lividus* ticks—obligatory parasites of the sand martin (*Riparia riparia*)—collected in August 1989 on islands of the Volga−Kama stretch of the Kuibyshev Reservoir in the Republic of Tatarstan in the central part of the Russian Plain (55°20′N, 49°40′E) (Figure 8.59).[24]

The complete genomes of TYUV and KAMV (GenBank ID: KF815939 and KF815940, respectively) were presented in a 1973 article in the *Journal of Medical Entomology*,[25] and it was established that KAMV was a new virus within the TYUV group of the *Flavivirus* genus.

Virion and Genome. TYUV is a prototypical virus of the Tyuleniy antigenic complex. The viruses of that complex belong to the ecological group of seabird tick-borne flaviviruses, which forms a distinct branch on the

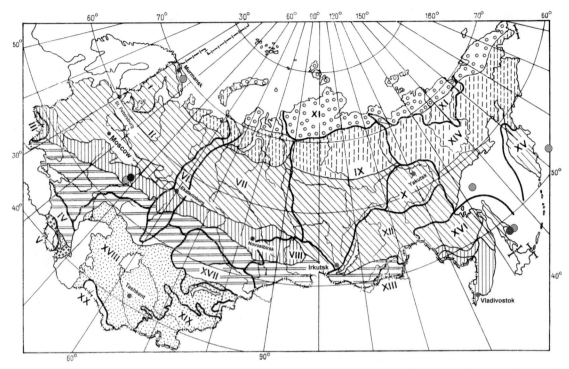

FIGURE 8.59   Places of isolation of TYUV and KAMV (family Flaviviridae, genus *Flavivirus*) in Northern Eurasia. Red circle: prototype strain TYUV/LEIV-6C identified by complete genome sequencing; Pink circles: strains of TYUV identified by serologic method; Dark-blue circle: prototype strain KAMV/LEIV - 20776 Taj identified by complete genome sequencing; Light-blue circle: strain of KAMV identified by serologic methods. (See other designations in Figure 1.1.)

phylogenetic tree.[26] Four species are known in the Tyuleniy antigenic complex: TYUV (in Russia and the United States), MEAV (in Europe), SREV (in Oceania) and KAMV (in Russia). The genetic similarity between the seabird tick-borne flaviviruses and the mammalian tick-borne flaviviruses is about 42% nt. A full-length genome comparison showed that the similarity among the four viruses in the Tyuleniy antigenic complex is 70% nt and 85% aa, on average. TYUV LEIV-61C, isolated in the Russian Far East, has 86% nt and 97% aa identities with TYUV isolated on the Pacific coast of the United States. Kama virus (strain LEIV-Tat20776) has 70% nt identity with the other viruses of the Tyuleniy antigenic complex (MEAV, SREV, TYUV). The similarity of the polyprotein precursor of KAMV is 74% aa

with each of TYUV and SREV, 78% aa with MEAV.[25]

Arthropod Vectors. TYUV is distributed over the basins of the Sea of Okhotsk and the Bering and Barents Seas. The infection rate of *Ixodes uriae* in the Pacific part of the virus's distribution is 4.5 times greater than in the Atlantic part (Table 8.31).[16,18–23] Outside of Northern Eurasia, TYUV is distributed over the west coasts of the United States (chiefly in Oregon) and Canada.[27,28] The infection rate of nymphs and larvae of *I. uriae* is one-twentieth to one-half the infection rate of the imago. The infection rates of *I. uriae* females and males (the males have only a rudimentary hypostome and do not feed) are practically the same.[21] These data testify to the transphase and transovarial transmission of TYUV. (The efficiency of this type of

TABLE 8.31　Infection Rate of TYUV (Family Flaviviridae, Genus *Flavivirus*) Among *I. Uriae* in Nesting Colonies of Alcidae Birds in the Basins of the Sea of Okhotsk and the Bering and Barents Seas

	Far East				Europe
	Okhotsk Sea Basin		Bering Sea Basin		Barents Sea Basin
	Sakhalin Region		Kamchatka Krai	Chukotka Okrug	Murmansk Region
Results of investigation	Tyuleniy Island (48°29′N, 144°38′E)	Iona Island (56°24′N, 143°23′E)	Ariy Kamen Island (Commander Islands) (55°13′N, 165°48′E)	Bering Strait Coast (64°50′N, 173°10′W)	Kharlov Island Near Kola Peninsula (68°49′N, 37°19′E)
Number of strains	9	4	22	0	2
Infection rate (%)	0.066	0.205	0.116	0	0.022
Total　Number of strains			35		2
Number of ticks tested			34,569		8,994
Infection rate (%)			0.101		0.022

transmission is about 5%.) Attempts to isolate TYUV from *I. signatus* ticks were unsuccessful.

The presence of antibodies to TYUV among local cows and indigenous people of the Commander Islands[19,21] indicates the possible role of sanguivorous mosquitoes (e.g., *Aedes communis*, *Ae. punctor*, and *Ae. excrucians*) in infection. Mosquitoes could also take part in virus circulation: Their infection rate from the end of July to the beginning of August reaches 0.3% in nesting colonies of seabirds and 0.1% on the seacoast.

Experimental infection of TYUV on the model of *Aedes aegypti* demonstrated the presence of the virus 4–31 days after inoculation, with 1.5–2.0 lg LD_{50}/10 mcL on days 4–17; 3.0–3.5 lg LD_{50}/10 mcL on days 23–27; and 1.5 lg LD_{50}/10 mcL on day 31. The transmission of TYUV during the feeding of infected mosquitoes on mice was established 7–19 days after infection of the mosquitoes. In *Culex pipiens molestus*, TYUV was detected 5–21 days (the period of observation) after infection, with 1.0–2.0 lg LD_{50}/10 mcL.[20]

Vertebrate Hosts. Migratory seabirds play a role in the exchange of TYUV group flaviviruses between the Northern and Southern Hemispheres.[18,29]

Investigation with the help of indirect complement-binding reactions of sera samples from 2,500 birds collected in the Far East revealed that the maximum TYUV infection rate takes place in Brünnich's guillemots (*Uria lomvia*), common murres (*U. aalge*), and tufted puffins (*Fratercula cirrhata*). Lower rates were seen in pelagic cormorants (*Phalacrocorax pelagicus*), red-faced cormorants (*Ph. urile*), glaucous-winged gulls (*Larus glaucescens*), kittiwakes (*Rissa tridactyla*), northern fulmars (*Fulmarus glacialis*), and sandpipers (Scolopacidae).[17,18,20,21,23,30] The presence of specific anti-TYUV antibodies among sandpipers—red-necked phalaropes

(*Pholaropus lobatus*) and ruffs (*Philomachus pugnax*)—testifies to the possibility that *Ixodes uriae* ticks have penetrated into the Southern Hemisphere, whereas Alcidae birds could transfer the virus through circumboreal routes.[27,28,31] Considering the annual migrations of these birds, TYUV can be found within the *I. uriae* area of distribution in nesting colonies of puffins.

About 90% of adult and 10% of juvenile northern fur seals (*Callorhinus ursinus*) on the Commander Islands have specific anti-TYUV antibodies, implying that these animals are involved in the circulation of that virus. A TYUV strain was isolated from the Arctic ground squirrel (*Citellus* (*Urocitellus*) *parryii*) on the southeastern coast of the Chukotka Peninsula (63°N, 180°E). This event is one more argument for virus splash into the continent, with rodents included in virus circulation.

In the tundra of the Kola Peninsula seacoast, antibodies specific to TYUV were detected among cattle (28.1%) as well as red-necked phalaropes (*Phalaropus lobatus*), snow buntings (*Plectrophenax nivalis*), ruffs (*Philomachus pugnax*), and rodents: tundra voles (*Microtus oeconomus*).[21] Thus, in the Atlantic part of its distribution, TYUV also tends to penetrate into the continent.

Experimental infection of kittiwakes (*Rissa tridactyla*), herring gulls (*Larus argentatus*), and Brünnich's guillemots (*Uria lomvia*) was followed by the development of clinical features with CNS lesions and lethal outcomes.[32]

Epidemiology. The indigenous population in the Far Eastern part of TYUV distribution has specific anti-TYUV antibodies: 8.4% in tundra on the coast of the Chukotka Peninsular, 4.2% in forest–tundra on the coasts of the Sea of Okhotsk and the Bering Sea, 7.4% —in taiga on Sakhalin island, and 9.1% in tundra on the coast of the Kola Peninsula.[21]

The development of fever in humans visiting nesting colonies of seabirds on the coast of the Barents Sea has been described in the literature.[33]

Ecological Peculiarities of TYUV and KAMV Distribution. Penetration of TYUV from the Northern to the Southern Hemisphere is carried out by about 20 species of birds, mostly turnstones (*Arenaria interpres*), that nest in the north of Asia and overwinter in Australia and New Zealand. Wedge-tailed shearwaters (*Puffinus pacificus*) nest in the Southern Hemisphere and carry out an annual migration along the coasts of the Pacific Ocean up to Northern Eurasia and North America.[23,34]

Close genetic relations found between TYUV and KAMV have not been explained yet because information is lacking about ecological links between Alcidae birds in the north and bank swallows in the central part of the Russian Plain. Nontheless, the closeness demonstrates an ancient link between the flaviviruses and Ixodidae ticks—obligatory parasites of colonial and burrow-shelter birds not only on the ocean coast, but also on the continental part of the distribution of those viruses.[19,20,23,35,36]

MEAV and SREV, which are genetically close to TYUV,[25,26] could be intermediate evolutionary branches between tick-borne viruses of seabirds and later mammalian viruses transmitted by ticks.[13,15]

The main vector of TYUV in subarctic regions—*Ixodes Uriae*, adapted to seabirds—is replaced by the *Ornithodoros capensis* complex or *Argas* spp. in the subtropics and tropics.[18,27] The northern boundary of the *Argas* genus distribution is limited by a July isotherm of 15–20°C and of the *Ornithodoros* genus by 20–25°C in Europe and 25–30°C in Asia.[37] The vector of KAMV—the *I. lividus* tick—has transpaleoarctic distribution, from the British Isles in the west to Japan in the east and from 62°N down to 43°S. This species of tick has an extrazonal distribution in the diggings of bank swallows (*Riparia riparia*) made in the soft ground of steeps along the banks of rivers and lakes in taiga, leaf forest, forest–steppe and steppe climatic belts. *I. lividus* ticks are typical parasites of—burrow-shelter birds and relate strictly to the life cycle of the host: After the appearance of birds in the nesting areas in May, larvae begin to feed. In June, nymphs feed on the nestlings; female imagoes also feed on the nestlings, but male imagoes do not.[38]

Given the presence of KAMV—a virus closely related to TYUV—in the central part of the Russian Plain, it is worthwhile, and even necessary, to carry out a wider search for TYUV analogues on the continental part of Northern Eurasia.

8.2.1.6 Dengue Virus (imported)

History. Dengue fever (DENF), etiologically linked to Dengue virus (DENV) (family Flaviviridae, genus *Flavivirus*), has been known in Asia, Africa, and America since the end of the eighteenth century.[1,2] Wide epidemics of DENF appeared in southeastern Asia after World War II.[3] According to WHO data, DENF morbidity, including imported cases, has been detected in more than 100 countries of Asia, Africa, and Europe. More than 2.5 billion people on Earth are under the threat of DENF. About 50 million people fall victim to DENF annually.[4] American armies sustained heavy losses as the result of DENF during World War II,[3] as well as during 1960–1990 in Vietnam, the Philippines, Somalia, and Haiti.[5] Simultaneous outbreaks of DENF and Chikungunya fever often occur.[6]

The virus etiology of DENF and its transmission by mosquitoes was established by P.M. Ashburn and C.F. Craig in experiments using volunteers at the beginning of the twentieth century.[7] DENV-1 was isolated in 1944 from the blood of patients with fever on the Hawaiian Islands,[8] DENV-2 in 1944 from the blood of patients with fever on New Guinea,[8] DENV-3 in 1956 from the blood of patients with fever in the Philippines,[9] and DENV-4 in 1956 from the blood of patient with fever during epidemics in Manila.[9]

Taxonomy. Four different serotypes of DENV form a distinct phylogenetic lineage on the mosquito-borne flavivirus lineage (Figure 8.47). Genetic variation among different strains suggested that DENV be divided into distinct genetic clusters considered as genotypes. The genetic diversity of DENV is best exemplified in DENV-2, the different strains of which are divided into four genotypes: Asian 1, Asian 2, American/Asian and so-called Cosmopolitan.[10] DENV-3 strains are divided into five genotypes (I–V),[11] and DENV-4 strains form three genotypes.[12] In general, a particular genotype is linked to specific geographical regions and that genotype may be used in describing imported cases of DENV infection.

Arthropod Vectors. DENF belongs to natural-foci diseases. Its vectors are anthropophilic species of mosquitoes: *Aedes aegypti* and *Ae. albopictus* in synantropic natural foci. Humans are the only vertebrate hosts in synantropic natural foci, whereas wild mammals are involved in virus circulation in sylvatic natural foci. Vectors in equatorial Africa are *Ae. furcifer*, *Ae. vittatus*, *Ae. tailori*, and *Ae. luteocephalus*.

Vertebrate Hosts. In southastern Asia, the vertebrate hosts of DENV are macaques (genus *Macaca*) and surilis (genus *Presbytis*) living in the rain forests of equatorial climatic belts; the main vector is *Aedes niveus*; a circulation of DENV-{1, 2, 4} has been identified. Natural foci of DENV were also found in the eastern part of equatorial Africa, in Senegal and Nigeria. The vertebrate hosts are patas monkeys (*Erythrocebus patas*); wild strains are considered possible precursors of epidemic ones. Among humans, wild strains provoke slight clinical forms of Dengue fever.[13–15]

Epidemiology. DENF has an epidemic character involving tens of thousands of people in southeastern Asia, Oceania, the Caribbean basin, Central and South America, and Africa. The transmission pathway is a mosquito bite, mainly by members of the *Aedes* genus. These mosquitoes are able to transmit DENV in 8−10 days after feeding on a sick person. About 60−70% of the human population falls victim to DENF during epidemics.[15]

DENV continues to circulate actively and to provoke wide epidemics. For example, all four types of DENV exist in Sri Lanka, with new clades replacing old ones, accompanied by a severe clinical picture.[16,17] In the 1980s, a new wave of DENF epidemics began to develop in Sri Lanka, India, Pakistan, and Central and South America.[18,19] These epidemics were linked mainly to the relatively new DENV-3, but to DENV-1 and DENV-2 as well.[20] In some cases—for instance, in Myanmar[21] and China[1]—all four types of DENV circulated simultaneously.

Clinical Features. The incubation period is 2−7 days. The start of the disease is quick, with fever and with frontal and retroorbital headache. Lymphadenopathia, rash in macule and papule forms (not always), leukopenia, skin hyperesthesia, changes in taste, loss of appetite, and muscle and joint pains gradually develop. Then, after 1−2 days of normal body temperature, the second wave of fever develops, accompanied by a measleslike rash. The palms and soles are rash free. Severe CNS complications have been described to arise in endemic regions (e.g., Brazil).[3]

The hemorrhagic clinical form of DENF, with shock and a high level of lethality (especially among children), was originally seen in the Philippines in 1953. Later, this clinical form was registered in India, Malaysia, Singapore, Indonesia, Vietnam, Cambodia, and Sri Lanka, as well as on islands in the Pacific. According to WHO data, more than 1.3 million patients had hemorrhagic DENF from 1956 to 1992, with 14,000 lethal outcomes. Starting from 1975, hemorrhagic DENF has become the main cause of hospitalization and deaths among children in the countries of southeastern Asia.[1]

The hemorrhagic form of DENF usually develops after a secondary infection by a type of DENV different from the primary one. The primary type of DENV is not neutralized, but fragments antigen binding (Fab)-associated enhancement of the infection occurs. For example, in French Polynesia in 2000, two years after epidemics of DENV-2, an outbreak etiologically linked to DENV-1 emerged and hemorrhagic DENF was detected among children 6−10 months and 4−11 years old.[16] Five symptoms are characteristic of the hemorrhagic clinical form of dengue: high temperature, rash, hemorrhagia, hepatomegalia, and insults to the circulatory system. Thrombocytopenia with blood condensation also occurs.[4] Hemorrhagic DENF can be without shock or can precede it. Shock develops in 3−7 days of the disease, wheninsults to the circulatory system appear: The skin becomes cold, sticky, and cyanochroic; the pulse rate increases; and drowsiness appears. In the absence of antishock actions, patients die within 12−24 h. The severity of the disease depends on a number of factors: the infection titer in the blood, the type of DENV, its biological properties, and more.[22−24]

Imported Cases of Dengue. There is a high risk of DENV infection for visitors to endemic regions, with consequent penetration of the virus into nonendemic regions.[1,25]

DENF has occurred in Spain in the past (e.g., in Cádiz in 1778). Several tens of human cases are introduced into the country each year from equatorial and subequatorial regions. DENV-1 and DENV-2 caused a huge outbreak in Greece in August−September of both 1927 and 1928: in those periods, about 650,000 of 700,000 inhabitants of Athens and Piraeus contracted DENF, including hemorrhagic forms and about 1,000 lethal outcomes.[26] Penetrations of DENV also took place in the Netherlands in 2006−2007[27] and in Japan,[28] France,[29] northern Italy,[30] and Germany in 2010.[31]

During 2002–2011 in Russia, among patients with fever from the risk group that visited tropical–equatorial countries, 48 cases of DENF were identified with the help of serological investigation (22 cases arrived from Indonesia; 11 from Thailand; 3 each from Vietnam and India; 2 each from Venezuela and the Dominican Republic; and 1 each from Sri Lanka, Malaysia, Singapore, Sierra Leone, and Costa Rica).[32–34] In 2013 in Russia, 30 cases of DENF were identified in Moscow, 8 in St. Petersburg, and 8 imported strains of DENV were isolated.

The risk of DENF for Europe has appeared again with the introduction of *Aedes albopictus* and *Ae. aegypti* mosquitoes in the countries of the Mediterranean and Black Sea basins.[35] Stable populations of both these species were found on the southeastern coast of the Black Sea (in Krasnodar Krai, Russia, as well as in Abkhazia).[36–38]

Control and Prophylaxis. The main approach to prophylaxis is to struggle against mosquito vectors. During the 1950s and 1960s, a program against *Ae. aegypti* mosquitoes that was unprecedented in terms of scale and expense was conducted in America, but it was stopped in 1970; as a result, in 1995 the number of *Ae. aegypti* mosquitoes was estimated to be same as that before the program began.[39] The struggle against mosquito vectors in Singapore turned out to be more successful, but still did not prevent DENF morbidity.[40]

Investigations into four-component vaccines are far from completion today.[22,41]

Express methods of DENF diagnostics are used in airports.[42] WHO issues a reference guide for the diagnosis, treatment, prophylaxis, and control of DENF.[43]

8.2.1.7 Sokuluk Virus

History. Prototypical strain LEIV-400K of Sokuluk virus (SOKV) (family Flaviviridae, genus *Flavivirus*, Entebbe bat antigenic complex) was isolated from internal parts of the bat *Vespertilio pipistrellus* collected in 1970 in the garret of a house in Sokuluk District, Kyrgyzstan (42°30′N, 74°30′E) (Figure 8.60).[1–4] Later, in 1971–1973, SOKV was isolated from Argasidae and Ixodidae ticks, as well as from birds, in Fergana Valley and Chuysky Valley in central Asia (Table 8.32, Figure 8.60). Further serological investigations with the help of HIT revealed that SOKV belongs to the Flaviviridae family, and with the help of complement-fixation testing (but not neutralization testing), to the Entebbe bat serogroup.[1–3] A prototypical strain of this serogroup was isolated from a Kenyan big-eared free-tailed bat (*Tadarida lobata*) collected near Entebbe, Uganda, in July 1957.[5]

Taxonomy. The genome of SOKV was sequenced, and genome analysis showed that the virus is related most closely (71% nt and 79% aa identities) to Entebbe bat virus (ENTV). SOKV has about 50% nt and 55% aa identities with other flaviviruses, except viruses of the Rio Bravo (RBV) and Modoc (MODV) groups (<50% similarity).[6] No arthropod vector of ENTV and SOKV has been established; however, phylogenetic analysis based on a full-length genome comparison placed SOKV and ENTV together on a distinct branch of mosquito-borne flaviviruses related to YFV and Sepik virus (SEPV) (Figure 8.47).

Arthropod Vectors. According to serological data, domestic animals do not take part significantly in SOKV circulation, although antibodies to SOKV were detected among cows and sheep. Isolation of SOKV from birds that were known not to have made contact with obligatory parasites of bats, as well as the presence of positive sera from humans and domestic animals, suggest the participation of mosquitoes in SOKV circulation. Transmission of the virus by bats could be carried out by *Argas vespertilionis* and *Ixodes vespertilionis*.[7–10]

Vertebrate Hosts. More than 20 flaviviruses were isolated from bats (order Chiroptera); about half are unique to these mammals.[11]

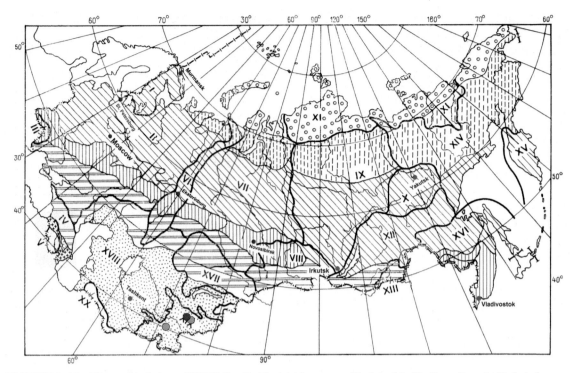

FIGURE 8.60 Places of isolation of SOKV (family Flaviviridae, genus *Flavivirus*) in Northern Eurasia. Red circle: prototype strain SOKV/LEIV-400K identified by complete genome sequencing; Pink circles: strains of SOKV identified by serological methods. (See other designations in Figure 1.1.)

TABLE 8.32 Isolation of SOKV on the Territory of Kyrgyzstan (1971)

	Source of isolation		Location of material collection		Number of strains
Class	Family	Species	Territory	Station	
Mammalia	Vespertilionidae	Bat (*Vespertilio pipistrellus*)	Fergana Valley	House garret	2
Avis	Motacillidae	White wagtail (*Motacilla alba*)	Chuysky Valley	Settlement	1
	Hirundinidae	Barn swallow (*Hirundo rustica*)	Chuysky Valley	Nesting colony in the house	1
	Alcedinidae	Common kingfisher (*Alcedo atthis*)	Chuysky Valley	Natural biocenosis, nest near the bank of the river	1
Arachnida	Argasidae	*Argas vespertilionis* ticks	Fergana Valley	Collected from *Vespertilio pipistrellus*	2
		Argas vulgaris ticks	Chuysky Valley	Digging colony of birds	2
	Ixodidae	*Hyalomma marginatum* ticks	Fergana Valley	Collected from the cattle	1

Bats of the suborder Microchiroptera are susceptible to a number of flaviviruses: JEV[12]; Dakar bat virus (DBV)[13–16]; Bukalasa bat virus (BBV)[17]; ENTV[18]; RBV,[19,20] closely related to MODV, from the deer mouse (*Peromyscus maniculatus*), in the north of California[21]; Montana Myotis leukoencephalitis virus (MMLV)[22,23]; Carey Island virus (CIV)[17]; Phnom Penh bat virus (PPBV)[24]; and Yokose virus (YOKV).[25]

The insectivorous bats *Vespertilio pipistrellus*, from which SOKV was isolated, belong to the evening bats family (Vespertilionidae), which is active during the evening and at night. Their daylight shelters are situated mostly in house garrets. *V. pipistrellus* is distributed over Europe, the Mediterranean, the Caucasus region, and central Asia. A part of the population overwinters in Africa, where infection by local viruses (e.g., BBV, DBV, ENTV) could occur.

Experimental infection of sparrows (*Passer montanus*) resulted in SOKV being detected in internal parts of infected birds on the 8th and 25th days after inoculation.[26]

Epidemiology. There are no laboratory-confirmed human cases of SOKV infection. Nevertheless, the proximity of SOKV hosts (bats) to human habitats, as well as the presence of encephalitis and hemorrhagic fever agents among the flaviviruses, suggest that SOKV may be dangerous to humans.

Complement-binding specific anti-SOKV antibodies were detected among humans in Kyrgyzstan and Turkmenistan (6.2% and 4.0%, respectively), testifying to recent infection events.[1–4,7,9,10,16,27–31]

8.2.1.8 West Nile Virus

History. WNV (family Flaviviridae, genus *Flavivirus*), theetiological agent of West Nile fever (WNF), was first isolated during research on YFV in 1937 from the blood of a native of Uganda who was suffering a mild fever.[1] The strain isolated, B956, belongs to genetic lineage II. (See "Taxonomy" next.) Strain Eg101, isolated from the sera of a child without clinical signs in Egypt,[2] is the prototype for African genetic lineage I, widely used for investigations. WNV belongs to the JEV group, has the broadest antigenic properties, and, on theoretical grounds, appears to be the most ancient member of the *Flavivirus* genus.[3] Low-passaged WNV strains are known by many investigators to be common causes of laboratory infection, apparent or inapparent.[4]

Taxonomy. Phylogenetic analysis revealed that different geographic isolates of WNV could be grouped into two major genetic lineages (Figure 8.61). Lineage I includes strains from Africa, southern and eastern Europe, India, and the Middle East. Lineage II includes isolates from west, central, and east Africa, as well as Madagascar. Lineage 1 can be subdivided into three clades: Clade 1a consists of strains from Europe, Africa, the United States, and Israel. The topotypic isolates of WNV in Australia—Kunjin virus (KUNV)—belong to clade 1b, and clade 1c is formed by isolates from India.[5] Subsequently, two genetically divergent Rabensburg strains—97–103 (isolated in the Czech Republic) and LEIV-Krnd88-190 (isolated in Russia)—were proposed to form novel lineages III and IV, respectively.[6–8] A fifth lineage was formed by strains from India.[9] Phylogenetic analyses based on complete genomic sequences revealed that the various lineages differed from each other by 20–25%. A putative novel sixth lineage has been detected in Spain in 2006, but only a partial sequence of the NS5 gene of this isolate is available in GenBank.[10]

World Distribution. The distribution of WNV in Northern Eurasia, and indeed, in the whole world, covers vast territories within the equatorial, tropical, and temperate (the southern part) climatic belts in Africa, Europe, Asia, Australia, and North America (the last starting from 1999).

In Africa, it is very difficult to find a country or landscape in which WNV has not been

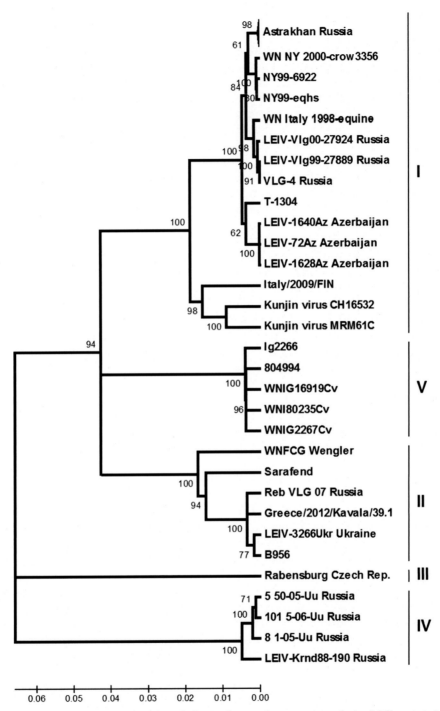

FIGURE 8.61   Phylogenetic tree constructed on the basis of a complete genome analysis of different strains of WNV.

detected by either a virological or serological approach. The isolation of this virus from a wide array of species of birds, mosquitoes, Ixodidae and Argasidae ticks, and domestic animals as well as humans testifies to the ecological plasticity of the virus and therefore to its ability to adapt to different ecological conditions. Two genetic lineages circulate in Africa: the first, which dominates, and the second.

Sporadic morbidity and epidemic outbreaks permanently take place in a number of African countries, especially the Republic of South Africa, where a wide outbreak with at least 3,000 human cases occurred in 1974 after an active period of rain. According to a report from the Pasteur Institute, during the last 10–15 years alone, epidemic outbreaks were registered in Algeria (in 1994, with more than 50 cases and 8 deaths, and in 1997, with 173 cases), in Tunisia (during 1997–2003, with 173 cases), Morocco (in 1996 and 2002; the epidemic reached both humans and horses), in Senegal (in 1993), and in Kenya (in 1998).[11] New centers of infection continue to be arise in Africa—for example, in 2009 in Morocco, where morbidity among people and horses was observed and 3.5% of birds had specific anti-WNV antibodies,[12] and in 2010 in the Republic of South Africa, where there were a number of lethal outcomes.[13]

The wide distribution of WNV in Africa and its circulation among populations of the majority of the continent's species of local and migrating birds indicates that the virus is able to penetrate to southern Europe and western Siberia through the birds' migration pathways. Most of the birds nesting in or migrating through the Volga delta overwinter in Africa.[14] Thus, Africa is the main source of penetration of WNV genotypes I and II into southern Europe and western Siberia.

In Asia, a peculiar third genotype of WNV appears to be circulating in the Indian subcontinent.[11] A prototypical strain of WNV genotype 3 was isolated from *xCulex vishnui* mosquitoes in southeastern India, and human morbidity was identified in India, Pakistan, and Israel. Taking into account the fact that most of the birds from western Siberia and many from eastern Siberia overwinter in India and other countries of southern Asia, there is a high probability that WNV genotype 3 has penetrated into Siberia. Also in Asia, both epidemics and sporadic cases etiologically linked with the first genotype of WNV have arisen regularly in Israel since at least1958. One such outbreak was observed in 1999–2000.[15] Surveillance in South Korea does not indicate any WNV circulation in that country.[16]

In Australia and Oceania, the Kunjin variant of the first genotype of WNV appears to be circulating.[17–19] KUNV could be introduced into Northern Eurasia (in eastern Siberia and the Far East) by migrating birds overwintering in southeastern Asia and Australia.[11,14] In 2011, an outbreak among horses in New South Wales, Australia, was identified.[20]

In central Europe, for a long time only two strains of WNV were known: one isolated in from *Aedes cantans* in 1972 in western Slovakia and the other isolated from *Ae. vexans*, *Ae. cinereus*, and *Culex pipiens* in 1997 in the Czech Republic, near the Austrian border. Anti-WNV antibodies were identified in the Czech Republic among 1.4–9.7% of birds, including crows, daws, turtle doves, common kestrels, ducks, coots, and thrushes. Later, two strains of the so-called Rabensburg genotype of WNV were isolated from *Cx. pipiens* in 1997 and 1999 in the Czech Republic.[21–23] The strain belonging to the second lineage of WNV was isolated from a goshawk in Hungary.[7]

In 1996 in Tuscany, Italy, Usutu virus (USUV), which is closely related to WNV, was isolated during an epizootic episode among birds, especially thrushes (*Turdus merula*), and then, again, in 2001 in Austria. Later, this virus was found in Hungary, Switzerland, and Germany.[24]

Practically all of the southern European countries are endemic for WNV.[25,26] Especially tragic events unfolded in Romania, where there was an epidemic in July–October 1996 with a peak at the end of August to the beginning of September in the southeastern part of the country, downstream of the Danube River. Six administrative units and Bucharest were affected, among other jurisdictions. Human morbidity reached 12.4%, and 835 patients with CNS insult were hospitalized. The number of patients with fever was at least 10 times more, and the number of infected individuals 100–300 times more. The outbreak, which dragged on until 2000,[27] testifies to the development of a city epidemic form of WNF. The virus belonged to the first genotype of WNV and probably was brought to Romania by birds from Africa.

WNV distribution in Europe indicates an especially high risk of a WVF outbreak in deltas of the large rivers—the Rhône in France and the Danube in Romania—through which the main migratory paths of birds overwintering in Africa lie.[14] In the recent past, WNV has been active in Europe in Italy,[28,29] Greece,[30,31] Spain,[10] Poland,[26] the Czech Republic,[3,22] and France.[22] Infected mosquitoes were imported into Great Britain from the United States by airplane travel.[32]

As for North America, before 1999 that continent was free of WNV. Penetration of WNV into America most likely happened by infected mosquitoes in the holds of ships from ports in the Mediterranean Sea or Black Sea.[11] Fifty-six cases of human WNF were revealed in New York City and its surroundings at the end of July–September 1999, with a peak in the second half of August. Seven cases (12.5%) had a lethal outcome. The virus was found in *Culex* sp. and *Aedes vexans* mosquitoes caught in September–October in New York City and in the states of New Jersey and Connecticut. Positive results were obtained by RT-PCR during an investigation of brain tissues of dead

birds: crows, seagulls, storks, herons, ducks, cuckoos, pigeons, jays, robins, hawks, and eagles. The genomes of the strains that were isolated were found to belong to the first genotype and were close to the strains isolated in 1996 in Romania and in 1998 in Israel.[33] In 1999, WNV was registered in the United States, probably translocated there by migrating birds or by infected mosquitoes inhabiting the holds of visiting ships. WNV was found in Florida and on the Cayman Islands. In the spring of 2002, WNV extended its coverage to the eastern part of the United States through the central and Mississippi migratory pathways. Also in 2002, the virus reached the central regions of the United States and southern Canada and, in the fall of that year, Mexico, Jamaica, other Caribbean countries and Central and South America (Cuba, Guatemala, and the archipelago of the Lesser Antilles islands). Then, in 2004, WNV penetrated to California through the Pacific migratory pathway.

By 2003–2004, practically all the territory of the United States, southern Canada, and Latin America became endemic with high morbidity and mortality.[34] The greatest morbidity in the United States was found in the states of NorthDakota, South Dakota, and Nebraska.[27,35] The number of diseased individuals reached 4,000–9,000 cases in separate years. During 1999–2006 in the United States, more than 16,000 WNF cases were identified, with more than 600 (4%) succumbing to the disease. The economic damage was estimated in billions of dollars.[36,37] Today, WNV continues to circulate in the United States.[38,39] Morbidity grew in the states of Louisiana and Mississippi after Hurricane Katrina.[40] In Montana, the infection rate of people living in close proximity to a colony of pelicans (*Pelecanus erythrorhynchos*) is five times higher than in other regions of the state.[41] In a sea park in Texas, grampuses (*Orcinus orca*) contracted encephalitis and died,[42] and previous episodes of polyencephalomyelitis were revealed among seals (*Phoca vitulina*). Also in Texas, three new genetic clades of WNV were found, testifying to rapid evolution of the virus on the American continent.[43] In 2012, an epidemic arose again, accompanied by a large number of lethal outcomes. In Texas, a state of emergency was declared.

Northern Eurasia. In Northern Eurasia, on the basis of the results of multiple investigations, the distribution of WNV includes Moldova, Ukraine, Belarus, Armenia, Azerbaijan, Georgia, Kazakhstan, Tajikistan, Kyrgyzstan, Uzbekistan, Turkmenistan, the south of the European part of Russia (the desert, semidesert, steppe, and forest–steppe landscape belts), and western Siberia.[11,35,44]

The first data on WNV isolation were obtained from *Hyalomma marginatum* ticks collected in the Astrakhan region in 1963. Data were also obtained in Azerbaijan from a blackbird (*Turdus merula*) and a European nuthatch (*Sitta europaea*) and, later, from a herring gull (*Larus argentatus*) and *Argasidae* ticks (*Ornithodoros coniceps*) parasitizing it.[14] WNF morbidity is now a permanent feature in the Astrakhan region, Kazakhstan, central Asian countries (republics of the former USSR), Ukraine, and Azerbaijan.

Virological, entomological, zoologico-ornithological, and epidemiological investigations of WNV in the Astrakhan region and the Kalmyk Republic were conducted especially actively.[8,39,45–61] Virus activity in the Volga

River delta was found at least as far as 50 years ago.[11,35,60] But interactions between WNV, on the one hand, and animal and vector populations, on the other, were not investigated in detail as well as genetic characteristics of the virus; indeed, the latter began to be studied well only during the first decade of twenty-first century, when suckling mice and Vero-E6 cell culture were used to isolate the virus and serological investigations were employed to detect viral RNA (neutralization testing, ELISA, HIT) and to sequence genes (RT-PCR).

WNV endemic territories in southern Russia were known from the moment the virus was isolated in the Astrakhan region in 1964. (The number of cases confirmed by ELISA in the southof the European part of Russia is presented in Table 8.33.) Sporadic cases with a moderate clinical picture and minor outbreaks were observed in the area practically annually, as well as in other southern regions of the former Soviet Union. The immune structure to WNV among humans in the USSR was also known, with the most immunity occurring in the south of Russia, mainly the Astrakhan region (Figure 8.62, Table 8.34).

All this familiarity with WNF is why an outbreak in 1999 in Volgograd was not exactly unexpected,[62] even though it originally was identified by regional experts as an enterovirus infection. Still, laboratory-confirmed WNF cases reached more than 500 that year, and according to our estimations, the number of

TABLE 8.33　WNF Cases Confirmed By ELISA in the South of the European Part of Russia (1968–2001)

Region		WNF morbidity/mortality (%)					
	1968	1990–1996	1997	1998	1999	2000	2001
Lower Volga (Volgograd region)	–	–	5	35	380/40 (10.5)	32/3 (9.4)	15
Volga–Akhtuba, Volga delta (Astrakhan region)	12	10	8	9	95/5 (5.3)	24	49/1 (2.0)
Total	12	10	13	44	475/45 (9.5)	56/3 (5.4)	64/1 (1.6)

FIGURE 8.62   Immune structure of human population to WNV on the territory of Russia.

infected patients exceeded 200,000 (Table 8.35). Mortality (about 10%) was also unusually high.

Large deltas of European rivers such as the Rhône, Danube, and Volga Rivers are known to be transit hubs for migrating birds and places of introduction of viruses linked with birds.[14] The main natural focus in Russia is the Volga delta.

The Volga delta and contiguous territories around the northern Caspian basin have been endemic for WNF for many years (Tables 8.33–8.35), and other arboviruses have been ecologically linked with aquatic and semiaquatic birds frequenting the region. Ninety percent of these species of birds overwinter on the African continent. Up to 100,000 birds pass over the region daily during their seasonal migrations via the Volga delta main line of the eastern Europe migratory route.

(See Figure 3.2.) The problem is that the Volga delta is the place from which viruses are introduced into anthropogenic biocenoses in close vicinity to human habitation. One consequence of this scenario was epidemic outbreaks in the Astrakhan and Volgograd regions in 1999–2001.

The Volga delta consists of three basic belts, each with its own unique ecosystem features (Figure 8.63).

The lower Volga delta borders the Caspian Sea and is characterized by extensive exposed spaces with water. The water depth usually does not exceed 1.0–1.5 m, a situation that is highly conducive to the mass propagation of mosquitoes and one that also provides nesting opportunities for aquatic and semiaquatic birds. Near where it empties into the Caspian Sea, the Volga bed turns significantly to the west, so the western part of the delta,

TABLE 8.34   Distribution of Antibodies to WNF Virus Among Humans and Domestic Animals in Russia (1986—1990)

Physicogeographical country	Landscape zone	Economic territory	Regions investigated	Category	Total	Positive number	%
Russian Plain	Northern, middle taiga, southern taiga	Northern	Vologodsk region; Komi Republic	Humans	1,740	9	0.5
				Cattle	325	0	0.0
				Deer	141	0	0.0
	Leaf-bearing forests	Central	Vladimir, Ivanov, Ryazan regions	Humans	692	12	1.7
	Forest—steppe, steppe	Central Chernozemny	Lipetsk, Tambov regions	Humans	694	5	0.7
		Povolzhsky	Volgograd, Samara, Saratov, Penza, Ul'yanov regions; Tatarstan Republic[a]	Humans	1,884	17	0.9
				Cattle	547	5	0.9
				Rodents	240	3	1.3
Central Asian Plain	Steppe, semidesert, desert	Povolzhsky	Astrakhan region; Kalmykiya Republic	Humans	722	61	8.4
				Cattle	178	6	3.4
				Camels	30	10	33.3
				Horses	41	1	2.4
				Sheep	60	0	0.0
Russian Plain	Steppe	Northern Caucasian	Rostov region; Stavropol, Krasnodar territories	Humans	1,132	8	0.7
Crimean—Caucasian mountain country	Mountain deciduous forests	Northern Caucasian	Kabardino-Balkaria, Chechnya Republics	Humans	461	1	0.2
				Cattle	211	1	0.5
				Sheep	59	0	0.0
Ural	Southern taiga	Ural	Orenburg region; Bashkortostan Republic	Humans	356	25	7.0
				Cattle	56	2	3.6
				Pigs	76	0	0.0
Western Siberia	Southern taiga, forest—steppe	Western Siberian	Altai territory; Omsk, Novosibirsk, Kemerovo, Tomsk regions	Humans	1,824	9	0.5
				Cattle	1,807	8	0.4
Eastern Siberia	Southern taiga, forest—steppe	Eastern Siberian	Chita region; Krasnoyarsk territory; Buryatiya Republic	Humans	936	4	0.4
				Cattle	25	0	0.0
				Deer	100	0	0.0
Amuro-Sakhalin country	Southern taiga, deciduous forests	Far Eastern	Amursk region; Primorsky Krai, Khabarovsk territory	Humans	2,917	21	0.7
				Cattle	1,539	6	0.4
				Pigs	1,667	23	1.4
Total	8 landscape belts from northern taiga to desert	9 economic territories	33 regions	Humans Animals	13,358 7,102	172 65	0.2—8.4 0—33.3

[a]Southern taiga and deciduous forests.

TABLE 8.35   Detection of Antibodies to WNV Before and After Epidemic Outbreak (1999)

| Region | Parameter | Year | | | Population (1999) | | Evaluation of WNV-Infected/WNF (1999) |
		1988–1989	1998	2000	Total	Urban population	
Lower Volga (Volgograd region)	Number tested	544	–	608	2,494,000	2,454,000	~95,700/~960
	Percentage positive	1.6	–	5.3 (+3.9)			
Volga–Akhtuba	Number tested	383	310	162	931,000	637,000	~117,000/~2,130
Volga delta (Astrakhan region)	Percentage positive	13.8	31.6 (+17.8)	50.0 (+18.4)			
Total	Percentage positive	6.7	31.6 (+24.9)	14.7	3,425,000	3,091,000	~212,700/~3,090

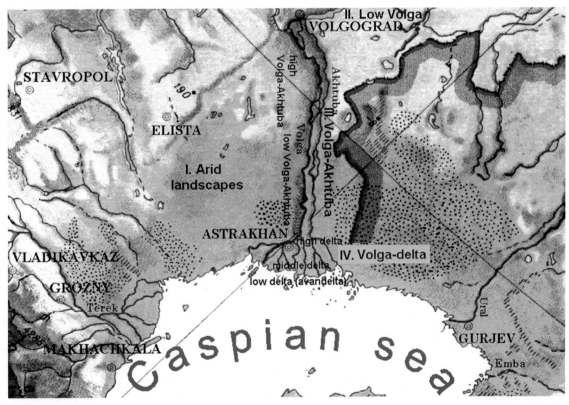

FIGURE 8.63   Ecosystem division of the northern Caspian Sea basin. I. Arid belt (semi-deserts and deserts); II. Low Volga (steppe); III. Volga Akhtuba (arthropodgenic-biocenosis); III.1. High Volga Akhtuba (semi-desert water ecosystems); III.2. Low Volga Akhtuba (desert) water ecosystems; IV. Volga delta; IV.1. High delta of Volga (anthropogenic biocenoses); IV.2. Middle delta of Volga (wild anthropogenic biocenoses); IV.3. Low delta of Volga (wild biocenoses).

including both the reed bed of the northwestern Caspian coast (up to Lagan in the Kalmyk Republic) and some flooded islands, is more extensive than the central and eastern parts. The extreme eastern part of the delta lies in Kazakhstan. A number of hunters and fishermen could be infected in the lower delta of the Volga.

The middle Volga delta is more distant from the sea, has powerful currents, and consists of shallow lake ecosystems with reeds and shrubs. Water ecosystems adjoin semidesert ones. Within the limits of this zone, wild biocenoses combine with anthropogenic areas around a number of settlements, whose inhabitants keep cattle, sheep, and camels. WNF is widely registered among the native population.

The upper Volga delta adjoins the Volga–Akhtuba lowlands and semideserts. Large cities, including Astrakhan, are located within the limits of this zone. Some species of wild birds that are common in the middle delta also occur in this zone, coming into close contact with domestic animals and synanthropic birds.

Analysis of retrospective data collected before 1999 revealed that the main locus of native-population morbidity by WNF is in the Volga delta (Table 8.35). Viruses could be introduced into the northern part of the Volga–Akhtuba lowland up to Volgograd and maybe even higher. Thus, in the future it will be necessary to control the introduction of the virus into the Volga–Akhtuba lowland from Astrakhan to Volgograd.

Arid landscapes occupying contiguous terrian to the west of the Volga–Akhtuba system and the Volga delta are situated within the boundaries of the Caspian Sea–Turanian Basin physicogeographical area (Figure 8.63).

Every year at the end of July, a group of specialists from the D.I. Ivanovsky Institute of Virology in Moscow has traveled to the Astrakhan region and the Kalmyk Republic to organize and conduct a joint scientific expedition with local Centers of Sanitary–Epidemiological Inspection for ecologo-virological monitoring of the northwestern Caspian region (Figures 8.64–8.66). The main goal of the expedition is to contain the ecological and epidemiological situation after suppression of WNV circulation in the previous epidemiological season as the result of a combination of natural factors.

The plan for the collection of field material took into account the results of previous expeditions, when key milestones and marker species of mosquitoes and wild and domestic animals were identified. In particular, the researchers planned to investigate the role of the Ixodidae tick *Hyalomma marginatum* (Figure 8.67) in WNV and other arbovirus circulation in anthropogenic and wild biotopes.

Both federal and local heads of various services, as well as virologists, epidemiologists, veterinarians, hunters, and frontier guards, were supplied with materials containing evaluations of ecologo-virological monitoring of their respective territories in the previous epidemiological season. Practical recommendations were given for prophylaxis of WNF, CCHF, and other arboviral diseases.

Field materials—bloodsucking mosquitoes, Ixodidae ticks, internals (blood, serum, liver, spleen, and brain) of wild birds and mammals, and sera from donors and domestic animals— were collected on the territory of the Astrakhan region and the Kalmyk Republic from the end of July to the beginning of August 2000–2004 within the boundaries of the Volga delta, the Volga–Akhtuba valley, and adjacent arid landscapes. Field materials were collected in the biotopes of the west Volga coast and the east Akhtuba coast, including internal water–meadows of the upper and lower Volga–Akhtuba zones, hydromorphic and adjacent meadow–steppe biotopes of the upper and meddle belts of the Volga, the Volga avandelta, the territory of the Sarpa Lakes, and the east side of Ergeny (see Figures 8.64–8.66).

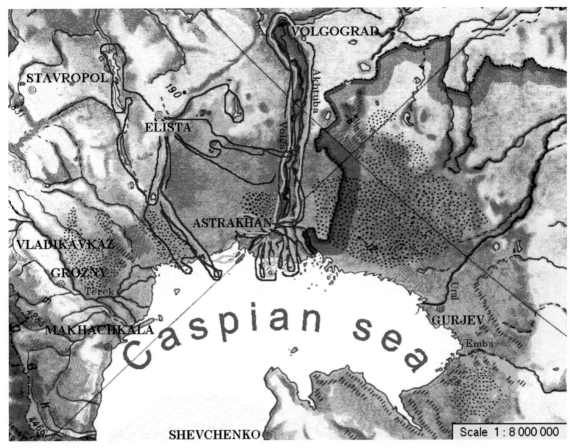

FIGURE 8.64 Pathway of the expedition mounted for the collection of field material for WNV investigation on the territory of the northwestern Caspian region during the 2004 epidemiological season. (Different-colored lines denote different expedition groups.)

During 2000–2004, the expedition collected 504,731 bloodsucking mosquitoes (of the order Diptera and family Culicidae: genera *Culex*, *Aedes*, *Coquillettidia*, and *Anopheles*); 11,266 Ixodidae ticks (of the taxon Acari and family Ixodidae: genera *Hyalomma*, *Rhipicephalus*, and *Dermacentor*), mainly *H. marginatum*; internal parts of 2,794 birds and 67 hares (*Lepus europaeus*); sera from 4,500 human donors (2,500 in the Astrakhan region and 2,000 in the Kalmyk Republic); and sera from 5,300 domestic animals (2,900 in the Astrakhan region and 2,400 in the Kalmyk Republic) (Figure 8.68).

The field materials that were collected were stored and transported to the D.I. Ivanovsky Institute of Virology in liquid nitrogen in dewars, in accordance with all requirements for the handling and transport of infectious samples.

Internal parts of 2,794 wild birds were investigated by virological methods (Table 8.36). Twelve WNV strains (Tables 8.36 and 8.37) were isolated. According to the bioprobe method used, the total WNV infection rate among wild birds is about 0.4%, with the highest level (0.7%) reached in the middle and

FIGURE 8.65   (A) D.K. Lvov, head of the scientific expedition. (B) Field laboratory deployed at the stern of the expedition ship. (C) One of the channels in the middle belt of the Volga. (D) Light traps at the head of the expedition ship. (E) Scientific researcher D.N. Lvov and senior technician E.I. Vakar are preparing mosquito-collecting materials on a small island in the lower belt of the Volga delta. (F) Scientific researcher M. Yu. Shchelkanov during the scientific collection of birds on the western hill—il'men territory in the upper Volga delta. A Ber hill can be seen in the distance on the right and a dug-up Ber hill on the left.

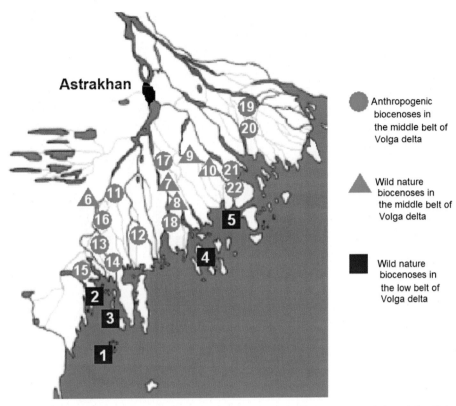

FIGURE 8.66 The places where field material was collected in the middle and lower belts of the Volga delta in 2000–2004.

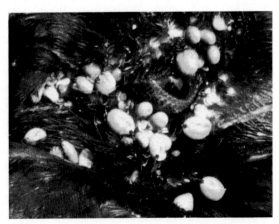

FIGURE 8.67 Preimago of the Ixodidae tick *Hyalomma marginatum* on the head of a rook (*Corvus frugilegus*).

upper belts of the Volga delta. The highest intensity of WNV circulation was among cormorants (3.4% in the middle and lower Volga delta) and Corvidae (3.3% in the upper and middle Volga delta); among other birds of this terrestrial ecological complex, the intensity was 1.4%, on average.

RT-PCR testing was performed on 5,080 pools containing 504,731 individual mosquitoes (order Diptera, family Culicidae) (Tables 8.37 and 8.38) and 892 pools containing 11,266 individual *Hyalomma marginatum* ticks (taxon Acari, family Ixodidae; 4,923 imagoes from cattle, horses, and sheep, as well as 6,343 preimagoes mainly from Corvidae birds) (Tables 8.37 and 8.39). Two WNV strains were

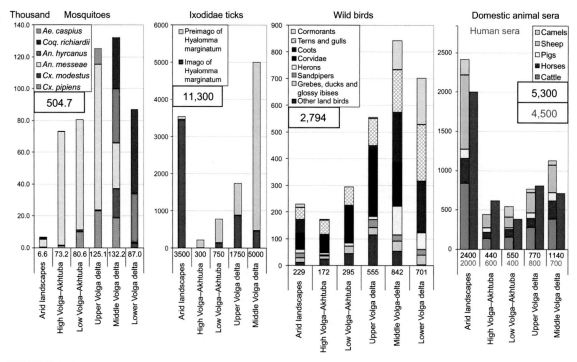

FIGURE 8.68   Field material collected during 2000–2004 in the lower Volga region for WNV investigations.

isolated from mosquitoes and 11 WNV strains from ticks.

With the help of RT-PCR, 2,794 samples of internal parts from wild birds were collected on the territory of the nothwestern Caspian region and were tested for any indication of WNV RNA. Positive results are presented in Tables 8.40 and 8.41. The total WNV infection rate among wild birds during 2001–2004 was established to be 4.8%. The highest WNV infection rate was found for cormorants (*Phalacrocorax carbo*) (10.6%; 22.0% in the middle belt of the Volga delta, 20.0% in the upper belt), Corvidae birds (6.1%; 12.0% in the upper belt of the Volga delta, 3.7% in the middle belt), and coots (*Fulica atra*) (5.0%; 6.4% in the middle belt of the Volga delta, 4.7% in the lower belt); the lowest WNV infection rate was discoverded in ducks and grebes (*Podiceps*): 1.3% (Figures 8.69–8.71). The most intensive WNV

circulation takes place in the middle (7.0%) and upper (7.4%) belts of the Volga delta. The highest portion of positive results was obtained from internal parts of wild birds in 2001 and was decreasing until 2004 (Figure 8.72).

RT-PCR testing for any indication of WNV RNA was carried out on 108 samples of internal parts collected from wild mammals on the territory of the northwestern Caspian region. Positive results are presented in Tables 8.42 and 8.43.

RT-PCR testing for WNV RNA was done on 3,066 pools containing 305,064 samples of mosquitoes (order Diptera, family Culicidae). Positive results are presented in Table 8.44. During 2001–2004, the highest WNV infection rate was detected in *Culex pipiens* (0.036%) living mainly in basements of human houses. A significant WNV infection rate was also

TABLE 8.36 Number of Internal Parts of Wild Birds Collected in Different Natural Belts of the Northwestern Caspian Region (2000−2004) and Investigated with a Bioprobe Approach

Kinds of birds	Arid landscapes	Upper Volga−Akhtuba	Lower Volga−Akhtuba	Upper Volga Delta	Middle Volga Delta	Lower Volga Delta	Total
Corvidae	59	63	135	249	164	2	672
				4 strains	3 strains		7 strains
Other land birds	10	22	43	114	52	2	243
	1 strain				1 strain		2 strains
Cormorants	13	3	0	5	109	173	303
					2 strains	1 strain	3 strains
Coots	52	5	6	15	188	191	457
Terns and gulls	45	54	70	103	160	213	645
Sandpipers	19	4	0	29	23	20	95
Herons	14	9	13	13	107	63	219
Ducks, podiceps, and glossy ibises	17	12	28	27	39	37	160
Total	229	172	295	555	842	701	2,794
	1 strain			4 strains	6 strains	1 strain	12 strains

TABLE 8.37 Number of Mosquitoes Collected in Different Natural Belts of the Northwestern Caspian Region (2001−2004) and Investigated by a Bioprobe Approach

Species of mosquitoes	Arid landscapes	Upper Volga−Akhtuba	Lower Volga−Akhtuba	Upper Volga Delta	Middle Volga Delta	Lower Volga Delta	Total
Culex pipiens	0	1,140	9,818	23,109	18,622	0	52,689
Culex modestus	300	200	1,200	380	18,504	2,933	23,517
Anopheles messeae	4,900	71,709	69,378	91,766	28,436	132	266,321
				2 strains			2 strains
Anopheles hyrcanus	900	0	0	0	34,216	31,149	66,265
Coquillettidia richiardii	0	0	0	0	31,528	52,834	84,362
Aedes caspius	500	100	214	9,830	933	0	11,577
Total	6,600	73,149	80,610	125,085	132,239	87,048	504,731
				2 strains			2 strains

TABLE 8.38 Number of *Hyalomma marginatum* Ticks Collected in Different Natural Belts of the Northwestern Caspian Region and Number of WNV Strains Isolated by a Bioprobe Method

Ontogenesis stage of *H. marginatum*	Arid landscapes	Upper Volga–Akhtuba	Lower Volga–Akhtuba	Upper Volga Delta	Middle Volga Delta	Lower Volga Delta	Total
Imago	3,448	0	140	877	458	0	**4,923**
Preimago	89	216	638	863	4,537	0	**6,343**
				9 strains	**2 strains**		**11 strains**
Total	3,537	216	778	**1,740**	**4,995**	0	**11,266**
				9 strains	**2 strains**		**11 strains**

TABLE 8.39 WNV Strains from Field Materials Collected in Different Natural Zones of the Northwestern Caspian Region (2000–2004)

Strains code	Source of isolation	Place of material collection	Identification approach
ARID LANDSCAPES (1 STRAIN)			
Ast02-2-558	Internals of pigeon (*C. livia*)	Narimanovsky department	ELISA, HIT, NT
UPPER VOLGA–AKHTUBA (0 STRAINS)			
LOWER VOLGA–AKHTUBA (0 STRAINS)			
UPPER VOLGA DELTA (15 STRAINS)			
Ast02-2-25	Internals of crow (*C. cornix*)	Vicinity of Astrakhan	ELISA, HIT, NT
Ast02-2-26	8 preimagoes of *H. marginatum* from crow (*C. cornix*)	Vicinity of Astrakhan	ELISA, HIT, NT
Ast02-2-173 (Dhori + WNV)	20 preimagoes of *H. marginatum* from rook (*C. frugilegus*)	Privolzhsky department	ELISA, HIT, NT
Ast02-2-176 (Dhori + WNV)	13 preimagoes of *H. marginatum* from rook (*C. frugilegus*)	Privolzhsky department	ELISA, HIT, NT
Ast02-2-188	10 preimagoes of *H. marginatum* from rook (*C. frugilegus*)	Privolzhsky department	ELISA, HIT, NT
Ast02-2-205	Internals of rook (*C. frugilegus*)	Privolzhsky department	ELISA, HIT, NT
Ast02-2-208	Internals of rook (*C. frugilegus*)	Privolzhsky department	ELISA, HIT, NT
Ast02-2-209 (Dhori + WNV)	12 preimagoes of *H. marginatum* from rook (*C. frugilegus*)	Privolzhsky department	ELISA, HIT, NT
Ast02-2-218	8 preimagoes of *H. marginatum* from crow (*C. cornix*)	Privolzhsky department	ELISA, HIT, NT
Ast02-2-239 (Dhori + WNV)	19 preimagoes of *H. marginatum* from rook (*C. frugilegus*)	Privolzhsky department	ELISA, HIT, NT

(Continued)

TABLE 8.39 (Continued)

Strains code	Source of isolation	Place of material collection	Identification approach
Ast02-2-298	Internals of rook (*C. frugilegus*)	Privolzhsky department	ELISA, HIT, NT
Ast02-2-326	6 preimagoes of *H. marginatum* from rook (*C. frugilegus*)	Privolzhsky department	ELISA, HIT, NT
Ast02-2-2045	10 preimagoes of *H. marginatum* from rook (*C. frugilegus*)	Krasnoyarsky department	ELISA, HIT, NT
Ast02-2-691	*An. messeae* mosquitoes (100 insects pooled)	Privolzhje city near Astrakhan	ELISA, RT-PCR
Ast02-2-692	*An. messeae* mosquitoes (100 insects pooled)	Privolzhje city near Astrakhan	ELISA, RT-PCR
MIDDLE VOLGA DELTA (8 STRAINS)			
Ast01-66	Internals of cormorants (*Ph. carbo*)	Ikryaninsky department	ELISA, HIT, RT-PCR
Ast01-182	20 preimagoes of *H. marginatum* from crow (*C. cornix*) and rook (*C. frugilegus*)	Limansky department	ELISA, HIT, NT
Ast01-183	20 preimagoes of *H. marginatum* from crow (*C. cornix*) and rook (*C. frugilegus*)	Limansky department	ELISA, HIT, NT
Ast01-187	Internals of crow (*C. cornix*)	Limansky department	ELISA, HIT, NT, RT-PCR
Ast02-3-146	Internals of pigeon (*C. livia*)	Ikryaninsky department	ELISA, HIT, NT
Ast02-3-165	Internals of cormorants (*Ph. carbo*)	Ikryaninsky department	ELISA, HIT, NT
Ast02-3-570	Internals of magpie (*P. pica*)	Limansky department	ELISA, HIT, NT
Ast04-2-824-A	Internals of crow (*C. cornix*)	Limansky department	ELISA, RT-PCR
LOWER VOLGA DELTA (1 STRAINS)			
Ast02-3-717	Internals of cormorants (*Ph. carbo*)	Limansky department	ELISA, HIT, NT
TOTAL: 25 WNV STRAINS			
	From the internals of wild birds		**12**
	From mosquitoes (Diptera, Culicidae)		**2**
	From preimago of ticks (Acari, Ixodidae)		**11**

detected in *Anopheles messeae* (0.028%), a common visitor to houses with domestic animals in anthropogenic biocenoses, as well as in *An. hyrcanus* (0.026%) in rushes in natural biocenoses. As is illustrated in Figure 8.73, the highest intensity of WNV circulation takes place among sanguivorous mosquito populations in anthropogenic biocenoses on the territory of the Volga delta (Figure 8.74).

RT-PCR testing was carried out for the detection of WNV RNA in 11,266 samples of *Hyalomma marginatum* ticks (taxon Acari,

TABLE 8.40 Positive Results Obtained from RT-PCR Testing for WNV RNA in Wild-Bird Internals Collected in the Northwestern Caspian Region (2001–2004)

Kinds of birds	Arid landscapes	Upper Volga–Akhtuba	Lower Volga–Akhtuba	Upper Volga Delta	Middle Volga Delta	Lower Volga Delta	Total
Corvidae	59	63	135	249	164	2	672
	1 (1.6%)	3 (2.2%)	30 (12.0%)	6 (3.7%)	1	41 (6.1%)
Other land birds	10	22	43	114	52	2	243
	1 (4.5%)		5 (4.4%)	2 (3.8%)		8 (3.3%)
Cormorants	13	3	0	5	109	173	303
			1 (20.0%)	24 (22.0%)	7 (4.0%)	32 (10.6%)
Coots	52	5	6	15	188	191	457
	2 (3.8%)			12 (6.4%)	9 (4.7%)	23 (5.0%)
Terns and gulls	45	54	70	103	160	213	645
	1 (2.2%)	2 (3.7%)	1 (1.4%)	4 (3.9%)	8 (5.0%)	3 (1.4%)	19 (2.9%)
Sandpipers	19	4	0	29	23	20	95
	1				1 (5.0%)	2 (2.1%)
Herons	14	9	13	13	107	63	219
			1 (7.7%)	5 (4.7%)	1 (1.6%)	7 (3.2%)
Ducks, podiceps, and glossy ibises	17	12	28	27	39	37	160
				2 (5.1%)		2 (1.3%)
Total	**229**	**172**	**295**	**555**	**842**	**701**	**2,794**
	3 (1.3%)	5 (2.9%)	4 (1.4%)	41 (7.4%)	59 (7.0%)	22 (3.1%)	134 (4.8%)

family Ixodidae). Positive results are presented in Table 8.45 and Figure 8.75.

The total WNV infection rate for *Hyalomma marginatum* during 2001–2004 on the territory of the northwestern Caspian region was 0.5% (0.22% for the imago, 0.71% for the preimago). The maximum value was detected in the upper delta of the Volga (1.84%) and the lower Volga–Akhtuba (1.16%).

Thirty-three laboratory-confirmed WNF cases were identified in the Astrakhan region in 2002, with 16 (48.5%) of the cases seen in the rural population and 17 (51.5%) in the urban population. The transmission pathway

was established in all cases: Patients confirmed that they had been bitten by mosquitoes numerous times. Inoculations took place in the fields where much of the rural population worked; however, 76.5% of infected patients from Astrakhan did not leave the town. The percentage of patients who were younger than 14 years old was 15.2%, while 30.3% of patients were 18–40 years old, 36.4% 41–60 years old, and 18.1% 61–76 years old. Of sick patients, 63.6% were male, 36.4% female.

WNF cases began to be registered starting in June 2001, with the maximum reached in August (Figure 8.76). Durint the first three

TABLE 8.41 Positive Results Obtained from RT-PCR Testing for WNV in Wild Birds Collected on the Territory of the Northwestern Caspian Region (2001–2004)

Ecological complex	Order	Family	Genus	Species	Positive Number	Positive Portion, %
Waterbirds	Podicipediformes	Podicipedidae	*Podiceps*	Great crested grebe (*P. cristatus*)	2	1.4
	Suliformes	Phalacrocoracidae	*Phalacrocorax*	Great cormorant (*Ph. carbo*)	33	23.7
				Pygmy cormorant (*Ph. pygmaeus*)	1	0.7
	Pelecaniformes	Ardeidae	*Ardea*	Grey heron (*A. cinerea*)	1	0.7
			Egretta	White egret (*E. alba*)	2	1.4
				Little egret (*E. garzetta*)	1	0.7
			Nicticorax	Night heron (*N. nicticorax*)	3	2.2
	Gruiformes	Rallidae	*Fulica*	Coot (*F. atra*)	26	18.7
	Charadriiformes	Charadriidae	*Philomachus*	Ruff (*Ph. pugnax*)	1	0.7
			Tringa	Redshank (*T. totanus*)	1	0.7
		Laridae	*Larus*	Herring gull (*L. argentatus*)	4	2.9
				Common gull (*L. canus*)	1	0.7
		Sternidae	*Chlidonias*	Whiskered tern (*Ch. hybrida*)	7	5.0
			Sterna	Common tern (*S. hirundo*)	5	3.6
				Little tern (*S. albifrons*)	2	1.4
Subtotal	5	7	*11*	15	90	64.7
Terrestrial birds	Columbiformes	Columbidae	*Columba*	Pigeon (*C. livia*)	2	1.4
	Galliformes	Phasianidae	*Perdix*	Grey partridge (*P. perdix*)	1	0.7
	Cuculiformes	Cuculidae	*Cuculus*	Common cuckoo (*C. canorus*)	1	0.7
	Coraciiformes	Upupidae	*Upupa*	Hoopoe (*U. epops*)	1	0.7
	Passeriformes	Laniidae	*Lanius*	Great grey shrike (*L. excubator*)	1	0.7
		Sylviidae	*Locustella*	Warbler (*L. fluviatilis*)	1	0.7
		Motacillidae	*Motacilla*	Pied wagtail (*M. alba*)	1	0.7
		Corvidae	*Corvus*	Hooded crow (*C. cornix*)	14	10.1
				Rook (*C. frugilegus*)	23	16.5
				Jackdaw (*C. monedula*)	1	0.7
			Pica	Magpie (*P. pica*)	1	0.7
			Garrulus	Jay (*G. glandarius*)	2	1.4
Subtotal	5	8	10	12	49	35.3
Total	10	15	21	27	139	100.0

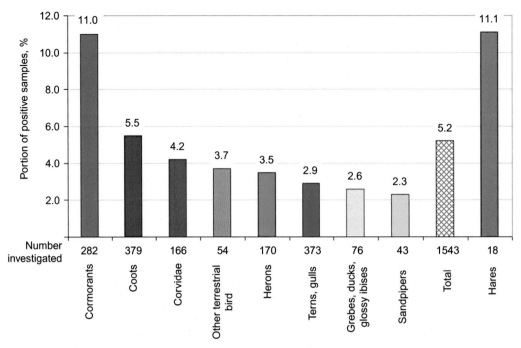

FIGURE 8.69 WNV infection rate obtained from RT-PCR testing of different species of wild vertebrates (2001–2004) in the middle and lower belts of the Volga River.

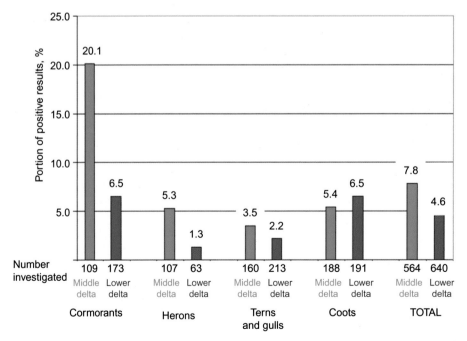

FIGURE 8.70 WNV infection rate obtained from RT-PCR testing of wild birds (2001–2004) in natural biocenoses of different ecosystems in the Volga delta.

II. ZOONOTIC VIRUSES OF NORTHERN EURASIA: TAXONOMY AND ECOLOGY

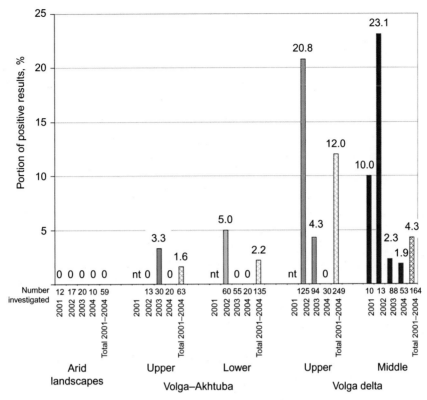

FIGURE 8.71 WNV infection rate obtained from RT-PCR testing of Corvidae birds (2001–2004) in different ecosystems.

days, 21 (63.4%) of patients appealed for medical aid; from the fourth to the sixth day, 9 patients (24.2%) did so. Twenty-six (78.8%) of the patients were hospitalized during the first five days of the disease.

Patients were hospitalized with the following initial diagnoses: adenovirus infection (33.3%), viral meningitis (36.4%), and a combined 30.3% for other diagnoses (acute respiratory viral infection, pneumonia, Q-fever, and Astrakhan fever). Thus, it would have been impossible to establish WNF without laboratory diagnostics. Seven patients (21.2%) demonstrated acute neuroinfection syndrome, whereas another 26 patients (78.8%) had no CNS insults. However, 27.0% of patients in the latter group had intracranial hypertension

syndrome. There were two cases of severe disease: a 71-year-old patient with seromeningitis and an 8-year-old child with neurotoxic syndrome during the acute period. All of the cases had a favorable result: No lethal cases were registered.

Sera from 2,884 farm animals collected in the Astrakhan region during 2001–2004 were tested by HIT and neutralization testing in order to detect specific anti-WNV antibodies. In addition, HIT-positive sera underwent neutralization testing. Anti-WNV antibodies were found by HIT in all species investigated: horses (mean positive result for the entire observation period, 9.8%; coincidence with neutralization testing, 94.1%), cattle (6.4%; 72.0%), camels (5.2%; 41.7%), pigs (3.1%;

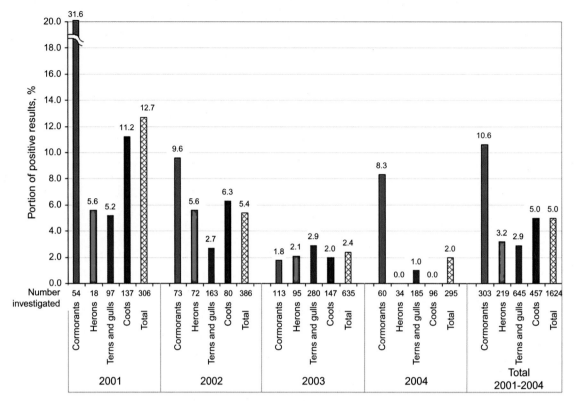

FIGURE 8.72   WNV infection rate obtained from RT-PCR testing of wild birds in natural biocenoses during different years (2001–2004).

TABLE 8.42   Positive Results Obtained from RT-PCR Testing for WNV RNA on Internal Parts Collected from Wild Mammals in the Northwestern Caspian Region (2001–2004)

Kinds of wild mammals	Arid landscapes	Upper Volga–Akhtuba	Lower Volga–Akhtuba	Upper Volga Delta	Middle Volga Delta	Lower Volga Delta	Total
Hare	2	13	13	9	14	4	55
					2 (14.3%)		2 (3.6%)
Hedgehog	1	1	3	0	0	0	5
Fox	0	1	0	0	0	0	1
Dog	0	0	1	0	0	0	1
Rodents	23	17	3	2	0	0	45
		1 (5.9%)					1 (2.2%)
Saiga	0	0	0	0	1	0	1
Total	26	32	20	11	15	4	108
							3 (2.8%)

TABLE 8.43   List of Positive Samples Obtained from
RT-PCR Testing for WNV on Internal Parts Collected
from Wild Mammals on the Territory of the
Northwestern Caspian Region (2001–2004)

Field material code	Species of mammals	Place of field material collection
UPPER VOLGA DELTA (1)		
Ast02-2-622	*Pygeretmus pumilio*	Narimanovsky department of Astrakhan region, western hill–il'men region
MIDDLE VOLGA DELTA (2)		
Ast02-3-564	*Lepus europaeus*	Limansky department of Astrakhan region
Ast02-3-920	*Lepus europaeus*	Limansky department of Astrakhan region

75.0%), and sheep (2.2%; 57.1%). Results obtained are presented in Table 8.46.

Cattle are the main host of *Anopheles messeae*, and cowsheds offer favorable conditions for the mosquitoes to reproduce. Cattle-specific antigens could often be found in the intestines of *Culex pipiens* females (but not *An. Messeae* females), which inhabitat damp basements. Town utilities adjoin with farm utilities in all settlements of the Astrakhan region, so cattle are the hosts both for *An. messeae* and for *Cx. pipiens*. Both species of mosquitoes are active vectors of WNV in anthropogenic biocenoses.

Horses were the only species of farm animals with clinically expressed WNF. In contrast to cattle, whose pastures are situated close to human settlements, horses browse far from settlements, often grazing in natural biocenoses. A significant portion of horse livestock in the Astrakhan region are of the Kushum breed, bred for meat and racing, and browse freely all year. Pedigree horses (Don, Akhaltekinsky and Arabian race horses) are kept in bloodstock farms in a stall, or they browse locally. Draft horses are kept in settlements. Horse-specific antigens have been found in the intestines of

replete females of all mosquitoes species (except for *Culiseta annulata*, which are relatively fewer). The total (2001–2004) distribution of HIT-positive horses increases from the upper Volga–Akhtuba to the lower, with the highest number found in the middle belt of the Volga delta (where the epicenter of the natural foci is located).

Pigs are the animals closest to human settlements, so pig-specific antigens are often found in the intestines of replete females of the anthropogenic mosquito species *Anopheles messeae* and *Culex pipiens*. Pigs are kept in individual yards or on pig farms. The latter are situated far from human settlements. As they are in cattle housing, *An. messeae* are the main mosquito species on the pig farm; nevertheless, all mosquitoes collected here by probe were negative for WNV. In 2003, we collected sera on the pig farms, and all probes were HIT negative. In 2004, we collected sera both on pig farms and in individual yards.

Sheep are the most numerous species of farm animal in the Astrakhan region. Sheep pastures are in the dry steppe, where conditions are favorable for the Ixodidae tick *Hyalomma marginatum*. Only a couple of species of mosquito could live in the saltish, dry steppe il'mens: *Aedes caspius* and *Cx. modestus*. The latter is an active vector for WNV. A stable and low level of infection rate among sheep (about 2%) reflects the low level of intensity of WNV circulation in arid landscapes of the Astrakhan region.

Kalmyk racing camels inhabit more arid landscapes than sheep inhabit; consequently, one might expect a lower level of seropositive camels. However, HIT often demonstrates a high percentage of positive results: 33.3% in 1989 and 13.9% in 2001. So, we instead collected sera from camels during 2002–2004 in semiwild pastures, and the percentage of seropositive results decreased. The coincidence between the results of HIT testing and neutralization testing is presented in Table 8.46.

TABLE 8.44   Results of Testing for the Detection of WNV RNA in Mosquitoes (Order Diptera, Family Culicidae) (2001−2004)[a]

Species of mosquitoes	Arid landscapes	Upper Volga−Akhtuba	Lower Volga−Akhtuba	Upper Volga Delta	Middle Volga Delta	Lower Volga Delta	Total
Aedes caspius	5/500	0	0	2/200	8/720	0	15/1,420
					1		1
Aedes flavescens	0	0	1/38	0	0	0	1/38
Aedes vexans	0	1/100	0	0	6/600	0	7/700
Anopheles hyrcanus	9/900	0	0	0	318/31,645	297/29,421	624/61,966
					10 (0.032%)	6 (0.020%)	16 (0.026%)
Anopheles messeae	49/4,900	161/16,100	121/12,100	406/40,672	216/21,335	1/132	954/95,239
	1 (0.020%)	1 (0.006%)	3 (0.025%)	13 (0.032%)	9 (0.042%)		27 (0.028%)
Culex modestus	3/300	2/200	11/1,100	3/240	60/5,970	151/14,933	230/22,743
			1 (0.091%)	1	1 (0.017%)	1 (0.007%)	4 (0.018%)
Culex pipiens	0	9/840	81/8,100	180/18,085	177/17,622	0	447/4,4647
		1	2 (0.025%)	6 (0.033%)	7 (0.040%)		16 (0.036%)
Culiseta annulata	0	0	0	1/4	0	0	¼
				1			1
Coquillettidia richiardii	0	0	0	0	294/29,300	493/49,007	787/78,307
					13 (0.044%)	14 (0.029%)	27 (0.034%)
Total	66/6600	173/17240	214/21338	592/59201	1079/107192	942/93493	3066/305064
	1 (0.015%)	2 (0.012%)	6 (0.028%)	21 (0.035%)	41 (0.038%)	21 (0.022%)	92 (0.030%)

[a]Data format: number of pools/number of mosquitoes.

Horses are the best marker of WNV circulation, because they have the largest percentage of HIT-positive results and the greatest coincidence between HIT and NT results. Kushum race horses are the most significant marker. Monitoring the infection rates among farm animals will be continued, taking into account the relationships and phenomena described.

It has been found that WNV can remain viable during interepidemiological periods in overwintering imagoes of sanguivorous mosquitoes (e.g., *Anopheles messeae*, *Culex pipiens* and *Culiseta annulata*) as well as overwintering imagoes of the Ixodidae tick *Hyalomma marginatum*. The scheme of WNV circulation on the territory of the northwestern Caspian region is presented in Figure 8.77.

After the 1999−2006 outbreak of WNF in four administrative units in southern Russia, a significant outbreak with more than 500 cases arose in the summer and autumn of 2010. The disease spread up to 500 km to the north and northeast from an earlier known endemic area and now includes an additional two administrative units (Tables 8.47 and 8.48, Figure 8.78).

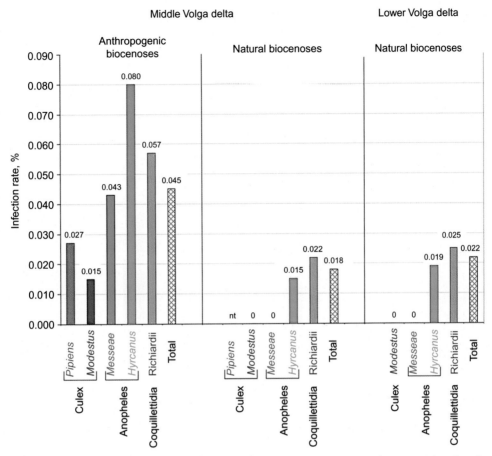

FIGURE 8.73 WNV infection rate of mosquitoes (order Diptera, family Culicidae) in natural and anthropogenic biocenoses of the middle and lower belts of the Volga delta.

8.3 FAMILY ORTHOMYXOVIRIDAE

The Orthomyxoviridae includes six genera of enveloped viruses with a segmented, negative-polarity ssRNA genome. The genome of the orthomyxoviruses consists of six (*Thogotovirus* and *Quaranjavirus*), seven (*Influenza C virus*), or eight (*Influenza A virus, Influenza B virus* and *Isavirus*) segments.[1,2] All orthomyxoviruses encode three enzymes formed of viral RdRp: PB1 (Figure 8.79), PB2, and PA. These proteins are about 30% similar among viruses of different genera. Common

structural proteins are NP, associated with genomic RNA; matrix protein; and two envelope proteins: hemagglutinin, or HA (possesses hemagglutinating activity) and neuraminidase, or NA (also called sialidase) in the influenza viruses.

Viruses of the *Thogotovirus* and *Quaranjavirus* genera are transmitted by arthropod vectors, predominantly Ixodidae and Argasidae ticks, respectively. Viruses of the *Influenza A virus, Influenza B virus* and *Influenza C virus* genera are important human pathogens transmitted by a respiratory route.[1,2]

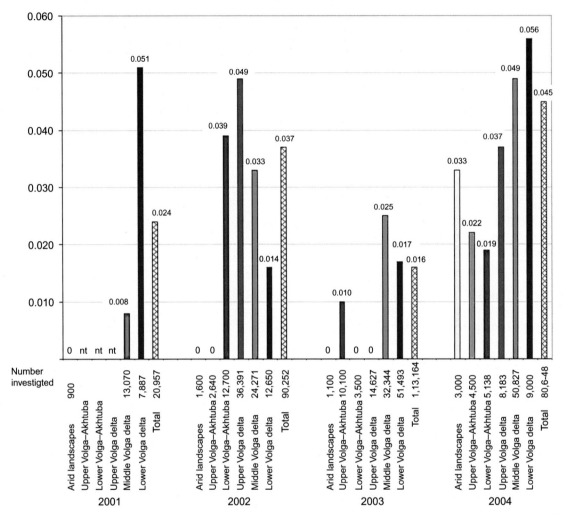

FIGURE 8.74 RT-PCR testing for WNV infection rate of mosquitoes in different ecosystems of the northern Caspian Sea basin in different years (2001–2004).

Genus *Isavirus* has only one species: infectious salmon anemia virus, which strikes fish in the Salmonidae family.

8.3.1 Genus *Influenza A Virus*

Genus *Influenza A virus* has just one named species: Influenza A virus, represented by numerous antigenic and genetic subtypes. The genome of Influenza A virus consists of 8 segments of ssRNA that encode 11 or more proteins.[1–5] Influenza A viruses are divided into distinct subtypes based on the antigenic and genetic properties of their HA and NA proteins. Sixteen subtypes of HA (HA1–16) and 9 subtypes of NA (NA1–9) have been found worldwide in aquatic birds. Two additional subtypes of HA (HA17 and HA18) and NA (NA10 and NA11) are seen in New World bats.[4,6,7] H17 and HA18 form a clade distinctly

TABLE 8.45 Positive Results Obtained from RT-PCR Testing for WNV RNA in the Ixodidae Tick *Hyalomma Marginatum* (2001–2004)

Stage of development of *H. Marginatum*	Arid landscapes	Upper Volga–Akhtuba	Lower Volga–Akhtuba	Upper Volga Delta	Middle Volga Delta	Lower Volga Delta	Total
Imago	3,448	0	140	877	458	0	4,923
	1 (0.03%)		1 (0.71%)	4 (0.46%)	5 (1.09%)		11 (0.22%)
Preimago	89	216	638	863	4,537	0	6,343
		1 (0.46%)	8 (1.25%)	28 (3.24%)	8 (0.18%)		45 (0.71%)
Total	3,537	216	778	1,740	4,995	0	11,266
	1 (0.03%)	1 (0.46%)	9 (1.16%)	32 (1.84%)	13 (0.26%)		56 (0.50%)

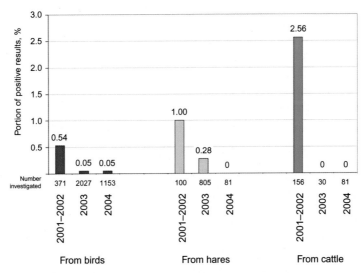

FIGURE 8.75 RT-PCR testing for WNV infection rate of *Hyalomma marginatum* ticks in the middle belt of the Volga delta (2001–2004).

related to the other Influenza A subtypes, but NA10 and NA11 form a new phylogenetic branch external to the Influenza A and Influenza B viruses (Figures 8.80 and 8.81).

8.3.1.1 Influenza A Viruses (H1–H18)

History. Influenza as a human disease was originally described in 412 B.C. by Hippocrates (Figure 8.82) in his book *Epidemics*, but the "father of medicine" did not consider influenza to be an infectious disease. Instead, the famous English physician Thomas Sydenham (Figure 8.83) was the first who suggested the infectious nature of the disease.[1,2]

The term "influenza" has been around since the first half of eighteenth century and derives

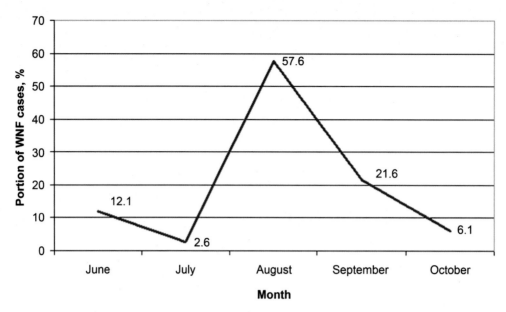

FIGURE 8.76   Seasonal dynamics of WNF in the Astrakhan region.

from the Italian "influenza di freddo" ("influence of the cold") or from Spanish "influencia de las estrellas" ("influence of the stars"), the latter reflecting the contemporaneous belief in astrological reasons for the emergence of disease.[3] Up to the nineteenth century, the archaic terms "catarrhus epidemicus," "cephalgia contagiosa," "febris catarrhalis" and "febris comatose" had wide currency.[4] The English word "grippe" (related to the Russian "грипп") is related to the German "greifen" ("to catch hold") and derived from the French "gripper" ("to catch hold," "paralyze"); the word gained currency at the beginning of nineteenth century. (Cf., e.g., the passage from Volume 1, Chapter 1 of Tolstoy's famous novel *War and Peace*: "She was, as she said, suffering from la grippe; grippe being then a new word in St. Petersburg, used only by the elite.").[5]

Before the nineteenth century, influenza A epidemics were described only qualitatively.

Subtypes of the etiological agent were retrospectively revealed for the 1889–1892 epidemic (H2N2), the 1897–1900 epidemic (H3N8), and the 1918–1919 pandemic (H1N1, the so-called Spanish flu)[5–9]—retrospectively only because Influenza A virus wasn't found until 1930 by Richard Shope (Figure 8.84) on the model of swine (*Sus scrofa*) flu.[10,11] Human flu was found two years later[12,13] by a group of English scientists: Wilson Smith (Figure 8.85), Christopher Andrewes (Figure 8.86) and Patrick Laidlaw (Figure 8.87). During the pandemic of 1918–1919, it was suggested that the etiological agent of influenza A was the so-called Afanasiev–Pfeiffer bacillus,"[14–16] named after the Russian bacteriologist Mikhail Afanasiev (Figure 8.88) and the German bacteriologist Richard Pfeiffer (Figure 8.89)—the modern *Haemophilus influenzae* bacillus.[17,18] Three Influenza A pandemics were described after the discovery of the etiological agent: the

Natural-territory unit	2001		2002		2003		2004		Total (2001–2004)	
	HIT	R	HIT	R	HIT	R	HIT	R	HIT	R
CATTLE										
Volga–Akhtuba Upper	6/44		2/30		1/30		0/29		9/133	
	13.6%	66.7%	6.7%	100%	10.0%	–	0%	–	6.8%	75.0%
Lower	1/41		6/35		0/40		0/40		7/156	
	2.4%	100%	17.1%	83.3%	0%	–	0%	–	4.5%	85.7%
Volga delta Upper	1/43		16/80		9/71		0/81		26/275	
	2.3%	100%	20.0%	62.5%	12.7%	–	0%	–	9.5%	64.7%
Middle	5/90		7/112		1/106		6/74		19/382	
	5.6%	80.0%	6.3%	85.7%	0.9%	–	8.1%	50.0%	5.0%	72.2%
Total	13/218		31/257		11/247		6/224	50.0%	61/946	
	6.0%	76.9%	12.1%	74.2%	4.5%	–	2.0%		6.4%	72.0%
HORSES										
Volga–Akhtuba Upper					3/40		1/40		4/80	
	–	–	–	–	7.5%	–	2.5%	100%	5.0%	100%
Lower			0/17		1/40		8/37		9/94	
	–	–	0%	–	2.5%	–	21.6%	75.0%	9.6%	75.0%
Volga delta Upper					0/40		0/73		0/113	
	–	–	–	–	0%	–	0%	–	0%	–
Middle			15/68		17/60		6/105		38/233	
	–	–	22.1%	93.3%	28.3%	–	5.7%	100%	16.3%	95.2%
Total			15/85		21/180		15/255		51/520	
	–	–	17.6%	93.3%	11.7%	–	5.9%	86.7%	9.8%	94.1%
PIGS										
Volga–Akhtuba Upper					0/30		0/30		0/60	
	–	–	–	–	0%	–	0%	–	0%	–
Lower							0/20		0/20	
	–	–	–	–	–	–	0%	–	0%	–
Volga delta Upper					0/42		2/30		2/72	
	–	–	–	–	0%	–	6.7%	100%	2.8%	100%
Middle					0/65		6/42		6/107	
	–	–	–	–	0%	–	14.3%	66.7%	5.6%	66.7%
Total					0/137	–		75.0%		
	–	–	–	–	0%	6.6%	8/122	3.1%	8/259	75.0%

(Continued)

TABLE 8.46　(Continued)

Natural-territory unit		2001 HIT	2001 R	2002 HIT	2002 R	2003 HIT	2003 R	2004 HIT	2004 R	Total (2001–2004) HIT	Total (2001–2004) R
SHEEP											
Volga–Akhtuba	Upper	1/37		1/71		0/30		0/30		2/168	
		2.7%	100%	1.4%	100%	0%	—	0%	—	1.2%	100%
	Lower	0/25		0/55		0/40		1/20		1/140	
		0%	—	0%	—	0%	—	5.0%	—	0.7%	—
Volga delta	Upper	2/53		4/75		1/91		0/40		7/259	
		3.8%	0%	5.3%	33.3%	1.1%	—	0%	—	2.7%	20.0%
	Middle	2/89		3/130		3/80		2/45		10/344	
		2.2%	100%	2.3%	66.7%	3.8%	—	4.4%	50.0%	2.9%	71.4%
Total		5/204		8/331		4/241		3/135		20/911	
		2.5%	60.0%	2.4%	57.1%	1.7%	—	2.2%	50.0%	2.2%	57.1%
CAMELS											
Volga–Akhtuba	Upper	—	—	—	—	—	—	—	—	—	—
	Lower	2/48		0/35		1/40		0/10		3/133	
		4.2%	50.0%	0%	—	2.5%	—	0%	—	2.3%	50.0%
Volga delta	Upper			0/22		0/20		1/9		1/51	
		—	—	0%	—	0%	—	11.1%	100%	2.0%	100%
	Middle	8/24		1/30		0/10				9/64	
		33.3%	37.5%	3.3%	0%	0%	—	—	—	14.1%	33.3%
Total		10/72		1/87		1/70		1/19		13/248	
		13.9%	40.0%	1.1%	0%	1.4%	—	5.3%	100%	5.2%	41.7%

[a] Abbreviations: HIT, hemagglutination inhibition test; R, coincidence coefficient between hemagglutination inhibition test (HIT) and neutralization test (NT); —, no data.

so-called Asian flu (1957–1959) (H2N2),[19–21] Hong Kong flu (1968–1970) (H3N2),[21–23] and swine flu (2009–2010) (H1N1 pdm09).[24–29] (The large epidemic of "Russian flu" (1977–1978) (H1N1)[21,23,30] did not reach pandemic scale.)

Avian flu has been known under the name "Lombardian disease" since the beginning of the nineteenth century.[31–34] In 1878, the Italian veterinarian Edoardo Perroncito (Figure 8.90) described a highly contagious disease (previously named "exsudative typhus of chickens") among chickens, with 100% lethality in the vicinity of Turin.[35] The terms "classic fowl plague" and "bird pest" came

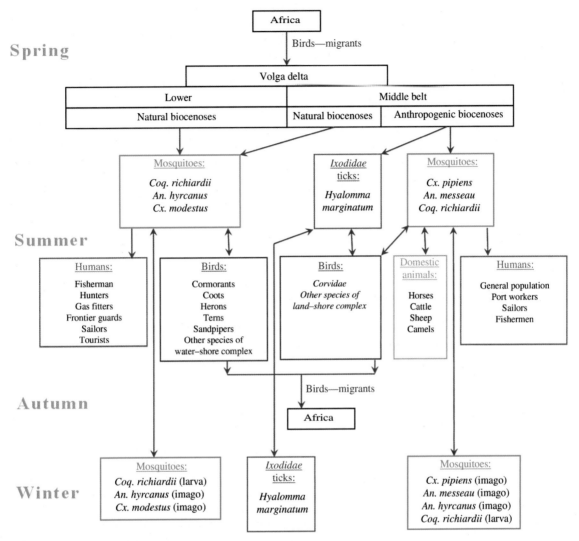

FIGURE 8.77   Circulation pattern for WNV on the territory of North-Western Caspian region.

into wide use in 1901, when a large epizootic outbreak in Tyrol province, Italy, did away with the population of farm birds there.[33] The term "Braunschweig disease" was used to identify an analogous disease among guinea fowls in Europe.

In 1901, the Italian scientists Eugenio Centanni and Ezio Savonuzzi demonstrated that the etiological agent of classic fowl plague is a filtrated substance.[34] Nevertheless, classic fowl plague wasn't identified as Influenza A virus until 1955, by Werner Shäfer (Figure 8.91) on the example of the historical strain A/chicken/Brescia/1/1902 (H7N7).[36,37] W.B. Becker was the first who identified Influenza A virus among wild birds when he

TABLE 8.47 WNF Morbidity in Russia (1999—2013)[a]

Administrative unit	Number of cases per year															Total
	1999	2000	2001	2002	2003	2004	2005	2006	2007	2008	2009	2010	2011	2012	2013	
Astrakhan region	95	24	49	31	11	25	73	16	1	1	5	22	18	67	70	508
Volgograd region	380	32	15	15			3	12	63	2	5	411	57	210	48	1,253
Voronezh region												27	34	36	6	103
Krasnodar Krai	85						2		7	2		5	3			104
Rostov region		7	5		3	7	18	13	18	1	2	63	16	48	11	212
Saratov region												2			28	30
Samara region															9	9
Lipetsk region														35	2	37
Belgorod region															2	2
Total	560	63	69	46	14	32	96	41	89	6	12	530	128	396	176	2,258

[a]WNF mortality in 1999 was 0.9%, in 2010, it was 1.2%.

TABLE 8.48 WNF Cases in Russia in 2013

Federal District	Number of cases	Portion, %	Infection rate per 100,000 population
Central	15	7.2	0.04
Northwestern	1	0.5	0.01
Southern	147	70.3	1.06
North Caucasian	0	0.0	0.00
Lower Volga	42	20.1	0.14
Ural	1[a]	0.5	0.01
Siberian	3[a]	1.4	0.02
Far Eastern	0	0.0	0.00
Total	209	100.0	

[a]Imported cases.

described antigenic relations between A/tern/South Africa/61 and A/chicken/Scotland/59 strains and developed the hypothesis that wild birds played a role in disseminating the virus.[38] Then, during 1960—1970, Gram Laver (Australia), Dmitry Lvov (Figure 2.36) (former USSR, Russia), and Robert Webster (Figure 2.20) (USA) independently formulated the more general idea that there were natural foci f Influenza A virus and that wild aquatic birds were a natural reservoir for the virus.[39—42]

Subtypes of Influenza A Virus in Northern Eurasia. At present, we know that numerous avian influenza viruses are abundant in the bird populations of Northern Eurasia. However, until the end of the 1960s, these data were absent. At that time in the former USSR, avian Influenza A virus was being isolated only from poultry. One of the first avian viruses isolated in the USSR—A/duck/Ukraine/1/1963—was destined to play an important role in the development of the theory of influenza virus evolution.[43]

In 1960—1964, a group of researchers in the Ukrainian Soviet Republic isolated several influenza virus strains from ducklings affected with sinusitis. The first three strains were isolated in 1960 in Crimea and in the Kharkov

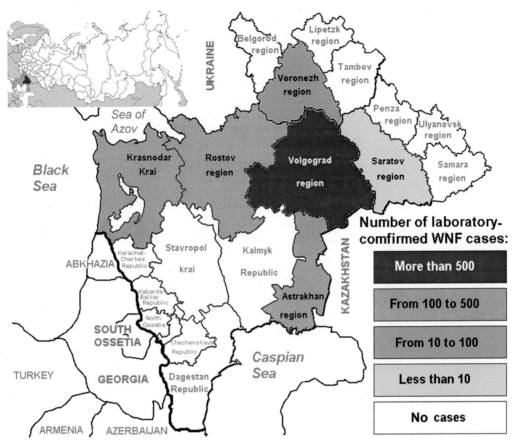

FIGURE 8.78   Present distribution of WNF in Russia.

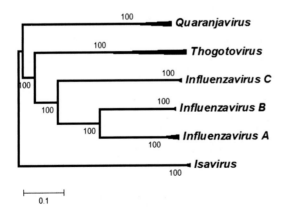

FIGURE 8.79 Phylogenetic structure of the Orthomyxoviridae family, constructed for PB1 amino acid sequences.

region.[44] These prototypical strains were Ya-60, B-60, and S-60. [45] Several other strains (Z-61, C-61, N-62, D-62, D-62, Z-62, S-64, and BV1) were isolated from ducks and chickens[46] in 1961–1964. The most peculiar features of these isolates were revealed in comparative studies performed at the D.I. Ivanovsky Institute of Virology in Moscow. As early as 1964, the duck strains Ya-60, B-60, Z-61, and C-61 were analyzed with respect to their antigenic specificity by HIT and were found to be antigenically distinct from the human H1 and H2 viruses.[47] After the appearance of the H3 pandemic virus in 1968, some of the Ukrainian duck strains were shown to be antigenically

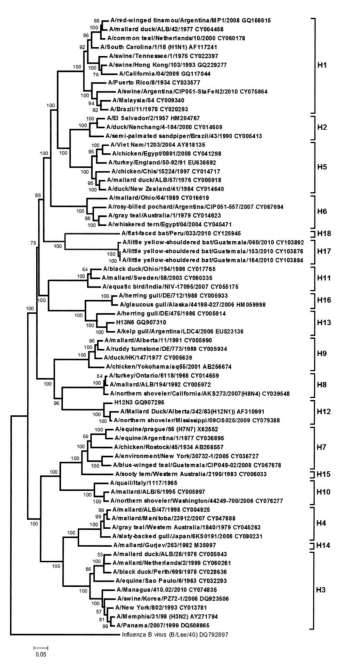

FIGURE 8.80   Phylogenetic structure of the *Influenza A virus* genus, constructed for HA amino acid sequences.

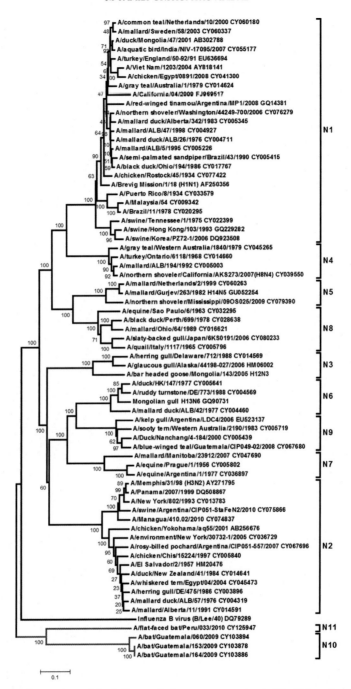

FIGURE 8.81   Phylogenetic structure of the *Influenza A virus* genus, constructed for NA amino acid sequences.

FIGURE 8.82   Hippocrates ('Ιπποκράτης) (460 B.C.−377 B.C.).

FIGURE 8.85   Wilson Smith (1897−1965).

FIGURE 8.83   Thomas Sydenham (1624−1689).

FIGURE 8.86   Christopher Andrewes (1896−1988).

FIGURE 8.84   Richard Shope (1901−1966).

FIGURE 8.87   Patrick Laidlaw (1881−1940).

II. ZOONOTIC VIRUSES OF NORTHERN EURASIA: TAXONOMY AND ECOLOGY

FIGURE 8.88 Mikhail Ivanovich Afanasiev (1850–1910).

FIGURE 8.91 Werner Shäfer (1913–2000).

FIGURE 8.89 Richard Pfeiffer (1858–1945).

FIGURE 8.90 Edoardo Perroncito (1847–1936).

related to the new subtype. In 1969, L.Ya. Zakstelskaya et al. used HIT to demonstrate that duck strains B-60 and BV1 cross-reacted with the A/Hong Kong/1/1968 pandemic strain and other human strains isolated in 1968, whereas the strain Ya-60 exhibited a negligible cross-reaction with the human viruses.[48,49] Moreover, HIT testing also showed that the B-60 and BV1 strains of the virus reacted with human sera, including those collected in 1881–1886 and in 1905–1908. On the basis of this phenomenon, the authors suggested that an avian virus similar to the strains B-60 and BV1 was the precursor of the human pandemic strain and that this antigenic variant had appeared in humans several times in the past.[48] Formerly known as Ya-60, strain A/duck/Ukraine/1/1960 was shown[50] to belong to the H11N2 subtype, whereas A/duck/Ukraine/2/1960 was identified as H3N6 and A/duck/Ukraine/1/1963 as H3N8.

The highly pathogenic H5N2 and H7N2 strains were isolated from chickens in the Moscow region.[51,52]

Several virus strains producing enteritis in chickens were isolated in 1972 and in 1974 in chicken farms and identified as H6N2 strains,[51,53,54] an unusual antigenic formula for a pathogenic virus affecting poultry.

Six H3N2 isolates were obtained in a chicken farm in Kamchatka from chickens affected with rhinitis, conjunctivitis, and laryngotracheitis.[51,55]

In 1977, isolates identified as H3N1 viruses were isolated from sick chickens and ducks in the Russian Federation[25] and Uzbekistan in the former USSR.[26]

In 1984, H8N4 strains were isolated in the western part of the Ukrainian Soviet Republic from the lungs of ducklings affected with pneumonia. The isolation was the only one of an H8 influenza virus in the USSR (Lvov DK, unpublished data).

In 1970, a large-scale series of virus isolations from wild birds, combined with some serological studies, was initiated as a part of the Coordinated Program of the National Committee on the Studies of Viruses Ecologically Linked to Birds together with the Virus Ecology Center of the D.I. Ivanovsky Institute of Virology. By the end of the 1970s, the pattern

of circulation of avian viruses in the territory of the USSR was identified.[3,11,26,30] In the ensuing years, the pattern of the Influenza A virus subtypes (including H15 and H16) circulating in Northern Eurasia was amplified (Figure 8.92).

Blood sera collected in the spring and autumn of 1970 near Lake Khanka and Peter the Great Bay (both in Primorsky Krai) from 262 birds—mainly mallards (*Anas platyrhynchos*), common teals (*An. crecca*), Baikal teals (*An. formosa*), garganeys (*An. querquedula*), falcated ducks (*An. falcata*), pintails (*An. acuta*), grey herons (*Ardea cinerea*), coots (*Fulica atra*), black guillemots (*Cepphus grylle*) and black-tailed gulls (*Larus crassirostris*)—were HIT-tested against H1, H4, H5, H6, H10, and H11 avian influenza viruses. No antibodies were found in the sera of grey herons and coots, nor were any found against H11 in any species. Antibodies against all the other subtypes tested were encountered occasionally

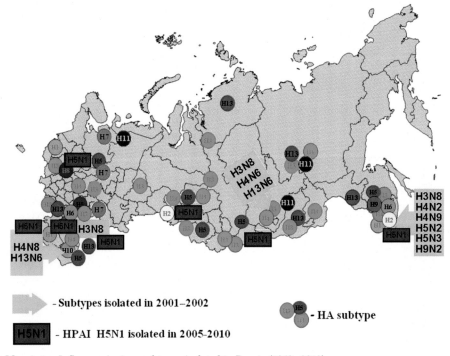

FIGURE 8.92   Avian Influenza A virus subtypes isolated in Russia (1962–2013).

in the sera of gulls, black guillemots, and ducks. In some species, such as teals, falcated ducks, and black guillemots, antibodies against several subtypes were detected.[27]

In 1972, sera were collected from gulls, cormorants, murres, and tufted puffins in the Commander Islands. Antibodies against H2, H3, H5, and H7 viruses were detected.[28] In 1970–1972, sera from gulls, cormorants, and murres were collected in the Kamchatka, Sakhalin, and Magadan regions and antibodies to H1, H2, H3, H5, H6, and H7 viruses were detected.[30] Antibodies against H1, H3, H4, H5, and H7 were identified in sera taken from Arctic terns (*Sterna paradisaea*), black-throated loons (*Gavia arctica*), mallards (*Anas platyrhynchos*), common teals (*Anas crecca*), tufted ducks (*Aythya fuligula*), greylag geese (*Anser anser*), skuas (*Stercorarius* sp.), and a blue whistling thrush (*Myophonus caeruleus*) collected in the White Sea Basin in the estuary of the Pechora River in the Arkhangelsk region of Russia in 1969–1972.[56] The serologic studies suggested a wide range of avian influenza viruses circulating in wild birds in Northern Eurasia. This suggestion was confirmed and extended by the isolation of virus strains from other wild birds.

Many avian species proved to be hosts of H1 viruses. A virus belonging to the H1N3 subtype was isolated in 1977 from a tern in the southern part of the Caspian Sea basin.[57] In 1978, an H1N4 strain was isolated from a common teal (*Anas crecca*) in the Russian Republic of Buryatia in eastern Siberia.[41] Several H1N1 viruses were isolated in Kazakhstan in 1979 from waterfowl, including the common teal (*An. crecca*), garganey (*An. querquedula*), shoveler (*Spatula clypeata*), and coot (*Fulica atra*),[58] as well as in 1980 from tree sparrows (*Passer montanus*) and hooded crows (*Corvus cornix*).[41] In 1979, an H1N1 virus was isolated from a hawfinch (*C. coccothraustes*) in Mongolia.[59] In the same year, an H1N2 strain was isolated from a black-headed gull (*Larus ridibundus*)

on an island in the northern part of the Caspian Sea.[41]

The avian viruses belonging to the H2 subtype seem not to be abundant in Russia. In fact, for a long time the only virological evidence of the presence of this subtype in Russia was the isolation of an H2N3 virus in 1976 from a pintail (*Anas acuta*) in Primorsky Krai.[60] However, serological data suggested that H2 viruses circulated in wild birds not only in Primorsky Krai, but also in other regions of the Far East, including the Commander Islands as well as the Kamchatka, Sakhalin, and Magadan regions.[54,61]

Avian influenza A viruses belonging to the H3 subtype are widespread in Northern Eurasia. An H3N2 virus was isolated from a common murre (*Uria aalge*) in 1974 on Sakhalin Island,[62] and another H3N2 strain was isolated in 1976 from a pintail (*Anas acuta*) in Primorsky Krai.[63] Two H3N2 strains were isolated in 1974 in the Ukrainian Soviet Republic from unusual hosts for avian viruses: the white wagtail (*Motacilla alba*) and the European turtle dove (*Streptopelia turtur*).[64] H3N2 strains were also isolated from grey crows (*Corvus cornix*) in 1972 in the Volga basin and from a shelducks (*Tadorna ferruginea*) in 1979 in Kazakhstan.[65] An H3N2 virus was isolated from a tree sparrow (*P. montanus*) in 1983 in the Ukrainian Soviet Republic.[66] In 1972–1973, H3N3 and H3N8 viruses were isolated from ducks and herons in Khabarovsk Krai. One of the viruses closely resembled a strain isolated a year later in central Asia. This resemblance demonstrated that H3N3 viruses circulated in regions fairly distant from one another.[67] In 1972–1973, H3N8 viruses were isolated in Khabarovsk Krai from wild ducks (*Anas* sp.), tufted puffins (*Fratercula cirrhata*), and horned puffins (*F. corniculata*)[65] and in the Arkhangelsk region in the Pechora River estuary (White Sea basin) from Arctic terns (*Sterna paradisaea*) and black-throated loons (*Gavia arctica*).[68] In 1978, H3N8 strains were isolated in the Republic of Buryatia from a mallard (*An. platyrhynchos*) and a pintail (*An. acuta*),[65] as well as in Khabarovsk Krai from the common murre (*U. aalge*)[67] and from black-headed gulls (*Larus ridibundus*).[69]

Avian viruses of the H4 subtype were isolated in 1970–1980 mostly in a narrow belt stretching from the lower Volga, through Kazakhstan, and on to the south of eastern Siberia. Several H4N6 strains were isolated in 1976 from slender-billed gulls (*Chroicocephalus genei*) in the Volga delta[70] and from great black-headed gulls (*Ichthyaetus ichthyaetus*) on the islands in the northern part of Caspian Sea.[41] In 1977, H4N8 virus was isolated from the black tern (*Chlidonias niger*) in Central Kazakhstan.[71] In the Republic of Buryatia, H4N6 strains were isolated in 1978 from the common goldeneye (*Bucephala clangula*).[41]

Isolations of H5 influenza viruses from wild birds were scarce. In 1976, several H5N3 strains were isolated from terns (common terns and little terns) and a slender-billed gull in the Volga River delta.[70] A detailed description of the penetration of the H5N1 strain of of highly pathogenic avian influenza (HPAI) A into Northern Eurasia and its further dissemination is presented shortly.

The strains belonging to the H6 subtype seem not to be abundant, but their geographic distribution is wide. An H6N2 strain was isolated in 1972 from the Arctic tern (*Sterna paradisaea*)[68] in the Arkhangelsk region (White Sea basin). One H6N4 strain was isolated in 1978 from the pintail (*Anas acuta*) in Primorsky Krai,[41] and an H6N8 strain was isolated from the common tern (*S. hirundo*) in 1977 in the Caspian Sea basin.[57] In 2010, two H6N2 strains were isolated on Kunashir Island (the southernmost of the Kuril Islands) and four were isolated on Sakhalin Island.

An H7N3 strain was isolated in 1972 from a sandpiper (a member of the Scolopacidae family) in the Arkhangelsk region of Russia.[68]

One strain of H8N4 was isolated in 2001 in the Republic of Buryatia, and one strain in 2003 in Mongolia.

An H9N2 strain was isolated from a mallard (*Anas platyrhynchos*)[72] in Primorsky Krai in 1982 and in Khabarovsk Krai in 2013.

Over 40 H10N5 strains were isolated from a wide array of bird species near Alakol Lake in east central Kazakhstan in 1979. The strains were isolated from several species of ducks (*Anas* sp.), from shorebirds (members of the order Charadriiformes), to passerine birds (members of the order Passeriformes), to coots (*Fulica atra*), plovers (members of the family Charadriidae, subfamily Charadriinae), and chukars (*Alectoris chukar*).[41] This situation is a rare case of an isolation of closely related viruses from an extremely wide array of avian species.

The viruses identified as H11N8 strains were isolated in 1972 from the Arctic tern (*Sterna paradisaea*) and the red-throated diver (*Gavia stellata*) in the estuary of the Pechora River in the northern part of European Russia.[54] Several H11N6 strains were isolated from the common teal (*Anas crecca*), the European widgeon (*An. penelope*), and the European golden plover (*Pluvialis apricaria*) in 1979 in eastern Siberia.[41] In 1987, H12N2 strains were isolated from mallards, a pintail, and European widgeons south of Issyk-Kul Lake in Kyrgyzstan.[72]

Two strains of H12N2 were isolated from wild ducks (subfamily Anatinae) in Kyrgyzstan.

The results of virus isolation and serological studies in the territory of the USSR in 1970–1980 suggested a wide circulation of avian influenza viruses in wild birds and enabled researchers to construct a map of avian influenza viruses encountered in different regions of Northern Eurasia. The general pattern of distribution of influenza virus subtypes in wild birds was fairly evident by the end of the decade. Virus isolation was continued in the ensuing years, and it brought several major results. Isolations were performed mostly in the central and southern parts of European Russia, in western and eastern Siberia, and in the Russian Far East.[72] Overall, 1,005 strains were isolated from wild birds in Russia in 1980−2013 (Table 8.49). About 250 samples were taken yearly from 50 to 100 birds in each geographic region. The mean percentage of successful isolations ranged from 3.5% to 5.7%. Over 50% of the isolates were H13 viruses (H13N2, H13N3, H13N6, and H13N8) isolated mostly from gulls and shorebirds in the northern part of the Caspian Sea. The viruses of the H3 subtype (over 25% of the total number of isolates) were isolated in several regions.

Many strains isolated in 1979−1985 from great black-headed gulls (*Ichthyaetus ichthyaetus*), herring gulls (*Larus argentatus*) and Caspian terns (*Hydroprogne caspia*) on the island of Maly Zhemchuzhny in the northern part of the Caspian Sea were not identified at the time of isolation with respect to the subtype of their HA. As it turned out, the strains belonged to the subtype H13, was first described in 1982,[73] and in 1989 the mysterious Caspian isolates were identified[74] as H13N2, H13N3, and H13N6. To characterize the H13 subtype molecularly and antigenically, the complete nucleotide sequence of the HA of the strain A/great black-headed gull/Astrakhan/277/84 was used for comparison with the HAs

TABLE 8.49   Isolation of Influenza A Strains from Birds in Northern Eurasia (1980−2014) (According to Data from the Russian State Collection of Viruses in the D.I. Ivanovsky Institute of Virology)

HA Subtype	NA subtype									Total	
	1	2	3	4	5	6	7	8	9	Number	%
1	59		6			4				69	6.87
2		2	1							3	0.30
3	2	18		3		38		177		238	23.68
4	3	2	3	2		36		7	1	54	5.37
5	57	3	9							69	6.87
6		8		1				1		10	1.00
7	8		1				2			11	1.09
8				1						1	0.10
9		10		4						14	1.39
10				12				7		19	1.89
11		1				6		1	5	13	1.29
12		2								2	0.20
13		99	78			311		10		498	49.55
14					3	1				4	0.40
Total Number	129	145	98	23	3	396	2	203	6	1,005	
%	12.84	14.43	9.75	2.29	0.30	39.40	0.20	20.20	0.60		

of two American strains isolated from a gull and a pilot whale.[75]

Virus isolation studies in the northern Caspian basin were continued in the 1990s and 2000s. Materials were collected from wild birds in the area of the northern coast of the Caspian Sea (including Maly Zhemchuzhny Island) from the delta of the Terek River in the north Caucasus region to the Emba River in western Kazakhstan. Most of the strains that were isolated belonged to the H13 subtype, including H13N2, H13N3, H13N6, and H13N8 isolates; besides these strains, only single isolates belonging to the H4N3, H4N6, H6N2, and H9N2 subtypes were isolated.[76,77]

In 1990, a new, previously unrecognized, subtype of influenza virus H14 HA was described[78] on the basis of the characterization of two strains isolated in 1982 from mallards (*Anas platyrhynchos*) in the Ural River delta. The H14N5 and H14N6 subtypes were isolated from mallards and gulls in Astrakhan.[76] A partial sequencing revealed that NS gene of the H14 strains isolated from the gulls was closely related to the NS gene of H9 and H13 strains isolated previously from gulls and terns in the Caspian Sea basin and to the H9N4 strain isolated in the Russian Far East. The NS gene of an H14N5 strain isolated from a mallard was much more distantly related to the NS gene of the viruses isolated from gulls.[76] The results suggest that reassortment events play a significant role in the evolution of H14 viruses, with the NS gene being an important determinant of the range of the host.

A large-scale isolation of avian influenza viruses from fecal samples was performed in 1995–1998 in eastern Siberia and the Far East by a group that included both Russian and Japanese researchers.[79] Scientific contacts between Russian and Japanese researchers of avian Influenza A virus were ongoing during the eighth Russian–Japanese Consultations at a conference titled "Protection of Migratory Wild Birds in the Asia–Pacific region" held at the

Russian Ministry of Natural Resources in Moscow April 01–05, 2011. At the conference, the D.I. Ivanovsky Institute of Virology took the initiative to renew the international meetings on medical ornithology at the level of experts of Asia–Pacific countries that had been taking place regularly during the 1970 and 1980s. As a result, the First International Meeting for Medical Ornithology in the Asia–Pacific Region was held in Tokyo, Japan, on June 23, 2011. The meeting was devoted to the topic of HPAI H5N1 distribution in Asia. A second meeting was conducted in Moscow at the D.I. Ivanovsky Institute of Virology March 15–16, 2012 (Figure 8.93).[80]

In the summer of 2000 in a valley in the Sayan Mountains in southeastern Siberia, the strains H3N8, H7N1, H7N8, H13N1, and H13N6 were isolated.[81] The H3N8 and H7N8 strains were isolated from ruddy shelducks (*Tadorna ferruginea*) and common redshanks (*Tringa totanus*), the H7N1 strains from common pochards (*Aythya ferina*), and the H13N1 strains from northern shovelers (*Anas clypeata*) and great crested grebes (*Podiceps cristatus*). The H13N6 strains were isolated from all of the aforementioned species, as well as from teals, ducks, and terns. In 2000–2002, the subtypes H3N8, H4N2, H4N6, H4N8, H4N9, H5N2, H5N3, H9N2, and H13N6 were isolated in the same region; 1,750 samples were taken from 48 bird species.[72] A strain isolated from the muskrat (*Ondatra zibethicus*)[81] in 2000 in the Republic of Buryatia was identified as an H4N6 virus closely resembling the H4N6 strains isolated from ducks in the same year and the same region.[72] The HAs of the H4 strains (including the muskrat strain) isolated in Buryatia formed a separate group of the Eurasian–Australian branch in the phylogenetic tree of H4 HA (Figure 8.94). They had a C-terminal proline residue in their HA1 subunit, in contrast to the serine residue of most Eurasian strains. The HA genes of the H5N2 isolates turned out[82] to have cleavage peptides LRNVPQRETR/GL identical to the ones of the

FIGURE 8.93   (A) Official emblem of the Second International Meeting for Medical Ornithology in the Asia—Pacific Region (Moscow, Russia, D.I. Ivanovsky Institute of Virology, March 15—16, 2012). (B) Chairman Dmitry Lvov opening the plenary session of the meeting. From left to right: Dr. Yasuko Neagari (Ministry of the Environment, Tokyo, Japan), Dr. Mikhail Shchelkanov (D.I. Ivanovsky Institute of Virology, Moscow, Russia), and Dr. Dmitry Lvov (D.I. Ivanovsky Institute of Virology, Moscow, Russia). (C) Organizing Committee of the Second International Meeting for Medical Ornithology in the Asia—Pacific Region, held at the D.I. Ivanovsky Institute of Virology. (D) Closing of the meeting. From left to right: Dr. Yoshihiro Sakoda (Hokkaido University, Sapporo, Japan), Dr. Mikhail Shchelkanov (D.I. Ivanovsky Institute of Virology, Moscow, Russia).

low-pathogenic strains isolated from ducks in Hong Kong and Malaysia. In contrast, the HAs of H3 and H4 strains isolated from teals in 2002 and from mallards in 2003 near Lake Chany in Novosibirsk Region western Siberia, were related to the HAs of the European H3 and H4 strains.[83,84] Interestingly, the HAs of the H3 strains were closely related to the HA of A/duck/Ukraine/1/1963 (H3N8).[83] However, unlike the Has of H3 and H4, the HAs of H2 strains isolated in the same area in 2003 from mallards resembled the HAs of H2 strains isolated in 2001 in Japan from mallards (*Anas platyrhynchos*).[84]

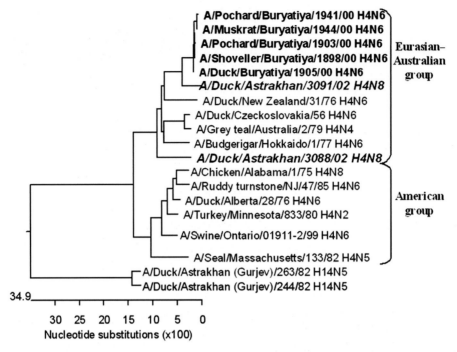

FIGURE 8.94 Phylogenetic tree for the HA gene of Influenza A H4 subtype.

In 2003, influenza A virus strains belonging to a rare subtype H8N6 were isolated in Mongolia from the great cormorant (*Phalacrocorax carbo*), white wagtail (*Motacilla alba*), and magpie (*Pica pica*).[85]

Penetration of HPAI H5N1 into Northern Eurasia: Reasons and Consequences. During longitudinal wide-scale monitoring of Influenza A viruses among wild bird populations in Northern Eurasia, several H5N2 and H5N3 strains were isolated in 1976 and 1981 in the Caspian Sea basin.[70,74] More recently, in 1991−2001, strains belonging to the same subtypes were isolated in Siberia, and their features proved to be relevant to H5 virus circulation. Onn the one hand, the HAs of the strains isolated from teals in 2001 in Primorsky Krai, as well as the HAs of strains isolated from a mallard in Lake Chany in western Siberia in 2003, were shown to be closely related to HAs of H5 strains isolated in 1997 in Italy from poultry.[79,82]

On the other hand, the HA of the H5N3 strain isolated from a wild duck as early as 1991 in Altai Krai in southwest Siberia was closely related to the HA of A/duck/Malaysia/F119-3/ 1997 (Figure 8.95). The HA of the Altai (1991) and Lake Chany (2003) viruses had a monobasic HA1−HA2 cleavage site, and, accordingly, it had a low-pathogenic avian influenza (LPAI) phenotype.[72,79,82,86]

Besides the amino acid sequence of the HA, the sequences of other genes of the H5 viruses isolated in Russia proved to be relevant. The NP genes of the H5N2 and H5N3 strains isolated in Primorsky Krai in 2001 formed a separate cluster in the phylogenetic tree, together with the NP genes of the H4N6 strains isolated from common shelducks (*Tadorna ferruginea*) and common pochards (*Aythya ferina*) in the Republic of Buryatia in 2000, the H2N3 strain isolated from the northern pintail (*Anas acuta*) in Primorsky Krai in 1976, and the

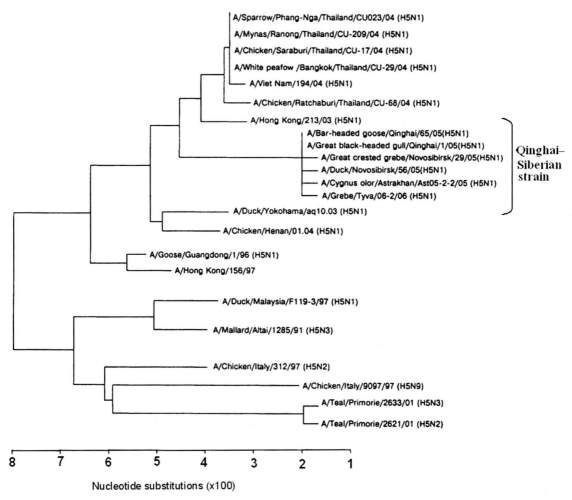

FIGURE 8.95 Phylogenetic tree for the HA gene of Influenza A H5 subtype.

H14N5 strain isolated from a wild duck in the Caspian Sea basin in 1982 (Figure 8.96).[43,72] However, they were very distantly related to the NP genes of H3N8, H6N1, and H5N1 strains isolated from poultry and humans in southeast Asia in 1996–2001 and to the NP genes of H4N8 viruses isolated from wild ducks in the Caspian Sea basin in the European Russia in 2002. By contrast, unlike the NP genes, NS genes of the strains from Primorsky Krai were closely related to the NS genes of the H5N1 and H4N8 viruses isolated in southeastern Asia in 1997–2001, as well as to the NS genes of an H4N8 virus isolated in the Caspian Sea basin in 2002 (Figure 8.97).[43,72]

An abundance of influenza A subtypes in the avian populations of Northern Eurasia provides excellent conditions for gene exchange. The extent of the exchange is demonstrated by the relatedness of different genes of the Russian isolates to the genes of European

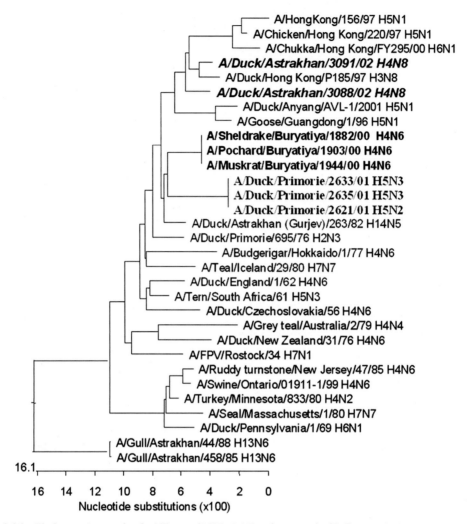

FIGURE 8.96   Phylogenetic trees for the NP gene (1,027–1,415 nt fragments) of Influenza A viruses.

strains, on the one hand, and South Asia isolates, on the other.[72,76,83,84] The exchange is to a certain extent restricted by host specificity, but this restriction is not rigid, and the virus genes frequently traverse interspecies barriers. Avian migration routes crossing Russian territory are an important factor in the gene flow. The extensive intra- and interspecies contacts in the natural habitats of wild birds in Russia stimulate rapid virus evolution and

the appearance of new variants through reassortment events and, presumably, through the postreassortment adjustment of genes, thereby restoring the functional intergenic match.[87,88] Another factor may be the occurrence of avian influenza viruses in lake water, first registered in 1979 in eastern Siberia.[41] This phenomenon might provide a means for the temporal as well as territorial transfer of genes, as suggested by the recent detection of influenza

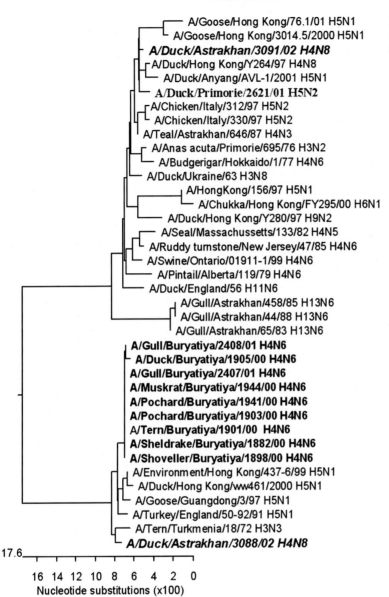

FIGURE 8.97 Phylogenetic trees for the NS gene (600–852 nt fragments) of Influenza A viruses.

viral RNA in the ice of high-latitude lakes in the Lena River basin in the Sakha−Yakutia Republic.[89]

Thus, the sequencing data suggest that there exists an extensive exchange of genes of the avian influenza viruses circulating in Europe, Siberia, and southeast Asia along the avian migration routes connecting Europe, through the Russian territory, with southeastern Asia, the cradle of potentially pandemic reassortant viruses. After the highly pathogenic H5N1 viruses began disseminating from southeastern

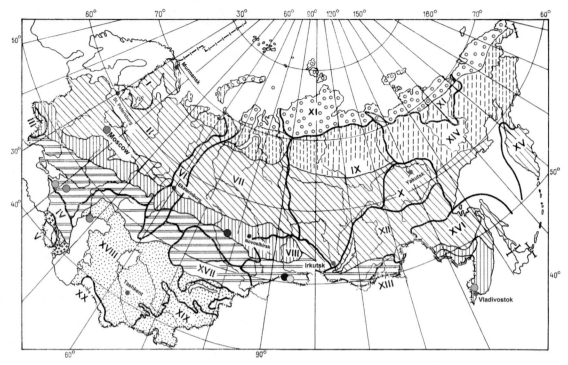

FIGURE 8.98   Places of isolation of HPAI H5N1 (family Orthomyxoviridae, genus *Influenza A virus*) in Northern Eurasia. (See other designations in Figure 1.1.)

Asia westward in 2004, their transfer to Russia by migrating birds was to be expected. On the eve of an HPAI H5N1 epizootic in southeastern Asia starting in the autumn of 2003, we warned about the likelihood of outbreaks with the Asian lineage HPAI H5N1 at the international conference titled "Options for the Control of Influenza V," held in Okinawa, Japan, October 04–13, 2003.[82]

Our second prediction was that overwintering migrating birds could transmit the HPAI virus into Northern Eurasia during their spring migration. We discussed two possible routes by which the birds might introduce the virus: the Dzungarian (Indian–Asian) migration route and the Asian–Pacific route. Preparing for these two possibilities, we increased our surveillance in the southern part of western Siberia (through the Russian

Foundation for Basic Research Project 03-a04-49158) and in Primorski Krai (through the International Science–Technical Center Project 2800) in the spring of 2004. In April of 2005, a wide epizootic outbreak emerged at Kukunor Lake (also called Qinghai Lake) in Qinghai Province, China, and from this location the virus could spread through the Dzungarian Gate, which links the northwestern mountain ranges of Tibet with the western Siberian lowland. Our second prediction was confirmed as well, when HPAI H5N1 first appeared in Northern Eurasia, in western Siberia (Novosibirsk Region, Russia) in the summer of 2005 (Figure 8.98). Although the official start of the epizootic among poultry was dated July 10, 2005 (Table 8.50), that one occurred among wild birds about 2 weeks before was retrospectively established.[5]   The   outbreak   spread

TABLE 8.50 Coverage by the Subjects of Russia of the Epizootic Provoked by HPAI H5N1 in the Southern Part of Western Siberia

Subject of Russia	Official start of the epizootic		Coverage of epizootic			Official mortality[a] among poultry	
	Date	Settlement	Districts	Settlements	Homesteads	Number of birds	Portion of the entire poultry population
Novosibirsk region	July 10, 2005	Suzdalka	11	37	268	5,031	1%
Omsk region	July 15, 2005	Pervotarovka	13	15	47	1,763	2%
Tyumen region	July 25, 2005	Peganovo	4	8	30	428	1%
Altai Krai	July 30, 2005	Titovka	12	22	78	2,454	1%

[a]In the period from October 7 to October 31, 2005.

quickly and caused over 90% lethality among poultry.

The virus isolations in the area were performed independently by two groups of researchers. A number of strains were isolated in Zdvinsky District, Novosibirsk Region, by a group of researchers from the D.I. Ivanovsky Institute of Virology in Moscow. The materials for isolation (cloacal and tracheal swabs, pools of internal organs, and blood) were taken from dead, sick, and healthy birds at the farm where the epizootic occurred and from wild birds in the vicinity.[90,91] Three strains were isolated from dead chickens (*Gallus gallus domesticus*), two strains from sick or dead ducks (*Anas platyrhynchos domesticus*), and one strain from a healthy great crested grebe (*Podiceps cristatus*). All of the strains were deposited into the Russian State Collection of Viruses functioning under the auspices of the D.I. Ivanovsky Institute of Virology (Table 8.51). Sequencing of the HA gene of two strains (Table 8.52)[92] revealed a close relationship to the HA of the H5N1 strains isolated at Qinghai Lake, China, and belonging to genetic group 2.2. Nevertheless, the Qinghai strains formed a so-called Qinghai genetic subgroup, whereas the strains isolated in Russia belonged to the western Siberian genetic subgroup within clade 2.2

clade (Figure 8.99). Still, the close relationship to H5N1 could be the consequence of a "bottleneck" type of genetic selection during transfer of the virus from Qinghai Lake to western Siberia by migrating wild birds.

The sequencing of the other genes of the western Siberian isolates (Table 8.52) confirmed their close relationship to the H5N1 viruses isolated in China: A/great black-headed gull/Qinghai/1/2005 (H5N1) and A/bar-headed goose/Qinghai/65/2005 (H5N1).[93] Several features of the primary structure of virus proteins, such as Lys627 residue in PB2 and Glu92 residue in NS1, characteristic of highly virulent variants of H5N1 viruses, correlated with the high pathogenicity of the Novosibirsk isolates. A deletion in the NA gene in amino acid positions 49–60 indicated that the strains belonged to the genotype Z, which dominated in 2004 in southeastern Asia.[94] The other group of strains was isolated by a team of researchers from the State Research Center of Virology and Biotechnology VECTOR (also known as the Vector Institute) in Koltsovo, Novosibirsk Region. Two strains were isolated from chickens and one strain from a turkey in the village of Suzdalka, Dovolnoe District, in July 2005. The viruses were isolated from homogenates

TABLE 8.51 Infection Activity *In Vitro* of HPAI H5N1 Strains Isolated in Natural and Anthropogenic Ecosystems of Northern Eurasia (2005−2010)

Month, Year	Region	Ecological group of birds	Strain[a]	Number of deposition certificate in Russian State Collection of Viruses	Clinical features[b]	Log_{10} $TCID_{50}/$ mL for SPEV
July 2005	Southern part of western Siberia (Novosibirsk region)	Wild	**A/grebe/Novosibirsk/29/2005**	2,372	∅	5.7
					Mean value	5.7
		Poultry	**A/duck/Novosibirsk/56/2005**	2,371	⊗	7.7
			A/duck/Novosibirsk/67/2005	2,376	⊕	10.2
			A/chicken/Novosibirsk/64/2005	2,373	⊕	11.2
			A/chicken/Novosibirsk/65/2005	2,374	⊕	10.7
			A/chicken/Novosibirsk/66/2005	2,375	⊕	10.7
					Mean value	10.1
November 2005	Mouth of Volga River (Astrakhan region, Kalmyk Republic)	Wild	**A/Cygnus olor/Astrakhan/Ast05-2-1/2005**	2,379	⊗	3.7
			A/Cygnus olor/Astrakhan/Ast05-2-2/2005	2,380	⊗	4.2
			A/Cygnus olor/Astrakhan/Ast05-2-3/2005	2,381	⊗	4.2
			A/Cygnus olor/Astrakhan/Ast05-2-4/2005	2,382	⊗	3.7
			A/Cygnus olor/Astrakhan/Ast05-2-5/2005	2,383	⊗	5.2
			A/Cygnus olor/Astrakhan/Ast05-2-6/2005	2,384	⊗	5.2
			A/Cygnus olor/Astrakhan/Ast05-2-7/2005	2,385	⊗	5.7
			A/Cygnus olor/Astrakhan/Ast05-2-8/2005	2,386	⊗	4.2
			A/Cygnus olor/Astrakhan/Ast05-2-9/2005	2,387	⊗	3.2
			A/Cygnus olor/Astrakhan/Ast05-2-10/2005	2,388	⊗	4.7
					Mean value	4.4
June 2006	Uvs-Nuur Lake (Tyva Republic)	Wild	**A/grebe/Tyva/Tyv06-1/2006**	2,393	⊗	8.0
			A/grebe/Tyva/Tyv06-2/2006	2,394	⊗	8.5
			A/cormorant/Tyva/Tyv06-4/2006	2,396	∅	5.0
			A/coot/Tyva/Tyv06-6/2006	2,397	⊗	5.0
			A/grebe/Tyva/Tyv06-8/2006	2,395	⊕	8.0
			A/tern/Tyva/Tyv06-18/2006	2,399	∅	5.0
					Mean value	6.6
February 2007	Vicinity of Moscow (Moscow and Kaluga regions)	Poultry	A/chicken/Moscow/1/2007	2,403	⊕	4.0
			A/chicken/Moscow/2/2007	2,404	⊕	4.5
			A/chicken/Moscow/3/2007	2,405	⊕	4.0

(*Continued*)

TABLE 8.51   (Continued)

Month, Year	Region	Ecological group of birds	Strain[a]	Number of deposition certificate in Russian State Collection of Viruses	Clinical features[b]	Log$_{10}$ TCID$_{50}$/mL for SPEV
February 2007 (Continued)			A/chicken/Moscow/4/2007	2,406	⊕	4.0
			A/goose/Moscow/5/2007	2,407	⊕	4.0
			A/chicken/Moscow/6/2007	2,408	⊕	4.5
			A/chicken/Moscow/7/2007	2,409	⊕	4.5
			A/chicken/Moscow/8/2007	2,410	⊕	4.0
			A/chicken/Moscow/9/2007	2,414	⊕	4.0
					Mean value	4.2
September 2007	North-Eastern part of Sea of Azov basin (Krasnodar krai)	Wild	**A/Cygnus cygnus/Krasnodar/329/2007**	2,421	⊗	3.5
					Mean value	3.5
		Poultry	**A/chicken/Krasnodar/300/2007**	2,418	⊗	3.5
			A/chicken/Krasnodar/301/2007	2,419	⊗	3.0
			A/chicken/Krasnodar/302/2007	2,420	⊗	3.5
					Mean value	3.3
December 2007	Southwestern part of Russian Plain (Rostov region)	Wild	**A/pigeon/Rostov-on-Don/6/2007**	2,423	∅	6.5
			A/pigeon/Rostov-on-Don/7/2007	2,424	∅	5.5
			A/heron/Rostov-on-Don/11/2007	2,425	∅	6.0
			A/pigeon/Rostov-on-Don/21/2007	2,426	∅	6.0
			A/rook/Rostov-on-Don/26/2007	2,427	∅	6.5
			A/rook/Rostov-on-Don/27/2007	2,428	∅	6.0
			A/tree sparrow/Rostov-on-Don/28/2007	2,429	∅	6.0
			A/starling/Rostov-on-Don/39/2007	2,435	∅	6.0
					Mean value	6.1
		Poultry	A/chicken/Rostov-on-Don/31/2007	2,430	⊗	7.5
			A/chicken/Rostov-on-Don/32/2007	2,431	⊗	7.0
			A/chicken/Rostov-on-Don/33/2007	2,432	⊕	7.0
			A/chicken/Rostov-on-Don/34/2007	2,433	⊕	7.5
			A/chicken/Rostov-on-Don/35/2007	2,434	⊕	7.0
			A/muscovy duck/Rostov-on-Don/51/2007	2,436	⊕	7.0
			A/chicken/Rostov-on-Don/52/2007	2,437	⊕	7.5
					Mean value	7.2

(Continued)

II. ZOONOTIC VIRUSES OF NORTHERN EURASIA: TAXONOMY AND ECOLOGY

TABLE 8.51   (Continued)

Month, Year	Region	Ecological group of birds	Strain[a]	Number of deposition certificate in Russian State Collection of Viruses	Clinical features[b]	Log$_{10}$ TCID$_{50}$/mL for SPEV
April 2008	Suifun-Khanka Lowland (Primorsky Krai)	Wild	A/*Anas crecca*/Primorje/8/2008	2,441	∅	4.0
					Mean value	4.0
		Poultry	**A/chicken/Primorje/1/2008**	2,440	⊕	4.5
			A/chicken/Primorje/11/2008	2,442	⊕	4.0
			A/chicken/Primorje/12/2008	2,443	⊕	4.5
					Mean value	4.3
June 2009	Uvs-Nuur Lake (Tyva Republic)	Wild	**A/grebe/Tyva/3/2009**	2,461	⊕	3.0
			A/grebe/Tyva/5/2009	2,462	⊕	2.0
			A/grebe/Tyva/8/2009	2,463	⊕	2.5
			A/bean goose/Tyva/10/2009	2,464	⊕	2.5
			A/grebe/Tyva/15/2009	2,465	⊕	3.0
			A/grebe/Tyva/16/2009	2,466	∅	2.0
					Mean value	2.5
June 2010	Uvs-Nuur Lake (Tyva Republic)	Wild	**A/grebe/Tyva/2/2010**		⊗	2.0
			A/grebe/Tyva/4/2010		⊕	1.7
			A/teal/Tyva/5/2010		⊕	1.8
			A/grebe/Tyva/6/2010		⊕	2.1
			A/grebe/Tyva/7/2010		⊕	1.9
			A/grebe/Tyva/8/2010		⊕	1.7
			A/grebe/Tyva/11/2010		⊕	2.0
			A/grebe/Tyva/13/2010		⊕	2.1
			A/grebe/Tyva/14/2010		⊕	1.5
					Mean value	1.9

[a]*Strains with complete genome sequences are marked by bold font.*
[b]*Description: ∅, Birds without clinical features; ⊗, birds with clinical features; ⊕, birds that died.*

of turkey spleen and chicken kidneys. The isolated viruses belonged to subtype H5N1, and their HA gene was closely related to the HA gene of the viruses isolated near Qinghai Lake in China in 2003. The sequences of the internal genes (PB1, PB2, PA, and NS) revealed a similarity to the avian H5N1 viruses isolated in Hong Kong in 2004 and in 2005 in Shantou,

TABLE 8.52 GenBank Identification Numbers of HPAI H5N1 Strains Isolated in Natural and Anthropogenic Ecosystems of Northern Eurasia (2005–2010) with Complete Nucleotide Sequences of Genomes

Strain	HA genotype	Source of isolation	Complete nucleotide sequences							
			PB2	PB1	PA	HA	NP	NA	M	NS
A/grebe/Novosibirsk/29/05	2.2	Great crested grebe (Podiceps cristatus)	DQ232607	DQ232605	DQ234075	DQ230521	DQ232609	DQ230523	DQ234077	DQ234073
A/duck/Novosibirsk/56/05	2.2	Domestic duck (Anas platyrhynchos domesticus)	DQ232608	DQ232606	DQ234076	DQ230522	DQ232610	DQ230524	DQ234078	DQ234074
A/Cygnus olor/Astrakhan/Ast05-2-1/05	2.2	Mute swan (Cygnus olor)	DQ389161	DQ394578	DQ394579	DQ389158	DQ394577	DQ389159	DQ394576	DQ389160
A/Cygnus olor/Astrakhan/Ast05-2-2/05	2.2	Mute swan (Cygnus olor)	DQ343506	DQ343505	DQ343504	DQ343502	DQ359694	DQ343503	DQ359692	DQ359693
A/Cygnus olor/Astrakhan/Ast05-2-3/05	2.2	Mute swan (Cygnus olor)	DQ358750	DQ358749	DQ358748	DQ358746	DQ358751	DQ358747	DQ358739	DQ358752
A/Cygnus olor/Astrakhan/Ast05-2-4/05	2.2	Mute swan (Cygnus olor)	DQ363916	DQ363915	DQ363917	DQ363918	DQ363929	DQ363919	DQ363925	DQ363926
A/Cygnus olor/Astrakhan/Ast05-2-5/05	2.2	Mute swan (Cygnus olor)	DQ365011	DQ365008	DQ365007	DQ365004	DQ365006	DQ365005	DQ365009	DQ365010
A/Cygnus olor/Astrakhan/Ast05-2-6/05	2.2	Mute swan (Cygnus olor)	DQ365001	DQ365000	DQ364999	DQ364996	DQ364998	DQ364997	DQ365002	DQ365003
A/Cygnus olor/Astrakhan/Ast05-2-7/05	2.2	Mute swan (Cygnus olor)	DQ363921	DQ363920	DQ363922	DQ363923	DQ363930	DQ363924	DQ363928	DQ363927
A/Cygnus olor/Astrakhan/Ast05-2-8/05	2.2	Mute swan (Cygnus olor)	DQ386305	DQ386304	DQ399537	DQ399540	DQ399539	DQ399541	DQ399542	DQ399538
A/Cygnus olor/Astrakhan/Ast05-2-9/05	2.2	Mute swan (Cygnus olor)	DQ399543	DQ406738	DQ406737	DQ399547	DQ399545	DQ399546	DQ400912	DQ399544
A/Cygnus olor/Astrakhan/Ast05-2-10/05	2.2	Mute swan (Cygnus olor)	DQ434890	DQ423612	DQ434891	DQ434889	DQ440580	DQ440579	DQ434888	DQ434887
A/grebe/Tyva/Tyv06-1/06	2.2	Great crested grebe (Podiceps cristatus)	DQ914807	DQ914810	DQ978999	DQ914808	DQ916293	DQ914809	DQ914805	DQ914806
A/grebe/Tyva/Tyv06-2/06	2.2	Great crested grebe (Podiceps cristatus)	DQ852607	DQ852606	DQ852603	DQ852600	DQ852602	DQ852601	DQ852604	DQ852605
A/grebe/Tyva/Tyv06-8/06	2.2	Great crested grebe (Podiceps cristatus)	DQ863510	DQ863509	DQ863508	DQ863503	DQ863506	DQ863507	DQ863504	DQ863505
A/chicken/Moscow/2/07	2.2	Chicken (Gallus gallus domesticus)	EF474443	EF474444	EF474445	EF474450	EF474447	EF474448	EF474449	EF474446
A/chicken/Krasnodar/300/07	2.2	Chicken (Gallus gallus domesticus)	EU163436	EU163435	EU163434	EU163431	EU163432	EU163433	EU163429	EU163430
A/Cygnus cygnus/Krasnodar/329/07	2.2	Whooper swan (Cygnus cygnus)	EU257707	EU257636	EU257637	EU257631	EU257635	EU257632	EU257633	EU257634
A/pigeon/Rostov-on-Don/6/07	2.2	Rock dove (Columba livia)	EU441930	EU441931	EU441932	EU441937	EU441933	EU441936	EU441935	EU441934
A/rook/Rostov-on-Don/26/07	2.2	Rook (Corvus frugilegus)	EU814510	EU814509	EU814508	EU814503	EU814506	EU814505	EU814504	EU814507
A/starling/Rostov-on-Don/39/07	2.2	Starling (Sturnus vulgaris)	EU486848	EU486849	EU486850	EU486855	EU486851	EU486854	EU486853	EU486852
A/chicken/Rostov-on-Don/35/07	2.2	Chicken (Gallus gallus domesticus)	EU414265	EU420032	EU408333	EU401751	EU401754	EU401753	EU401752	EU401755
A/muscovy duck/Rostov-on-Don/51/07	2.2	Muscovy duck (Cairina moschata)	EU441922	EU441923	EU441924	EU441929	EU441925	EU441928	EU441927	EU441926
A/chicken/Primorje/1/08	2.3.2.1	Chicken (Gallus gallus domesticus)	EU672455	EU672456	EU672457	EU676174	EU672458	EU672460	EU676173	EU672459
A/grebe/Tyva/3/2009	2.3.2.1	Great crested grebe (Podiceps cristatus)	GQ386146.1	GQ386147.1	GQ386145.1	GQ386142.1	GQ386144.1	GQ386143.1	GQ386148.1	GQ386149.1
A/bean goose/Tyva/10/2009	2.3.2.1	Taiga bean goose (Anser fabalis)	GQ386154.1	GQ386155.1	GQ386153.1	GQ386150.1	GQ386152.1	GQ386151.	GQ386156.1	GQ386157.1
A/grebe/Tyva/2/2010	2.3.2.1	Great crested grebe (Podiceps cristatus)	HQ630841.1	HQ630842.1	HQ630840.1	HQ630838.1	HQ630839.1	HQ630837.1	HQ630843.1	HQ630844.1

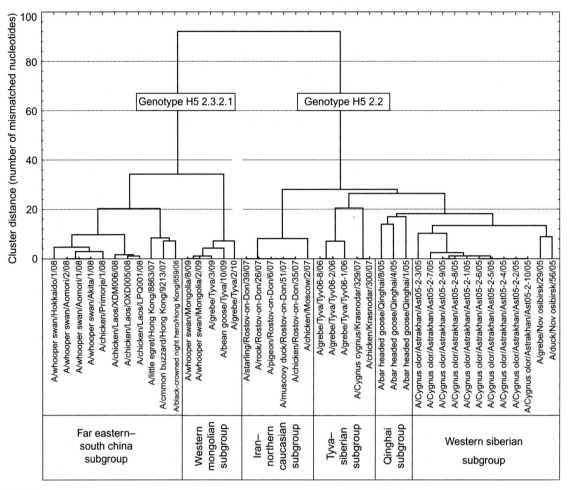

FIGURE 8.99   Phylogenetic analysis of complete nucleotide sequences of ORFs for HPAI H5N1 HA.

Guangdong Province, China.[95] The viruses were highly pathogenic to chickens in a laboratory test.[96]

Our third prediction was that the virus would move with the migrating birds to their overwintering locations. As it turned out, coincident with this prediction, epizootic outbreaks occurred along the main migration routse in the Urals, the Russian Plain, Europe, Africa, central Asia, and India (Figure 8.100).[97–103] In November 2005, an epizootic with mass deaths emerged in the downstream part of the mouth of the Volga River (Figure 8.98) among a local population of mute swans (*Cygnus olor*).[102] Many of the swans contracted neurological disorders, including the inability to keep their neck or head raised, paralysis of the extremities (mainly the legs), and depression. Because there are no human settlements in this part of the Volga, no infections were detected among poultry. Migrating tufted ducks (*Aythya fuligula*), among which no clinical features were detected, are suspected to be the source of the infection, because the start

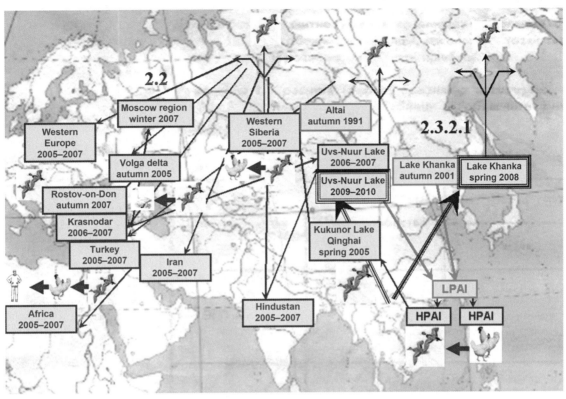

FIGURE 8.100 Reasons and consequences of HPAI H5N1 (clades 2.2 and 2.3.2) virus penetration into Northern Eurasia.

of the epizootic coincided with the appearance of the ducks. Sequencewise, the swan viruses, including A/Cygnus olor/Astrakhan/Ast05-2/2005, were closely related to the western Siberian strains (Tables 8.51 and 8.52, Figure 8.99),[102] indicating the distribution of the virus through the eastern European flyway of birds (Figure 8.100), connecting western Siberia, the Russian Plain, eastern Europe, the Middle East, and Africa.[54]

Our fourth prediction was that the virus would return in birds migrating from their overwintering places to Northern Eurasia in the spring of 2006, with a widening of the epizootic. Dramatic events occurred June 10–28, 2006, at Uvs-Nuur Lake, which is situated on the boundary between the Great Lakes Depression of Mongolia and the Tyva Republic of Russia (Figure 8.98). An estimated 3,000-plus birds died in the Russian part of this lake, which is only about 1% of the total area of the lake. The species most affected was the great crested grebe (Podiceps cristatus); as also affected were coots (Fulica atra) and cormorants (Phalacrocorax carbo). Terns and gulls were involved in the epizootic to a significantly less extent. The absence of poultry farms in the vicinity of Uvs-Nuur Lake precluded outbreaks among poultry. The Tyva strains appeared to be the beginning of a new genetic lineage in the Qinghai–Siberian genotype 2.2. The lineage was designated as a Tyva–Siberian sub-group[104] (Figure 8.99) that was isolated not only in Siberia, but also in Europe. It is believed

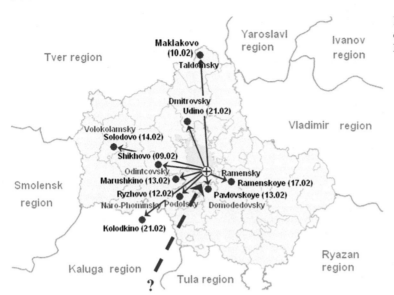

FIGURE 8.101 Dynamics of epizootic provoked by HPAI H5N1 in the Moscow region in February 2007.

that the Tyva—Siberian subgroup emerged in 2006 on nesting grounds of wild ducks in western Siberia. These birds were thought to have been infected in 2005 at the Great Lakes Depression in Mongolia and at their overwintering grounds in India. After the virus was introduced into the nesting places of Northern Eurasia, it spread through the region.

A series of nine outbreaks occurred on the outskirts of Moscow beginning in February 2007 (Figures 8.98 and 8.101). The occurrence of the outbreak at that time of the year seems to preclude the participation of wild waterfowl in introducing or spreading the virus; in addition, terrestrial wild birds tested negative by RT-PCR. The virus was isolated (Table 8.51) from dead and sick poultry, and all the isolates were identified as HPAI H5N1 (Table 8.52) with a high level of sequence similarity to the Qinghai—Suberian genotype 2.2 (Figure 8.99). This outcome implied a common source of infection for all the local outbreaks (Figure 8.101), and subsequent epidemiologic investigation demonstrated a link to live-bird markets in Moscow, where the affected farmers had purchased poultry several days before.

A complete genome analysis of the prototype A/chicken/Moscow/2/2007 revealed[105] that the highest similarity occurred for the strains isolated in the Caucasus region during the winter of 2005—2006: A/cat/Dagestan/87/2006, A/Cygnus cygnus/Iran/754/2006, A/chicken/Krasnodar/01/2006, A/chicken/Adygea/203/2006 (the similarity for nucleotide sequences of PB2 was 99.5%; PB1, 99.3%; PA, 99.7%; HA, 99.1—99.4%; NP, 99.4%; NA, 99.1—99.6%; M1, 99.5—99.9%; and NS1, 99.9%). The closest neighbor for A/chicken/Moscow/2/2007 was found to be A/Cygnus cygnus/Iran/754/2006. Later, this genetic subgroup of HPAI/H5N1/2.2 was designated as "Iran—North Caucasian" (Figure 8.99). Nevertheless, A/chicken/Moscow/2/2007 differed significantly from other Qinghai—Siberian strains: In four genes—PB2, PB1, HA, and NP—there were 12 unique amino acid substitutions (Table 8.53), and all 8 amino acid substitutions in PB1 were unique at that time, affording evidence of active circulation of the virus before 2007. However, the specific origin of A/chicken/Moscow/2/2007 has not been officially identified, and it is suspected that the virus was circulating in a small intermountain

TABLE 8.53 Point Mutations in the Proteins of HPAI/H5N1/2.2 Strains Isolated in Northern Eurasia (2005–2007)

Strains [a]	NP	NA	M1	M2	NS1	NS2
A/{*}/05	Y_{10} N_{397}				D_{202}	M_{50} I_{60}
↓	↓ ↓	no	no	no	↓	↓ ↓
A/{*}/{06–07}	H_{10} S_{397}				G_{202}	V_{50} T_{60}
A/duck/Novosibirsk/56/05	T_{403}	M_{29} L_{143}				
↓	↓	↓ ↓	no	no	no	no
A/grebe/Novosibirsk/29/05	A_{403}	I_{29} V_{143}				
A/duck/Novosibirsk/56/05	T_{403}	L_{143}				
↓	↓	↓	no	no	no	no
A/Cygnus olor/Astrakhan/Ast05-2-{1-10}/05	A_{403}	V_{143}				
A/duck/Novosibirsk/56/05	Y_{10} N_{397} T_{403}	L_{143} P_{320}			D_{202}	M_{50} I_{60}
↓	↓ ▼ ↓	↓ ▼	no	no	▼	▼ ▼
A/grebe/Tyva/Tyv06-{1,2,8}/06	H_{10} S_{397} A_{403}	V_{143} L_{320}			G_{202}	V_{50} T_{60}
A/duck/Novosibirsk/56/05	Y_{10} K_{90} A_{373} N_{397} T_{403}	R_{44} L_{143}			D_{202}	M_{50} I_{60}
↓	▼ ▼ ▼ ▼ ↓	↓ ↓	no	no	▼	▼ ▼
A/chicken/Moscow/2/07	H_{10} R_{90} T_{373} S_{397} A_{403}	C_{44} V_{143}			G_{202}	V_{50} T_{60}
A/duck/Novosibirsk/56/05	Y_{10} N_{397} T_{403}	Q_{39} G_{41} L_{143} P_{320} K_{412}			I_{64} D_{202}	M_{50} I_{60}
↓	↓ ▼ ▼	↓ ↓ ↓ ▼ ↓	no	no	↓ ▼	▼ ▼
A/{*}/Krasnodar/{*}/07	H_{10} S_{397} A_{403}	L_{39} R_{41} V_{143} L_{320} E_{412}			M_{64} G_{202}	V_{50} T_{60}
A/duck/Novosibirsk/56/05	Y_{10} K_{90} A_{323} A_{373} N_{397} T_{403}	A_{46} V_{63} I_{102} L_{143} Q_{288}	no	V_{68} Q_{81}	D_{202}	M_{50} I_{60}
↓	▼ ▼ ↓ ▼ ▼ ↓	↓ ↓ ↓ ↓ ↓	no	↓ ↓	▼	▼ ▼
A/{*}/Rostov-on-Don/{*}/07	H_{10} R_{90} T_{323} T_{373} S_{397} A_{403}	V_{46} L_{63} V_{102} V_{143} R_{288}		I_{68} R_{81}	G_{202}	V_{50} T_{60}

[a] Group of strains is shown with the use of braces: Designations common to all strains in the given group are shown outside the braces; the variable part of the designations is cited inside the braces; the asterisk "*" means "any designation." Only mutations that are found in all the strains of the given group are listed in the table.

[b] **Bold font** indicates substitutions with respect to HPAI/H5N1/2.2 consensus; the frame —substitutions unique to Northern Eurasian strains (Tables 1–2)—that is, they did not occur among Northern Eurasian strains previously; the frame with grey background —substitutions unique to all HPAI/H5N1/2.2 genotypes (strains isolated in both Northern Eurasia and other places); « ↓ »—substitution that takes place in the strains of the given epizootic outbreak only; « ▼ »—substitution that takes place in the strains of both the given and later or previous epizootic outbreaks.

valley ecosystem in the north or south Caucasus in the winter of 2007 and was introduced into the live-bird market through contaminated poultry cages or contaminated grain.

In September 2007, an outbreak was detected in the northeastern part of the basin of the Sea of Azov on a chicken farm called "Lebyazhje-Chepiginskaya" in the Krasnodar region of Russia (Figure 8.98). The virus isolates—A/chicken/Krasnodar/300/2007 from poultry and A/Cygnus cygnus/Krasnodar/329/2007 from a sick whooper swan (*Cygnus cygnus*) found in a "liman" (shallow gulf) near the farm—were closely related to each other (they had two synonymous nucleotide substitutions in PB1, two synonymous in PB2, one nonsynonymous in M1, two nonsynonymous in NA, and one nonsynonymous in NS1) and belonged to the Iran—North Caucasian subgroup of Qinghai—Siberian genotype 2.2 (Figure 8.99). The isolated strains contained 10 unique amino acid substitutions with respect to a Qinghai—Siberian consensus in PB2, PA, HA, NA, and NS1, suggesting that regional variants were continuing to emerge.[106]

In December 2007, a poultry farm called "Gulyai-Borisovskaya" in the Rostov region became infected (Figure 8.98). Unfortunately, the infection was not reported in time, and infected poultry manure was spread on adjacent fields, where wild terrestrial birds could be infected.[107] This exposure is thought to have contributed to the infection of a number of species. including rooks (*Corvus frugilegus*), jackdaws (*Corvus monedula*), rock doves (*Columba livia*), common starlings (*Sturnus vulgaris*), tree sparrows (*Passer montanus*), house sparrows (*Passer domesticus*), and more. Surveillance of these species by RT-PCR detected H5 virus in 60% of pigeons and crows, in around 20% of starlings, and in 10% of tree sparrows, all without clinical features. These results were confirmed by viruses isolated from wild birds and poultry (Table 8.51). Birds whose infection was confirmed by RT-PCR and virus isolation

seemed reluctant to move and had ruffled feathers. On necropsy, the birds were observed to have had conjunctivitis; hemorrhages on the lower extremities and in muscle, adipose, intestine, mesentery, and brain tissue; and changes in the structure of the pancreas and liver. Wide involvement of wild terrestrial birds in virus circulation, presumably from the exposure to infected chicken manure, distinguished this outbreak from others. Genome analysis (Table 8.52) revealed that the strains which were isolated belonged to the Iran—Northern Caucasian subgroup of the Qinghai—Siberian genotype (Figure 8.99). They were phylogenetically similar to A/chicken/Moscow/2/2007 and 13 unique amino acid substitutions with respect to Qinghai—Siberian consensus in PB2, PA, HA, NP, NA, and M2 (Table 8.53).

The main genetic characteristics of the Qinghai—Siberian clade[5,43,92,108] persisted as the virus lineage extended into the western part of Northern Eurasia and Africa (2005—2008). The other genes of the Qinghai—Siberian genotype are associated with group Z, which has dominated among poultry in southeastern Asia since 2003—2004.[93,94] The group Z genotype has several unique genetic markers, including a 20-mer amino acid fragment deletion $C_{49}NQSIITYENNTWVNQTYVN_{68}$ in the N1 gene, compared with the genotype of A/goose/Guangdong/1996. The HA cleavage site—$P_{337}QGERRRKKRGLF_{349}$—has multiple basic amino acid insertion. Three types of silent nucleotide substitutions are known in the coding region of the cleavage site among members of the Qinghai—Siberian clade: $G \overset{1020}{\rightarrow} A$ (A/chicken/Volgograd/236/2006), $G \overset{1028}{\rightarrow} A$ (A/pied magpie/Liaoning/7/2005), and $A \overset{1044}{\rightarrow} G$ (A/chicken/Crimea/04/2005). In addition, four types of amino acid substitutions are known: $G \overset{339}{\rightarrow} R$ (A/bar-headed goose/Qinghai/{1,3}/2005; A/brown-headed gull/Qinghai/1/2005; A/great black-headed gull/Qinghai/3/2005; A/great cormorant/Qinghai/3/2005; A/whooper swan/ Mongolia/13/2005), $R\overset{341}{\to}G$ (A/chicken/Sudan/ 1784-{7,10}/2006), $R\overset{343}{\to}K$ (A/pied magpie/ Liaoning/7/2005), and $K\overset{344}{\to}R$ (A/whooper swan/Mongolia/7/2005). The presence of consensus G339 makes the Qinghai−Siberian clade different from other HPAI/H5N1 variants from southeastern Asia containing consensus R339. The highest portion of R339, 5/15 (33.3%), was detected in the initial outbreak at Kukunor Lake, suggesting that the Qinghai−Siberian strains which originated from southeastern Asia were under a "bottleneck" selection at an early stage of their evolution. The NS1 protein has E92, instead of the more common D92 in virus variants from birds,[109,110] and a 5 aa deletion.

The Qinghai−Siberian clade includes viruses that have infected and caused severe disease and mortality in humans, but currently they do not appear to be transmitted efficiently in humans. Upon analyzing representative viruses in our collection for their potential to replicate in mammals, we found that isolated strains replicated effectively in mammalian cell culture lines BHK-21, LECH, Vero E6, MDCK, and SPEV.[5,108,111] PB2 has consensus K627 that promotes virulence in mammalian cells.[93,112] Six representative isolates from the Qinghai−Siberian clade have E627: A/bar-headed goose/Qinghai/2/2005, A/ ruddy shelduck/Qinghai/1/2005, A/duck/ Novosibirsk/02/2005, A/duck/Kurgan/08/ 2005, A/Cygnus olor/Astrakhan/Ast05-2-4/ 2005, and A/Cygnus olor/Italy/808/2006. These strains are uniformly distributed over time and territory as the result of the stochastic nature of E627 and the absence of any tendency for K627 elimination. Amino acid substitutions that are correlated with virus tropism in mammals include $D\overset{701}{\to}N$ and $S\overset{714}{\to}R$ in PB2, $L\overset{13}{\to}P$ and $S\overset{678}{\to}N$ in PB1, and $K\overset{615}{\to}N$ in PA. In the viruses of the Qinghai−Siberian clade, proline is present in all the genomes in the 13th aa position of PB1 and asparagine at position 701 of the PB2

protein was found only in A/ruddy shelduck/ Qinghai/1/2005.

On the basis of the amino acid sequence of HA receptor-binding sites of Qinghai−Siberian isolates containing E202, Q238, and G240, its affinity of Qinghai Siberian isolates for α2′-′ 3′-sialic acids was predicted. However, a double mutation $Q\overset{238}{\to}L$ and $G\overset{240}{\to}S$ or just a single mutation $E\overset{202}{\to}D$ could switch HA receptor-binding affinity from avian to human receptors.[113]

All the Qinghai−Siberian isolates are sensitive to amantadine, rimantadine, and oseltamivir, as has been confirmed by both direct biological experiments in vitro[114] and the presence[5,92,102,104−108] of marker substitutions in M and NA virus proteins.

Genetic stratification of the Qinghai−Siberian (2.2) genotype of HPAI/H5N1 virus in Northern Eurasia appeared to occur in accordance with the following ecological model: In the summer of 2005, western Siberian cluster variants selected during an epizootic outbreak at Kukunor Lake, Qinghai province, China, spread to the summer nesting places of birds. In winter, 2005−2006, HPAI/ H5N1/2.2 was under selection in two main overwintering areas: (1) Africa, Transcaucasia, and the Middle East (penetrating along eastern and western European flyways); and (2) India and central Asia (penetrating along the Indian−Asian flyway).[54] The first overwintering area could be the source for the Iran−North Caucasian subgroup, the second for the Tyva−Siberian subgroup. Returning to their nesting areas in Northern Eurasia in the spring of 2006, wild birds afforded a mixed virus population the opportunity to spread (Figure 8.100).[5,24,28,42,43,108] A decrease in the potential of isolated strains to reproduce in vitro (Figure 8.102) is more evident in poultry ($TCID_{50} = 11.847−0.272 \times t$) than in to wild birds ($TCID_{50} = 6.185−0.066 \times t$)[115] (Time t is time expressed in months starting from the beginning of 2005.) Tables 8.53 and 8.54 facilitate a comparison of phenotypical changes

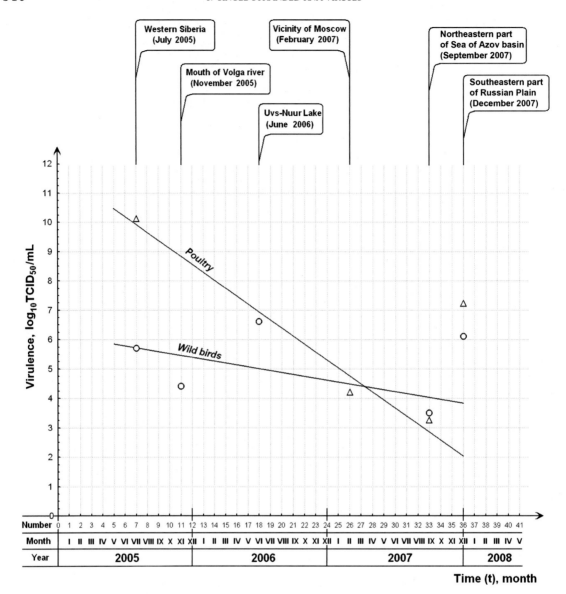

FIGURE 8.102 Virulence dynamics of HPAI/H5N1/2.2 strains isolated in Northern Eurasia (July 2005–September 2007; epizootic episode in December 2007 was excluded from the calculation of the trend because of its ecologically distinctive features described in the text).

with point mutations in the proteins of HPAI/H5N1/2.2 strains isolated in Northern Eurasia.

Although HPAI H5N1 has penetrated into Northern Eurasia through the Dzungarian flyway of wild birds, this fact did not exclude the possibility of the virus transferring through other flyways —(e.g., through the Far East–Pacific flyway).[54] Indeed, in April

TABLE 8.54 Possible Influence of Amino Acid Substitutions on the Virulence of HPAI/H5N1/2.2 Strains Isolated in Northern Eurasia (2005–2007)

| Protein | Position | Virulence | | Function region of the protein | Influenced process of virus life cycle |
		Increased	Decreased		
PB2	69	E	A	Domain of uncovalent binding with C-terminus of PB1	Formation and functioning of polymerase complex PB2–PB1–PA
	73	Q	R		
	221	A	T		
	473	M	T	NLS	Import of RNP into the nucleus
PB1	212	L	M	NLS	Import of RNP into the nucleus
	294	Q	H	Enzyme polymerase center	Functioning of polymerase complex PB2–PB1–PA
	451	V	L		
	618	E	K	Domain of uncovalent binding with N-terminus of PB2	Formation and functioning of polymerase complex PB2–PB1–PA
	678	S	G		
	741	A	N		
PA	213	R	K	NLS	Import of RNP into the nucleus
HA	512	A	S	Elongation fragment during fusion peptide disengage	Virion and endosome membrane fusion
	545	L	M	C-terminus transmembrane domain	
NP	10	Y	H	NLS	Import of RNP into the nucleus
	323	A	T		
	373	A	T		
	397	N	S		
	403	T	A		
NA	29	M	I	N-terminus transmembrane domain	Assembling of new virions
M2	68	I	V	C-terminus cytoplasm domain	Interaction between M2 and M1
	81	R	Q		
NS1	64	I	M	Unknown	Interaction with signal systems of infected cell
	202	D	G		
NS2	50	M	V	Terminus of α-helix N2	Export of RNP from the nucleus
	60	I	T	Base of α-helix C1	

2008, a second breach of HPAI H5N1 into Northern Eurasia emerged in the Russian Far East[116] and was linked with another genotype: 2.3.2.1 (Figure 8.99, Table 8.52). The epizootic originated from unvaccinated poultry in the outermost backyard of Vozdvizhenka, a village in Primorsky Krai surrounded by a small river and meadow where poultry often interacted with wild waterfowl. One initial theory of the introduction of the virus to poultry was from the birds' exposure to hunted ducks, but the direct interaction of wild birds with poultry seems more likely. The isolates (see Table 8.51) from dead chickens and the common teal (*Anas crecca*) collected in the vicinity of epizootic farms were identical and indicated a direct role of migrating birds in the introduction of the virus. The teal, which appeared to be the most likely source of infection of poultry, had no obvious behavior changes but did have hemorrhagic lesions in the intestines on necropsy. It is interesting to underline the fact that common teals were the source of isolation of H5 strains in Primorski Krai in autumn 2001. The teals migrate for long distances, so, the operative hypothesis is that HPAI/H5N1/2.3.2.1 variants may have migrated from the Far East to southern China, Vietnam, or Laos. The HPAI/H5N1 virus was widely distributed in Primorski Krai in the spring of 2008: According to RT-PCR testing, 26% of wild ducks in the Suifun River—Lake Khanka lowland were infected.[116] However, during monitoring in the autumn of 2008, we were not able to find HPAI/H5N1. Nevertheless, the emergence of the HPAI/H5N1 virus in Primorski Krai creates a new type of HPAI stratification in Northern Eurasia (genotype 2.2 in the eastern portion of the subcontinent, genotype 2.3.2.1 in the western portion), as well as the threat of introducing HPAI into North America. The closest neighbors of Primorski Krai 2008 strains are A/chicken/Viet Nam/10/2005 (the nucleotide sequence similarity for the HA gene is 97.5%), A/chicken/Guandong/178/2004 (97.3%), and A/duck/Viet Nam/12/2005 (97.2%). The cleavage site of HA—P_{337}QRERRRKRGLF$_{348}$—contains a multiple basic amino acid motif, which is typical for HPAI but differs from the Qinghai—Siberian HA cleavage site. The 2008 isolates also belong to group Z for internal genes, have avian-type receptor specificity, and are sensitive to M2-channel formation

and neuraminidase inhibitors. Nevertheless, in contrast to the HPAI/H5N1/2.2 strains, which contain K627 (an amino acid that promotes virulence in mammalian cells), the Far Eastern 2008 isolates contains E627, typical for avian-adaptive variants. Finally, direct experiments *in vitro* verified that the 2008 isolates had a reduced tropism for mammalian cell lines.[116]

A comparison of the biological microchip of a representative isolate from the clade 2.2 and 2.3.2.1 isolates shows a clear difference in hybridization pattern in the HA and neuraminidase genes (Figure 8.103). The sequence similarities of the HA/H5/2.2 strains to A/chicken/Primorje/1/2008 and A/*Anas crecca*/Primorje/8/2008 are 92.9—95.3% nt for the HA gene and 94.1—95.3% nt for the NA gene. The nucleotide differences lead to different hybridization patterns for HA/H5/2.2 and 2.3.2.1on the biological microchip, as shown in Figure 8.103.

In June 2009 and 2010, epizootics provoked by HPAI H5N1 reoccurred in Uvs-Nuur Lake in the Tyva Republic, Russia (Figures 8.98 and 8.100). Although the HPAI H5N1 viruses that affected the area in June 2006 belonged to the 2.2 clade, the viruses that appeared in 2009 and 2010 belonged to the 2.3.2.1 clade. (See Tables 8.51 and 8.52.) Nevertheless, these viruses differed from viruses isolated in Primorsky Krai in 2008. Thus, taking into account the fact that the epizootic included Great Lake Depression territory in western Mongolia, researchers classified the Uvs-Nuur strains of 2009—2010 into a new so-called western Mongolian genetic subgroup[5] (Figure 8.99).[117,118]

Fortunately, both clades (2.2 and 2.3.2.1) of HPAI H5N1 that had penetrated into Northern Eurasia had low epidemic potential because their receptor specificity did not switch from $\alpha2'-3'$- to $\alpha2'-6'$-sialoside affinity, a fact that was revealed by the primary structure of the HA receptor-binding region and direct testing in sialoside-based experiments *in vitro*.[5,80]

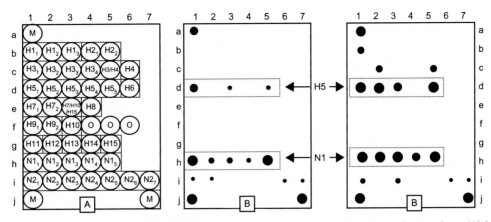

FIGURE 8.103 Subtyping of the HPAI/H5/2.2 and 2.3.2 virus strains by using biological microchips: (A) biological microchip structure; (B) hybridization pattern for A/chicken/Novosibirsk/64/2005 belonging to the HA/H5/2.2 (Qinghai—Siberian) genotype; (C) hybridization pattern for A/chicken/Primorje/1/2008 belonging to the HA/H5/2.3.2 genotype.

Thus, we discuss the epizootic event provoked by HPAI H5N1 in Northern Eurasia during 2005–2010 as a model of an emerging—reemerging situation in need of permanent ecologo-virological monitoring.

Influenza A Viruses Among Mammals. The circulation of Influenza A viruses among swine (order Artiodactyla: family Suidae, genus *Sus*) was originally established in 1930 by Richard Shope (Figure 8.84): His investigations not only established the viral etiology of swine flu and isolated the first historical strain A/swine/Iowa/15/1930 (H1N1), but also serologically demonstrated the close relation between human infection agents and those of swine.[11]

Shope's findings gave rise to a number of isolations of swine respiratory disease agents. Many of these agents later turned out not to be Influenza A virus; for example, "Köbe porcine influenza virus," isolated in Germany;[119] "infectious pneumonia of pigs;"[120,121] "Beveridge—Betts virus"[122] (more often, these pathogens belonged to *Chlamydia* sp.); and "Hemagglutinating virus of Japan,"[123,124] which initially was named "Influenza D virus" and was later identified as

Sendai virus (SeV) (family Paramyxoviridae, genus *Respirovirus*).[125] Nevertheless, a number of strains isolated at the end of 1940s in Korea (strain Oti),[126] and in the 1950s and 1960s in Lithuania (prototype A/swine/Kaunas/353/1959),[127] Estonia,[128] Poland,[129] and Russia[130] were identified as Influenza A (H1N1) virus. Also, in the middle of twentieth century, Influenza A strains closely related to A/swine/Iowa/15/1930 (H1N1) were isolated in Czechoslovakia[131,132] and Hungary.[133] Finally, after the beginning of the "Asian flu" pandemic in 1957, swine Influenza A (H2N2) virus strains were isolated initially in China[134] and later in Czechoslovakia[135] and Moldova (prototype A/swine/Moldova/1/1964).

For a long time, the close relationship between swine and human Influenza A virus strains was held to be the consequence of laboratory contamination. A change in this situation came after the "Hong Kong flu" (H3N2) pandemics in 1968, when a great number of Influenza A (H3N2) strains were isolated all over the world. It then became clear that Influenza A virus is able to penetrate from humans to pigs and, conversely, from pigs to humans.

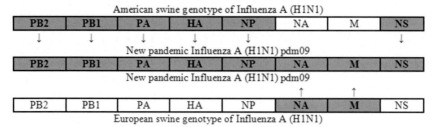

FIGURE 8.104   Reassortation scheme of the emergence of pandemic Influenza A (H1N1) pam09.

The principal peculiarity of pigs is the presence of both $\alpha2'-6'$-sialosides (typical of human cells) and $\alpha2'-3'$-sialosides (typical of avian cells) on the surface of respiratory tract cells. This feature permits both human (or adapted swine) and bird Influenza A virus strains to circulate simultaneously, giving rise to conditions favorable to the reassortment and emergence of virus variants with suddenly appearing new properties.[42,136–143] Avian Influenza A virus strains have been demonstrated to initiate productive infection in swine under experimental conditions.[31,144–147]

The great number of reassortment forms of Influenza A viruses isolated from swine constitute evidence of the extremely high reassortment potential of the swine viral population. Thus, A/swine/England/191973/1992, isolated from nasal swabs of sick pigs in Great Britain in 1992, belongs to the unique H1N7 subtype, which was formed by the reassortment of A/USSR/90/1977 (H1N1) (the source of PB2, PB1, PA, HA, NP, and NS segments) and A/equine/Prague/1/1956 (H7N7) (the source of NA and M segments).[148,149] Another swine reassortant virus, of the subtype H1N2 (A/swine/Ehime/1/1980),[150] was originally isolated in Ehime Prefecture, Japan, in 1980. It was formed by the strains A/swine/Hong Kong/1/1974 (H1N1) and A/swine/Taiwan/7310/1970 (H3N2), a close relative of A/Taiwan/1/1969 (H3N2).[150,151] According to serological data, subtype H1N2 did not circulate in Japan before 1980,[150] but it was isolated after 1980 from A/swine/Miyagi/5/2003 (H1N2) and A/swine/Miyazaki/1/2006 (H1N2).[151]

The most evident illustration of the reassortment potential of swine populations is the emergence of the pandemic "swine flu" H1N1 pdm09 in 2009 as the result of the reassortment of two swine genotypes of the H1N1 subtype: the "American swine genotype" (the source of PB2, PB1, PA, HA, NP, and NS segments) and the "European swine genotype" (the source of NA and M segments) (Figure 8.104).[24–29] Using different receptor-mimicking sialosides (Table 8.55), we investigated the evolution of receptor specificity in Influenza A (H1N1) pdm09 virus during pandemic and postpandemic epidemiological seasons. Different types of sialoside specificity spectra are presented in Figure 8.105.

To compare $\alpha2'-3'$- and $\alpha2'-6'$-sialoside specificities, we introduced the special parameter $W_{3/6}$, which is the ratio of the optical density for flat $\alpha2'-3'$-sialosides (3'SL and 3'SLN) to the optical density for flat $\alpha2'-6'$-sialosides (6'SL and 6'SLN): $W_{3/6} = (d[3'SL] + d[3'SLN])/(d[6'SL] + d[6'SLN])$. If $W_{3/6}$ is <1 ($W_{3/6} < 1.00$), then $\alpha2'-6'$-specificity dominates. In contrast, if $W_{3/6} > 1.00$, then $\alpha2'-3'$-specificity dominates. (Strains with $W_{3/6} \approx 1.00$ have approximately equal $\alpha2'-3'$- and $\alpha2'-6'$-specificities.)[152]

The sialoside specificity of the first pandemic strains isolated in our study, A/California/04/2009 (H1N1) pdm09,

TABLE 8.55 Receptor-Mimicking Oligosaccharides (Sialosides) Used in the Detection of Receptor Specificity of the Influenza A (H1N1) pdm09 Virus

Type of covalent bond	Brief designation	Chemical composition
2'−3'	3'SL	3'-sialyllactose: Neu5Acα2-3Galβ1-4Glcβ
	3'SLN	3'-sialyllactosamine: Neu5Acα2-3Galβ1-4GlcNAcβ
	6Su-3'SLN	6-Su-3'-sialyllactose: 6-Su-Neu5Acα2-3Galβ1-4Glcβ
	SLeᵃ	Neu5Aα2-3Galβ1-3(Fucα1-4)GlcNAcβ
	SLeˣ	Neu5Acα2-3Galβ1-4(Fucα1-3)GlcNAcβ
	SLeᶜ	Neu5Acα2-3Galβ1-3GlcNAcβ
2'−6'	6'SL	6'-sialyllactose: Neu5Acα2-6Galβ1-4Glcβ
	6'SLN	6'-sialyllactosamine: Neu5Acα2-6Galβ1-4GlcNAcβ
	6Su-6'SLN	6-Su-6'-sialyllactose: 6-Su-Neu5Acα2-6Galβ1-4Glcβ

demonstrates dual affinity to both α2'−3'- and α2'-6'-sialosides (Figure 8.106). Therefore, such strains might be able to effect swine−human and human−human transmission, and their pathogenicity is higher than that of seasonal influenza viruses ($W_{3/6} \approx 1$). The other strains isolated during the epidemic seasons of 2009−2011 in Russia have a different value for the W parameter. Most strains were isolated from nasopharyngeal swabs and had the value $W_{3/6} < 1$. These strain had D222 and Q223 in the receptor-binding site of HA1 and usually did not cause any severe complication of the disease, such as pneumonia. The strains with $W_{3/6} \approx 1$ were isolated from autopsies of patients who died of primary viral pneumonia. About 10% of these strains had the substitutions D222N/G and Q223R in HA1 (Tables 8.56 and 8.57). The third group of

strains had a $W_{3/6} > 1$ multiple substitution in the receptor-binding site and caused the majority of viral pneumonias and deaths of patients (Tables 8.56 and 8.57).[153−160]

Pigs could be the source of Influenza A virus not only in humans, but also in synantropic animals. S. Agapov published an article on the pathogenic properties of Influenza A virus specimens isolated from brown rats (*Rattus norvegicus*) in pigsties.[161] Experimental infection of swine influenza A virus strains in rodents— mice (subfamily Murinae) and hamsters (subfamily Cricetinae)—has been described in a number of publications.[3,133,146,161−163]

Rodents have become a widely used laboratory model for Influenza A virus. Productive infection in laboratory mice (order Rodentia: family Muridae, genus *Mus*) was revealed in a pioneer publication[13] of W. Smith (Figure 8.85), C. Andrewes (Figure 8.86) and P. Laidlaw (Figure 8.87). Adapted to mice, Influenza A virus strains are widely used to investigate infectious process, pathology, and the efficiency of antivirals.[161,164−168]

In 2000, the strain Influenza A/muskrat/Buryatia/1944/2000 (H4N6) was isolated from muskrat (*Ondatra zibethicus*) hunted in the Selenga River delta, near where it empties into Lake Baikal. Despite mountain relief along the lake coast, the delta represents a sandbank wedge overgrown with low reeds where the conditions are conducive to a mass nesting of ducks and a high density of population of muskrats. As a result, there is a high level of interaction between the populations of aquatic birds and muskrats. In particular, A/muskrat/Buryatia/1944/2000 (H4N6) has the highest homology with A/pochard/Buryatia/1903/2000 (H4N6). The strain from muskrat turned out to be virulent to mice without any preliminary adaptation, like the majority of H4 strains from Siberian ducks. It was suggested that virulence was promoted by an R220G mutation in HA.[72,81]

The Russian State Collection of Viruses contains the Influenza A/*Sciurus vulgaris*/

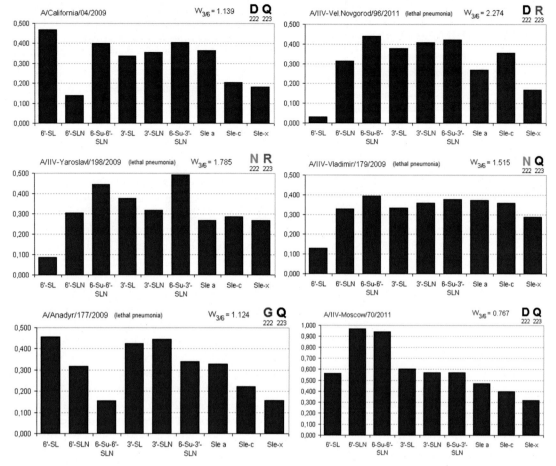

FIGURE 8.105   Different types of sialoside specificity spectra with $W_{3/6}$ parameter value and amino acid substitution in the limits of the receptor binding site of HA.

Primorje/1004/1979 strain with an undetermined subtype isolated from a red squirrel (*Sciurus vulgaris*).[5]

Weasels (order Carnivora: family Mustelidae) are another sensitive group of hosts for Influenza A viruses. The sensitivity of the domestic ferret (*Mustela putorius furo*), an albino form of the forest polecat (*Mustela putorius*), to the virus was explored even in the earliest scientific publications devoted to Influenza A virus.[13,14] Today, ferrets are the best animal model of Influenza A virus

infection. In particular, sera of infected ferrets (as well as infected rats) are widely utilized for Influenza A virus subtype identification.

In 1985, Japanese scientists demonstrated that the epidemic strain A/Kumamoto/22/1977 (H3N2) was able to provoke disease in the European mink (*Mustela lutreola*),[169] and perhaps it was this virus that caused a respiratory disease epizootic on Japanese fur farms during 1977–1978. In 1984–1985, during an epizootic among minks in Sweden, six strains of Influenza A (H10N4) virus (prototype A/

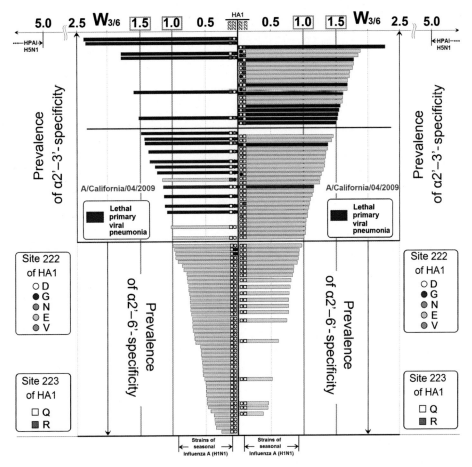

FIGURE 8.106 Receptor specificity toward α2'−3'/α2'−6'-sialosides and the structure of 222−223 sites of HA in pandemic Influenza A (H1N1) pdm09 isolated from hospitalized patients during epidemiological seasons 2009−2010 and 2010−2011.

mink/Sweden/E12665/1984) were isolated and turned out to have an avian origin.[170] In 2007, an Influenza A/stone marten/Germany/R747/06 (H5N1) strain was isolated from the internals of a stone marten (*Martes foina*) that was found dead in a place where there was mass mortality of birds in Germany.[171,172]

The circulation of Influenza A virus among cats (order Carnivora: family Felidae) was originally established in 1942 by the Japanese virologists J. Nakamura and T. Iwasa: Strain A/cat/Fusan/1/1942 (known as "Chiba

virus")[173] turned out to be an avian strain of the H7N7 subtype.[168] In 1970, C.K. Paniker and C.M. Nair described the successful experimental infection of adult cats and eight-month-old kittens by A/Hong Kong/1/1968 (H3N2), of the "Hong Kong flu" pandemic strain.[174] A number of H5N1 strains from Felidae members—tigers (*Panthera tigris*),[175−177] leopards (*P. pardus*),[176] and domestic cats (*Felis catus*)[178−180]—were described after 2005.

The first experiment involving the infection of dogs (order Carnivora: family

TABLE 8.56   Correlation Between Receptor Specificity of Pandemic Influenza A (H1N1) pdm09 Strains from Hospitalized (2009−2011) Patients, the Structure of Those Strains' Receptor-Binding Site of HA, and the Probability of Patients Developing Lethal Primary Viral Pneumonia (LPVP)

Strains			Prevalence of $\alpha2'-6'$-specificity	Prevalence of $\alpha2'-3'$-specificity			Total
			$W_{3/6} \leq 1.0$	$1.0 < W_{3/6} \leq 1.5$	$1.5 < W_{3/6}$		
Number			67	37	25		129
From patients with LPVP			0	11 (29.7%)	15 (60.0%)		26 (20.2%)
Amino acid substitutions in the receptor-binding site	D222	G	2 (3.0%) 4 (6.0%)	10 (27.0%) 14 (37.8%)	9 (36.0%) 11 (44.0%)	21 (16.3%) 29 (22.5%)	
		N	0	3 (8.1%)	2 (8.0%)	5 (3.9%)	
		E	2 (3.0%)	0	0	2 (1.6%)	
		V	0	1 (2.7%)	0	1 (0.8%)	
	Q223	R	0	2 (5.4%)	13 (52.0%)	15 (11.6%)	
	D222Q223		4 (6.0%)	15 (40.5%)	20 (80.0%)	39 (30.2%)	

TABLE 8.57   Detection of Amino Acid Substitutions in the Limits of the HA Receptor-Binding Site of Influenza A (H1N1) pdm09 Virus Strains from Hospitalized Patients (2009−2011) with Lethal and Favorable Outcomes

Outcome of the disease	Amino acid substitutions in the receptor-binding site of HA										Q223		Strains with amino acid substitutions in the limits of the receptor-binding site	
	D222													
	G		N		E		V		Total		R			
	Number	%	Number	%	Number	%	Number	%	Number	%	Number	%	Number	%
Lethal (26 patients)	13	50.0	4	15.4	0	0	1	3.8	18	69.2	8	30.8	23	88.5
Nonlethal (103 patients)	8	7.8	1	1.0	2	1.9	0	0	71	10.7	7	6.8	16	15.5

Canidae, genus *Canis*) was carried out by S.M. Titova[181] in 1954. Then, in 1959, an analogous experiment was conducted by A.D. Ado and S.M. Titova.[182] In 1968, J.D. Todd and D. Cohen[183] repeated Ado and Titova's 1959 experiment. Isolation of Influenza A virus from dogs—A/dog/Vladivostok/1/1970 (H3N2)—under natural conditions was originally performed by T.V. Pysina[31,184,185] in Vladivostok, Primorsky Krai, Russia, in the winter of 1970. A year later, one more canine strain—A/dog/Vladivostok/110/1971 (H3N2) —was isolated in Vladivostok.[31,184,185] The Russian State Collection of Viruses in the D.I. Ivanovsky Institute of Virology contains the strain Influenza A/Canis lupus albus/ Chukotka/1320/1976 (H6N2), isolated from a tundra wolf (*Canis lupus albus*) in Chukotka in the Russian Far East. The strain A/dog/Thailand/KU-08/2004 was isolated from a dog in Thailand;[178] this strain had an avian origin, but provoked lethal pneumonia in dogs.[186] It is noteworthy that Influenza A virus can be isolated from nasal swabs of dogs during inapparent infection,[187] so this virus might be more widely distributed among dogs than is usually considered. Influenza A virus is often the cause of pericarditis in dogs.[188]

The circulation of Influenza A viruses among horses (order Perissodactyla: family Equidae, genus *Equus*) was originally explored in 1956 by a group of Czechoslovakian scientists headed by Bella Tumova (Figure 8.107). In that year, a widespread epizootic emerged among horses (*Equus ferus caballus*) and the historical strain A/equine/Prague/1/1956 was isolated.[189] A subtype of this strain was given an initial designation $H_{eq1}N_{eq1}$ and later was identified as H7N7 (but, for a long time, veterinarians designated this subtype as equine influenza type 1).[146] Later, Influenza A (H7N7) strains were isolated in other European countries[190] and the United States.[191]

During the "Asiatic flu" pandemic of 1958–1961, a number of strains of Influenza A (H2N2) were isolated from sick horses in the Moscow region of the former USSR[192] and in

Hungary.[133,163] It was shown that these strains were significantly different from A/equine/Prague/1/1956 (H7N7), belonged to the H2N2 subtype, and had a human origin.

Equine Influenza A type 2 was originally found in 1963 in Miami, Florida, in the United States, when the prototypical strain A/equine/Miami/1963 was isolated and designated as subtype $H_{eq2}N_{eq2}$.[193] Later, this subtype was identified as H3N8 and was multiply isolated[194–196] in both North and South America. In the former USSR, Influenza A (H3N8) virus strains were isolated from horses in the Ukrainian Soviet Republic during a widespread epizootic in 1970 in the vicinity of Kiev.[31]

The Russian State Collection of Viruses contains the Influenza A/equine/Mongolia/3/1975 (H5N3) strain, which originates from birds and over came the interspecies barrier to penetrate into the equine population.

The circulation of Influenza A virus among camels (suborder Tylopoda: family Camelidae, genus *Camelus*) was originally established by D. K. Lvov[59] (Figure 2.36) in 1980. In December 1979, an epizootic of "contagious cough" among Bactrian camels (*Camelus bactrianus*) emerged in Mongolia. Thirteen strains were isolated from nasal swabs;[59] prototypical strains were A/camel/Mongolia/1/1980, A/camel/Mongolia/1/1981, and A/camel/Mongolia/7/1983. Later, these strains were identified[197] as belonging to H1N1, a virus closely related to the epidemic strain A/USSR/90/1977 (H1N1).

The first Influenza A virus strains from cattle (order Artiodactyla: family Bovidae, subfamily Bovinae) were isolated by J. Romváry[133,163] in 1958 from domestic cows (*Bos taurus taurus*). During 1958–1961, about a hundred epizootic outbreaks emerged in Hungary. Together with his coauthors, Romváry isolated Influenza A2 virus strains (the modern H2N2) from calves. During the "Hong Kong flu" (H3N2) pandemic of 1968–1970, a number of strains were isolated from cattle in the Russian Federation,[145,198] Tajikistan,[199] and the Ukrainian Soviet Republic in the former USSR.[31] The circulation of Influenza A viruses among cattle has been confirmed by multiple serological data.[31,200–204]

The first isolation of Influenza A strain from sick sheep (*Ovis aries*) (order Artiodactyla: family Bovidae, subfamily Caprinae) was carried out in 1959 by a group of Hungarian scientists under the direction of G. Takatsy during an epizootic among farm animals.[133,163] The Strain A/sheep/Hungary/B111/59 (H2N2) isolated by Takatsy was later utilized by J.L. McQueen and F.M. Davenport for experimental infection in lambs, but they observed no clinical symptoms.[205]

The circulation of Influenza A viruses among deer (order Artiodactyla: family Cervidae) was originally established by T.V. Pysina and D.K. Lvov when they isolated the A/*Rangifer tarandus*/Chukotka/1254/77 (H6N2) strain from slowed reindeer (*Rangifer tarandus*) in the Chukotka Peninsula.[206] The Russian State Collection of Viruses in the D.I. Ivanovsky Institute of Virology contains the strains A/deer/Primorje/1201/78 (H1N1), isolated from red deer (*Cervus elaphus*) in Primorsky Krai, and A/*Rangifer tarandus*/Yamal/865/90 (H13N1), isolated from reindeer (*R. tarandus*) on the coast of the Barents Sea. Specific antibodies towards Influenza A (H1N1) and A (H3N2) were detected in the sera of red deer (*C. elaphus*) and elks (*Alces alces*) in the north of Germany.[207,208] S.Q. Li established the presence of about a 10% immune layer toward Influenza A (H1N1) and A (H3N2) among Cervidae in the northeastern provinces of China.[209]

The strain Influenza A/whale/Pacific Ocean/19/1976 (H1N3) (or, alternatively, A/whale/PO/19/1976) from a whale belonging to the Balaenopteridae family (order Cetacea, suborder Mysticeti) and bagged in the South Pacific Ocean was isolated by a group of Soviet virologists under the direction of D.K. Lvov[210] (Figure 2.36) in 1976. This strain turned out to be reassortant between human and avian virus variants.[211]

Two strains of Influenza A virus were isolated by a group of American virologists under the direction of R. Webster[212] (Figure 2.20) from slowed long-finned pilot whales (*Globicephala melaena*) near Portland, Maine, in the United States in 1984: A/whale/Maine/1/84 (H13N9) (from periapical lymph nodes in the lungs) and A/whale/Maine/2B/84 (H13N2) (from the lungs). Further molecular genetic investigation, carried out by a Russian–American group of scientists, revealed that Influenza A variants in gulls (family Laridae) were the source of these strains.[75]

A number of Influenza A virus strains were isolated on the coast of North America: H4N5,[213] H4N6,[214] and H7N7.[215,216] Thus, one could expect to find Influenza A viruses among seals in Northern Eurasia as well.

Pathogenesis. Epithelial cells of mucous membranes are the main targets of Influenza A viruses. Degeneration, necrosis, and further apoptosis, followed by tearing away of the epithelial cell layer take place as a result of the infection. Nevertheless, the main element of Influenza A virus–induced pathogenesis is lesions on the system of vessels; the lesions emerge as the result of the toxic effect of the virus, an effect that includes the multiple formation of active oxygen forms. The latter provoke the generation of hydroperoxides, which interact with lipids and phospholipids of the cell wall to oxidize their peroxide, thereby hindering transport across the cell membrane.[217–219] A subsequent increase in the permeability of vessels, the fragility of their walls, and a violation of the body's microcirculation result in hemorrhagic manifestations, from nasal bleeding to hemorrhagic hypostasis of the lungs and hemorrhages in the substance of the brain.[219,220] Frustration of the circulation, in turn, defeats the nervous system. The pathomorphological picture is characterized by the existence of lymphomonocytic infiltrates around small and average-size veins, hyperplasia of glial elements, and a focal demyelinization that testifies to the toxic and allergic nature of the pathological process in the CNS during influenza.[219,221,222]

The most significant factors involved in cell tropism of the Influenza A virus are the receptor assembly on the surface of the potential target cell and the ability of cell proteases to cleave HA into two subunits (HA1and HA2) followed by fusion peptide rescue.[223–227] For example, for avian Influenza A virus variants, there is an obvious threshold in the virulence level: so-called LPAI and HPAI. HPAI strains strike vascular endothelial and perivascular parenchymal cells as well as the cardiovascular system, quickly reproduce high titers in practically all internal organs, and cause systemic disease leading to death of a bird 1−7 days after infection. LPAI strains, to the contrary, reproduce in low titers, have a narrow tropism toward mucous in the digestive and respiratory tracts (Figure 8.108), and cause enteritis or rhinitis with low mortality. (However, bird diseases connected with LPAI also cause significant damage to agriculture and can break the interspecies barrier, resulting in diseases that are dangerous to people). Wild aquatic and semiaquatic birds, which are natural reservoirs of Influenza A viruses, can have inapparent disease during either LPAI or HPAI infection.[5,24,27,28,39,41–43,53,226,228–230]

The ability of HA to be cleaved by proteases depends on the amino acid composition of the proteolytic cleavage site: LPAI strains contain only one or two positively charged basic amino acids (K or R), whereas HPAI strains have an enriched amount of basic amino acids.[5,24,27,28,39,41,228–230] Nevertheless, pandemic strains with extremely high virulence in humans have only single basic amino acids within the limits of the proteolytic cleavage site (Table 8.58). Still, it is noteworthy that LPAI could provoke human disease as well.

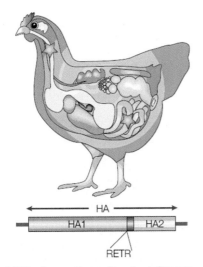

LPAI—low-pathogenic avian influenza

HA cleavage could be performed by attachment to mucous cells of the digestive and respiratory tracts

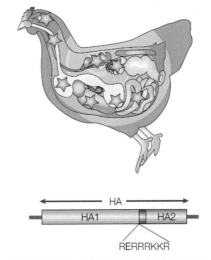

HPAI—highly pathogenic avian influenza

HA cleavage could be performed by cells of practically all tissues

FIGURE 8.108 Comparative characteristics of LPAI and HPAI. Blue stars designate localization of proteases, which cleave HA.

TABLE 8.58   Amino Acid Sequences of the Proteolytic Cleavage Site of HA in LPAI and HPAI (Positively Charged Residues are Marked in Bold on the Blue Backspace; "."—Deletions for the Alignment; "|"—Point of Proteolytic Cleavage)

AVIAN STRAINS OF INFLUENZA A VIRUS

Nonvirulent (LPAI H5)	PQ....RETR|G
Nonvirulent (LPAI H7)	PEXP...KXR|G
Virulent (HPAI H5)	PQ..RKRKKR|G
Virulent (HPAI H7)	PEPSKKRKKR|G

PANDEMIC STRAINS OF INFLUENZA A VIRUS

«Spanish flu» (1918–1919) (H1N1)	PS....IQSR|G
«Asian flu» (1957–1958) (H2N2)	PQ....IESR|G
«Hong Kong flu» (1968–1970) (H3N2)	PE....KQTR|G
«Russian flu» (1977–1978) (H1N1)	PS....IQSR|G
«Swine flu» (2009–2010) (H1N1 pdm09)	PS....IQSR|G

AVIAN STRAINS OF INFLUENZA A VIRUS ISOLATED FROM HUMANS

Hong Kong 1997 (HPAI H5N1)	PQRERRRKKR|G
Hong Kong 1999 (LPAI H9N2)	PQ....RSSR|G
Netherlands 2003 (HPAI H7N7)	PEIP.KRRRR|G
Southeastern Asia 2003 (HPAI H5N1)	PQRERRRKKR|G

Except for the amino acid composition of the proteolytic cleavage site of HA, the efficiency of the cleavage process depends on glycosylation of HA in the vicinity of this site.[231,232]

Amino acid substitutions that switch virus tropisms from avian to mammalian cells in different Influenza A virus proteins have been described: E627K,[112,233] D701N,[100,233] S714R[234] in PB2; L13P,[234] K615N,[234] and S678N[234] in PA.

Clinical features. The clinical features of Influenza A virus infection among birds vary with the virulence level, the subtype of the virus, and the species of the host. Symptoms of disease in poultry have been investigated more than those in wild birds. Four basic clinical forms (listed in order of increasing severity) are distinguishable in birds: sinusitis, enteritis, catarrhal disease, and systemic disease. The last is linked with HPAI and with only two type of HA: H5 and H7. It was the systemic form of the disease that was historically designated "classical avian plaque." A comparison of clinical characteristics of the different forms of avian influenza is presented in Table 8.59. The most massive epizootics provoked by HPAI systemic disease are given in Table 8.60. Clinical symptoms of sick birds are shown in Figure 8.109.

TABLE 8.59   Comparison of Clinical Characteristics of Different Forms of Avian Influenza

Clinical form	Subtype of Influenza A virus	Level of virulence	Type of transmission	Typical level of lethality (with forced slaughter)
Sinusitis	H3N1, H3N2, H3N6, H3N8, H4N1, H4N8, H6N1, H6N3, H6N6, H6N7, H10N1, H10N5, H10N9, H11N2, H11N6, H11N9	LPAI	Aerosol	1–%
Enteritis	H6N2, H6N4, H6N8, H8N2, H8N4, H11N3, H11N8, H12N2	LPAI	Alimentary	2–30%
Catarrhal disease	H5N2, H7N1, H7N2, H7N3, H7N9, H8N4, H9N2, H9N4, H10N7	LPAI, HPAI	Aerosol	5–80%
Systemic disease (classical avian plague)	H5N1, H5N2, H5N3, H5N8, H5N9, H7N1, H7N3, H7N4, H7N7	HPAI	Aerosol and alimentary	100%

TABLE 8.60 Large Epizootics Provoked by HPAI

Year	Country	Subtype	Prototype strain	Scale of epizootic
1959	Scotland	H5N1	A/chicken/Scotland/59	5,000 chickens died on two farms
1963	Great Britain	H7N3	A/turkey/England/63	29,000 turkeys became ill
1966	Canada	H5N9	A/turkey/Ontario/7732/66	8,000 turkeys became ill
1967	USSR	H5N1	A/chicken/USSR/314/67	5,000 chickens became ill
1976	Australia	H7N7	A/chicken/Victoria/76	40,000 chickens and 16,000 ducks became ill
1979	Germany	H7N7	A/chicken/Germany/79 A/goose/Leipzig/187-7/79	600,000 chickens and geese became ill
	Great Britain	H7N7	A/turkey/England/199/79	3 farms with turkeys
1983	Ireland	H5N8	A/turkey/Ireland/1378/83	270,000 ducks, 28,000 chickens, 9,000 turkeys
1983–1985	USA	H5N2	A/chicken/Pennsylvania/1370/83	17,000,000 chickens and turkeys
1985	Australia	H7N7	A/chicken/Victoria/85	220,000 chickens
1991	Great Britain	H5N1	A/turkey/England/50-92/91	8,000 turkeys
1992	Australia	H7N3	A/chicken/Victoria/1/92	13,000 chickens, 6,000 ducks
1994	Australia	H7N3	A/chicken/Queensland/667-6/94	22,000 chickens
	Pakistan	H7N3	A/chicken/Pakistan/447/95	3,200,000 chickens
1994–1995	Mexico	H5N2	A/chicken/Puebla/8623-607/94	360 farms with poultry
1997	Hon Kong	H5N1	A/chicken/Hong Kong/220/97	1,400,000 chickens
	Italy	H5N2	A/chicken/Italy/330/97	6,000 poultry
	Australia	H7N4	A/chicken/New South Wales/1651/97	160,000 chickens, 300 emus
1999–2000	Italy	H7N1	A/turkey/Italy/1265/99	14,000,000 poultry
2002	Chili	H7N3	A/chicken/Chile/1/02	15,000 chickens
2003	Netherlands, Belgium, Denmark, Germany	H7N7	A/chicken/Netherlands/1/03	255 farms in Netherlands (30,000,000 chickens), 8 farms in Belgium (3,000,000 chickens), 1 farm in Germany (80,000 chickens)
2004	Canada	H7N3	A/chicken/Canada-BC/1/04	19,000,000 chickens
	USA	H5N2	A/chicken/USA-TX/1/04	7,000 chickens
	South Africa	H5N2	A/ostrich/South Africa/1/04	24,000 emus, 5,000 chickens
2003–2009	Southeastern Asian countries	H5N1	A/duck/China/E319-2/03	More than 300,000,000 poultry
2005–2010	Russia and European countries	H5N1	A/duck/Novosibirsk/56/05	3,000,000 chickens
	Transcaucasian countries	H5N1	A/chicken/Turkey/986/06	500,000 chickens and turkeys
	African countries	H5N1	A/chicken/Egypt/1/06	500,000 chickens, turkeys, and ducks
	Middle East countries	H5N1	A/chicken/Gaza/450/06	300,000 chickens and turkeys
	India	H5N1	A/chicken/Navapur/Nandurbar/India/7966/06	150,000 chickens and ducks

FIGURE 8.109 Sick birds as the result of HPAI/H5N1/2.2 virus infection. A - sick domestic duck (Anas platyr-hynchos domesticus) (south of Western Siberia, July, 2005); B - sick mute swan (Cygnus olor) (mouth of Volga river, November, 2005); C - sick great crested grebe (Podiceps cristatus) (Uvs-Nur Lake; June, 2006); D - sick coot (Fulica atra) (Uvs-Nur Lake; June, 2006); E - rooks (Corvus frugilegus) on the mixed fodder ground in the poultry farm "Gulyai-Borisovskay" (Rostov region; December, 2007); F - intestine vessel plethohora and changes in pancreas structure of infected rook (Corvus frugilegus) from poultry farm "Gulyai-Borisovskay" (Rostov region; December, 2007).

Diagnostics. Avian Influenza A (family Orthomyxoviridae, genus *Influenza A virus*) is to be differentiated from Newcastle disease (family Paramyxoviridae, genus *Avulavirus*), avian rhinotracheitis (family Herpesviridae, genus *Iltovirus*), avian bronchitis (family Coronaviridae, genus *Coronavirus*), ornithosis (*Chlamydia psittaci*), and mycotoxicosis[32,144,146,235−238]; similarly, mammalian Influenza A is to be distinguished from other respiratory diseases, mainly parainfluenza (family Paramyxoviridae, genus *Paramyxovirus*), coronavirus disease (family Coronaviridae, genus *Betacoronavirus*), torovirus disease (family Coronaviridae, genus *Torovirus*), respiratory syncytial disease (family Paramyxoviridae, genus *Pneumovirus*), and arboviruses with influenzalike symptoms.[144,146,219,221]

The classic diagnostic approach is to isolate the virus with the use of sensitive biological models (ferrets, developing chicken embryoa, and cell lines). Influenza A virus infection could be retrospectively detected by HIT[239] or neutralization testing, but the most effective diagnostic methods are RT-PCR and biological microchips.

Control and Prophylaxis. Vaccination, together with the forced slaughter of livestock. is the most effective and accessible approach to Influenza A prophylaxis among domestic animals. Each country chooses its own strategy for combining these methods. For example, in Russia only livestock in small and individual farms is to be vaccinated whereas birds in poultry farms are not vaccinated, but are killed if either HPAI or LPAI is detected.[32,144,146,235]

8.3.2 Genus *Quaranjavirus*

In 2013, the Quaranfil group, which includes Quaranfil virus (QRFV), Johnston Atoll virus (JAV), and Lake Chad virus (LCV), was allocated to the separate *Quaranjavirus* genus of the Orthomyxoviridae family. Tyulek virus (TLKV) (see Section 8.3.2.1) is a new member of this genus.

The genome of the quaranjaviruses consists of six segments of negative ssRNA. Segments 1−3 encode the proteins of a replicative polymerase complex (polymerase basic protein 2, or PB2; polymerase acidic protein, or PA; and polymerase basic protein 1, or PB1, respectively). The PB1 protein (polymerase 1 basic protein, RdRp) is one of the most conservative proteins of all viruses with a segmented RNA genome. The amino acid sequence similarity of the PB1 protein among the viruses of different genera in the Orthomyxoviridae family is 25−30%, on average, but the similarity of the functional domains of RdRp (pre-A, A, B, C, D, and E motifs) is 40−50% (Figure 8.110).

The envelope glycoprotein GP (HA, segment 5) of the quaranjaviruses has a very low similarity to the homologous protein (HA, segment 4) of influenza viruses. However, it has some similarities tgo the surface glycoprotein of the baculoviruses.[1] The amino acid sequences of *Thogotovirus* genus members have about 20% identity with QRFV and TLKV.

Two other segments of the genome (segments 4 and 6) of the quaranjaviruses encode two proteins whose function is unknown. These proteins are probably structural proteins, which act as nucleocapsid (N) and matrix protein (M), respectively, but currently their function is not well known.

Other viruses of the *Quaranjavirus* genus have been found in South Africa, Nigeria, Egypt, Iran, Afghanistan, and Oceania. The quaranjaviruses are associated with Argasidae ticks (*Argas arboreus*, *A. vulgaris*, *Ornithodoros capensis*), which are obligate parasites of birds.[2−7]

8.3.2.1 Tyulek Virus

History. Tyulek virus (TLKV), prototypical strain LEIV-152Arg, was isolated from Argasidae ticks *Argas vulgaris* collected in the burrows of a colony of birds in the floodplain of Aksu River near the village of Tyulek (43°N, 74°E), in the northern part of the Chu Valley in Kyrgyzstan; Figure 8.111) in 1978.[1,2] Subsequently, a total of

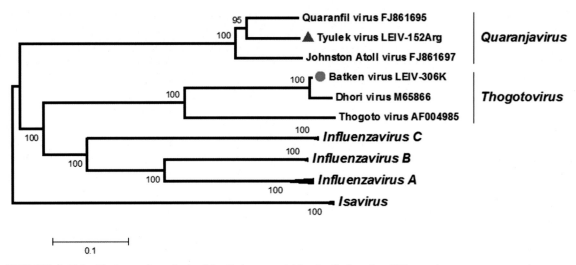

FIGURE 8.110   Phylogenetic analysis of the Orthomyxoviridae family based on PB1 protein sequence comparison.

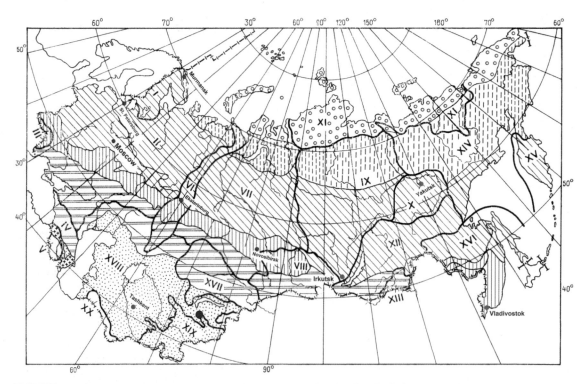

FIGURE 8.111   Place of isolation (red circle) of Tyulek virus (TLKV) (family Orthomyxoviridae, genus *Quaranjavirus*) in Northern Eurasia. (See other designations in Figure 1.1.)

II. ZOONOTIC VIRUSES OF NORTHERN EURASIA: TAXONOMY AND ECOLOGY

42 strains of TLKV was isolated in 1981, 1984, and 1986.[3] TLKV has been classified into the Quaranfil group of the Orthomyxoviridae family on the basis of its antigenic reactions.[4–14]

Taxonomy. Like the other members of the *Quaranjavirus* genus, TLKV has a genome that consists of six ssRNA segments.[13] The PB1 protein amino acid sequence of TLKV has 86% and 84% identities with QRFV and JAV, respectively (Table 8.61). The similarity of the PB2 and PA proteins of TLKV to those of Orf virus (ORFV) is 70%, on average.

The envelope glycoprotein (GP, segment 5) of the quaranjaviruses has very low similarity to the homologous protein (HA) of influenza viruses. However, it has some similarities to the surface glycoprotein of the baculoviruses.[4] The similarity of the GP of TLKV to that of QRFV is 72% nt and 80% aa (Table 8.62). Segment 5 of TLKV has one ORF and encodes a protein with unknown function (524 aa). Its similarity to the same protein of QRFV is 85% aa. Segment 6 encodes a protein 266 aa long, which has no homology with any of the virus's proteins that are deposited in the database

GenBank. The similarity of this protein in TLKV and the same protein in QRFV is 60%.

Figures 8.110 and 8.112 show the results of phylogenetic analysis based on a comparison of PB1 and the envelope protein (GP and HA, respectively). On the phylogenetic trees, TLKV is grouped with QRFV and JAV within the *Quaranjavirus* genus.[13]

Arthropod Vectors. Natural foci of TLKV associated with *Argas vulgaris* ticks in Kyrgyzstan are located below the northern border of the area of distribution of Argasidae ticks (43₀N). This boundary coincides with the line of a frost-free period of 150–180 days a year and an average daily temperature above 20° for no less than 90–100 days per year. The ability of these ticks to withstand prolonged starvation (up to 9 years), as well as their long life cycle (25–30 years), polyphagia, and ability to transfer viruses transovarially, provides stability of the virus's natural foci.[1–3,15–18]

Animal Hosts. TLKV was isolated from Argasidae ticks collected in the nesting burrows of birds. Complement-fixation testing of the birds from these colonies revealed that

TABLE 8.61   Divergence of PB1 (Segment 3) of the Orthomyxoviridae

Genus	Virus		Differences (%)									
			1	2	3	4	5	6	7	8	9	10
Quaranjavirus	Tyulek LEIV-152Arg	1		0.28	0.33	0.67	0.65	0.60	0.74	0.68	0.75	0.71
	Quaranfil virus	2	0.20		0.32	0.67	0.66	0.61	0.75	0.70	0.76	0.69
	Johnston Atoll virus	3	0.31	0.27		0.65	0.64	0.60	0.75	0.68	0.74	0.72
Thogotovirus	Batken LEIV-306K	4	0.82	0.85	0.82		0.20	0.59	0.71	0.72	0.74	0.74
	Dhori virus 1313/61	5	0.85	0.87	0.85	0.09		0.60	0.73	0.73	0.74	0.73
	Thogoto virus	6	0.82	0.81	0.81	0.71	0.71		0.73	0.68	0.71	0.73
Influenza A virus	Influenza A virus PR8	7	0.91	0.92	0.95	0.89	0.90	0.90		0.61	0.72	0.71
Influenza B virus	Influenza B virus	8	0.94	0.95	0.94	0.93	0.93	0.89	0.79		0.67	0.68
Influenza C virus	Influenza C virus	9	0.95	0.95	0.93	0.93	0.93	0.90	0.88	0.89		0.69
Isavirus	Infectious salmon anemia virus	10	0.95	0.92	0.95	0.92	0.92	0.92	0.91	0.90	0.90	

Nucleotide differences (%) are shown above the diagonal, and amino acid differences (%) are shown below the diagonal.

TABLE 8.62   Divergence of HA (Segment 4) of the Orthomyxoviridae

Genus	Virus		Differences (%)									
			1	2	3	4	5	6	7	8	9	10
Quaranjavirus	Tyulek LEIV-152Arg	1		0.28	0.33	0.67	0.65	0.60	0.74	0.68	0.75	0.71
	Quaranfil virus	2	0.20		0.32	0.67	0.66	0.61	0.75	0.70	0.76	0.69
	Johnston Atoll virus	3	0.31	0.27		0.65	0.64	0.60	0.75	0.68	0.74	0.72
Thogotovirus	Batken LEIV-306K	4	0.82	0.85	0.82		0.20	0.59	0.71	0.72	0.74	0.74
	Dhori virus 1313/61	5	0.85	0.87	0.85	0.09		0.60	0.73	0.73	0.74	0.73
	Thogoto virus	6	0.82	0.81	0.81	0.71	0.71		0.73	0.68	0.71	0.73
Influenza A virus	Influenza A virus PR8	7	0.91	0.92	0.95	0.89	0.90	0.90		0.61	0.72	0.71
Influenza B virus	Influenza B virus	8	0.94	0.95	0.94	0.93	0.93	0.89	0.79		0.67	0.68
Influenza C virus	Influenza C virus	9	0.95	0.95	0.93	0.93	0.93	0.90	0.88	0.89		0.69
Isavirus	Infectious salmon anemia virus	10	0.95	0.92	0.95	0.92	0.92	0.92	0.91	0.90	0.90	

Nucleotide differences (%) are shown above the diagonal, and amino acid differences (%) are shown below the diagonal.

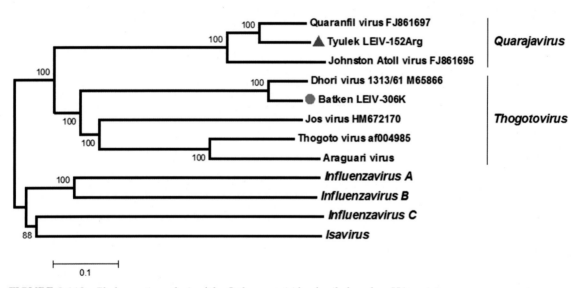

FIGURE 8.112   Phylogenetic analysis of the Orthomyxoviridae family based on HA protein sequence comparison.

some of them had anti-TLKV antibodies. But all attempts to isolate TLKV from birds, Ixodidae ticks, and mosquitoes ended in failure.[2,3,17,18] Antibodies to QRFV have been found in camels, cows, swine, and donkey.

Human Pathology. No cases of disease caused by TLKV have been registered. Two strains of QRFV were isolated from children bitten by ticks in the village of Quaranfil, Egypt. In and around this village, antibodies to

QRFV have been found in 2.6% of the human population.[11]

8.3.3 Genus *Thogotovirus*

The genus *Thogotovirus* currently includes four viruses: Thogoto virus (THOV), Dhori virus (DHOV), Araguari virus (ARGV), and Jos virus (JOSV).[1,2] The viruses of *Thogotovirus* are arboviruses, transmitted mainly by Ixodidae ticks; therefore, the genus had previously been called *Orthoacarivirus*, to emphasize these viruses' association with ixodids (taxon Acari: order Parasitiformes, family Ixodidae). THOV was originally isolated from the ticks *Rhipicephalus* (*Boophilus*) *decoloratus* and *Rh. evertsii* collected from cattle in Thogoto forest, Nairobi, Kenya, in 1960. Subsequently, it was isolated from human, cows, camels, and ticks in many countries in Africa.[3,4]

The genome of the thogotoviruses consists of six segments of negative-polarity ssRNA that encode seven proteins. (Segment 6 encodes two forms of matrix protein.)[1,2] The most conservative proteins of the replicative complex (PB1, PB2, PA) of thogotoviruses have 25−30% identity with those of the *Influenza A virus* genus.

8.3.3.1 *Dhori Virus and Batken Virus (var. Dhori virus)*

History. Dhori virus (DHOV) was originally isolated from *Hyalomma dromedarii* ticks collected from camels in India.[1] DHOV has also been isolated in Egypt, Portugal, Russia, and Transcaucasia.[2−7] In Russia, several strains of DHOV were isolated from *H. plumbeum* ticks, *Anopheles hyrcanus* mosquitoes, and *Lepus europaeus* hares, all in the Volga River estuary.[5,7] One strain of DHOV was isolated from the cormorant *Phalacrocorax carbo* in Maly Zhemchuzhnyi Island in the Caspian Sea (45°00′N, 48°18′E; Figures 8.113 and 8.114).[4]

The prototypical strain of Batken virus (BKNV), LEIV-K306, was isolated from *Hyalomma marginatum* ticks collected from sheep near the town of Batken in Kyrgyzstan in April 1970.[8] Other strains of BKNV were isolated from a mixed pool of *Aedes caspius* and *Culex hortensis* mosquitoes in Kyrgyzstan[9] and from *Ornithodoros lahorensis* and *Dermacentor marginatus* ticks in Transcaucasia.[10] Antigenic studies showed that BKNV is closely related to DHOV, but differs from it.[8]

Taxonomy. The similarity of the structural homologous proteins of the thogotoviruses (THOV, DHOV, ARGV, and JOSV) ranges from 25% (M-protein, segment 6) to 45% (NP, segment 5). The envelope protein HA (segment 4) has 35−45% identity, on average. The similarity of the nonstructural proteins (PB1, PB2, and PA) ranges from 60% to 74%.

BKNV has a high similarity to DHOV. The proteins are 96% (PB2, PA, NP, M) and 98% (PB1) identical. The similarity of the envelope protein HA of BKNV to that of DHOV is 90%, a percentage that explains the antigenic differences between these two closely related viruses. Because the homology of the other structural and nonstructural proteins of BKNV and DHOV is 96−98%, it can be concluded that BKNV is a variant of DHOV, typical to central Asia and Transcaucasia. A phylogenetic analysis based on a comparison of the PB1 and HA proteins is presented in Figures 8.110 and 8.112.

Arthropod Vectors. Apparently, the main arthropod vector of DHOV and BKNV is *Hyalomma* sp. ticks—in particular, *H. marginatum*. DHOV has also been isolated from *H. dromedarii*, *Dermacentor marginatus*, and *Ornithodoros lahorensis* ticks. Rare isolations of DHOV and BKNV from mosquitoes (*Anopheles hyrcanus*, *Aedes caspius*, and *Culex hortensis*) suggest that they play some role in the circulation of these viruses.[9]

Vertebrate Hosts. Antibodies to DHOV were found in 100% of camels, 19% of horses, and 2% of goats in the Indian state of Gujarat, where the virus was first isolated. Antibodies

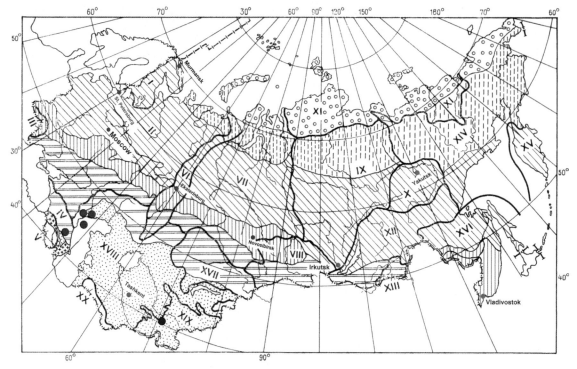

FIGURE 8.113   Places of isolation of Dhori virus (DHOV) (red circles) and Batken virus (BKNV) (dark-blue circle) (family Orthomyxoviridae, genus *Thogotovirus*) in Northern Eurasia. (See other designations in Figure 1.1.)

to BKNV were found in 1.0% of sheep and 1.3% of cattle in Kyrgyzstan.[9] Two strains of DHOV were isolated from a hare (*Lepus europaeus*) and a cormorant (*Phalacrocorax carbo*) in natural foci of the virus.[4,5] The bird, from which DHOV was isolated on Maly Zhemchuzhnyi Island, was ill with respiratory failure, inability to fly, and loss of coordination[4] (Figure 8.114C).

Human Disease. Several cases of disease caused by DHOV have been registered.[11] The disease occurred with fever, encephalitis (40%), headache, and weakness. Antibodies to DHOV were found in 4−9% of the population in the Volga River delta (in the south of Russia) and in 0.8% in the south of Portugal.[12] Antibodies to BKNV were found in the sera of 0.3% of the human population of Kyrgyzstan.

Five cases of laboratory infection were identified.[11]

8.4 FAMILY TOGAVIRIDAE

The Togaviridae family consists of two genera (*Alphavirus* and *Rubivirus*) of enveloped RNA viruses. The virion of the togaviruses (70 nm) contains a core particle (40 nm) formed by a capsid protein and comprising a single-stranded, positive-sense genomic RNA 11,400−11,800 nt long. The lipid bilayer contains the heterodimers of two surface glycoproteins E1 and E2, which form an icosahedral surface of the virion. The genomic RNA has a cap structure at the 5′- and poly-A tail at the 3′-end, as well as two ORFs encoding

FIGURE 8.114   Maly Zhemchuzhny Island (45°00′N, 48°18′E) in the northwestern part of the Caspian Sea: (A) Terns and gulls in a nesting colony; (B) A group of fledglings of Pallas's gull (*Larus ichthyaetus*); (C) A sick great cormorant (*Phalacrocorax carbo*), a source of DHOV isolation; (D) The expedition team (from left to right): Stepan Lvov (researcher); Dmitry Lvov (director); captain and sailor of the expedition ship; (E) Chief sanitary physician officer of the Kalmyk Republic Konstantin Yashkulov during scientific hunting; (F) Mikhail Shchelkanov (researcher) during sample collection near the mortheastern extremity of the island; in the background is a group of Caspian seals (*Phoca caspica*) surrounded by terns and gulls.

nonstructural and structural proteins. The non-structural proteins are encoded by the 5'-ORF (which occupies two-thirds of the genome), whereas the structural proteins are encoded by the subgenomic 3'-ORF.[1]

Most viruses of the *Alphavirus* genus are arboviruses and can replicate in either a vertebrate host and or an invertebrate vector.[2,3] The *Rubivirus* genus consists of one species—*Rubella virus*—that is transmitted by aerosol and is the causative agent of disease known as rubella.[4,5]

8.4.1 Genus *Alphavirus*

The genome of the alphaviruses is a single-stranded RNA with positive polarity about 11,500 nt in length. The viral RNA has a cap at the 5'-end and a poly-A tail at the 3'-end. A large part of the genome of the alphaviruses (about two-thirds, beginning from one-third into the genome and extending to the 5'-end) encodes nonstructural proteins that form the viral replicative complex nsP1, nsP2, nsP3, and nsP4). Structural proteins (core, E3, E2, 6K, and E1) are translated from subgenomic RNA (26S RNA), which is formed in the process of replicating the virus and corresponds to the other one-third of the genome (Figure 8.115).[1]

The alphaviruses can infect a wide range of vertebrates. Most of the alphaviruses are arbo-viruses and are associated with mosquitoes (genera *Culex*, *Culiseta*, *Aedes*, *Coquillettidia*, and *Haemogogus*) and birds, the latter of which can transfer viruses during migration.[2–4] Other vertebrate hosts of the alphaviruses are ruminants, reptiles, amphibians, and fish.[5,6] The

alphaviruses are divided into 10 antigenic complexes. Among the alphaviruses are dangerous pathogens of humans or animals, such as Eastern equine encephalitis virus (EEEV), Western equine encephalitis virus (WEEV), Sindbis virus (SINV), Chikungunya virus (CHIKV), and others.[7,8]

8.4.1.1 *Chikungunya Virus (imported)*

History. CHIKV (family Togaviridae, genus *Alphavirus*, Semliki Forest group) is the etiological agent of a fever that is mortally dangerous to humans. This disease is accompanied by joint and muscle pains (right up to complete immobilization of the patient) and a two-wave course of the fever, together with a macular–papular rash emergency (usually during the second wave).[1] The etymology of the name "Chikungunya" is «chee-kungunyala», which, in the Makonde local language, means "doubled up," owing to the severe joint pains.

CHIKV was originally isolated by R.W. Ross from the serum of a patient with fever during the decoding of an epidemic outbreak in Tanzania in February–March 1956.[2–4] The close relation of CHIKV to Mayaro virus (MAYV), from the Semliki Forest group, was demonstrated in 1957 by serological methods.[5,6]

Distribution. CHIKV was also isolated in Cambodia in southeastern Asia in 1963,[7] in Hindustan in 1964,[8,9] and in the eastern part of New Guinea in 2012.[10] The basic area over which CHIKV is distributed (Table 8.63) comprises (1) the sub-Saharan region of Africa bounded by the equatorial and subequatorial climatic belts (only the southern parts of Africa and Madagascar are in the tropical belt) (Angola, Burundi, the Central African Republic, Kenya, Namibia, Nigeria, Senegal, the Republic of South African, Tanzania, Uganda, Zimbabwe, Madagascar, Sierra Leone, and Guinea); (2) southeastern Asia (Myanmar, Malaysia, Cambodia, India, Hong Kong, Laos, Sri Lanka, Thailand, Vietnam, the southern provinces of China, and Pakistan); and (3)

FIGURE 8.115 FIGURE 8.115 Scheme of the genome organization of SINV (*Togaviridae*, *Alphavirus*). *Drawn by Tanya Vishnevskaya.*

TABLE 8.63   Laboratory-Confirmed Epidemic Outbreaks of Chikungunya Fever Since the Middle of the 1980s

Year	Country	Region	Genotype [a]	Relation to the basic area of distribution
1985	Uganda[18]	Africa	CESA	Basic area
1985	Philippines[19]	Malay Archipelago	A	Basic area
1987	Malawi[20]	Africa	CESA	Basic area
1988	Thailand[21]	Southeastern Asia	A	Basic area
1990	Australia[22]	Australia	A	**Imported cases**
1991	Thailand[21]	Southeastern Asia	A	Basic area
1992	Guinea[23]	Africa	WA	Basic area
1995	Thailand[21]	Southeastern Asia	A	Basic area
1996	Senegal[24]	Africa	WA	Basic area
1998	Indonesia[25]	Southeastern Asia	A	Basic area
1998	Malaysia[26]	Southeastern Asia	A	Basic area
1999	Congo[27]	Africa	CESA	Basic area
1999	Central African Republic[13]	Africa	CESA	Basic area
2000	Indonesia[25]	Southeastern Asia	A	Basic area
2003	Timor[13]	Malay Archipelago	A	Basic area
2004	Kenya[28]	Africa	CESA	Basic area
2005	United States[29]	North America	CESA	**Imported cases**
2005	Reunion[30]	Islands near the eastern coast of Madagascar	CESA	Basic area
2005	Mauritius[30]		CESA	Basic area
2005	Seychelles[30]		CESA	Basic area
2006	Comoro Islands[30]	Islands in the Mozambique Channel	CESA	Basic area
2006	Madagascar[30]	Madagascar	CESA	Basic area
2006	Cameroon[31]	Africa	CESA	Basic area
2006	India[32]	Hindustan	CESA	Basic area
2006	Australia[33]	Australia	CESA	**Imported cases**
2006	Malaysia[34]	Southeastern Asia	CESA	Basic area
2006	Canada[13]	North America	CESA	**Imported cases**
2006	Belgium[35]	Europe	CESA	**Imported cases**
2006	Czech Republic[35]	Europe	CESA	**Imported cases**
2006	Germany[35]	Europe	CESA	**Imported cases**

(Continued)

TABLE 8.63　(Continued)

Year	Country	Region	Genotype [a]	Relation to the basic area of distribution
2006	Norway[35]	Europe	CESA	**Imported cases**
2006	Switzerland[35]	Europe	CESA	**Imported cases**
2006	France[35]	Europe	CESA	**Imported cases**
2006	Hong Kong[36]	Southeastern Asia	CESA	**Imported cases**
2006	United States[35]	North America	A	**Imported cases**
2007	Japan[37]	Eastern Asia	CESA	**Imported cases**
2007	Italy[38]	Europe	CESA	**Imported cases**
2007	Spain[39]	Europe	CESA	**Imported cases**
2009	South Korea[40]	Eastern Asia	CESA	**Imported cases**
2009	Malaysia[41]	Southeastern Asia	A	Basic area
2009	Japan[42]	Eastern Asia	CESA	**Imported cases**
2010	France[43]	Europe	CESA	**Imported cases**
2010	Brazil[44]	South America	Unknown	**Imported cases**
2011	Japan[45]	Eastern Asia	CESA	**Imported cases**

[a]*Chikungunya virus (CHKV) genotypes: A, Asian genotype; CESA, centre, east, and south African genotype; WA, west African genotype.*

Oceania (Indonesia, Malaysia, the Philippines, Singapore, and a number of Pacific Ocean islands).[1−4,10−17]

Taxonomy. CHIKV belongs to the Togaviridae family, Alphavirus genus, Semliki Forest group. On the basis of comparative analysis of the E1 gene, CHIKV was classified into three genotypes: A (Asian), CESA (centre, east, and south African), and WA (west African)[1,12−14] (Table 8.64).

Vertebrate Hosts. Rodents, bats, and monkeys are the natural reservoir of CHIKV.[1,11−14,46] There is substantial evidence, that, in Africa, wild primates play an important role in the natural transmission cycle, but it is not clear whether infection in primates is incidental to or necessary for the maintenance of the virus. In Uganda, CHIKV was frequently isolated from *Aedes africanus*

mosquitoes, which preferto feed on monkeys in the forest canopy.[47] Specific anti-CHIKV antibodies were found among chimpanzees (*Pan troglodytes*) in equatorial and savanna forests in the Democratic Republic of the Congo (Kinshasa)[48] and in savannas in southern Africa. Antibodies were found over a wide area in vervet monkeys (*Cercopithecus aethiops*) and baboons (*Pipio ursinus*), and in both species the virus could circulate in the blood for two to three days at high concentrations without evidence of illness.[49] So, wild animals could play an important role as amplifying hosts.[49] CHIKV was isolated in Dakar < Senegal, from bats, which developed viremia after experimental infection. But in India, inoculation of the virus into two species of fruit-eating bats was followed by low virulence.[50,51] Antibodies were found among donkeys, bats,

TABLE 8.64   Members of the Semliki Forest Group (Family Togaviridae, Genus *Alphavirus*), to which CHIKV Belongs

Group	Virus		Distribution of natural foci	Prototype strain	GenBank ID
Semliki Forest	Bebaru virus (BEBV)		Malaysia	MM2354	AF339480
	Getah virus (GETV)		Asia	M1	EU015061
	Semliki Forest virus		Africa	42S	X04129
	Mayaro virus (MAYV)		South America, Trinidad	Brazil	AF237947
	O'nyong-nyong virus (ONNV)		Africa	SG650	AF079456
	Ross River virus (RRV)		Australia, Oceania	NB5092	M20162
	Una virus (UNAV)		South America	BeAr 13136	AF339481
	Chikungunya virus (CHIKV)	A	Asia	Gibbs 63-263	AF192901
		CESA	Centre, east, and south Africa	Ross	AF192905
		WA	West Africa	37997	AY726732

and wild rodents in Africa[52] and among domestic animals in Asia.[49,50]

Inoculation of African strains into cattle, sheep, goats, and horses failed to produce viremia. Apart from chickens, adult fowl and several species of wild birds did not develop viremia after experimental infection. But experimental infection of vervet monkeys and baboons led to high viremia (up to 8 $\log_{10}$ PFU/mL) during six days, which is sufficient for the infection of mosquitoes.[53]

Arthropod Vectors. CHIKV is transmitted by bloodsucking mosquitoes. The main vectors for this virus during epidemics are *Aedes aegypti* and *Ae. albopictus* in urban regions and mosquitoes from the *Aedes*, *Culex*, and *Coquillettidia* genera in rural landscapes.[1,11–14,46] CHIKV has been multiply isolated from *Ae. africanus*, *Ae. luteocephalus*, *Ae. furcifer–taylori*, *Cx. fatigans*, and *Coq. fuscopenatta*, all of which could preserve the virus and realize virus circulation in natural foci.[1,54,55]

Epidemiology. A high level of viremia in humans (up to 8 $\log_{10}$ PFU/mL) makes it possible for mosquitoes to transmit CHIKV from human to human[1]—a plausible reason that large epidemic outbreaks have been known in big cities of southern and southeastern Asia since the 1960s.[11,13,56–58] Beginning in the middle of the 1980s, epidemiological processes linked to CHIKV have intensified (Table 8.63), although this fact could be explained by improvements in laboratory diagnostics: Previously, Chikungunya fever was often confused with dengue. In any event, CHIKV-provoked lethality has increased, in some cases up to 4.5%).[1,59]

Increases in the frequency of imported Chikungunya fever cases seen at the beginning of the twenty-first century (Table 8.63) are most dangerous, especially when the possibility of CHIKV penetration into local mosquito populations is taken into account. Since 2006, imported cases of Chikungunya fever have become more frequent in Europe (Italy,[15,38,60,61]

Spain,[39] France,[35,44,62] Belgium,[35] Switzerland,[35] Germany,[35] the Czech Republic,[35] Norway[35]); the Americas (Canada,[13] the United States,[35,63] Brazil[44]); eastern Asia (Hong Kong,[36] South Korea,[40] Japan[37,45]); and Australia.[33] Outbreaks in Brazilian cities emerged with infections from *Aedes aegypti*, whereas in rural regions *Aedes albopictus* was the vector, introduced from southeastern Asia,[44] including Japan.[64]

Imported Cases of Chikungunya Fever in Russia. A 59-year-old patient arrived in Russia September 22, 2013, and suddenly fell ill, with a body temperature of 38.7°C. Antipyretic drugs were not effective. Early in the morning on September 24, 2013, the patient was delivered to a Moscow infection hospital with a diagnosis of "fever with unknown etiology." The fever had mid-level severity, and the patient complained of shivering, headache, and asthenia. Hyperemia of the conjunctivae, papular—hemorrhagic rash on the abdomen, and cruses were found.

A medical radiograph (Figure 8.116) of the lungs of the patient revealed decreased clarity at the back of the lung field and diffuse reticular pneumosclerosis in the right lower lobe pyramid, as well as local changes with expressed peribronchial and perivascular alterations. A round shadow was detected near (i.e., peribronchially to) the intermediate bronchus. The roots were intensified. The heart was enlarged at the left. Thus, the medical radiography portrait was consistent with right-side pneumonia with lymphadenopathy. Several peculiarities of the case were the bareness of clinical symptoms (pneumonia was diagnosed only via medical radiography), a rapid progression of changes in the lungs, and the absence of inflammation markers in the peripheral blood. Three days later, positive dynamics were detected: The basal parts of the right lung were restored to their previous level of clarity, although the shadow indicating a hypertrophic lymph node and right root broadening remained.

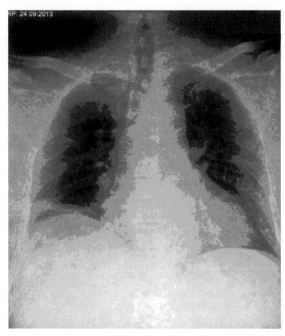

FIGURE 8.116   Medical radiological photograph of the lungs of a patient with Chikungunya fever on the first day of the disease.

Bioprobes (blood swabs and nasopharyngeal swabs) were delivered to the D.I. Ivanovsky Institute of Virology. The absence of Influenza A and B viruses was established by RT-PCR. The strain CHIKV/LEIV-Moscow/1/2013 was isolated with the use of intracerebrally inoculated newborn mice and was identified with the help of a complete-genome (GenBank ID: KF872195) next-generation sequence approach. Phylogenetic analysis (Figure 8.117, Table 8.64) revealed that the CHIKV/LEIV-Moscow/1/2013 strain belonged to an Asian genotype. This strain was deposited into the Russian State Collection of Viruses (deposition certificate N 1239 with a priority of November 11, 2013).[65]

Serological methods revealed eight cases of imported Chikungunya fever that had previously been described in Russia:[66] from Indonesia, Singapore, India, the island of Réunion, and the

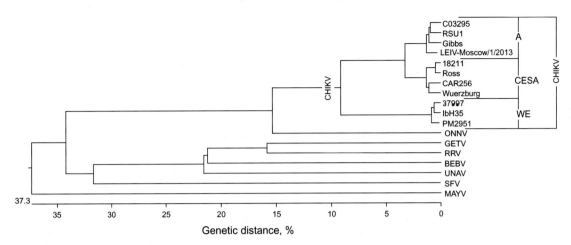

FIGURE 8.117   Phylogenetic tree for nucleotide sequences of the E1 gene (1,044 nt) of the Semliki Forest group viruses. Three genotypes of CHIKV are presented: A (Asian), CESA (centre, east, and south African), and WA (west African).

Maldives Islands. The CHIKV/LEIV-Moscow/1/2013 strain was found to belong to the A-genotype, whereas most of the cases imported into Europe belong to the CESA genotype, reflecting the "bridge" role of Russia between Europe and Asia. The modern-day intensification of both international links and transport flows among countries increases the probability of imported cases of infection emerging. The penetration of *Aedes aegypti* and *Aedes albopictus* to the Russian Black Sea coast[1,67,68] suggests the emergence of seasonal outbreaks in the dynamically developing greater Sochi region as well.

8.4.1.2 Getah Virus

History. Getah virus (GETV) was originally isolated in western Malaysia from *Culex gelidus* and *Cx. tritaeniorhynchus* mosquitoes.[1-3] This virus is widespread in southeastern Asia and in Australia.[3-5] The first isolation of GETV in Northern Eurasia was carried out by M.P. Chumakov[6] (Figure 2.10) in 1972.

Taxonomy. GETV belongs to the Togaviridae family, Alphavirus genus, Semliki Forest group, which also includes subtypes of GETV—Ross River virus (RRV) (Australia, Oceania) and Sagiyama virus (SAGV) (Japan)—as well as

other viruses: Bebaru virus (BEBV), CHIKV, MAYV, O'nyong-nyong virus (ONNV), Semliki Forest virus (SFV), and Una virus (UNAV)[7].

The Genome of GETV is 11,598 nt long. The strains of GETV, circulating in different geographical regions of northeastern and southeastern Asia, have a high level of similarity.[8-11] A pairwise comparison of complete genome sequences revealed that isolates from Malaysia, South Korea, China, Mongolia, Japan, and Russia have 96—98% nt identities, suggesting that the rate of GETV evolution is low. Phylogenetic analysis of the E2 gene (Figure 8.118) is not conducive to dividing the GETV strains into distinct clusters.

Analyses of numerous strains isolated in Japan showed that genetic differences were determined by the time of isolation more than the place of isolation.[8] An analysis of 21 strains of GETV isolated in different regions of Russia revealed their high degree of similarity, but still, they could be divided into three groups on the basis of minimal differences. The first group comprises strains from tundra and forest—tundra in the Magadan region and the Sakha—Yakutia Republic in the north of Asia. The second group encompasses strains

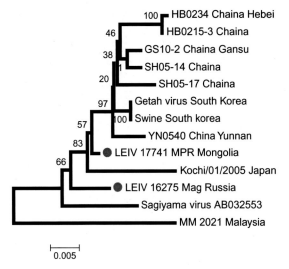

FIGURE 8.118 Phylogenetic tree based on the E2 sequences of different isolates of GETV.

from leaf-bearing forests of Khabarovsk Krai. The third group consists of isolates from forest—steppe and steppe landscape belts of Khabarovsk Krai, the Republic of Buryatia, and Mongolia.[10,12]

Distribution. According to our data,[6,10,12–22] GETV is distributed over eastern Siberia and North Pacific physicogeographical lands (Figure 8.119). The most intensive virus circulation was revealed in the steppe landscape belt of Mongolia, as well as in the mixed forests of Khabarovsk Krai and in the northern taiga of the Magadan region and the Sakha—Yakutia Republic. GETV circulation intensity is significantly lower in tundra and forest—tundra landscapes, a phenomenon that could be explained by the temperature there.

GETV is the only member of the *Alphavirus* genus whose distribution extends to the rough

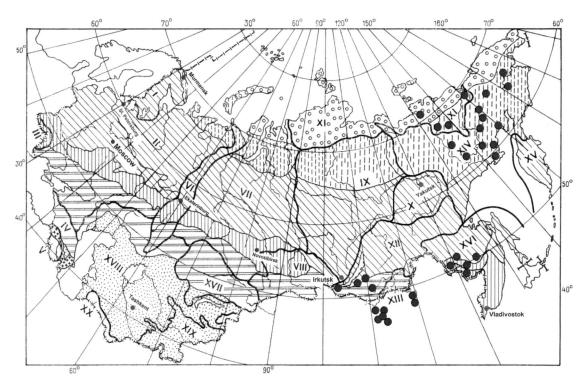

FIGURE 8.119 Places of isolation (red circles) of GETV (family Togaviridae, genus *Alphavirus*) in Northern Eurasia. (See other designations in Figure 1.1.)

climatic conditions of the high latitudes of Northern Eurasia.[18,19] GETV has penetrated to the north of Asia from the overwintering places of birds, which regularly migrate by the east Asian flyway[17,23] (Figure 3.2). The distribution of the virus in the north coincides with that of *Aedes* mosquitoes, which are the effective vector of GETV.

GETV and closely related viruses are known outside of Northern Eurasia in Japan, various countries in southeastern Asia, and Australia.[1−3,5,24−29]

Human Infection. The pathogenicity of GETV to humans has not yet been described. Nevertheless, the antigenically close RRV has been associated with large epidemic outbreaks of polyarthritis in Australia and Sarawak.[2,4]

Vertebrate Animal Infection. Symptomatic and subclinical infections of animals were reported in 1998 in Japan, where there was a large outbreak involving 722 racehorses.[30,31] Among the clinical features seen were fever, rash on various parts of the body, and edema on the hind legs. Virus isolates were more similar to the prototypical Malaysian strain than to the Japanese Sagiyama strain. GETV has been implicated in illness and abortion or stillbirths in pigs.[32,33] Disease among horses was described in India.[34] Infection in cattle is usually subclinical.[3]

Arthropod Vectors. GETV has been isolated from *Culex gelidus*, *Cx. tritaeniorhynchus* (Malaysia, Cambodia, China), *Cx. bitaeniorhynchus*, *Anopheles amictus* (Australia), *Cx. vishnui* (Philippines); the Sagiyama subtype of GETV was isolated from *Cx. tritaeniorhynchus* and *Aedes vexans*, as well as from pigs with fever, in Japan.[27,35]

Although their natural transmission cycle is not known in details, mosquitoes acquire GETV mainly while feeding on domestic mammals and fowl. There may also be a jungle cycle involving wild vertebrates.[5] The Bebaru subtype was isolated from *Culex lophoceratomyia* and *Aedes* spp. mosquitoes collected in mangrove swamp forests of western Malaysia.[32]

The main vectors in Russia (i.e., in Northern Eurasia) are *Aedes nigripes*, *Ae. communis*, *Ae. impiger*, *Ae. punctor*, and *Ae. excrucians*.[18,21]

8.4.1.3 Sindbis Virus and a Set of Var. Sindbis Virus: Karelian Fever Virus, Kyzylagach Virus ()

History. SINV was first isolated from the ornithophilic mosquitoes *Culex univittatus* collected in 1952 in the Sindbis district near Cairo, Egypt.[1] Subsequently, SINV has been found in many regions of Africa, Europe, the Middle East, central and southeastern Asia, Australia, New Zealand, and the Philippines.[2−10] In the Old World, SINV is widely distributed and has several geographical variants: Karelian fever virus (KFV), Ockelbo virus (OCKV), Babanki virus (BBKV), Kyzylagach virus (KYZV), and Whataroa virus.[4,11−14] SINV was categorized into the western equine encephalomyelitis complex.[4,13]

KFV was first noted in the summer of 1981 in the central and southwestern parts of Fennoscandia, including Russia, Finland, Sweden, and southern Norway (Figure 8.120).[14]

The prototypical strain LEIV-65A of KYZV was first isolated from *Culex modestus* mosquitoes collected in a colony of Ardeidae birds (herons) in Kyzylagach Reservation, located on the coast of Kyzylagach Bay in the Caspian Sea (39°10′N, 48°58′E; Figure 8.120).[15]

Taxonomy. On the basis of a comparison of a partial sequence of the E2 gene, isolates of SINV can be divided into five genotypes (Figure 8.121).[9] Genotype I includes viruses from Europe and Africa, genotype II isolates from Australia and Oceania, and genotype III viruses from India and the Philippines. Together with the Chinese strain SINV XJ-160, KYZV was assigned to genotype IV. Genotype V consists of only the strain M78 from New Zealand.

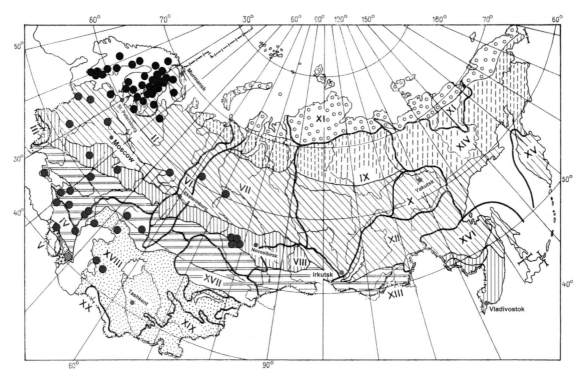

FIGURE 8.120 Places of isolation of SINV (red circles), KFV (blue circles), and KYZV (green circle) (family Togaviridae, genus *Alphavirus*) in Northern Eurasia. (See other designations in Figure 1.1.)

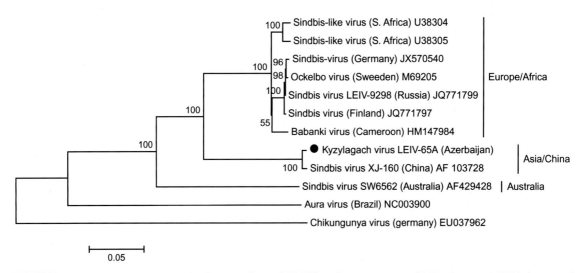

FIGURE 8.121 Phylogenetic tree of different isolates of SINV based on sequences of 26S subgenomic RNA (structural proteins) rooted in Aura virus (AURAV) and CHIKV sequences.

The strains of genotype I form two subclusters, one of which comprises SINVs from northern Europe and sub-Saharan Africa and the second of which consists of strains from the Mediterranean region (southern Europe, northern Africa, and the Middle East).[9] The genetic distance between the viruses of the different genotypes of SINV (e.g., between the European and Australian isolates) is not more than 23% nt (Table 8.65). At the same time, SINVs isolated in the same geographic region are characterized by a high degree of similarity (Figure 8.122). Thus, SINV strains isolated in Russia, Germany, Sweden(OCKV), and Finland have about 99% similarity (Table 8.65).[3,5,6,11] Babanki virus, which is from Cameroon, has 98% similarity to the European strains of SINV.

Despite the high degree of similarity among the different genotypes of SINV, known cases of human disease are caused only by strains of the European–African subcluster of genotype I (Karelian fever, a disease of Ockelbo, a disease of Babanki). KYZV has a high similarity (99%) to the Chinese isolate SINV XJ-160, isolated from *Anopheles* sp. mosquitoes in the Xinjiang Uighur Autonomous Region in the northwest of China.[16] The divergence of KYZV and XJ-160 from the European isolates of SINV is 18% nt and 7% aa of the entire genome sequence (Table 8.65). The geographic isolation of KYZV and XJ-160 and their genetic divergence from the European and Australian isolates suggest that KYZV is a variant of SINV that is typical to Central Asia.

Distribution. SINV has been isolated in many regions of southern Europe, the Middle East, Africa, southeastern Asia, the Philippines, and Australia.[2,17,18] The African continent is almost all endemic for SINV: Strains are known from Egypt, the Republic of South Africa, Uganda, the Central African Republic, Sudan, Nigeria, and Zimbabwe. As for Asia, there are strains from Turkey, India, Malaysia, and the Philippines. In Australia, SINV strains were multiply isolated in the north of the continent. In Europe, SINV has been isolated in Sicily (Italy) and Slovenia. On

TABLE 8.65 Genetic Divergence (Percent of Differences in the Entire Genome Sequences) Among Different Isolates of SINV

SINV strains		1	2	3	4	5	6	7	8	9	10	11
Kyzylagach_virus_LEIV-65A	1		0.01	0.19	0.19	0.19	0.19	0.18	0.19	0.25	0.37	0.42
Sindbis_virus_XJ-160_AF103728	2	0.01		0.19	0.18	0.18	0.18	0.18	0.18	0.25	0.37	0.42
Sindbis_virus_LEIV-9298_(Russia)_JQ771799	3	0.07	0.07		0.01	0.01	0.01	0.03	0.03	0.23	0.36	0.41
Sindbis_virus_(Germany)_JX570540	4	0.07	0.07	0.01		0.01	0.01	0.03	0.03	0.23	0.36	0.41
Ockelbo_virus_(Sweden)_M69205	5	0.07	0.07	0.01	0.01		0.01	0.03	0.03	0.23	0.36	0.41
Sindbis_virus_(Fennoscandia)_JQ771797	6	0.07	0.07	0.01	0.01	0.01		0.03	0.03	0.23	0.36	0.41
Babanki_virus_(Cameroon)_HM147984	7	0.07	0.07	0.01	0.01	0.01	0.01		0.03	0.23	0.37	0.41
Sindbis-like_virus_(S._Africa)_U38305	8	0.07	0.07	0.01	0.01	0.01	0.01	0.01		0.23	0.36	0.41
Sindbis_virus_SW6562_(Australia)_AF429428	9	0.12	0.13	0.11	0.11	0.11	0.11	0.11	0.11		0.36	0.42
Aura_virus_(Brazil)_NC003900	10	0.32	0.32	0.31	0.31	0.31	0.31	0.31	0.31	0.31		0.43
Chikungunya_virus_EU037962	11	0.40	0.40	0.40	0.40	0.40	0.40	0.40	0.40	0.41	0.41	

Nucleotide differences (%) are shown above the main diagonal; amino acid differences (%) are shown below the main diagonal.

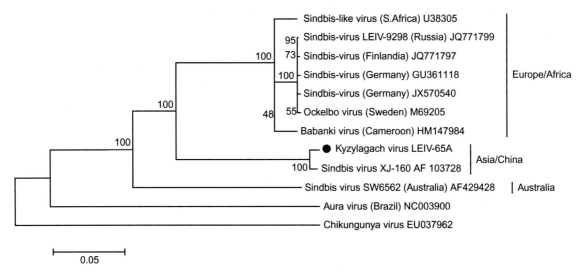

FIGURE 8.122 Phylogenetic tree of different isolates of SINV based on sequences of genomic RNA encoding structural proteins, rooted in AURAV and CHIKV sequences.

the territory of the former USSR, SINV strains were multiply isolated in Belarus, Ukraine, Azerbaijan, Tajikistan, and western Siberia (in the areas around the central region of the Ob River valley).[17–19].

Vertebrate Hosts. The main vertebrate hosts of SINV are different species of birds, predominantly of the orders Passeriformes, Pelecaniformes, Ciconiiformes, and Anseriformes. SINV infection in birds can chronic, allowing them to transfer the virus during their seasonal migration.[17–20] Migratory birds play an important role in the wide distribution of this virus. SINV has been known to persist for as much as two months after experimental infection.

SINV strains have been multiply isolated from aquatic and semiaquatic birds in the delta of the Nile River in Egypt, from the white wagtail (*Motacilla alba*) and the common hill myna (*Gracula religiosa*) in India, and from the reed warbler (*Acrocephalus scirpuceus*) in the western part of Slovakia.

In Zimbabwe, SINV has been isolated from insectivorous bats of the Rhinolophidae and Hipposideridae families.[2] Occasionally,

SINV has been isolated from rodents and amphibians.

On the territory of the former Soviet Union, SINV was originally isolated from a yellow herons (*Ardeola ralloides*) caught out of a bird colony in the southeastern part of Azerbaijan in 1968. Serological methods have revealed SINV circulation in the Astrakhan region among aquatic and semiaquatic birds, especially those of the orders Pelecaniformes (18%), Ciconiiformes (15%), and Anseriformes (11%). Neutralizing antibodies to SINV were found in coots (*Fulica atra*) (16.7%) from natural foci of the middle belt of the Volga River delta. In the Kuban River delta in Krasnodar Krai, specific anti-SINV antibodies were found among eight species of aquatic and semiaquatic birds, most frequently mallards (*Anas platyrhynchos*) and purple herons (*Ardea purpurea*). In Belarus, anti-SINV antibodies were detected in 4% of birds in the summer and in 0.4% in the fall.[21]

Antibodies to SINV have been detected among farm animals (Table 8.66). For example, neutralizing antibodies were found among

TABLE 8.66 Detection of Antihemagglutinating Antibodies to SINV Among Humans and Farm Animals in the Russian Federation (1982–1992)

Federal district	Federal subject	Species	Tested	Positive Number	%
Northern	Volgograd, Arkhangelsk regions, Komi Republic, Karelia Republic	Humans	2,278	21	0.9
		Farm animals	3,042	493	16.2
Northwestern	Novgorod, Kaliningrad, Leningrad, Pskov regions	Humans	894	2	0.2
		Farm animals	2,291	37	1.6
Central	Moscow, Tver, Vladimir, Ivanovo, Kostroma, Bryansk, Kaluga, Tula, Ryazan, Orel, Smolensk regions	Humans	2,417	0	0
		Farm animals	415	2	0.5
Central–Chernozem	Belgorod, Kursk, Lipetsk, Tambov regions	Humans	1,152	0	0
North Caucasian	Krasnodar, Rostov regions, Stavropol krai, Kabardino-Balkar Republic	Humans	1,312	1	0.1
		Farm animals	281	6	2.1
Volga	Astrakhan, Volgograd, Penza, Saratov, Samara, Ulyanovsk regions, Kalmyk Republic, Tatarstan Republic	Humans	1,152	41	3.6
		Farm animals	2,213	1	0.05
Volga–Vyatka	Kirov, Nizhny Novgorod regions, Chuvash Republic, Mordva Republic, Udmurtia Republic, Mari El Republic	Humans	1,384	4	0.3
		Farm animals	681	10	1.5
Ural	Chelyabinsk, Orenburg, Perm, Sverdlovsk regions, Bashkortostan Republic	Humans	1,665	5	0.3
		Farm animals	175	18	10.3
Western Siberian	Tyumen, Novosibirsk, Kemerovo, Omsk, Tomsk, Kurgan regions, Altai Krai	Humans	4,826	3	0.06
		Farm animals	660	8	1.2
Eastern Siberian	Irkutsk, Chita regions, Krasnoyarsk Krai, Sakha-Yakutia Republic	Humans	3,308	4	0.1
		Farm animals	1,130	17	1.5
Far Eastern	Amur region, Kamchatka Krai, Khabarovsk Krai, Primorsky Krai	Humans	2,554	4	0.2
		Farm animals	638	22	3.4

cattle (17.5%) and horses (15.0%) in the middle belt of the Volga River delta.

Arthropod Vectors. SINV is closely associated with ornithophilic mosquitoes. In Egypt, this virus was isolated from *Culex univittatus*, *Cx. antennatus*, and *Anopheles pharoensis*; in Uganda, from *Coquillettidia* spp.; in Sarawak, (Malaysia), from *Cx. bitaeniorhynchus*; in Australia, from *Cx. annulirostris*, *Aedes normanensis*, and *Ae. vigilax*; in India, from *Coq. fuscopennata*; in Sudan, from *Cx. quinquefasciatus*; and in Europe, from *Cx. pipiens*, *Cx. torrentium*, *Culiseta morsitans*, *Coq. richiardii*, *Ochlerotatus communis*, *Oc. excrucians*, *Ae. cinereus*, and *An. hyrcanus*.[22,23]

According to our data, in the Volga River delta SINV is transferred by *Culex pipiens* in anthropogenic biocenoses and by *Anopheles hyrcanus* and *Coquillettidia richiardii* in natural ones. In the natural foci of the middle belt of the Volga delta, 1 strain can be isolated from approximately 3,800 *An. hyrcanus* or 3,300 *Coq. Richiardii* mosquitoes; in the low belt of the delta the ratio is 1 in about in a power less, and in anthropogenic biocenoses it is 1 strain per 1,500 *Cx. pipiens* mosquitoes.

SINV strains from Gamasidae ticks (*Ornithonyssus bacoti*) in India and from Ixodidae ticks (*Hyalomma marginatum*) in Sicily (Italy) are known.[2] Productive experimental infections were described in the Argasidae ticks *Ornithodoros savignyi* and *Argas persicus* (although infected ticks did not transmit the virus during feeding).[23] Most likely, ticks do not play an important role in SINV circulation or as a reservoir for this virus.

Human Pathology. SINV causes acute fever in humans but has a favorable outcome. Antibodies to SINV are widely detected in human sera (Table 8.66), although in eastern Siberia and the Far East cross-reactions with GETV (another member of the Semliki Forest serogroup) can take place.

The start of the disease is sudden. Clinical symptoms include fever, muscle and joint pain, and rash. Severe progressive arthritis of large joints could develop several years after the disease and could lead to disability. This pathology appears in 6–20% of citizens of endemic territories. Outbreaks of Sindbis fever in Egypt and Israel emerged at the same time as West Nile fever; hence, it is necessary to distinguish these infections in the laboratory.

References

8.1 Family Bunyaviridae

1. Smithburn KC, Haddow AJ, Mahaffy AF. A neurotropic virus isolated from *Aedes* mosquitoes caught in the Semliki forest. *Am J Trop Med Hyg* 1946;**26**:189–208.
2. Plyusnin A, Beaty BJ, Elliot RM, et al. Family Bunyaviridae. In: King AM, Adams MJ, Carstens EB, Lefkowitz EJ, editors. *Virus taxonomy: classification and nomenclature of viruses: Ninth Report of the International*

Committee on Taxonomy of Viruses. 1st ed. London: Elsevier;2012. p. 725–41.
3. Schmaljohn CS, Nichol ST. Bunyaviridae. In: Knipe DM, Howley PM, editors. *Fields virology*. 5th ed. Philadelphia, PA: Lippincott Williams and Wilkins;2007. p. 1741–89.
4. Bishop DH, Calisher CH, Casals J, et al. Bunyaviridae. *Intervirology* 1980;**14**(3–4):125–43.
5. Elliott RM. *The Bunyaviridae*. New York, NY: Plenum Press;1996.
6. Guu TS, Zheng W, Tao YJ. Bunyavirus: structure and replication. *Adv Exp Med Biol* 2012;**726**:245–66.

8.1.1 Genus *Hantavirus*

1. Plyusnin A, Beaty BJ, Elliot RM, et al. Family Bunyaviridae. In: King AM, Adams MJ, Carstens EB, et al., editors. *Virus taxonomy: classification and nomenclature of viruses: Ninth Report of the International Committee on Taxonomy of Viruses.* 1st ed. London: Elsevier;2012. p. 725–41.
2. Henttonen H, Buchy P, Suputtamongkol Y, et al. Recent discoveries of new hantaviruses widen their range and question their origins. *Ann NY Acad Sci* 2008;**1149**:84–9.
3. Maes P, Klempa B, Clement J, et al. A proposal for new criteria for the classification of hantaviruses, based on S and M segment protein sequences. *Infect Genet Evol* 2009;**9**(5):813–20.
4. Plyusnin A. Genetics of hantaviruses: implications to taxonomy. *Arch Virol* 2002;**147**(4):665–82.

8.1.2 Hemorrhagic Fever with Renal Syndrome Virus and Related Viruses

1. Drozdov SG, Tkachenko EA. Hemorrhagic fever with renal syndrome. In: Lvov DK, Klimenko SM, Gaidamovich SY, editors. *Arboviruses and arbovirus infection*. Moscow: Meditsina;1989. p. 289–307 [in Russian].
2. Lee HW, Dalrympl JM. *Manuel of hemorrhagic fever with renal syndrome.* Seoul: Korea Univ;1989.
3. Liang M, Li D, Xiao S-Y, et al. Antigenic and molecular characterization of hantavirus isolates from China. *Virus Res* 1994;**31**:219–33.
4. Tkachenko EA, Dzagurova TK, Tkachenko PE. Hantaviruses: ecology, molecular biology, morphology, pathogenesis and diagnostics of hantavirus infections. *Molekulyarnaya Meditsina* 2009;(5):36–41 [in Russian].
5. Hofmann J, Meisel H, Klempa B, et al. Hantavirus outbreak, Germany, 2007. *Emerg Infect Dis* 2008;**14**(5):850–2.
6. Plyusnin A, Vapalahti O, Vaheri A. Hantaviruses: genome structure, expression and evolution. *J Gen Virol* 1996;**77**(11):2677–87.
7. Schwarz AC, Rauft U, Piechotowski I, et al. Risk factors for human infection with Puumala virus, southwestern Germany. *Emerg Infect Dis* 2009;**15**(7):1032–9.

8. Tkachenko EA, Dzagurova TK. Hemorrhagic fever with renal syndrome in Russia—the problem of XXI century. In: Lvov DK, Uryvaev LV, editors. *Investigation of virus evolution in the limits of the problems of biosafety and socially significant infections.* Moscow: D.I. Ivanovsky Institute of Virology of RAMS; 2011. p. 162—74 [in Russian].

9. Avsic-Zupanc T, Xiaos Y, Stojanovic R, et al. Characterization of Dobrava virus: a hantavirus from Slovenia, Yugoslavia. *J Med Virol* 1992;**38**:132—7.

10. Gligic A, Obradovic M, Stojanovich R, et al. Epidemic hemorrhagic fever with renal syndrome in Yugoslavia, 1986. *Am J Trop Med Hyg* 1989;**41**(1):102—8.

11. Lokugamage K, Kariwa H, Hayasaka D, et al. Genetic characterization of hantaviruses transmitted by the Korean field mouse (*Apodemus peninsulae*), Far East Russia. *Emerg Infect Dis* 2002;**8**(8):768—76.

12. Yao LS, Zhao H, Shao LJ, et al. Complete genome sequence of an Amur virus isolated from *Apodemus peninsulae* in Northeastern China. *J Virol* 2012;**86**(24):13816—17.

13. Plyusnin A, Vapalahti O, Lehvaslaiho H, et al. Genetic variation of wild Puumala viruses within the serotype local rodent populations and individual animals. *Virus Res* 1995;**58**:25—41.

14. Yakimenko VV, Tyulko JS, Dekonenko AE. Phylogenetic relations of Western Siberian hantaviruses belonging to Tula and Puumala genotypes. In: *Hantaviruses and hantavirus infections.* Vladivostok: OAO Primpoligraphkombinat; 2003. p. 161—72 [in Russian].

15. Horling J, Chizhikov V, Lundkvist A, et al. Khabarovsk virus: a phylogenetically and serologically distinct hantavirus isolated from *Microtus fortis* trapped in Far-Eastern Russia. *J Gen Virol* 1996;**77**(4):687—94.

16. Okumura M, Yoshimatsu K, Kumperasart S, et al. Development of serological assays for Thottapalayam virus, an insectivore-borne Hantavirus. *Clin Vaccine Immunol* 2007;**14**(2):173—81.

17. Hugot JP, Plyusnina A, Herbreteau V, et al. Genetic analysis of Thailand hantavirus in *Bandicota indica* trapped in Thailand. *Virol J* 2006;**3**:72.

18. Song JW, Kang HJ, Song KJ, et al. Newfound hantavirus in Chinese mole shrew, Vietnam. *Emerg Infect Dis* 2007;**13**(11):1784—7.

19. Kuchuloria T, Clark DV, Hepburn MJ, Tsertsvadze T, Pimentel G, Imnadze P. Hantavirus infection in the Republic of Georgia. *Emerg Infect Dis* 2009;**15**(9):1489—91.

20. Kim GR, Lee YT, Park CH. A new natural reservoir of hantavirus: isolation of hantaviruses from lung tissues of bats. *Arch Virol* 1994;**134**(1—2):85—95.

21. Slonova RA, Tkachenko EA, Kushnarev EL, Dzagurova TK, Astakova TI. Hantavirus isolation from birds. *Acta Virol* 1992;**36**(5):493.

22. Lvov DK. Hemorrhagic fever with renal syndrome. In: Lvov DK, editor. *Handbook of virology: Viruses and viral infection of human and animals.* Moscow: MIA;2013p. 797—802 [in Russian].

23. Gor'kova TI, Shamak DR. Hemorrhagic fever with the renal syndrome. *Klinic Med* 1973;**51**(2):137—8 [in Russian].

24. Weiss S, Witkowski PT, Auste B, et al. Hantavirus in bat, Sierra Leone. *Emerg Infect Dis* 2012;**18**(1):159—61.

25. Yashina LN, Slonova RA, Oleinik OV, et al. A new genetic variant of the PUUV virus from the Maritime Territory and its natural carrier red-grey vole *Clethrionomys rufocanus.* *Vopr Virusol* 2004;**49**(6):34—7 [in Russian].

26. Hardestam J, Karlsson M, Falk KI, et al. Puumala hantavirus excretion kinetics in bank voles (*Myodes glareolus*). *Emerg Infect Dis* 2008;**14**(8):1209—15.

27. Olsson GE, White N, Hjalten J, et al. Habitat factors associated with bank voles (*Clethrionomys glareolus*) and concomitant hantavirus in northern Sweden. *Vector Borne Zoon Dis* 2005;**5**(4):315—23.

28. Chen LF, Chen SH, Wang KL, et al. Characterization of S gene of a strain of hantavirus isolated from *Apodemus peninsulae* in Heilongjiang Province. *Bing Du Xue Bao* 2012;**28**(5):517—21 [in Chinese].

29. Dekonenko AE, Tkachenko EA. Hantaviruses and hantavirus infections. *Vopr Virusol* 2004;**49**(3):52—6 [in Russian].

30. Klempa B, Tkachenko EA, Dzagurova TK, et al. Hemorrhagic fever with renal syndrome caused by 2 lineages of Dobrava hantavirus, Russia. *Emerg Infect Dis* 2008;**14**(4):617—25.

31. Dzagurova TK, Tkachenko EA, Iunicheva Iu V, et al. Discovery, clinical and etiological characteristic of hemorrhagic fever with renal syndrome in the subtropical zone of Krasnodar region. *Zh Mikrobiol Epidemiol Immunobiol* 2008;**1**:12—16 [in Russian].

32. Dzagurova TK, Yunicheva YuV, Morozov VG, et al. Ecological differences between different genovariants of Dobrava-Belgrade virus. *Meditsinskaya Virusologiya* 2007;**24**:109—11 [in Russian].

33. Kuklev EV, Minin GD, Korobov LI, et al. Natural foci infections in low Volga Federal district and morbidity dynamics. Report 1. Morbidity of hemorrhagic fever with renal syndrome. *Problems Specially Dangerous Infections* 2004;**1**:28—30 [in Russian].

34. Kuchuloria T, Clark DV, Hepburn MJ, et al. Hantavirus infection in the Republic of Georgia. *Emerg Infect Dis* 2009;**15**(9):1489—91.

35. Tarasov MA, Kuklev EV, Velichko LN, et al. Quantity evaluation of link between hemorrhagic fever with renal syndrome morbidity among humans and

epizootic potential dynamics. *Problems Especially Dangerous Infections* 2004;**7**:37−9 [in Russian].

36. Bernstein AD, Gavrilovskaya IN, Apekina NS, et al. Peculiarities of natural foci hantavirus zoonoses. *Epidemiologiya i vakcinoprofilaktika* 2010;**2**:5−13 [in Russian].

37. Tkachenko EA, Okulova NM, Yunicheva YuV, et al. The epizootological and virological characteristics of a natural hantavirus infection focus in the subtropic zone of the Krasnodarsk Territory. *Vopr Virusol* 2005;**50**(3):14−29 [in Russian].

38. Pettersson L, Klingstrom J, Hardestam J, et al. Hantavirus RNA in saliva from patients with hemorrhagic fever with renal syndrome. *Emerg Infect Dis* 2008;**14**(3):406−11.

39. Hjertqvist M, Klein SL, Ahlm C, et al. Mortality rate patterns for hemorrhagic fever with renal syndrome caused by Puumala virus. *Emerg Infect Dis* 2010;**16**(10):1584−6.

40. Olsson GE, Hjertqvist M, Lundkvist A, Hornfeldt B. Predicting high risk for human hantavirus infections, Sweden. *Emerg Infect Dis* 2009;**15**(1):104−6.

41. Tkachenko EA, Dzagurova TK, Tkachenko PE. Hantaviruses: ecology, molecular biology, morphology, pathogenesis and diagnostics of hantavirus infections. *Molekul Med* 2009;**5**:36−41 [in Russian].

42. Tkachenko EA, Dzagurova TK, Nabatnikov PA, et al. *Development of the vaccine against hemorrhagic fever with renal syndrome*. Moscow: D.I. Ivanovsky Institute of Virology of RAMS; 2010 [in Russian].

8.1.3 Genus *Nairovirus*

1. Plyusnin A, Beaty BJ, Elliot RM, et al. Family Bunyaviridae. In: King AM, Adams MJ, Carstens EB, et al., editors. *Virus taxonomy: classification and nomenclature of viruses: Ninth Report of the International Committee on Taxonomy of Viruses*. 1st ed. London: Elsevier;2012. p. 725−41.

2. Schmaljohn CS, Nichol ST. Bunyaviridae. In: Knipe DM, Howley PM, editors. *Fields Virology*. 5th ed. Philadelphia, PA: Lippincott Williams and Wilkins;2007. p. 1741−89.

3. Casals J, Tignor GH. The Nairovirus genus: serological relationships. *Intervirology* 1980;**14**(3−4):144−7.

4. Clerx JP, Bishop DH. Qalyub virus, a member of the newly proposed Nairovirus genus (Bunyavividae). *Virology* 1981;**108**(2):361−72.

5. Clerx JP, Casals J, Bishop DH. Structural characteristics of nairoviruses (genus *Nairovirus*, *Bunyaviridae*). *J Gen Virol* 1981;**55**(1):165−78.

6. Gould EA, Chanas AC, Buckley A, et al. Immunofluorescence studies on the antigenic

interrelationships of the Hughes virus group (genus *Nairovirus*) and identification of a new strain. *J Gen Virol* 1983;**64**(3):739−42.

7. Chastel C, Main AJ, Richard P, et al. Erve virus, a probable member of Bunyaviridae family isolated from shrews (*Crocidura russula*) in France. *Acta Virol* 1989;**33**(3):270−80.

8. Zeller HG, Karabatsos N, Calisher CH, et al. Electron microscopic and antigenic studies of uncharacterized viruses. II. Evidence suggesting the placement of viruses in the family Bunyaviridae. *Arch Virol* 1989;**108**(3−4):211−27.

8.1.3.1 Crimean−Congo Hemorrhagic Fever Virus

1. Butenko AM, Chumakov MP, Bashkirtsev VN, et al. Isolation and investigation of Astrakhan strain ("Drozdov") of Crimean hemorrhagic fever virus and data on serodiagnostics of this infection. In: *Proceedings of the XV-th Scientific Session of the Institute of poliomyelitis and viral encephalitis*, Moscow, vol. 1; 1968. p. 88−90 [in Russian].

2. Chumakov MP, Smirnova SE, Shalunova NV, et al. Isolation and study of the virus from Crimean hemorrhagic fever patient in Samarkand region of Uzbek SSR, strain Khodzha. In: *Viral hemorrhagic fevers. Proceedings of the Institute of poliomyelitis and viral encephalitis*, Moscow, vol. 19; 1971. p. 19−21 [in Russian].

3. Hoogstraal H. The epidemiology of tick-borne Crimean-Congo hemorrhagic fever in Asia, Europe, and Africa. *J Med Entomol* 1979;**15**(4):307−417.

4. Simpson DI, Knight EM, Courtois G, et al. Congo virus: a hitherto undescribed virus occurring in Africa. I. Human isolations—clinical notes. *East Afr Med J* 1967;**44**(2):86−92.

5. Carroll SA, Bird BH, Rollin PE, et al. Ancient common ancestry of Crimean-Congo hemorrhagic fever virus. *Mol Phylogenet Evol* 2010;**55**(3):1103−10.

6. Chamberlain J, Cook N, Lloyd G, et al. Co-evolutionary patterns of variation in small and large RNA segments of Crimean-Congo hemorrhagic fever virus. *J Gen Virol* 2005;**86**(12):3337−41.

7. Deyde VM, Khristova ML, Rollin PE, et al. Crimean-Congo hemorrhagic fever virus genomics and global diversity. *J Virol* 2006;**80**(17):8834−42.

8. Kuhn JH, Seregin SV, Morzunov SP, et al. Genetic analysis of the M RNA segment of Crimean-Congo hemorrhagic fever virus strains involved in the recent outbreaks in Russia. *Arch Virol* 2004;**149**(11):2199−213.

9. Seregin SV, Samokhvalov EI, Petrova ID, et al. Genetic characterization of the M RNA segment of Crimean-Congo hemorrhagic fever virus strains isolated in Russia and Tajikistan. *Virus Genes* 2004;**28**(2):187−93.

10. Yashina L, Petrova I, Seregin S, et al. Genetic variability of Crimean-Congo haemorrhagic fever virus in Russia and Central Asia. *J Gen Virol* 2003;**84**(5):1199–206.

11. Yashina L, Vyshemirskii O, Seregin S, et al. Genetic analysis of Crimean-Congo hemorrhagic fever virus in Russia. *J Clin Microbiol* 2003;**41**(2):860–2.

12. Barr DA, Aitken C, Bell DJ, et al. First confirmed case of Crimean-Congo haemorrhagic fever in the UK. *Lancet* 2013;**382**(9902):1458.

13. Hubalek Z, Rudolf I. Tick-borne viruses in Europe. *Parasitol Res* 2012;**111**(1):9–36.

14. Maltezou HC, Papa A. Crimean-Congo hemorrhagic fever: risk for emergence of new endemic foci in Europe? *Travel Med Infect Dis* 2010;**8**(3):139–43.

15. Berezin VV, Chumakov MP, Stolbov DN, et al. About natural hosts of Crimean-Congo hemorrhagic fever virus in Astrakhan region. In: *Proceedings of the Institute of poliomyelitis and viral encephalitis*, Moscow, vol. 19; 1971. p. 210–8 [in Russian].

16. Butenko AM, Leshchinskaya EV, Lvov DK. Crimean-Congo hemorrhagic fever virus. *Vestnik Rossiiskoi Academii Estestvenykh Nauk* 2002;**2**:41–9 [in Russian].

17. Zeller HG, Cornet JP, Camicas JL. Experimental transmission of Crimean-Congo hemorrhagic fever virus by west African wild ground-feeding birds to *Hyalomma marginatum* rufipes ticks. *Am J Trop Med Hyg* 1994;**50**(6):676–81.

18. Zgurskaya GN, Berezin VV, Smirnova SE, et al. Investigation of the problems on the transmission and inter-epidemiological safety of Crimean-Congo hemorrhagic fever virus by ticks *Hyalomma plumbeum plumbeum* Panz. In: *Proceedings of the Institute of poliomyelitis and viral encephalitis*, Moscow, vol. 12; 1971. p. 217–20 [in Russian].

19. Lvov DK, Deryabin PG, Aristova VA, et al. *Atlas of distribution of natural foci virus infections on the territory of Russian Federation*. Moscow: SMC MPH RF Publ;2001.

20. Lvov DN, Dzharkenov AF, Aristova VA, et al. The isolation of Dhori viruses (Orthomyxoviridae, Thogotovirus) and Crimean-Congo hemorrhagic fever virus (Bunyaviridae, Nairovirus) from the hare (*Lepus europaeus*) and its ticks *Hyalomma marginatum* in the middle zone of the Volga delta, Astrakhan region, 2001. *Vopr Virusol* 2002;**47**(4):32–6 [in Russian].

21. Zimina YuV, Birulya V, Berezin VV, et al. Materials of zoologic-parasitological characteristics of Crimean hemorrhagic fever (CHF) in Astrakhan region. In: *Proceedings of the Institute of poliomyelitis and viral encephalitis*, Moscow, vol. 7; 1965. p. 288–96 [in Russian].

22. Zeller HG, Cornet JP, Camicas JL. Crimean-Congo haemorrhagic fever virus infection in birds: field investigations in Senegal. *Res Virol* 1994;**145**(2):105–9.

23. Lindeborg M, Barboutis C, Ehrenbord C, et al. Migratory birds, ticks and Crimean-Congo hemorrhagic fever virus. *Emerg Infect Dis* 2012;**18**(12):2095–7.

24. Palomar AM, Portillo A, Santibanez P, et al. Crimean-Congo hemorrhagic fever virus in ticks from migratory birds, Morroco. *Emerg Infect Dis* 2013;**19**(2):260–3.

25. Christova I, Di Caro A, Papa A, et al. Crimean-Congo hemorrhagic fever, southwestern Bulgaria. *Emer Infect Dis* 2009;**15**(6):983–5.

26. Karti SS, Odabasi Z, Korten V, et al. Crimean-Congo hemorrhagic fever in Turkey. *Emerg Infect Dis* 2004;**10**(8):1379–84.

27. Yilmaz GR, Buzgan T, Irmak H, et al. The epidemiology of Crimean-Congo hemorrhagic fever in Turkey, 2002–2007. *Int J Infect Dis* 2009;**13**(3):380–6.

28. Sang R, Lutomiah J, Koka H, et al. Crimean-Congo hemorrhagic fever virus in Hyalommid ticks, northeastern Kenya. *Emerg Infect Dis* 2011;**17**(8):1502–5.

29. Leblebicioglu H. Crimean-Congo haemorrhagic fever in Eurasia. *Int J Antimicrob Agents* 2010;**36**(Suppl. 1):S43–6.

30. Grard G, Drexler JF, Fair J, et al. Re-emergence of Crimean-Congo hemorrhagic fever virus in Central Africa. *PLoS Negl Trop Dis* 2011;**5**(10):e1350.

31. Vorou R, Pierroutsakos IN, Maltezou HC. Crimean-Congo hemorrhagic fever. *Curr Opin Infect Dis* 2007;**20**(5):495–500.

32. Onishchenko GG, Tumanova IYu, Vyshemirskii OI, et al. ELISA and RT-PCR-based research of viruses in the ticks collected in the foci of Crimean-Congo fever in Kazakhstan and Tajikistan in 2001–2002. *Vopr Virusol* 2005;(1):23–6 [in Russian].

33. Yevchenko YuM, Lvov DK, Sysolyatina GV, et al. Crimean-Congo hemorrhagic fever virus infection rate in *Hyalomma marginatum* Koch, 1844 ticks in the Stavropol region during the epidemic season of 2000. *Vopr Virusol* 2001;(4):18–19 [in Russian].

34. Bodur H, Akinci E, Ascioglu S, et al. Subclinical infections with Crimean-Congo hemorrhagic fever virus, Turkey. *Emerg Infect Dis* 2012;**18**(4):640–2.

35. Gunes T, Engin A, Poyraz O, et al. Crimean-Congo hemorrhagic fever virus in high-risk population, Turkey. *Emerg Infect Dis* 2009;**15**(3):461–4.

36. Chinikar S, Goya MM, Shirzadi MR, et al. Surveillance and laboratory detection system of Crimean-Congo haemorrhagic fever in Iran. *Transbound Emerg Dis* 2008;**55**(5–6):200–4.

37. Knust B, Medetov ZB, Kyraubayev KB, et al. Crimean-Congo hemorrhagic fever, Kazakhstan, 2009–2010. *Emerg Infect Dis* 2012;**18**(4):643–5.

38. Mustafa ML, Ayazi E, Mohareb E, et al. Crimean-Congo hemorrhagic fever, Afghanistan, 2009. *Emerg Infect Dis* 2011;**17**(10):1940−1.

39. Aradaib IE, Erickson BR, Karsany MS, et al. Multiple Crimean-Congo hemorrhagic fever virus strains are associated with disease outbreaks in Sudan, 2008−2009. *PLoS Negl Trop Dis* 2011;**5**(5):e1159.

40. Aradaib IE, Erickson BR, Mustafa ME, et al. Nosocomial outbreak of Crimean-Congo hemorrhagic fever, Sudan. *Emerg Infect Dis* 2010;**16**(5):837−9.

41. Zakhashvili K, Tsertsvadze N, Chikviladze T, et al. Crimean-Congo hemorrhagic fever in man, Republic of Georgia, 2009. *Emerg Infect Dis* 2010;**16**(8):1326−8.

42. Trusova IN, Butenko AM, Larichev VF, et al. Dynamics of specific IgM and IgG antibodies in patients with Crimean hemorrhagic fever. *Epidemiol Infect Dis (Moscow)* 2013;**2**:17−24 [in Russian].

43. Butenko AM. Modern status of the problem of Crimean hemorrhagic fever, West Nile fever and other arboviral infections in Russia. In: Lvov DK, Uryaev LV, editors. *Investigation of virus evolution in the limits of the problems of biosafety and socially significant infections.* Moscow: D.I. Ivanovsky Institute of Virology of RAMS; 2011, p. 175−9 [in Russian].

44. Sannikova IV, Pasechnikov AD, Maleev VV, et al. Crimean-Congo hemorrhagic fever in Stavropol region: peculiarities of severe clinical forms and predictors of the lethality. *Infectsionnye Bolezni* 2007;(4):25−8 [in Russian].

45. Moskvitina EA, Vodyanitskaya SYu, Pichyurina NL, et al. Investigation of modern status of natural foci of Crimean-Congo hemorrhagic fever in Rostov region. In: *Problems of specially dangerous infections.* Saratov; 2004. p. 34−7 [in Russian].

46. Podsvirov AV, Podsvirova VV, Sanjiev VB, et al. Epizootological and epidemiological characteristics of the Kalmyk Republic in respect to Crimean-Congo hemorrhagic fever during 2000−2006. In: Butenko AM, Barinsky IF, Karganova GG, editors. *Arboviruses and arboviral.* Moscow; 2007. p. 155−7 [in Russian].

47. Kovtunov AI, Yustratov VI, Nikeshina NN, et al. Epidemiology of Crimean hemorrhagic fever in Astrakhan region. In: Butenko AM, Barinsky IF, Karganova GG, editors. *Arboviruses and arboviral infections.* Moscow; 2007. p. 108−13 [in Russian].

48. Lobanov AP, Savchenko ST, Smelyansky VV, et al. Crimean hemorrhagic fever in Volgograd region. In: Butenko AM, Barinsky IF, Karganova GG, editors. *Arboviruses and arboviral.* Moscow; 2007. p. 132−5 [in Russian].

49. Lvov DK, Kolobukhina LV, Shchelkanov MYu, et al. Clinical profile and diagnostics algorithm of Crimean-Congo hemorrhagic fever and West Nile fever. Methodical handbook. Moscow; 2006.

50. Sannikova IV. Nosocomial infection of Crimean hemorrhagic fever. *Zh Microbiol Epidemiol Immunobiol* 2005;**4** (Suppl. 1):38−42.

51. Sannikova IV. Crimean-Congo hemorrhagic fever (clinical aspect, diagnostics, treatment, organization of medical help). *Handbook.* Stavropol; 2007.

52. Kolobukhina LV, Lvov DN. Crimean-Congo hemorrhagic fever. In: Lvov DK, editor. *Handbook of virology: Viruses and viral infection of human and animals.* Moscow: MIA;2013p. 772−9 [in Russian].

53. Krasovskaya TYu, Naidenova EV, Sharova IN, et al. Development and introduction of PCR-test-systems for the detection of West Nile virus and Crimean-Congo hemorrhagic fever virus. In: Butenko AM, Barinsky IF, Karganova GG, editors. *Arboviruses and arboviral infections.* Moscow; 2007. p. 97−9 [in Russian].

54. Drosten C, Kummerer BM, Schmitz H, Gunther S. Molecular diagnostics of viral hemorrhagic fevers. *Antiviral Res* 2003;**57**(1−2):61−87.

55. Duh D, Saksida A, Petrovec M, et al. Novel one-step real-time RT-PCR assay for rapid and specific diagnosis of Crimean-Congo hemorrhagic fever encountered in the Balkans. *J Virol Methods* 2006;**133** (2):175−9.

56. Ibrahim SM, Aitichou M, Hardick J, et al. Detection of Crimean-Congo hemorrhagic fever, Hanta, and sandfly fever viruses by real-time RT-PCR. *Methods Mol Biol* 2011;**665**:357−68.

57. Vanhomwegen J, Alves MJ, Avsvic Zupanc T, et al. Diagnostic assays for Crimean-Congo hemorrhagic fever. *Emerg Infect Dis* 2012;**18**(12):1958−65.

58. Elaldi N, Bodur H, Ascioglu S, et al. Efficacy of oral ribavirin treatment in Crimean-Congo haemorrhagic fever: a quasi-experimental study from Turkey. *J Infect Dis* 2009;**58**(3):238−44.

59. Izadi S, Salehi M. Evaluation of the efficacy of ribavirin therapy on survival of Crimean-Congo hemorrhagic fever patients: a case−control study. *Jpn J Infect Dis* 2009;**62**(1):11−15.

60. Mardani M, Jahromi MK, Naieni KH, et al. The efficacy of oral ribavirin in the treatment of Crimean-Congo hemorrhagic fever in Iran. *Clin Infect Dis* 2003;**36** (12):1613−18.

61. Buttigieg KR, Dowall SD, Findlay-Wilson S, et al. A novel vaccine against Crimean-Congo haemorrhagic fever protects 100% of animals against lethal challenge in a mouse model. *PLoS One* 2014;**9**(3):e91516.

62. Mousavi-Jazi M, Karlberg H, Papa A, et al. Healthy individuals' immune response to the Bulgarian Crimean-Congo hemorrhagic fever virus vaccine. *Vaccine* 2012;**30**(44):6225−9.

63. Tkachenko EA, Butenko AM, Badalov ME, et al. In: *Proceedings of the Institute of poliomyelitis and viral encephalitis*, Moscow, vol. 19; 1971, p. 119−29 [in Russian].
64. Aristova VA, Kolobukhina LV, Shchelkanov MYu, et al. Ecology of Crimean-Congo hemorrhagic fever virus and clinical picture of the disease in Russia and its neighboring countries. *Vopr Virusol* 2001;**45**(4):7−15 [in Russian].

8.1.3.2 Artashat Virus

1. Lvov DK. *Natural foci of arboviruses in the USSR. Soviet medical reviews. Virology*, 1. UK: Harwood Academic Publishers GmbH;1987153−96.
2. Lvov DK. Arboviral infections in subtropics and on south of temperate zone in USSR. In: Lvov DK, Klimenko SM, Gaidamovich SY, editors. *Arboviruses and arboviral infections*. Moscow: Meditsina;1989. p. 235−49 [Chapter 8] [in Russian].
3. Shakhazaryan SA, Oganesyan AS, Manukyan DV, et al. Isolation of arboviruses in Armyan SSR. In: Lvov DK, editor. *Uspekhi nauki i tekhniki. Ch. « Virology Arboviruses and arboviral infection »*. Moscow: : Academy of Science of USSR;199227(1): 57−60. [in Russian].
4. Alkhovsky SV, Lvov DK, Shchelkanov MYu, et al. Taxonomic status of Artashat virus (ARTSV) (Bunyaviridae, Nairovirus), isolated from ticks *Ornithodoros alactagalis* Issaakjan, 1936 и *O. verrucosus* Olenev, Sassuchin et Fenuk, 1934 (Argasidae Koch, 1844), collected in Transcaucasia. *Vopr Virusol* 2014;**59** (3):24−8 [in Russian].
5. Honig JE, Osborne JC, Nichol ST. The high genetic variation of viruses of the genus *Nairovirus* reflects the diversity of their predominant tick hosts. *Virology* 2004;**318**(1):10−16.
6. Plyusnin A, Beaty BJ, Elliot RM, Goldbach R, et al. Family Bunyaviridae. In: King AM, Adams MJ, Carstens EB, Lefkowitz EJ, editors. *Virus taxonomy: Ninth Report of the International Committee of Taxonomy of Viruses*. 1st ed. London: Elsevier;2012. p. 725−41.
7. Dilcher M, Koch A, Hasib L, et al. Genetic characterization of Erve virus, a European Nairovirus distantly related to Crimean-Congo hemorrhagic fever virus. *Virus Genes* 2012;**45**(3):426−32.
8. Woessner R, Grauer MT, Langenbach J, et al. The Erve virus: possible mode of transmission and reservoir. *Infection* 2000;**28**(3):164−6.
9. Chastel C, Main AJ, Richard P, et al. Erve virus, a probable member of Bunyaviridae family isolated from shrews (*Crocidura russula*) in France. *Acta Virol* 1989;**33** (3):270−80.
10. Shchelkanov MYu, Gromashevsky VL, Lvov DK. The role of ecovirological zoning in prediction of the influence of climatic changes on arbovirus habitats.

Vesstnik Rossiiskoi Academii Meditsinskikh Nauk 2006;**2**: 22−5 [in Russian].
11. Filippova NA. *Fauna of USSR*. Moscow-Leningrad: USSR Academy of Sciences;1966vol. 4(3). Arachnoidea. Argas ticks (Argasidae). [in Russian].

8.1.3.3 Caspiy Virus

1. Bergeron E, Albarino CG, Khristova ML, et al. Crimean-Congo hemorrhagic fever virus-encoded ovarian tumor protease activity is dispensable for virus RNA polymerase function. *J Virol* 2010;**84**(1):216−26.
2. Clifford CM. Tick-borne viruses in sea-birds. In: Kurstak E, editor. *Arctic and tropical arboviruses*. NY/San-Francisco/London: Harcourt Brace Jovanovich Publ.;1979p. 83−100 [Chapter 6].
3. Filippova NA. *Birds parasitizing ticks Ornithodoros Koch*. Digest of Parasitology of the Zoological Institute USSR Academy of Sciences;196321: 16−27. [in Russian].
4. Filippova NA. *Fauna of USSR*. Moscow-Leningrad: USSR Academy of sciences;1966vol. 4(3). Arachnoidea. Argas ticks (Argasidae). [in Russian].
5. Lvov DK. Natural foci of arboviruses, related with the birds in USSR. In: Lvov DK, Il'ichev VD, editors. *Migration of the birds and transduction of infection agents*. Moscow: Nauka;1979p. 37−101 [Chapter 2] [in Russian].
6. Lvov DK, Gromashevsky VL, Sidorova GA, et al. Preliminary dates about isolation of three of new arboviruses in Caucuses and Central Asia. In: Lvov DK, editor. *Ecology of the viruses*, vol. 2. Moscow: USSR Academy of Sciences;1974. p. 80−1 [in Russian].
7. Lvov DK. *Natural foci of arboviruses in the USSR. Soviet medical reviews. Virology*, vol. 1. UK: Harwood Academic Publishers GmbH;1987. p. 153−96.
8. Lvov DK, Timofeeva AA, Gromashevsky VL. The new arboviruses, isolated in USSR in 1969−1975. In: *Proceedings of 18th session of the Institute of Poliomyelitis and Viral Encephalitis of USSR Academy of Medical Sciences*, Moscow; 1975. p. 322−324 [in Russian].
9. Lvov DK, Timopheeva AA, Smirnov VL, et al. Ecology of tick-borne viruses in colonies of birds in the USSR. *Med Biol* 1975;**53**:325−30.
10. Sidorova GA, Andreev VP. Some features of the new arboviruses, isolated in Uzbekistan and Turkmenia. In: Lvov DK, editor. *Ecology of the viruses*. Moscow: USSR Academy of Medical Sciences;1980. p. 108−14 [in Russian].
11. Lvov DK, Alkhovsky SV, Shchelkanov MYu, et al. Genetic characterization of Caspiy virus (CASV) (Bunyaviridae, Nairovirus), isolated from seagull *Larus argentatus* (Laridae Vigors, 1825) and ticks *Ornithodoros capensis* Neumann in eastern and western cost of Caspian sea. *Vopr Virusol* 2014;**59**(1):24−9 [in Russian].

12. Wang Y, Dutta S, Karlberg H, et al. Structure of Crimean-Congo hemorrhagic fever virus nucleoprotein: superhelical homo-oligomers and the role of caspase-3 cleavage. *J Virol* 2012;**86**(22):12294–303.

13. Zhirnov OP, Syrtzev VV. Influenza virus pathogenicity is determined by caspase cleavage motifs located in the viral proteins. *J Mol Genet Med* 2009;**3**(1):124–32.

14. Frias-Staheli N, Giannakopoulos NV, Kikkert M, et al. Ovarian tumor domain-containing viral proteases evade ubiquitin- and ISG15-dependent innate immune responses. *Cell Host Microbe* 2007;**2**(6):404–16.

15. Honig JE, Osborne JC, Nichol ST. The high genetic variation of viruses of the genus Nairovirus reflects the diversity of their predominant tick hosts. *Virology* 2004;**318**(1):10–16.

16. Gould EA, Chanas AC, Buckley A, et al. Immunofluorescence studies on the antigenic interrelationships of the Hughes virus group (genus *Nairovirus*) and identification of a new strain. *J Gen Virol* 1983;**64**(3):739–42.

17. Hoogstraal H, Korser MN, Taylor MA, et al. Ticks (Ixodoidea) on birds migrating from Africa to Europe and Asia. *Bull WHO* 1961;**24**:197–212.

8.1.3.4 Chim Virus

1. Lvov DK, Gromashevsky VL, Sidorova GA, et al. Preliminary data about isolation of three novel arboviruses in Caucasus and Central Asia. In: Lvov DK, editor. *Ecology of viruses*. Moscow: USSR Academy of Medical Sciences;1974. p. 80–1 [in Russian].

2. Lvov DK, Sidorova GA, Gromashevsky VL, et al. Chim virus, a new arbovirus isolated from ixodid and argasid ticks collected in the burrows of great gerbils on the territory of the Uzbek SSR. *Vopr Virusol* 1979;(3):286–9 [in Russian].

3. Lvov DK. Isolation viruses from natural foci in the USSR. In: Lvov DK, Klimenko SM, Gaidamovich SY, editors. *Arboviruses and arboviral infections*. Moscow: Meditsina;1989p. 220–35 [Chapter 7].

4. Chim (CHIMV). In: Karabatsos N, editor. *International catalogue of arboviruses and some others viruses of vertebrates*. San Antonio, TX: American Society of Tropical Medicine and Hygiene;1985. p. 331–32.

5. Meliev A, Shermukhamedova DA. Results of searching of the arboviruses in Uzbek SSR. In: *Proceedings of 11th All-Union conference for natural foci infection*, Moscow; 1984. p. 107–8 [in Russian].

6. Sidorova GA, Andreev VL. Some features of the ecology of novel arboviruses, isolated in Uzbekistan and Turkmenia. In: Lvov DK, editor. *Ecology of viruses*. Moscow: USSR Academy of Medical Sciences;1980p. 108–14 [in Russian].

7. Drobichshenko NI, Karimov CK, Rogovaya SG. Isolation of Chim virus and Karshi virus from rodent in Kazakhstan. In: Lvov DK, editor. *Ecology of viruses*.

Moscow: USSR Academy of Medical Scienses;1980p. 133–5 [in Russian].

8. Lvov DK. Arboviruses in the USSR. In: Vesenjak-Hirjan J, editor. *Arboviruses in the mediterranean countries*. Stuttgart/New York: Gustav Fischer Verlag;1980*Zentralbl Bakt* 1980;(Suppl 9): 35–48. [in German].

9. Lvov DK, Alkhovsky SV, Shchelkanov MYu, et al. Taxonomic status of Chim virus (CHIMV) (Bunyaviridae, Nairovirus, Qalyub group) isolated from Ixodidae and Argasidae ticks collected in the great gerbil (*Rhombomys opimus* Lichtenstein, 1823) (Muridae, Gerbillinae) burrows in Uzbekistan and Kazakhstan. *Vopr Virusol* 2014;**59**(3) [in Russian].

10. Honig JE, Osborne JC, Nichol ST. The high genetic variation of viruses of the genus *Nairovirus* reflects the diversity of their predominant tick hosts. *Virology* 2004;**318**(1):10–16.

11. Alkhovsky SV, Lvov DK, Shchelkanov MYu, et al. Genetic characterization of Geran virus (GERAV) (Bunyaviridae, Nairovirus, Qalyub group) isolated from ticks *Ornithodoros verrucosus* (Olenev, Zasukhin et Fenyuk, 1934) (Argasidae), collected in the burrow of *Meriones libycus* (Lichtenstein, 1823) (Muridae, Murinae) in Azerbaijan. *Vopr Virusol* 2014;**59**(5):13–18 [in Russian].

12. Bandia (BDAV). In: Karabatsos N, editor. *International catalogue of arboviruses and some others viruses of vertebrates*. San Antonio, TX: American Society of Tropical Medicine and Hygiene;1985. p. 205–6.

13. Clerx JP, Bishop DH. Qalyub virus, a member of the newly proposed Nairovirus genus (Bunyavividae). *Virology* 1981;**108**(2):361–72.

14. Qalyub (QYBV). In: Karabatsos N, editor. *International catalogue of arboviruses and some others viruses of vertebrates*. San Antonio, TX: American Society of Tropical Medicine and Hygiene;1985. p. 847–8.

15. Alkhovsky SV, Lvov DK, Shchelkanov MYu, et al. Taxonomy of Issyk-Kul virus (ISKV, Bunyaviridae, Nairovirus), the etiologic agent of Issyk-Kul fever, isolated from bats (Vespertilionidae) and ticks *Argas (Carios) vespertilionis* (Latreille, 1796). *Vopr Virusol* 2013;**58**(5):11–15 [in Russian].

16. Filippova NA. Fauna of USSR. Moscow/Leningrad: 4 (3) Arachnoidea. Argas ticks (Argasidae) [in Russian]

17. Abdel-Wahab KS, Kaiser MN, Williams RE. Serological studies of Qalyub virus in certain animal sera in Egypt. *J Hyg Epidemiol Microbiol Immunol* 1974;**18**(2): 240–5.

8.1.3.5 Geran Virus

1. Lvov DK, Alkhovsky SV, Shchelkanov MYu, et al. Genetic characterization of the Geran virus (GERV,

Bunyaviridae, Nairovirus, Qalyub group), isolated from ticks *Ornithodoros verrucosus* Olenev, Zasukhin and Fenyuk, 1934 (Argasidae), collected in burrow of *Meriones erythrourus* Grey, 1842 in Azerbaijan. *Vopr Virusol* 2014;**59**(5) [in Russian].

2. Alkhovsky SV, Lvov DK, Shchelkanov MYu, et al. Taxonomy of Issyk-Kul virus (ISKV, Bunyaviridae, Nairovirus), the etiologic agent of Issyk-Kul fever, isolated from bats (Vespertilionidae) and ticks *Argas (Carios) vespertilionis* (Latreille, 1796). *Vopr Virusol* 2013;**58**(5):11−15 [in Russian].

3. Lvov DK, Alkhovsky SV, Shchelkanov MYu, et al. Taxonomy of previously unclassified Chim virus (CHIMV—Chim virus) (Bunyaviridae, Nairovirus, Qalyub group), isolated in Uzbekistan and Kazakhstan from ixodes (Acari: Ixodidae) and argas (Acari: Argasidae) ticks, collected in a burrows of great gerbil *Rhombomys opimus* Lichtenstein, 1823 (Muridae, Gerbillinae). *Vopr Virusol* 2014;**59**(3) [in Russian].

4. Honig JE, Osborne JC, Nichol ST. The high genetic variation of viruses of the genus Nairovirus reflects the diversity of their predominant tick hosts. *Virology* 2004;**318**(1):10−16.

5. Clerx JP, Bishop DH. Qalyub virus, a member of the newly proposed *Nairovirus* genus (Bunyavividae). *Virology* 1981;**108**(2):361−72.

6. Miller BR, Loomis R, Dejean A, et al. Experimental studies on the replication and dissemination of Qalyub virus (Bunyaviridae: Nairovirus) in the putative tick vector, *Ornithodoros (Pavlovskyella) erraticus. Am J Trop Med Hyg* 1985;**34**(1):180−7.

7. Qalyub virus. In: Karabatsos N, editor. *International catalogue of arboviruses including certain other viruses of vertebrates.* San Antonio, TX: American Society of Tropical Medicine and Hygiene;1985. p. 847.

8. Bishop DH, Calisher CH, Casals J, et al. Bunyaviridae. *Intervirology* 1980;**14**(3−4):125−43.

9. Bandia virus. In: Karabatsos N, editor. *International catalogue of arboviruses including certain other viruses of vertebrates.* San Antonio, TX: American Society of Tropical Medicine and Hygiene;1985. p. 205−6.

10. Bres P, Cornet M, Robin Y. The Bandia Forest virus (IPD-A 611), a new arbovirus prototype isolated in Senegal. *Ann Inst Pasteur (Paris)* 1967;**113**(5):739−47.

8.1.3.6 Issyk-Kul Virus

1. Lvov DK, Karas FR, Timofeev EM, et al. "Issyk-Kul" virus, a new arbovirus isolated from bats and *Argas (Carios) vespertilionis* (Latr., 1802) in the Kirghiz S.S.R. Brief report. *Arch Gesamte Virusforsch* 1973;**42**(2):207−9.

2. Issyk-Kul. In: Karabatsos N, editor. *International catalogue of arboviruses incuding certain other viruses of vertebrates.* San Antonio, TX: American Society of Tropical Medicine and Hygiene;1985. p. 461.

3. Lvov DK. Issyk-Kul virus disease. In: Monath TP, editor. *The Arboviruses: ecology and epidemiology.* Boca Raton, FL: CRS Press;1988. p. 53−62.

4. Lvov DK. Arbovirus infections in the subtropics and on the South of temperate zone of USSR. In: Lvov DK, Klimenko SM, Gaidamovich SY, editors. *Arboviruses and arbovirus infections.* Moscow: Meditsina;1989. p. 235−49 [in Russian].

5. Lvov DK, Ilyichev VD. *Migration of birds and the transfer of the infectious agents.* Moscow: Nauka;1979 [in Russian].

6. Lvov DK, Zakaryan VA, Abelyan KE. Viruses, ecologically associated with birds and Chiroptera in Transcaucasian. In: Lvov DK, editor. *Trends in Science and Technology. Virology. Arbovirus and arboviral infection*, vol. II. VINITI;1992. p. 99−100 [in Russian].

7. Karas FR. Ecology of arboviruses of mountain system in Central Asia region of USSR [Doctor of Med. Sci. thesis]. Moscow: D.I. Ivanovsky Institute of Virology of USSR Academy of Medical Sciences; 1979 [in Russian].

8. Karimov SK. Results and prospects of study of arboviral infection in Kazakhstan. In: Lvov DK, editor. *Ecology of Viruses in Kazakhstan and Central Asia.* Alma-Ata; 1980. p. 3−7 [in Russian].

9. Pak TP. The problems of arbovirus conservation in the interepidemic period. In: Lvov DK, editor. *Ecology of viruses.* Moscow: USSR Aademy of Sciences;1980. p. 118−21 [in Russian].

10. Pak TP, Lvov DK, Kostyukov MA, et al. The results of the virus scan in Tajikistan. In: Lvov DK, editor. *Ecology of Viruses in Kazakhstan and Central Asia.* Alma-Ata; 1980. p. 7−10 [in Russian].

11. Kostyukov MA. Study of Issyk-Kul virus ecology. In: Gaidamovich SY., editor. *Arboviruses.* Moscow; 1981. p. 78−82 [in Russian].

12. Kostyukov MA, Gordeeva ZE, Rafiyev ChK, et al. Immunological structure of the population to Issyk-Kul and Wad Medani virus in Tajikistan. In: Lvov DK, editor. *Ecology of viruses.* Moscow: D.I. Ivanovsky Institute of Virology of USSR Academy of Medical Sciences;1976. p. 108−12 [in Russian].

13. Nemova NV, Bylychev VP, Gordeeva ZE, Kostiukov MA. Isolation of the Issyk-Kul virus from blood-sucking biting flies *Culicoides schultzei* Enderlein in southern Tadzhikistan. *Meditsinskaia Parazitologiia i Parazitarnye Bolezni* 1989;**6**:25−6.

14. Keterah. In: Karabatsos N, editor. *International catalogue of arboviruses incuding certain other viruses of vertebrates.*

San Antonio, TX: American Society of Tropical Medicine and Hygiene;1985.

15. Varma MG, Converse JD. Keterah virus infections in four species of Argas ticks (Ixodoidea: Argasidae). *J Med Entomol* 1976;**13**(1):65–70.

16. Plyusnin A, Beaty BJ, Elliot RM, et al. Family Bunyaviridae. In: King AM, Adams MJ, Carstens EB, Lefkowitz EJ, editors. *Virus taxonomy: Ninth Report of the International Committee of Taxonomy of Viruses.* 1st ed. London: Elsevier;2012. p. 725–41.

17. Alkhovsky SV, Lvov DK, Shchelkanov MYu, et al. Taxonomy of Issyk-Kul virus (ISKV, Bunyaviridae, Nairovirus), the etiologic agent of Issyk-Kul fever, isolated from bats (Vespertilionidae) and ticks *Argas (Carios) vespertilionis* (Latreille, 1796). *Vopr Virusol* 2013;**58**(5):11–15 [in Russian].

18. Dacheux L, Cervantes-Gonzalez M, Guigon G, et al. A preliminary study of viral metagenomics of French bat species in contact with humans: identification of new mammalian viruses. *PLoS One* 2014;**9**(1): e87194.

19. Honig JE, Osborne JC, Nichol ST. The high genetic variation of viruses of the genus Nairovirus reflects the diversity of their predominant tick hosts. *Virology* 2004;**318**(1):10–16.

20. Lvov DK, Kostiukova MA, Pak TP, et al. Isolation of an arbovirus antigenically related to Issyk-Kul virus from the blood of a human patient. *Vopr Virusol* 1980; (1):61–2 [in Russian].

21. Lvov DK, Kostiukov MA, Daniyarov OA, et al. Outbreak of arbovirus infection in the Tadzhik SSR due to the Issyk-Kul virus (Issyk-Kul fever). *Vopr Virusol* 1984;**29**(1):89–92 [in Russian].

22. Kostiukov MA, Bulychev VP, Lapina TF. Experimental infection of *Aedes caspius caspius* Pall. mosquitoes on dwarf bats, *Vespertilio pipistrellus*, infected with the Issyk-Kul virus and its subsequent transmission to susceptible animals. *Medicinskaya Parazitologiya i Parazitarnye Bolezni* 1982;**51**(6):78–9 [in Russian].

8.1.3.7 Uzun-Agach Virus

1. Lvov DK. Arboviral zoonoses of Northern Eurasia (Eastern Europe and the commonwealth of independent states. Boca Raton AMArbar In: Beran GW, editor. *Handbook of zoonoses. Section B: Viral.* London/Tokyo: CRC Press;1994. p. 237–60.

2. Lvov DK. *Ecological soundings of the former USSR territory for natural foci of arboviruses. Soviet medical reviews. Section E. Virology reviews.* USA: Harwood Academic Publishers GmbH;1993. p. 1–47.

3. Lvov DK. Isolation of the viruses from natural sources in the USSR. In: Lvov DK, Klimenko SM, Gaidamovich SY, editors. *Arboviruses and arboviral infections.* Moscow: Meditsina;1989. p. 220–35 [Chapter 7]. [in Russian].

4. Karimov CK. Arboviruses in Kazakhstan region. Alma-Ata/Moscow: D.I. Ivanovsky Institute of Virology RAMS; 1983 [in Russian].

5. Lvov DK. *Natural foci of arboviruses in the USSR. Soviet medical reviews. Section E. Virology reviews.* UK: Harwood Academic Publishers GmbH;1987. p. 153–96.

6. Alkhovsky SV, Lvov DK, Shchelkanov MYu, et al. Genetic characterization of the Uzun-Agach virus (UZAV, Bunyaviridae, Nairovirus), isolated from bat *Myotis blythii oxygnathus* Monticelli, 1885 (Chiroptera; Vespertilionidae) in Kazakhstan. *Vopr Virusol* 2014;**59** (5) [in Russian].

7. Alkhovsky SV, Lvov DK, Shchelkanov MYu, Shchetinin AM, et al. Taxonomy of Issyk-Kul virus (ISKV, Bunyaviridae, Nairovirus), the etiologic agent of Issyk-Kul fever, isolated from bats (Vespertilionidae) and ticks *Argas (Carios) vespertilionis* (Latreille, 1796). *Vopr Virusol* 2013;**58**(5):11–15 [in Russian].

8. Calisher CH, Childs JE, Field HE, et al. Bats: important reservoir hosts of emerging viruses. *Clin Microbiol Rev* 2006;**19**(3):531–45.

9. Calisher CH, Ellison JA. The other rabies viruses: the emergence and importance of lyssaviruses from bats and other vertebrates. *Travel Med Infect Dis* 2012;**10**(2):69–79.

10. Drexler JF, Corman VM, Muller MA, et al. Bats host major mammalian paramyxoviruses. *Nat Commun* 2012;**3**:796.

11. Leroy EM, Kumulungui B, Pourrut X, et al. Fruit bats as reservoirs of Ebola virus. *Nature* 2005;**438**(7068):575–6.

8.1.3.8 Sakhalin Virus and Paramushir Virus

1. Lvov DK, Timofeeva AA, Gromashevski VL, et al. "Sakhalin" virus—a new arbovirus isolated from *Ixodes (Ceratixodes) putus* Pick.-Camb. 1878 collected on Tuleniy Island, Sea of Okhotsk. *Arch Gesamte Virusforsch* 1972;**38**(2):133–8.

2. Lvov DK, Gromashevsky VL. LEIV-71C strain of Sakhalin virus. Deposition certificate of Russian State Collection of viruses. 01.12.1969 [in Russian].

3. Sakhalin virus. In: Karabatsos N, editor. *International catalogue of arboviruses and certain other viruses of vertebrates.* San Antonio, TX: American Society of Tropical Medicine and Hygiene;1985. p. 881–2.

4. Lvov SD. *Natural virus foci in high latitudes of Eurasia. Soviet medical reviews. Section E. Virology Reviews,* 5. UK: Harwood Academic Publishers GmbH;1993. p. 137–85.

5. Lvov DK, Timofeeva AA, Gromashevski VL, et al. The ecology of Sakhalin virus in the north of the Far East of the USSR. *J Hyg Epidemiol Microbiol Immunol* 1974;**18** (1):87–95 [in Russian].

6. Lvov DK, Timopheeva AA, Smirnov VA, et al. Ecology of tick-borne viruses in colonies of birds in the USSR. *Med Biol* 1975;**53**(5):325–30.

7. Lvov DK, Timofeeva AA, Lebedev AD, et al. Foci of arboviruses in the north of the Far East (building hypothesis and its experimental verification). *Vesstnik Academii Meditsinskikh Nauk SSSR* 1971;**2**:52–64 [in Russian].

8. Bishop DH, Calisher CH, Casals J, et al. Bunyaviridae. *Intervirology* 1980;**14**(3–4):125–43.

9. Lvov DK, Alkhovsky SV, Shchelkanov MYu, et al. Genetic characterization of Sakhalin virus (SAKV), Paramushir virus (PMRV) (Sakhalin group, Nairovirus, Bunyaviridae) and Rukutama virus (RUKV) (Uukuniemi group, Phlebovirus, Bunyaviridae), isolated from the obligate parasites of colonial sea-birds—ticks *Ixodes (Ceratixodes) uriae*, White 1852 and *I. signatus* Birulya, 1895 in water area of sea of Okhotsk and Bering sea. *Vopr Virusol* 2014;**59**(3) [in Russian].

10. Paramushir virus. In: Karabatsos N, editor. *International catalogue of arboviruses and certain other viruses of verte-brates*. San Antonio, TX: American Society of Tropical Medicine and Hygiene;1985. p. 797–8.

11. Lvov DK. Arboviral zoonoses of Northern Eurasia (Eastern Europe and the commonwealth of independent states). Boca Raton AMArbar In: Beran GW, editor. *Handbook of zoonoses. Section B: Viral*. London/Tokyo: CRC Press;1994. p. 237–60.

12. Lvov DK. *Natural foci of arboviruses in the USSR. Soviet medical reviews. Section E. Virology reviews*. UK: Harwood Academic Publishers GmbH;1987. p. 153–96.

13. Lvov DK, Gromashevski VL, Skvortsova TM, et al. Arboviruses of high latitudes in the USSR. In: Kurstak E, editor. *Arctic and Tropical Arboviruses*. NY/San Francisco/London: Harcourt Brace Jovanovich Publ;1979. p. 21–38.

14. Lvov DK, Sazonov AA, Gromashevsky VL, et al. "Paramushir" virus, a new arbovirus, isolated from ixodid ticks in nesting sites of birds on the islands in the north-western part of the Pacific Ocean basin. *Arch Virol* 1976;**51**(1–2):157–61.

15. Avalon virus. In: Karabatsos N, editor. *International cat-alogue of arboviruses and certain other viruses of verte-brates*. San Antonio, TX: American Society of Tropical Medicine and Hygiene;1985. p. 193–4.

16. Clo Mor virus. In: Karabatsos N, editor. *International catalogue of arboviruses and certain other viruses of verte-brates*. San Antonio, TX: American Society of Tropical Medicine and Hygiene;1985. p. 335–6.

17. Taggert virus. In: Karabatsos N, editor. *International catalogue of arboviruses and certain other viruses of verte-brates*. San Antonio, TX: American Society of Tropical Medicine and Hygiene;1985. p. 975–6.

18. Labuda M, Nuttall PA. Tick-borne viruses. *Parasitology* 2004;**129**(Suppl. 1):S221–45.

19. Quillien MC, Monnat JY, Le Lay G, et al. Avalon virus, Sakhalin group (Nairovirus, Bunyaviridae) from the seabird tick *Ixodes (Ceratixodes) uriae* White 1852 in France. *Acta Virol* 1986;**30**(5):418–27.

20. Main AJ, Downs WG, Shope RE, et al. Avalon and Clo Mor: two new Sakhalin group viruses from the North Atlantic. *J Med Entomol* 1976;**13**(3):309–15.

21. Lvov DK, Gromashevsky VL, Skvortsova TM. LEIV-6269C strain of Rukutama virus. Deposition certificate of Russian State Collection of viruses. 1976. N 641 [in Russian].

22. Lvov DK. Natural foci of arboviruses, related with the birds in USSR. In: Lvov DK, Il'ichev VD, editors. *Migration of the birds and transduction of infection agents*. Moscow: Nauka;1979. p. 37–101 [Chapter 2] [in Russian].

23. Timofeeva AA, Pogrebenko AG, Gromashevsky VL, et al. Natural foci of infection on the Iona island in the Sea of Okhotsk. *Zoologicheskii Zhurnal* 1974;**53** (6):906–11 [in Russian].

24. Clifford CM. Tick-borne viruses of seabirds. In: Kurstak E, editor. *Arctic and tropical arboviruses*. NY/San Francisco/London: Academic Press Harcourt Brace Jovanovich Publ;1979. p. 83–100.

25. Artsob H, Spence L. Arboviruses in Canada. In: Kurstak E, editor. *Arctic and tropical arboviruses*. NY/San Francisco/London: Academic Press Harcourt Brace Jovanovich Publ;1979. p. 39–65.

8.1.3.9 Tamdy Virus

1. Lvov DK. Tamdy virus strain LEIV-1308Uz. *Am J Trop Med Hyg* 1978;**27**:411–12.

2. Lvov DK, Sidorova GA, Gromashevsky VL, et al. Virus "Tamdy"—a new arbovirus, isolated in the Uzbek S.S.R. and Turkmen S.S.R. from ticks *Hyalomma asiaticum* asia-ticum Schulce et Schlottke, 1929, and *Hyalomma plum-beum plumbeum* Panzer, 1796. *Arch Virol* 1976;**51** (1–2):15–21.

3. Tamdy virus (TAMV). In: Karabatsos N, editor. *International Catalogue of arboviruses and some others viruses of vertebrates*. San Antonio, TX: American Society of Tropical Medicine and Hygiene;1985. p. 979–80.

4. Lvov DK. Arboviral infections in subtropics and on south of temperate zone in USSR. In: Lvov DK, Klimenko SM, Gaidamovich SY, editors. *Arboviruses and arboviral infections*. Moscow: Meditsina;1989. p. 235–49 [Chapter 8] [in Russian].

5. Lvov DK. *Ecological sounding of the USSR territory for natural foci of arboviruses. Soviet medical reviews. Section E: Virology Reviews*, 5. USA: Harwood Academic Publishers GmbH;1993. p. 1–47.

6. Shakhazaryan SA, Oganesyan AS, Manukyan DV, et al. Isolation of arboviruses in Armyan SSR. In: Lvov DK, editor. *Uspekhi nauki I tekhniki: Virology. Arboviruses and arboviral infection*, vol. 24. Moscow: USSR Academy of Sciences;1991. p. 22 [in Russian].

7. Sidorova GA, Andreev VP. Some features of the new arboviruses, isolated in Uzbekistan and Turkmenistan. In: Lvov DK, editor. *Ecology of the viruses*. Moscow: USSR Academy of Sciences;1980. p. 108–14 [in Russian].

8. Gromashevsky VL, Skvortsova TM, Nikiforov LR, et al. Isolation of arboviruses in Turkmen SSR and Azerbaijan SSR. In: Lvov DK, editor. *Biology of the viruses*. Moscow: D.I. Ivanovsky Institute of Virology RAMS; 1975. p. 91–94 [in Russian].

9. Meliev A, Kadyrov AM, Shermukhamedova DA, et al. Ecology of Tamdy virus in Uzbekistan. In: Lvov DK, editor. *Uspekhi nauki i tekhniki: virology. Arboviruses and arboviral infection*. Moscow: USSR Academy of Sciences;1991p. 21 [in Russian].

10. Skvortsova TM, Gromashevsky VL, Sidorova GA, et al. Results of viral surveillance of arthropods vectors in Turkmenia. In: Lvov DK, editor. *Ecology of the viruses*. Moscow: USSR Academy of Sciences;1982. p. 139–44 [in Russian].

11. Meliev A, Shermukhamedova DA. *Search of arboviruses in Turkmenia. Proceedings of All-Union Conference for natural foci infections*. Moscow: USSR Academy of sciences;1984. p. 107–8. [in Russian].

12. Karas FR, Vargina SG, Steblyanko EN. About ecology of Tamdy virus in Kirgizia. In: Lvov DK, editor. *Uspekhi nauki i tekhniki; Virology. Arboviruses and arboviral infection*. Moscow: USSR Academy of Sciences;1976. p. 87–8 [in Russian].

13. Vargina SG, Breininger IG, Gershtein VI. Dynamics of circulation of arboviruses in Kirgizia. In: Lvov DK, editor. *Uspekhi nauki i tekhniki: Virology. Arboviruses and arboviral infection*. Moscow: USSR Academy of Sciences;1992. p. 38–45 [in Russian].

14. Karimov SK, Drobishchenko NI, Kiryushchenko TV. Isolation of Tamdy virus from ticks *Hyalomma asiaticum asiaticum* in Kazakh SSR. In: Lvov DK, editor. *Ecology of the viruses*. Moscow: USSR Academy of Sciences;1982. p. 151–4 [in Russian].

15. Meliev A, Kadyrov AM, Shermukhamedova DA, et al. Ecology of Tamdy virus in Uzbekistan. In: Lvov DK, editor. *Uspekhi nauki i tekhniki: Virology. Arboviruses and arboviral infection*. Moscow: USSR Academy of Sciences;1991. p. 1–21 [in Russian].

16. Lvov DK, Sidorova GA, Gromashevsky VL, et al. Isolation of Tamdy virus (Bunyaviridae) pathogenic for man from natural sources in Central Asia, Kazakhstan and Transcaucasia. *Vopr Virusol* 1984;**29**:487–90 [in Russian].

17. Lvov DK. Natural foci of arboviruses, related with the birds in USSR. In: Lvov DK, Il'ichev VD, editors. *Migration of the birds and transduction of infection agents*. Moscow: Nauka;1979. p. 37–101 [Chapter 2] [in Russian].

18. Lvov DK. Arboviral zoonoses of Northern Eurasia (Eastern Europe and the Commonwealth of Independent States). In: Beran GW, editor. *Handbook of zoonoses. Section B: Viral*. Boca Raton/London/Tokyo: CRC Press;1994. p. 237–60.

19. Elkina NYu, Gromashevsky VL, Skvortsova TM. Tamdy virus, strain LEIV-10224Az from ticks *Hyalomma anatolicum* from Apsheron district of Azerbaijan SSR. Deposition certificate of Russian State Collection of viruses. N 792 [in Russian].

20. Lvov DK, Alkhovsky SV, Shchelkanov MYu, et al. Taxonomy of previously unclassified Tamdy virus (TAMV) (Bunyaviridae, Nairovirus) isolated from *Hyalomma asiaticum asiaticum* Schülce et Schlottke, 1929 (Ixodidae, Hyalomminae) in Central Asia and Transcaucasia. *Vopr Virusol*. 2014;**59**(2):15–22 [in Russian].

21. Pomerantsev BI. *Fauna of USSR*. Moscow-Leningrad: USSR Academy of Sciences;1966. vol. 4(2) Arachnoidea. Argas ticks (Argasidae) [in Russian].

8.1.3.10 Burana Virus (BURV)

1. Karas FR. Arboverises in Kirgizia. In: Gaidamovich SY, editor. *Arboviruses*. Moscow: D.I. Ivanosky Institute of Virology;1978p. 40–4 [in Russian].

2. Karas FR, Lvov DK, Vargina SG, et al. Isolation of new virus—Burana virus, from ticks *Haemaphysalis punctata* in the north climatic region of Kirgizia. In: Lvov DK, editor. *Ecology of viruses*. Moscow: Nauka;1976. p. 94–7 [in Russian].

3. Lvov DK, Alkhovsky SV, Shchelkanov MYu, et al. Taxonomic status of Burana virus (BURV) (Bunyaviridae, Nairovirus, Tamdy group), isolated from ticks *Haemaphysalis punctata* Canestrini et Fanzago, 1877 and *Haem. concinna* Koch, 1844 (Ixodidae, Haemaphysalinae) in Kyrgyzstan. *Vopr Virusol* 2014;**59**(4):10–15 [in Russian].

4. Plyusnin A, Beaty BJ, Elliot RM, et al. Family Bunyaviridae. In: King AM, Adams MJ, Carstens EB, Lefkowitz EJ, editors. *Virus taxonomy: Ninth Report of the International Committee of Taxonomy of Viruses*. London: Elsevier;2012. p. 725–41.

5. Altamura LA, Bertolotti-Ciarlet A, Teigler J, et al. Identification of a novel C-terminal cleavage of Crimean-Congo hemorrhagic fever virus PreGN that leads to generation of an NSM protein. *J Virol* 2007;**81**(12):6632—42.
6. Vincent MJ, Sanchez AJ, Erickson BR, et al. Crimean-Congo hemorrhagic fever virus glycoprotein proteolytic processing by subtilase SKI-1. *J Virol* 2003;**77**(16):8640—9.
7. Wang Y, Dutta S, Karlberg H, et al. Structure of Crimean-Congo hemorrhagic fever virus nucleoprotein: superhelical homo-oligomers and the role of caspase-3 cleavage. *J Virol* 2012;**86**(22):12294—303.
8. Vargina SG, Breininger IG, Gerstein VI. Dynamics of arbovirus circulation in Kirgizia. In: Lvov DK, editor. *Itogi nauki i techniki. Virology*. Moscow: USSR Academy of sciences;1992p. 38—45 [in Russian].

8.1.4 Genus *Orthobunyavirus*

1. Schmaljohn CS, Nichol ST. Bunyaviridae. In: Knipe DM, Howley PM, editors. *Fields virology*. 5th ed. Philadelphia, PA: Lippincott Williams and Wilkins;2007. p. 1741—89.
2. Elliott RM. Nucleotide sequence analysis of the large (L) genomic RNA segment of Bunyamwera virus, the prototype of the family Bunyaviridae. *Virology* 1989;**173**(2):426—36.
3. Elliott RM. Identification of nonstructural proteins encoded by viruses of the Bunyamwera serogroup (family Bunyaviridae). *Virology* 1985;**143**(1):119—26.
4. Gentsch JR, Bishop DL. M viral RNA segment of bunyaviruses codes for two glycoproteins, G1 and G2. *J Virol* 1979;**30**(3):767—70.
5. Van Knippenberg I, Carlton-Smith C, Elliott RM. The N-terminus of Bunyamwera orthobunyavirus NSs protein is essential for interferon antagonism. *J Gen Virol* 2010;**91**(8):2002—6.
6. Habjan M, Pichlmair A, Elliott RM, et al. NSs protein of Rift Valley fever virus induces the specific degradation of the double-stranded RNA-dependent protein kinase. *J Virol* 2009;**83**(9):4365—75.
7. Hart TJ, Kohl A, Elliott RM. Role of the NSs protein in the zoonotic capacity of Orthobunyaviruses. *Zoonoses Public Health* 2009;**56**(6—7):285—96.
8. Plyusnin A, Beaty BJ, Elliot RM, et al. Family Bunyaviridae. In: King AM, Adams MJ, Carstens EB, Lefkowitz EJ, editors. *Virus taxonomy : classification and nomenclature of viruses : Ninth Report of the International Committee on Taxonomy of Viruses*. London: Elsevier;2012. p. 725—41.

8.1.4.1 Batai Virus and Anadyr Virus

1. Dunn EF, Pritlove DC, Elliott RM. The S RNA genome segments of Batai, Cache Valley, Guaroa, Kairi, Lumbo, Main Drain and Northway bunyaviruses: sequence determination and analysis. *J Gen Virol* 1994;**75**(3):597—608.
2. Hubalek Z. Mosquito-borne viruses in Europe. *Parasitol Res* 2008;**103**(Suppl. 1):29—43.
3. Hunt AR, Calisher CH. Relationships of bunyamwera group viruses by neutralization. *Am J Trop Med Hyg* 1979;**28**(4):740—9.
4. Director of the Institute of Medical Research (Kuala Lumpur, Malaysia). Personal communication. 1957.
5. Karabatsos N. *International catalogue of arboviruses including certain other viruses of vertebrates*. San Antonio, TX: American Society of Tropical Medicine and Hygiene;1985.
6. Bardos V, Chupkova E. The Chalovo virus—the second virus isolated from mosquitoes in Czechoslovakia. *J Hyg Epidemiol Microbiol Immunol* 1962;(6):186—92.
7. Smetana A, Danielova V, Kolman JM, et al. The isolation of the Chalovo virus from the mosquitoes of the group *Anopheles maculipennis* in Southern Moravia. *J Hyg Epidemiol. Microbiol Immunol* 1967;**11**(1):55—9.
8. Vinograd IA, Gaidamovich SY, Obukhova R, et al. Investigation of biological properties of Olyka virus isolated from mosquitoes (Culicidae) in Western Ukraine. *Vopr Virusol* 1973;**17**:714—19 [in Russian].
9. Vinograd IA, Obukhova R. Isolation of arboviruses from birds in Western Ukraine. In: *Proceedings of D. I. Ivanovsky Institute of Virology*, Moscow, vol. 3; 1975. p. 84—7 [in Russian].
10. Vinograd IA, Rogochiy EG. Persistence of Batai virus, strain Olyka, from bird. *Virusy i Virusnye Zabolevaniya* 1978;**6**:191—203 [in Russian].
11. Vinograd IA, Vigovskiy AI, Beletskaya GB, et al. The ecology of Batai virus in Danube delta. In: Lvov DK, editor. Ecology of viruses. Moscow; 1980: 93—7. [in Russian].
12. Yadav PD, Sudeep AB, Mishra AC, et al. Molecular characterization of Chittoor (Batai) virus isolates from India. *Indian J Med Res* 2012;**136**(5):792—8.
13. Briese T, Bird B, Kapoor V, et al. Batai and Ngari viruses: M segment reassortment and association with severe febrile disease outbreak in East Africa. *J Virol* 2006;**80**:5627—30.
14. Shchetinin AM, Lvov DK, Alkhovsky S, et al. Complete genome analysis of Batai virus (BATV) and a new Anadyr virus (ANADV) of Bunyamwera group (Bunyaviridae, Ortho-bunyavirus) isolated in Russia. *Vopr Virusol* 2014;**59**(6):16—22 [in Russian].

15. Nashed NW, Olson JG, el-Tigani A. Isolation of Batai virus (Bunyaviridae, Bunyavirus) from the blood of suspected malaria patients in Sudan. *Am J Trop Med Hyg* 1993;**48**(5):676–81.

16. Geevarghese G, Prasanna NY, Jacob PG, et al. Isolation of Batai virus from sentinel domestic pig from Kolar district in Karnataka State, India. *Acta Virol* 1994;**38**(4):239–40.

17. Brummer-Korvenkontio M. Bunyamwera arbovirus supergroup in Finland. A study on Inkoo and Batai viruses. *Commentat Biol* 1974;**76**:152.

18. Lvov DK, Klimenko SM, Gaidamovich SY. *Arboviruses and arbovirus infections*. Moscow: Meditsina;1989 [in Russian].

19. Shope RE. Arboviral zoonoses in Western Europe. In: Beran GW, Steel JH, editors. *Handbook of zoonoses. Section B: Viral*. Boca Raton, FL: CRC Press;1994. p. 227–35.

20. Aspock H, Kunz C. Uberwintering des Chalovo Virus in experimentall infizierten Weibach von *Anopheles maculipennis* messeae Fall. *Zentralbl Bacteriol* 1970;**213**:429–33 [in German].

21. Galkina I. Distribution of arboviruses in the Volga territory [Ph.D. thesis]. Moscow: D.I. Ivanovsky Institute of Virology; 1991 [in Russian].

22. Kolobukhina L, Lvov DN. Batai fever. In: Lvov DK, editor. *Handbook of virology: Viruses and viral infection of human and animals*. Moscow: MIA;2013p. 736–9 [in Russian].

23. Lvov DN. Populational interactions of West Nile and other arbovirus infections with arthropod vectors, vertebrates and humans in the middle and low belts of Volga delta [Ph.D. thesis]. Moscow: D.I. Ivanovsky Institute of Virology; 2008 [in Russian].

24. Shchelkanov MYu, Ananyev VYu, Lvov DN, et al. Complex environmental and virological monitoring in the Primorje Territory in 2003–2006. *Vopr Virusol* 2007;**52**(5):37–48 [in Russian].

25. Beletskaya G, Alekseev AN. Experimental demonstration of the Batai virus maintenance in hibernating females of *Anopheles maculipennis*. *Med Parasitol* 1988;**66**(4):23–7 [in Russian].

26. Bardos V, Sluka F, Chupkova E. Serological study on the medical importance of Chalovo virus. In: Bardos V, editor. *Arboviruses of the California complex of the Bunyamwera group*. Bratislava: Slovak Academy of Sciences;1969p. 333–6 [in Slovak].

27. Sluka F. The clinical picture of the Chalovo virus infection. In: Bardos V, editor. *Arboviruses of the California complex of the Bunyamwera group*. Bratislava: Slovak Academy of Sciences;1969. p. 337–9.

28. Abramova LN, Terskikh II, Skvortsova TM, et al. Pathogenic properties of the LEIV-23 Astrakhan strain of Batai virus for primates. *Vopr Virusol* 1983;**28**(6):660–3.

29. Chippaux A, Chippaux-Hyppolite C, Clergeaud P, et al. Isolation of 2 human strains of Ilesha virus in the Central African Republic. *Bull Soc Med Afr* 1969;**14**(1):88–92 [in French].

30. Morvan JM, Digoutte JP, Marsan P, et al. Ilesha virus: a new aetiological agent of haemorrhagic fever in Madagascar. *Trans R Soc Trop Med Hyg* 1994;**88**(2):205.

31. Terekhin SA, Grebennikova T, Khutoretskaya N, et al. Molecular-genetic analysis of the Batai virus strains isolated from mosquitoes in Volgograd Region of the Russian Federation, West Ukraine, and Czech Republic. *Molekul Gen Mikrobiol Virusol* 2010;**1**:27–9 [in Russian].

8.1.4.2 California Encephalitis Complex Viruses: Inkoo Virus, Khatanga Virus, Tahyna Virus

1. Casals J. Immunological relationship between Tahyna and California encephalitis viruses. *Acta Virol* 1962;**6**:140–2.

2. Karabatsos N. *International catalogue of arboviruses including certain other viruses of vertebrates*. San Antonio, TX: American Society of Tropical Medicine and Hygiene;1985.

3. Kokernot RH, McIntosh BM, Worth CB, et al. Isolation of viruses from mosquitoes collected at Lumbo, Mozambique. I. Lumbo virus, a new virus isolated from *Aedes (Skusea) pembaensis* Theobald. *Am J Trop Med Hyg* 1962;**11**:678–82.

4. Bardos V, Danielova V. The Tahyna virus a virus isolated from mosquitoes in Czechoslovakia. *J Hyg Epidemiol Microbiol Immunol* 1959;**3**:264–76.

5. Brummer-Korvenkontio M, Saikku P. Mosquito-borne viruses in Finland. *Med Biol* 1975;**53**(5):279–81.

6. Brummer-Konvenkontio M, Saikku P, Korhonen P, et al. Arboviruses in Finland. IV. Isolation and characterization of Inkoo virus, a Finnish representative of the California group. *Am J Trop Med Hyg* 1973;**22**(3):404–13.

7. Burgdorfer W, Newhouse VF, Thomas LA. Isolation of California encephalitis virus from the blood of a snowshoe hare (*Lepus americanus*) in western Montana. *Am J Hyg* 1961;**73**:344–9.

8. Lavrent'ev MV, Prilipov AG, Lvov SD, et al. Phylogenetic analysis of the nucleotide sequences of Khatanga virus strains, the new representative of California encephalitis serocomplex, isolated in different regions of the Russian Federation. *Vopr Virusol* 2008;**53**(6):25–9 [in Russian].

9. Lvov DK, Gromashevsky VL, Skvortsova TM, et al. Circulation of viruses of the California serocomplex (Bunyaviridae, Bunyavirus) in the central and southern

parts of the Russian plain. *Vopr Virusol* 1998;**43** (1):10−14 [in Russian].

10. Poglazov AB, Shlyapnikova OV, Prilipov AG. Characteristic of the new Khatanga virus genotype. *Vestnik Rossiiskoi Akademii Meditsinskih Nauk* 2011;**5**:15−9 [in Russian].

11. Schmaljohn CS, Nichol ST. Bunyaviridae. In: Knipe DM, Howley PM, editors. *Fields virology*. Philadelphia, PA: Lippincott Williams and Wilkins;2007. p. 1741−89.

12. Arcan P, Topcin V, Rosin N, et al. Isolation of Tahyna virus from *Culex pipiens* mosquitoes in Romania. *Acta Virol* 1974;**18**:175.

13. Aspock H, Kunz C. Izolierung des Tahyna Virus aus Stechmucken in Isterreich. *Arch Gesamte Virusforsch* 1966;**18**:8−15 [in German].

14. Aspock H, Kunz C. Untersuchungen uber die Okologie des Tahyna Virus. *Zentralbl Bakt* 1967;**203**:1−24 [in German].

15. Aspock H, Kunz C. Feldutersuchungen uber die Bedentung des Igels (*Erinaceus europaeus roumanicus* Barret-Hamilton) im Zykles des Tahyna Virus. *Zentralbl Bakt* 1970;**213**:304−10 [in German].

16. Bardos V, Ryba J, Hubalek Z. Isolation of Tahyna virus from field collected *Culiseta annulata* (Schrk.) larvae. *Acta Virol* 1975;**19**(5):446.

17. Demikhov VG, Chaitsev VG, Butenko AM, et al. California serogroup virus infections in the Ryazan region of the USSR. *Am J Trop Med Hyg* 1991;**45**:371.

18. Godsey MSJr, Amoo F, Yuill TM, et al. California serogroup virus infections in Wisconsin domestic animals. *Am J Trop Med Hyg* 1988;**39**(4):409−16.

19. Hannoun C, Panthier R, Corniou B. Isolation of Tahyna virus in the south of France. *Acta Virol* 1966;**10** (4):362−4.

20. Kolman JM, Malkova D, Nemec A, et al. The isolation of the Tahyna virus from the mosquito *Aedes vexans* in Southern Moravia. *J Hyg Epidemiol Microbiol Immunol* 1964;**8**:380−6.

21. Malkova D, Holubova J, Marhoul Z, et al. Study of arboviruses in the region of Most from 1981 to 1982. Isolation of Tahyna viruses. *Cesk Epidemiol Mikrobiol Immunol* 1984;**33**(2):88−96 [in Czech].

22. Pilaski J, Mackenstein H. Nachweis des Tahyna Virus beistechmucken in zwei verschiedenen europaischen Naturherden. *Zentralbl Bakt* 1985;**180**:394−420 [in German].

23. Rosicky B, Malkova D, editors. *Tahyna virus natural focus in Southern Moravia*. Prague: Rozpravy CSAV;1980.

24. Traavik T, Mehl R, Wiger R. California encephalitis group viruses isolated from mosquitoes collected in Southern and Arctic Norway. *Acta Pathol Microbiol Scand B* 1978;**86B**(6):335−41.

25. Chandler LJ, Hogge G, Endres M, et al. Reassortment of La Crosse and Tahyna bunyaviruses in *Aedes triseriatus* mosquitoes. *Virus Res* 1991;**20**(2):181−91.

26. Le Duc JW. Epidemiology and ecology of the California serogroup viruses. *Am J Trop Med Hyg* 1987;**37**(Suppl. 3):60−8.

27. Gerhardt RR, Gottfried RL, Apperson CS, et al. First isolation of La Crosse virus from naturally infected *Aedes albopictus*. *Emerg Infect Dis* 2001;**7**(5):807−11.

28. Kitron U, Swanson J, Crandell M, et al. Introduction of *Aedes albopictus* into a La Crosse virus-enzootic site in Illinois. *Emerg Infect Dis* 1998;**4**(4):627−30.

29. Danielova V, Ryba J. Laboratory demonstration of transovarial transmission of Tahyna virus in *Aedes vexans* and the role of this mechanism in overwintering of this arbovirus. *Folia Parasitol* 1979;**26**:361−8.

30. Lvov SD, Gromashevsky VL, Morozova TN, et al. Distribution of viruses from the Californian encephalitis serogroup (Bunyaviridae, Bunyavirus) in the northern expanses of Russia. *Vopr Virusol* 1997;(5):229−35 [in Russian].

31. Lvov DK, Aristova VA, Butenko AM, et al. *Viruses of Californian serogroup and etiologically linked diseases: clinic-epidemiological characteristics, geographical distribution, methods of virological and serological diagnostics. Methodological manual*. Moscow: RAMS;2003 [in Russian].

32. Lvov DK, Deryabin PG, Aristova VA, et al. *Atlas of distribution of natural foci virus infections on the territory of Russian Federation*. Moscow: SMC MPH RF Publ;2001 [in Russian].

33. Lvov DK, Gromashevsky VL, Sidorova GA, et al. Isolation of Tahyna virus from *Anopheles hyrcanus* mosquitoes in Kyzylagach preserve, South-Eastern Azerbaijan. *Vopr Virusol* 1972;**17**(1):18−21 [in Russian].

34. Black SS, Harrison LR, Pursell AR, et al. Necrotizing panencephalitis in puppies infected with La Crosse virus. *J Vet Diagn Invest* 1994;**6**(2):250−4.

35. Rodl P, Bardos V, Ryba J. Experimental transmission of Tahyna virus (California group) to wild rabbits (*Oryctolagus cuniculus*) by mosquitoes. *Folia Parasitol (Praha)* 1979;**26**(1):61−4.

36. Lvov SD. Californian encephalitis. In: Lvov DK, Klimenko SM, editors. *Gaidamovich SYa. Arboviruses and arbovirus infections*. Moscow: Meditsina;1989p. 212−14 [in Russian].

37. Grimstad PR. California group viral infection. In: Beran GW, Steel JH, editors. *Handbook of zoonoses. Section B: Viral*. Boca Raton, FL: CRC Press;1994. p. 71−9.

38. Haddow AD, Odoi A. The incidence risk, clustering and clinical presentation of La Crosse virus infections in the eastern United States, 2005−2007. *PLoS One* 2009;**4**:e6145.

II. ZOONOTIC VIRUSES OF NORTHERN EURASIA: TAXONOMY AND ECOLOGY

39. Lambert AJ, Blair CD, D'Anton M, et al. La Crosse virus in *Aedes albopictus* mosquitoes, Texas, USA, 2009. *Emerg Infect Dis* 2010;**16**(5):856–8.

40. Chun RWM. Clinical aspects of La Crosse encephalitis: neurological and psychological sequelae. In: Calisher CH, Thompson WH, editors. *California serogroup viruses.* New York: Alan R. Liss; 1983. p. 193–212.

41. Gligic A, Adamovic ZR. Isolation of Tahyna virus from Aedes vexans mosquitoes in Serbia. *Mikrobiologiya* 1976;**12**:119–29 [in Russian].

42. Tatum LM, Pacy JM, Frazier KS, et al. Canine La Crosse viral meningoencephalomyelitis with possible public health implications. *J Vet Diagn Invest* 1999;**11**(2):184–8.

43. Kolobukhina LV, Lvov SD. Californian encephalitis. In: Lvov DK, editor. *Handbook of virology: Viruses and viral infection of human and animals.* Moscow: MIA;2013p. 741–9 [in Russian].

44. Lvov DK, Klimenko SM, Gaidamovich Sy. *Arboviruses and arbovirus infections.* Moscow: Meditsina;1989 [in Russian].

45. Malkova D, Danielova V, Kolman JM, et al. Natural focus of Tahyna virus in South Moravia. *J Hyg Epidemiol Microbiol Immunol* 1965;**9**(4):434–40.

46. Brummer-Korvenkontio M. Arboviruses of the California complex and Bunyamwera group in Finland. In: Bardos V, editor. *Arboviruses of the California complex and the Bunyamwera group.* Bratislava: SAS;1969. p. 131–3.

47. Lvov DK, Kolobukhina LV, Gromashevsky VL, et al. Isolation of California antigenic group viruses from patients with acute neuroinfection syndrome. *Arbovirus Inf Exed* 1996;**6**:16–18.

48. Lvov DK, Kostiukov MA, Pak TP, et al. Isolation of Tahyna virus (California antigenic group, family Bunyaviridae) from the blood of febrile patients in the Tadzhik SSR. *Vopr Virusol* 1977;**6**:682–5 [in Russian].

49. Lvov SD. *Natural virus foci in high latitudes of Eurasia. Soviet medical reviews. Section E. Virology reviews,* 5. London: Harwood Academic Publishers GmbH;1993. p. 137–85.

50. Lvov SD, Gromashevsky VL, Skvortsova TM, et al. Circulation of California encephalitis complex viruses in the northwestern Russian plain. *Med Parazitol (Mosk)* 1989;**2**:74–7 [in Russian].

51. Bardos V, Medek M, Kania V, et al. Tasclische Bild der Tahina Virus (California Gruppe) infectionenbei. *Kindern Pediat Grenigeb* 1980;**19**:11–23 [in German].

52. Bardos V, Senxa F. Acute human infection caused by Tahyna virus. *Cas Lek Ces* 1963;**52**:394–402 [in Slovak].

53. Hubalek Z, Bardos M, Medek M, et al. Tahyna virus-neutralization antibodies in patients in Southern Moravia. *Ces Epidemiol Microbiol Immunol* 1979;**28**: 87–96 [in Czech].

54. Janbon M, Bertran A, Hannoun C, et al. Meningoencephalit virus Tahyna. *J Med Mon* 1974;**9**:7–10.

55. Mittermayer T, Bilcikova M, Jass J, et al. Clinical manifestation of Tahyna virus infection in patients in Eastern Slovakia. *Bratisl Lek Lis* 1964;**44**:636–9 [in Slovak].

56. Traavik T, Mehl R, Wiger R. Mosquito-borne arboviruses in Norway: further isolations and detection of antibodies to California encephalitis viruses in human, sheep and wildlife sera. *J Hyg (London)* 1985;**94**(1): 111–22.

57. Gu HX, Artsob H. The possible presence of Tahyna (Bunyaviridae, California serogroup) virus in the People's Republic of China. *Trans R Soc Trop Med Hyg* 1987;**81**:693.

58. Gu HX, Artsob H, Lin YZ, et al. Arboviruses as aethiological agents of encephalitis in the Peoples Republic of China. *Trans R Soc Trop Med Hyg* 1992;**86**:198–201.

59. Lu Z, Lu X-J, Zang S, et al. Tahyna virus and human infection, China. *Emerg Infect Dis* 2009;**15**(2):306–9.

60. Heath SE, Artsob H, Bell RJ, et al. Equine encephalitis caused by snowshoe hare (California serogroup) virus. *Can Vet J* 1989;**30**(8):669–71.

61. Henderson BE, Coleman PH. The growing importance of California arboviruses in the etiology of human disease. *Prog Med Virol* 1971;**13**:404–60.

62. Calisher CH. Medically important arboviruses of the United States and Canada. *Clin Microbiol Rev* 1994;**7**(1):89–116.

63. Kolobukhina LV, Lvov DK, Butenko AM, et al. Significance of viruses of antigen complex of California encephalitis in pathology. *Klinic Med* 1989;**67**(9):61–4 [in Russian].

64. Kolobukhina LV, Lvov DK, Butenko AM, et al. The clinico-laboratory characteristics of cases of diseases connected with viruses of the California encephalitis complex in the inhabitants of Moscow. *Zhurnal Mikrobiol, Epidemioli Immunobiol* 1989;**10**:68–73 [in Russian].

65. Kolobukhina LV, Lvov DK, Butenko AM, et al. Signs and symptoms of infections caused by California serogroup viruses in humans in the USSR. *Arch Virol* 1990;(Suppl. 1):243–7.

66. Kolobukhina LV, Lvov DK, Skvortsova TM, et al. Diseases associated with viruses of the California encephalitis serogroup, in Russia. *Vopr Virusol* 1998;**43**(1):14–17 [in Russian].

8.1.4.3 Khurdun Virus

1. Galkina IV, Lvov DN, Gromashevsky VL, et al. Khurdun virus, a presumably new RNA-containing virus associated with coots (*Fulica atra*), isolated in the Volga river delta. *Vopr Virusol* 2005;**50**(4):29–31 [in Russian].

2. Alkhovskiĭ SV, Lvov DN, Samokhvalov EI, et al. Screening of birds in the Volga delta (Astrakhan region, 2001) for the West Nile virus by reverse transcription-polymerase chain reaction. *Vopr Virusol* 2003;**48** (1):14−17 [in Russian].

3. Lvov DK, Butenko AM, Gromashevsky VL, et al. West Nile virus and other zoonotic viruses in Russia: examples of emerging-reemerging situations. *Arch Virol Suppl* 2004;**18**:85−96.

4. Plyusnin A, Beaty BJ, Elliot RM, et al. Family Bunyaviridae. In: King AM, Adams MJ, Carstens EB, Lefkowitz EJ, editors. *Virus taxonomy: classification and nomenclature of viruses : Ninth Report of the International Committee on Taxonomy of Viruses.* London: Elsevier;2012. p. 725−41.

5. Alkhovsky SV, Shchetinin AM, Lvov DK, et al. The Khurdun virus (KHURV): a new representative of the Orthobunyavirus (Bunyaviridae). *Vopr Virusol* 2013;**58** (4):10−13 [in Russian].

6. Muller R, Poch O, Delarue M, et al. Rift Valley fever virus L segment: correction of the sequence and possible functional role of newly identified regions conserved in RNA-dependent polymerases. *J Gen Virol* 1994;**75** (6):1345−52.

7. Strandin T, Hepojoki J, Vaheri A. Cytoplasmic tails of bunyavirus Gn glycoproteins-Could they act as matrix protein surrogates?. *Virology* 2013;**437**(2):73−80.

8. Hepojoki J, Strandin T, Wang H, et al. Cytoplasmic tails of hantavirus glycoproteins interact with the nucleocapsid protein. *J Gen Virol* 2010;**91**(9): 2341−50.

9. Mohamed M, McLees A, Elliott RM. Viruses in the Anopheles A, Anopheles B, and Tete serogroups in the *Orthobunyavirus* genus (family Bunyaviridae) do not encode an NSs protein. *J Virol* 2009;**83**(15): 7612−18.

8.1.5 Genus *Phlebovirus*

1. Schmaljohn CS, Nichol ST. Bunyaviridae. In: Knipe DM, Howley PM, editors. *Fields virology.* 5th ed. Philadelphia, PA: Lippincott Williams and Wilkins;2007. p. 1741−89.

2. Plyusnin A, Beaty BJ, Elliot RM, et al. Family Bunyaviridae. In: King AM, Adams MJ, Carstens EB, Lefkowitz EJ, editors. *Virus taxonomy: classification and nomenclature of viruses : Ninth Report of the International Committee on Taxonomy of Viruses.* London: Elsevier;2012. p. 725−41.

3. Ronnholm R, Pettersson RF. Complete nucleotide sequence of the M RNA segment of Uukuniemi virus encoding the membrane glycoproteins G1 and G2. *Virology* 1987;**160**(1):191−202.

8.1.5.1 Bhanja Virus and Razdan Virus (var. Bhanja virus)

1. Shah KV, Work TH. Bhanja virus: a new arbovirus from ticks *Haemaphysalis intermedia* Warburton and Nuttall, 1909, in Orissa, India. *Indian J Med Res* 1969;**57** (5):793−8.

2. Pavlov P, Rosicky B, Hubalek Z, et al. Isolation of Bhanja virus from ticks of the genus *Haemaphysalis* in southeast Bulgaria and presence of antibodies in pastured sheep. *Folia Parasitol (Praha)* 1978;**25**(1):67−73.

3. Verani P, Balducci M, Lopes MC, et al. Isolation of Bhanja virus from Haemaphysalis ticks in Italy. *Am J Trop Med Hyg* 1970;**19**(1):103−5.

4. Vesenjak-Hirjan J, Calisher CH, Brudnjak Z, et al. Isolation of Bhanja virus from ticks in Yugoslavia. *Am J Trop Med Hyg* 1977;**26**(1):1003−8.

5. Filipe AR, Alves MJ, Karabatsos N, et al. Palma virus, a new Bunyaviridae isolated from ticks in Portugal. *Intervirology* 1994;**37**(6):348−51.

6. Boiro I, Lomonossov NN, Malenko GV, et al. Forecariah virus, a new representative of the Bhanja antigenic group, isolated in the Republic of Guinea. *Bull Soc Pathol Exot Filiales* 1986;**79**(2):183−6.

7. Butenko AM, Gromashevskii VL, Lvov DK, et al. Kisemayo virus, a representative of the Bhanja antigenic group. *Vopr Virusol* 1979;**6**:661−5 [in Russian].

8. Hubalek Z, Halouzka J. Numerical comparative serology of the Bhanja antigenic group (Bunyaviridae). *Arch Virol* 1985;**84**:175−80.

9. Marchette NJ. Arboviral zoonoses of Asia. In: Beran GW, Steel JH, editors. *Handbook of zoonoses. Section B: Viral.* Boca Raton, FL: CRC Press;1994. p. 275−88.

10. Lvov DK, Gromashevsky VL, Zakaryan VA, et al. Razdan virus, a new ungrouped bunyavirus isolated from *Dermacentor marginatus* ticks in Armenia. *Acta Virol* 1978;**22**(6):506−8.

11. Razdan virus. In: Karabatsos N, editor. *International Catalogue of Arboviruses Including Certain other Viruses of Vertebrates.* San Antonio, TX: American Society of Tropical Medicine and Hygiene;1985. p. 851−2.

12. Hubalek Z. Geographic distribution of Bhanja virus. *Folia Parasitol (Praha)* 1987;**34**(1):77−86.

13. Hubalek Z. Experimental pathogenicity of Bhanja virus. *Zentralbl Bakt* 1987;**266**(1−2):284−91.

14. Alkhovsky SV, Lvov DK, Shchelkanov MYu, et al. Molecular-genetic characterization of the Bhanja virus (BHAV) and the Razdan virus (RAZV) (Bunyaviridae, Phlebovirus) isolated from the ixodes ticks *Rhipicephalus bursa* (Canestrini and Fanzago, 1878) and *Dermacentor marginatus* (Sulzer, 1776) in Transcaucasus. *Vopr Virusol* 2013;**59**(4):14−19 [in Russian].

15. Dilcher M, Alves MJ, Finkeisen D, et al. Genetic characterization of Bhanja virus and Palma virus, two tick-borne phleboviruses. *Virus Genes* 2012;**45**(2):311−15.

16. Matsuno K, Weisend C, Travassos da Rosa AP, et al. Characterization of the Bhanja serogroup viruses (Bunyaviridae): a novel species of the genus *Phlebovirus* and its relationship with other emerging tick-borne phleboviruses. *J Virol* 2013;**87**(7):3719—28.

17. Bao CJ, Qi X, Wang H. A novel bunyavirus in China. *N Engl J Med* 2011;**365**(9):862—3.

18. Yu XJ, Liang MF, Zhang SY, et al. Fever with thrombocytopenia associated with a novel bunyavirus in China. *N Engl J Med* 2011;**364**(16):1523—32.

19. Palacios G, Savji N, Travassos da Rosa A, et al. Characterization of the Uukuniemi virus group (Phlebovirus: Bunyaviridae): evidence for seven distinct species. *J Virol* 2013;**87**(6):3187—95.

20. Muller R, Poch O, Delarue M, et al. Rift Valley fever virus L segment: correction of the sequence and possible functional role of newly identified regions conserved in RNA-dependent polymerases. *J Gen Virol* 1994;**75**(6):1345—52.

21. Marklewitz M, Handrick S, Grasse W, et al. Gouleako virus isolated from West African mosquitoes constitutes a proposed novel genus in the family Bunyaviridae. *J Virol* 2011;**85**(17):9227—34.

22. McMullan LK, Folk SM, Kelly AJ, et al. A new phlebovirus associated with severe febrile illness in Missouri. *N Engl J Med* 2012;**367**(9):834—41.

23. Hubalek Z. Biogeography of tick-borne Bhanja virus (Bunyaviridae) in Europe. *Int Persp Infect Dis* 2009;**11**:372691.

24. Sikutova S, Hornok S, Hubalek Z, et al. Serological survey of domestic animals for tick-borne encephalitis and Bhanja viruses in northeastern Hungary. *Vet Microbiol* 2009;**135**(3—4):267—71.

25. Camicas JL, Denbel V, Heme G, et al. Ecological and nosological study of tick-borne arboviruses in Senegal. II. Experimental study of the pathogenicity of the Bhanja virus in small domestic ruminants. *Rev Elev Med Trop* 1981;**34**:257—61.

26. Madr V, Hubalek Z, Zendulkova D. Experimental infection of sheep with Bhanja virus. *Folia Parasitol (Praha)* 1984;**31**(1):79—84.

27. Semashko IV, Chuikov MP, Tsykin LB, et al. Experimental pathogenecity of Bhanja virus for lambs of different infection routes. In: *Ecology of viruses*. Baku; 1976. p. 184—6 [in Russian].

28. Balducci M, Verani P, Lopes MC, et al. Experimental pathogenicity of Bhanja virus for white mice and Macaca mulatta monkeys. *Acta Virol* 1970;**14**(3):237—43.

29. Calisher CH, Goodpasture HC. Human infection with Bhanja virus. *Am J Trop Med Hyg* 1975;**24**(1):1040—2.

30. Punda V, Ropac D, Vesenjak-Hirjan J. Incidence of hemagglutination-inhibiting antibodies for Bhanja virus in humans along the north-west border of Yugoslavia. *Zentralbl Bakt* 1987;**265**(1—2):227—34 [in German].

31. Vesenjak-Hirjan J, Calisher CH, Bens I, et al. First natural clinical human Bhanja virus infection. *Zentralbl Bakt* 1980;(Suppl.):297—301 [in German].

8.1.5.2 Gissar Virus

1. Gordeeva ZE, Kostyukov MA, Kuyma AU, et al. Gissar virus—a new virus of family Bunyaviridae, isolated from argases ticks *Argas vulgaris* Fil. in Tajikistan. *Med Parasitol Parasit Bolez* 1990;**6**:34—5 [in Russian].

2. Lvov DK, Skvortsova TM, Kostyukov MA, et al. *The strain 5595 of Gissar virus. Deposition certificate of Russian State Collection of viruses*. 1982. N 796.

3. Lvov DK. Arboviral zoonoses of Northern Eurasia (Eastern Europe and the Commonwealth of Independent States). Boca Raton, FL In: Beran GW, editor. *Handbook of zoonoses. Section B: Viral*. London/Tokyo/: CRC Press;1994. p. 237—60.

4. Lvov DK, Alkhovsky SV, Shchelkanov My, et al. Molecular genetic characterization of Gissar virus (GSRV) (Bunyaviridae, Phlebovirus, Uukuniemi group), isolated from ticks *Argas reflexus* Fabricius, 1794 (Argasidae), collected in dovecote in Tajikistan. *Vopr Virusol* 2014;**59**(4):20—4 [in Russian].

5. Hannoun C, Corniou B, Rageau J. Isolation in Southern France and characterization of new tick-borne viruses related to Uukuniemi: grand Arbaud and Ponteves. *Acta Virol* 1970;**14**(2):167—70.

6. Palacios G, Savji N, Travassos da Rosa A, et al. Phlebovirus: Bunyaviridae): evidence for seven distinct species. Characterization of the Uukuniemi virus group *J Virol* 2013;**87**(6):3187—95.

7. Kostyukov MA, Gordeeva ZE, Nemova NV, et al. About ecology of Gissar virus. In: Lvov DK, editor. *Itogi nauki i techniki. Virology. Arboviruses and arboviral infections*. Moscow: VINITI;1991p. 18—19 [in Russian].

8. Filippova NA. *Fauna of USSR*, Moscow/Leningrad: USSR Academy of Sciences;1966. vol. 4(3). Arachnoidea. Argas ticks (Argasidae) [in Russian].

8.1.5.3 Khasan Virus

1. Lvov DK, Leonova GN, Gromashevsky VL, et al. Khasan virus, a new ungrouped bunyavirus isolated from *Haemaphysalis longicornis* ticks in the Primorie region. *Acta Virol* 1978;**22**(3):249—52.

2. Saikku P, von Bonsdorff CH. Electron microscopy of the Uukuniemi virus, an ungrouped arbovirus. *Virology* 1968;**34**(4):804—6.

3. Khasan virus. In: Karabatsos N, editor. *International catalogue of arboviruses including certain other viruses of*

vertebrates. San Antonio, TX: American Society of Tropical Medicine and Hygiene;1985. p. 567–8.

4. Alkhovsky SV, Lvov DK, Shchelkanov MYu, et al. The taxonomy of the Khasan virus (KHAV), a new representative of Phlebovirus genera (Bunyaviridae), isolated from the ticks *Haemaphysalis longicornis* (Neumann, 1901) in the Maritime territory (Russia). *Vopr Virusol* 2013;**58**(5):15–18 [in Russian].

5. Palacios G, Savji N, Travassos da Rosa A, et al. Characterization of the Uukuniemi virus group (Phlebovirus: Bunyaviridae): evidence for seven distinct species. *J Virol* 2013;**87**(6):3187–95.

6. Yu XJ, Liang MF, Zhang SY, et al. Fever with thrombocytopenia associated with a novel bunyavirus in China. *N Engl J Med* 2011;**364**(16):1523–32.

7. McMullan LK, Folk SM, Kelly AJ, et al. A new phlebovirus associated with severe febrile illness in Missouri. *N Engl J Med* 2012;**367**(9):834–41.

8. Hoogstraal HF, Roberts HS, Kohls GM, et al. Review of *Haemaphysalis (Kaiseriana) longicornis* Neumann (resurrected) of Australia, New zealand, New Caledonia, Fiji, Japan, Korea, and northeastern China and USSR, and its parthenogenetic and bisexual populations (Ixodoidea, Ixodidae). *J Parasit* 1968;**54**(6):1197–213.

9. Zhang YZ, Zhou DJ, Qin XC, et al. The ecology, genetic diversity, and phylogeny of Huaiyangshan virus in China. *J Virol* 2012;**86**(5):2864–8.

8.1.5.4 Sandfly Fever Naples Virus and Sandfly Fever Sicilian Virus

1. Plyshin A, Beaty BJ, Elliot RM, et al. Bunyaviridae. In: King AM, Adams MJ, Carstens EB, et al., editors. *Virus taxonomy. Ninth Report of International Committee of Taxonomy of viruses*. London: Elsevier Academic Press;2012. p. 725–41.

2. Theiler M, Downs WG. *The arthropod-borne viruses of Vertebrate*. New Haven/London: Yale University Press;1973.

3. Tesh RB, Peters CJ, Meegan JM. Studies on the antigenic relationship among phleboviruses. *Am J Trop Med Hyg* 1982;**31**(1):149–55.

4. Doerr R, Russ VK. Weitere unterschunget uber das pappataci-frieber. *Arch F Schiffs Tropen-Hyg* 1909;**13**: 693–706 [in German].

5. Sabin A. Phlebotomus fever. In: Rivers TM, editor. *Viral on Rickettsae infections of man*. Philadelphia, PA: Lippincott;1948 [Chapter 31].

6. Sabin AB. Experimental studies on Phlebotomus (pappataci, sandfly) fever during World War II. *Arch Gesamte Virusforsch* 1951;**4**(4):367–410.

7. Peralta PH, Shelokov A. Isolation and characterization of arboviruses from Almirante, Republic of Panama. *Am J Trop Med Hyg* 1966;**15**(3):369–78.

8. Nicoletti L, Verani P, Caciolli S, et al. Central nervous system involvement during infection by Phlebovirus toscana of residents in natural foci in central Italy (1977–1988). *Am J Trop Med Hyg* 1991;**45** (4):429–34.

9. Karimabad virus. In: Karabatsos N, editor. *International catalogue of arboviruses including certain other viruses of vertebrates*. San Antonio, TX: American Society of Tropical Medicine and Hygiene;1985. p. 519–50.

10. Salehabad virus. In: Karabatsos N, editor. *International catalogue of arboviruses including certain other viruses of vertebrates*. San Antonio, TX: American Society of Tropical Medicine and Hygiene;1985. p. 887–8.

11. Papa A, Konstantinou G, Pavlidou V, et al. Sandfly fever virus outbreak in Cyprus. *Clin Microbiol Infect* 2006;**12**(2):192–4.

12. Anagnostou V, Pardalos G, Athanasion-Metaxa M, et al. Novel Phlebovirus in febrile child, Greece. *Emerg Infect Dis* 2011;**17**(3):940–1.

13. Charrel RN, Izri A, Temmam S, et al. Cocirculation of 2 genotypes of Toscana virus, southeastern France. *Emerg Infect Dis* 2007;**13**(3):465–8.

14. Gaidamovich SY, Khutoretskaya NV, Asyamov YV, et al. Sandfly fever in Central Asia and Afganistan. *Arch Virol* 1991;**1**:287.

15. Lvov DK. Arboviral zoonoses of Northern Eurasia (Eastern Europe and the common wealth of Independent States). In: Beran GW, editor. *Handbook of zoonoses. Section B: Virol*. Boca Raton, FL: CRC Press;1994. p. 237–60.

16. Palacios G, Tesh RB, Savji N, et al. Characterization of the Sandfly fever Naples species complex and description of a new Karimabad species complex (genus *Phlebovirus*, family Bunyaviridae). *J Virol* 2014;**95**:292–300.

17. Bichad L, Dachraoi K, Piorkowski G, et al. Toscana virus isolated from sandflies, Tunisia. *Emerg Infect Dis* 2013;**19**(2):322–4.

18. Izri A, Temmam S, Moureau G, et al. Sandfly fever Sicilian virus, Algeria. *Emerg Infect Dis* 2008;**14** (5):795–7.

19. Shope R. Arboviral zoonoses of Western Europe. In: Beran GW, editor. *Handbook of zoonoses. Section B: Virol*. 2nd ed. Boca Raton, FL: CRC Press;1994. p. 227–35.

20. Tesh RB. Phlebotomus fever. In: Monath TP, editor. *The Arboviruses: epidemiology and ecology*, vol. IV. Boca Raton, FL: CRC Press;1988 [Chapter 37].

21. Tesh RB, Saidi S, Gajdamovic SJ, et al. Serological studies on the epidemiology of sandfly fever in the Old World. *Bull World Health Organ* 1976;**54**(6):663–74.

22. Tesh R, Saidi S, Javadian E, et al. Studies on the epidemiology of sandfly fever in Iran. I. Virus isolates

obtained from Phlebotomus. *Am J Trop Med Hyg* 1977;**26**(2):282−7.

23. Charrel RN, Gallian P, Navarro-Mari JM, et al. Emergence of Toscana virus in Europe. *Emerg Infect Dis* 2005;**11**(11):1657−63.

24. Eitrem R, Stylianou M, Niklasson B. High prevalence rates of antibody to three sandfly fever viruses (Sicilian, Naples, and Toscana) among Cypriots. *Epidemiol Infect* 1991;**107**(3):685−91.

25. Gabriel M, Resch C, Günther S, et al. Toscana virus infection imported from Elba into Switzerland. *Emerg Infect Dis* 2010;**16**(6):1034−6.

26. Hemmersbach-Miller M, Parola P, Charrel RN, et al. Sandfly fever due to Toscana virus: an emerging infection in southern France. *Eur J Intern Med* 2004;**15**(5):316−17.

27. Kay MK, Gibney KB, Riedo FX, et al. Toscana virus infection in American traveler returning from Sicily, 2009. *Emerg Infect Dis* 2010;**16**(9):1498−500.

28. Papa A, Andriotis V, Tzilianos M. Prevalence of Toscana virus antibodies in residents of two Ionian islands, Greece. *Travel Med Infect Dis* 2010;**8**(5):302−4.

29. Peyrefitte CN, Devetakov I, Pastorino B, et al. Toscana virus and acute meningitis, France. *Emerg Infect Dis* 2005;**11**(5):778−80.

30. Saidi S, Tesh R, Javadian E, et al. Studies on the epidemiology of sandfly fever in Iran. II. The prevalence of human and animal infection with five phlebotomus fever virus serotypes in Isfahan province. *Am J Trop Med Hyg* 1977;**26**(2):288−93.

31. Sanbonmatsu-Gamez S, Perez-Ruiz M, Collao X, et al. Toscana virus in Spain. *Emerg Infect Dis* 2005;**11**(11):1701−7.

32. Sonderegger B, Hachler H, Dobler G, et al. Imported aseptic meningitis due to Toscana virus acquired on the island of Elba, Italy, August 2008. *Euro Surveill* 2009;**14**(1) pii: 19079.

33. Goverdhan MK, Dhanda V, Modi GB, et al. Isolation of phlebotomus (sandfly) fever virus from sandflies and humans during the same season in Aurangabad District, Maharashtra State, India. *Indian J Med Res* 1976;**64**(1):57−63.

34. Venturi G, Ciccozzi M, Montieri S, et al. Genetic variability of the M genome segment of clinical and environmental Toscana virus strains. *J Gen Virol* 2007;**88**(4):1288−94.

35. Whittinghan HE. The etiology of phlebotomus fever. *J State Med* 1924;**32**:461.

36. Pick A. Zur Pathologie und Therapie einer Eigenthumlichen endemischen Krankheits-form. *Wein Med Wehnschr* 1896;**36**:1141 [in German].

37. Bartelloni PJ, Tesh RB. Clinical and serologic responses of volunteers infected with phlebotomus fever virus (Sicilian type). *Am J Trop Med Hyg* 1976;**25**(3):456−62.

8.1.5.5 Uukuniemi virus and Zaliv Terpeniya virus

1. Oker-Blom N, Salminen A, Brummer-Korvenkontio M, et al. Isolation of some viruses other than typical tick-borne encephalitis viruses from *Ixodes ricinus* ticks in Finland. *Ann Med Exp Biol Fenn* 1964;**42**:109−12.

2. Uukuniemi virus. In: Karabatsos N, editor. *International catalogue of arboviruses and some others viruses of vertebrates*. San Antonio, TX: American Society of Tropical Medicine and Hygiene;1985. p. 1063−4.

3. Gaidamovich SY, Nikiforov LP, Gromashevsky VL, et al. Isolation and study of Sumakh virus, a member of the Uukuniemi group, in the U.S.S.R. *Acta Virol* 1971;**15**(2):155−60.

4. Gromashevsky VL, Chervonsky VI, Nikiforov LP. *Features of Uukuniemi group Sumah virus, isolated from trush* Turdus merula *in southern-eastern part of Azerbaijan. Problems of medical virology*, XI. Moscow: USSR Academy of Medical Sciences;197198−9. [in Russian].

5. Nikiforov LP, Gromashevsky VL, Veselovskaya OV. About natural foci of Uukuniemi virus in southern of Azerbaijan. In: Lvov DK, editor. *Ecology of Viruses.*, 1. Moscow: USSR Academy of Medical Sciences;1973. p. 122−5 [in Russian].

6. Lvov DK, Gromashevski VL, Skvortsova TM, et al. Arboviruses of high latitudes in the USSR. In: Kurstak E, editor. *Arctic and tropical arboviruses*. NY/San-Francisco/London: Harcourt Brace Jovanovich Publ;1979p. 21−38 [Chapter 3].

7. Lvov SD. *Natural virus foci in high latitudes of Eurasia. Soviet medical reviews. Section E: Virology reviews.*, 3. USA: Harwood Academic Publishers GmbH;1993. p. 137−85.

8. Isotov VK, Zhuravleva MG, Kuzminskaya OP, et al. Isolation of three Uukuniemi virus strains from ticks in Volgograd district. In: Chumakov MP, editor. *Proceedings of the Institute of Poliomyelitis and Viral Encephalitis. Actual problems of virology and natural-foci viral infection*. Moscow: USSR Academy of Medical Sciences;1972p. 283−4 [in Russian].

9. Lvov DK. Natural foci of arboviruses, associated with the birds in USSR. In: Lvov DK, Il'ichev VD, editors. *Migration of the birds and transduction of contagium*. Moscow. Nauka;1979p. 37−101 [Chapter 2]. [in Russian].

10. Samoilova TI, Votyakov VI, Mishaev NP, et al. Detection of Uukuniemi virus in Belorussia. *Vopr Virusol* 1973;**1**:111−12 [in Russian].

11. Vinograd IA, Gaidamovich SY, Marushchak OG. Isolation of Uukuniemi group viruses from arthropods and birds in Chernovitskay district. In: *Problems of medical virology*. Moscow, vol. 2; 1991. p. 116−21 [in Russian].

12. Bochkov MV, Sarmanova ES, Michailova IS, et al. Studies of Uukuniemi virus strains, isolated in

Lithuanian SSR and Estonian SSR. In: Chumakov MP, editor. *Proceedings of the Institute of Poliomyelitis and Viral Encephalitis. Actual problems of virology and natural-foci viral infection.* Moscow: USSR Academy of Medical Science;1972p. 300 [in Russian].

13. Lvov DK. *Ecological sounding of the USSR territory for natural foci of arboviruses. Soviet medical reviews. Section E: Virology reviews, 5.* USA: Harwood Academic Publishers GmbH;1993. p. 1–47.

14. Moteyunas LI, Karaseova PS, Vargin VV. Immunological struture of the population of Lithuanian SSR for Uukuniemi virus. In: Chumakov MP, editor. *Proceedings of the Institute of Poliomyelitis and Viral Encephalitis. Actual problems of virology and natural-foci viral infection.* Moscow: Academy of Medical Science USSR;1972. p. 302–3 [in Russian].

15. Samoilova TI, Voinov IN. Experimental study of relationships between Uukuniemi virus and *Ixodes ricinus* ticks. In: Lvov DK, editor. *Ecology of viruses.* Moscow: Nauka;1980. p. 76–80 [in Russian].

16. Voinov IN. Investigation of arbovirus infections in Belarus and other western areas of the USSR. In: Gaidamovich SY, editor. *Arboviruses.* Moscow; 1978. p. 20–4 [in Russian].

17. Kolman JM, Malkova D, Smetana A. Isolation of a presumably new virus from unengorged *Ixodes ricinus* ticks. *Acta Virol* 1966;**10**(2):171–2.

18. Kozuch O, Greshicova M, Nosek J, et al. Uukuniemi virus in western Slovakia and northern Moravia. *Folia Parasitol* 1970;**17**:337–40.

19. Malkova D, Danielova V, Holubova J, et al. Less known arboviruses of Central Europe. *Rozpravy CSAV (Praha) Mat Prir Vedy* 1986;**96**(5):1–75 [in Czech].

20. Wroblewska-Mularczykowa Z, Sadowski W, Zukowski K. Isolation of arbovirus strains of Uuruniemi type in Poland. *Folia Parasitol* 1970;**17**:375–8.

21. Lvov DK, Timopheeva AA, Gromashevski VL, et al. "Zaliv Terpeniya" virus, a new Uukuniemi group arbovirus isolated from *Ixodes (Ceratixodes) putus* Pick.-Camb. 1878 on Tyuleniy Island (Sakhalin region) and Commodore Islands (Kamchatsk region). *Arch Gesamte Virusforsch* 1973;**41**(3):165–9 [in German].

22. Zaliv Terpeniya virus. In: Karabatsos N, editor. *International Catalogue of arboviruses and some others viruses of vertebrates.* San Antonio, TX: American Society of Tropical Medicine and Hygiene;1985. p. 1119–20.

23. Lvov DK, Smirnov VA, Gromashevsky VL, et al. Isolation of Terpeniye Gulf arbovirus from *Ixodes (Ceratixodes) putus* Pick-Cambr., 1878 ticks in Murmansk Province. *Med Parazitol* 2000;**1973**(6):728–30 [in Russian].

24. Lvov DK, Timopheeva AA, Smirnov VA, et al. Ecology of tick-borne viruses in colonies of birds in the USSR. *Med Biol* 1975;**53**(5):325–30.

25. Traavik T. *Tick- and mosquito-associated viruses in Norway.* Oslo: Nat Inst Publ Hlth;1979. p. 1–55.

26. Palacios G, Savji N, da Rosa AT, et al. Characterization of the Uukuniemi virus group (Phlebovirus: Bunyaviridae): evidence for seven distinct species. *J Virol* 2013;**87**(6):3187–95.

27. Filippova NA. About the species of *Ixodes persulcatus* (Parasitiformes, Ixodidae) group. *Parazitologiya* 1973;**7**:3–13 [in Russian].

28. Kozuch O, Rajcani J, Sekeyova M, et al. Uukuniemi virus in small rodents. *Acta Virol* 1970;**14**(2):163–6.

29. Vinograd IA, Vigovskiy AI, Chumachenko SS, et al. New isolates of arboviruses in Ukraine and their biological characteristics. In: Gaidamovich SY, editor. *Arboviruses.* Moscow; 1981. p. 45–9 [in Russian].

30. Vinograd IA, Vigovskiy AI, Obukhova VR, et al. Isolation of Uukuniemi arbovirus from *Coccothraustes coccothraustes* and *Streptopelia turtur* in the Carpathian area and western Ukrainian Polesye. In: Gaidamovich SY, editor. *Arboviruses.* Moscow; 1975. p. 84–7 [in Russian].

31. Saikku P. Passerine birds in the ecology of Uukuniemi virus. *Med Biol* 1974;**52**(2):98–103.

32. Saikku P, Brummer-Korvenkontio M. Arboviruses in Finland. II. Isolation and characterization of Uukuniemi virus, a virus associated with ticks and birds. *Am J Trop Med Hyg* 1973;**22**(3):390–9.

33. Malkova D, Holubova J, Kolman JM, et al. Antibodies against some arboviruses in personswith various neuropathies. *Acta Virol* 1980;**24**:298.

34. Vasilenko VA, Yiks SP, Tamm OM. Investigation of the role of Uuruniemi virus human pathology. In: Gaidamovich SY, editor. *Arboviruses.* Moscow; 1975. p. 110–4 [in Russian].

35. Kolman JM, Kopecky K, Wokounova D. Serological examination of the population from the endemic area of the tick-borne encephalitis virus (TBEV) and Uukuniemi virus (UKV). *Cesk Epidemiol Mikrobiol Imunol* 1973;**22**(3):153–9 [in Czech].

8.1.5.6 Komandory Virus and Rukutama Virus (var. Komandory virus)

1. Lvov DK. Arboviral zoonoses of Northern Eurasia (Eastern Europe and the commonwealth of independent states). In: Beran GW, editor. *Handbook of zoonoses. Section B: Viral.* London/Tokyo: CRC Press Boca Raton;1994. p. 237–60.

2. Alkhovsky SV, Lvov DK, Shchelkanov My, et al. Genetic characteristic of a new Komandory virus (KOMV; Bunyaviridae, Phlebovirus) isolated from ticks *Ixodes uriae*, collected in guillemot (*Uria aalge*) nesting sites on Komandorski Islands, Bering Sea. *Vopr Virusol* 2013;**6**:18–22 [in Russian].

3. Lvov DK, Alkhovsky SV, Shchelkanov MYu, et al. Genetic characterization of Sakhalin virus (SAKV), Paramushir virus (PMRV) (Sakhalin group, Nairoviridae, Bunyaviridae) and Rukutama virus (RUKV) (Uukuniemi group, Phlebovirus, Bunyaviridae), isolated from the obligate parasites of colonial sea-birds—ticks *Ixodes (Ceratixodes) uriae*, White 1852 and *I. signatus* Birulya, 1895 in water area of sea of Okhotsk and Bering sea. *Vopr Virusol* 2014;**59**(3):11−17 [in Russian].

4. Darwish MA, Hoogstraal H, Roberts TJ, et al. Seroepidemiological survey for Bunyaviridae and certain other arboviruses in Pakistan. *Trans R Soc Trop Med Hyg* 1983;**77**(4):446−50.

5. Palacios G, Savji N, Travassos da Rosa A, et al. Characterization of the Uukuniemi virus group (phlebovirus: bunyaviridae): evidence for seven distinct species. *J Virol* 2013;**87**(6):3187−95.

6. Lvov DK, Timofeeva AA, Gromashevski VL, et al. "Zaliv Terpeniya" virus, a new Uukuniemi group arbovirus isolated from *Ixodes (Ceratixodes) putus* Pick.-Camb. 1878 on Tyuleniy Island (Sakhalin region) and Commodore Islands (Kamchatsk region). *Arch Gesamte Virusforsch* 1973;**41**(3):165−9.

7. Lvov DK, Alkhovsky SV, Shchelkanov MYu, et al. Genetic characterization of Zaliv Terpeniya virus (ZTV, Bunyaviridae, Phlebovirus, Uukuniemi serogroup) strains, isolated from ticks *Ixodes (Ceratixodes) uriae* White, 1852, obligate parasites of Alcidae birds, in high latitudes of Northern Eurasia and mosquitoes *Culex modestus* Ficalbi, 1889 in subtropics Transcaucasus. *Vopr Virusol* 2014;**59**(1):12−18 [in Russian].

8. Lvov DK, Timopheeva AA, Chervonski VI, et al. Tuleniy virus. A new Group B arbovirus isolated from *Ixodes (Ceratixodes) putus* Pick.-Camb. 1878 collected on Tuleniy Island, Sea of Okhotsk. *Am J Trop Med Hyg* 1971;**20**(3):456−60.

9. Lvov DK, Timofeeva AA, Gromashevski VL, et al. "Sakhalin" virus—a new arbovirus isolated from *Ixodes (Ceratixodes) putus* Pick.-Camb. 1878 collected on Tuleniy Island, Sea of Okhotsk. *Arch Gesamte Virusforsch* 1972;**38**(2):133−8 [in German].

10. Lvov DK, Timopheeva AA, Gromashevsky VL, et al. "Okhotskiy" virus, a new arbovirus of the Kemerovo group isolated from *Ixodes (Ceratixodes) putus* Pick.-Camb. 1878 in the Far East. *Arch Gesamte Virusforsch* 1973;**41**(3):160−4 [in German].

11. Lvov DK, Timopheeva AA, Smirnov VA, et al. Ecology of tick-borne viruses in colonies of birds in the USSR. *Med Biol* 1975;**53**(5):325−30.

8.2 Family Flaviviridae

1. Simmonds P, Becker P, Collet MS, et al. Flaviviridae. In: King AMQ, Adams MJ, Carstens EB, Lefkowitz EJ, editors. *Virus taxonomy: 9th Rep Intern Comm Taxonomy of Viruses*. UK/USA: Elsevier Academic Press;2012. p. 1003−20.

2. Lindenbach BD, Thief HJ, Rice CM. Flaviviridae: the viruses and their replication. In: Knipe DM, Howley PM, editors. *Fields virology*. 5th ed. Philadelphia, PA: Lippincott Williams, Wilkins;2007. p. 1101−52.

3. Kuno G, Chang GJ, Tsuchiya KR, Karabatsos N, Cropp CB. Phylogeny of the genus Flavivirus. *J Virol* 1998;**72**(1):73−83.

4. Chambers TJ, Hahn CS, Galler R, Rice CM. Flavivirus genome organization, expression, and replication. *Ann Rev Microbiol* 1990;**44**:649−88.

5. Calisher CH. Persistent emergence of dengue. *Emerg Infect Dis* 2005;**11**(5):738−9.

6. Tomori O. Yellow fever: the recurring plague. *Crit Rev Clin Lab Sci* 2004;**41**(4):391−427.

7. Kilpatrick AM. Globalization, land use, and the invasion of West Nile virus. *Science* 2011;**334**(6054):323−7.

8.2.1.1 Omsk Hemorrhagic Fever Virus

1. Akhrem-Akhremovich RM. Spring-summer fever in Omsk region. In: *Proceedings of Omsk Medical Institute*, vol. 13. Omsk; 1948. p. 3−16 [in Russian].

2. Vorobjeva NN, Kharitonova NN, Radykova OA. Latent Omsk hemorrhagic fever infection among animals in natural focies. In: *Biological and epizootological characteristics of natural focies of Omsk hemorrhagic fever in Western Siberia*. Novosibirsk; 1979. p. 73−81 [in Russian].

3. Gavrilovskaya AA. Materials for epidemiology, etiology and prophylaxis of spring-summer fever in some districts of Omsk region. In: *Proceedings of Omsk Medical Institute*, vol. 13. Omsk; 1948. p. 27−43 [in Russian].

4. Mazhbich IB, Netsky GI. Three years of investigation of Omsk hemorrhagic fever (1946−1948). In: *Proceedings of the Omsk Institute for epidemiology and microbiology*, vol. 1. Omsk; 1952. p. 51−67 [in Russian].

5. Chumakov MP. For the results of expedition of the Institute of neurology on the investigation of Omsk hemorrhagic fever (OHF). *Vestnik Academii Meditsinskikh Nauk SSSR* 1948;(2):19−26 [in Russian].

6. Chumakov MP, Belyaeva AP, Gagarina AV, et al. Isolation and investigation of strains of the causative agent of Omsk haemorrhagic fever. In: *Endemic viral infections (haemorrhagic fever). Proceedings of Institute of Poliomyelitis and Viral Encephalitis of AMS USSR*, vol. 7; 1965. p. 327−45 [in Russian].

7. Fedorova TN. *Virological characteristics of Omsk haemorrhagic fever and tick-borne encephalitis natural foci in Western Siberia* [Thesis of medical doctor dissertation]. Tomsk: Tomsk University Press; 1969 [in Russian].

8. Gagarina AV. Epidemiological and virological characteristics of natural focies of Omsk hemorrhagic fever (OHF). In: *Proceedings of the Omsk Institute for epidemiology and microbiology*, vol. 2. Omsk; 1954. p. 21–8 [in Russian].

9. Gagarina AV. Transmission of Omsk haemorrhagic fever by ticks. In: *Endemic viral infections (haemorrhagic fever). Proceedings of Institute of Poliomyelitis and Viral Encephalitis of AMS USSR*, vol. 7; 1965. p. 422–29 [in Russian].

10. Grard G, Moureau G, Charrel RN, et al. Genetic characterization of tick-borne flaviviruses: new insights into evolution, pathogenetic determinants and taxonomy. *Virology* 2007;**361**:80–92.

11. Li L, Rollin PE, Nichol ST, et al. Molecular determinants of antigenicity of two subtypes of the tick-borne flavivirus haemorrhagic fever virus. *J Gen Virol* 2004;**85**:1619–24.

12. Lin D, Li L, Dick D, et al. Analysis of the complete genome of the tick-borne flavivirus Omsk hemorrhagic fever virus. *Virology* 2003;**313**:81–90.

13. Busygin FF. Omsk haemorrhagic fever (current status of the problem). *Vopr Virusol* 2000;**45**(3):4–9 [in Russian].

14. Busygin FF, Tsaplin IS, Lebedev EP. The modern state of epidemic process and epidemiological characteristics of Omsk haemorrhagic fever foci. In: *The problems of infectious pathology*. Omsk; 1975. p. 32–45 [in Russian].

15. Yakimenko VV, Karan LS, Malkova MG, et al. On the possibility of mixed infection of vertebrates and arthropods of tick-borne encephalitis virus and Omsk haemorrhagic fever virus. In: *Modern scientific and applied aspects of tick-borne encephalitis*, Moscow; 2007. p. 143-4 [in Russian].

16. Lvov DK. Omsk hemorrhagic fever. In: Lvov DK, editor. *Handbook of virology. Viruses and viral infection of human and animals*. Moscow: MIA;2013p. 766–8 [in Russian].

17. Yastrebov V.K., Yakimenko V.V. Omsk hemorrhagic fever: sum of data (1946–2013). Vopr Virusol 2014; (6):5-11 [in Russian].

18. Vorobjeva NN, Kukharchuk LP, Strizhak VM. Isolation of Omsk hemorrhagic fever virus from *Aedes* mosquitoes in Barabinsk lowland (Zdvinsk district). In: *Natural foci infections in the Far East*, Khabarovsk; 1973. p. 70–2 [in Russian].

19. Busygin FF, Tarasevich LN, Lebedev EP. Taxons of blood-sucking mosquitoes in the ecology of arboviruses in Western Siberia. In: *Arboviruses. Materials of Session of Problem Commission*. Tallinn; 1984. p. 25–6 [in Russian].

20. Kalmin OB, Busygin FF, Yakimenko VV. Blood-sucking mosquitoes as vectors for Omsk hemorrhagic fever virus. In: *Natural foci human diseases*. Omsk; 1996. p. 123–7 [in Russian].

21. Volynets LV, Bogdanov II, Fedorov TN. About possible role of blood-sucking mosquitoes in the transmission of Omsk hemorrhagic fever virus. Problems of medical virology. *Arboviruses*, vol. 11. Moscow; 1971. p. 154–6 [in Russian].

22. Isachenko AG, Shlyapnikov AA. *Nature of the world: landscapes*. Moscow: Mysl;1989 [in Russian].

23. Lvov DK. Omsk hemorrhagic fever. In: Monath TP, editor. *The arboviruses: epidemiology and ecology*, 3. Boca Raton, FL: CRC Press;1988. p. 205–16 [Chapter 34].

24. Lebedev EP, Sizemova GA, Busygin FF. Clinic and epidemiological characteristics of Omsk hemorrhagic fever. *J Microbiol (Moscow)* 1975;**11**:132–3 [in Russian].

25. Sizemova GA. Clinic and epidemiological characteristics of Omsk hemorrhagic fever. Endemic virus infections. In: *Proceedings of the Institute of Poliomyelitis*, Moscow; 1965. p. 430–9 [in Russian].

26. Yakimenko VV. *Omsk haemorrhagic fever virus: epidemiological and clinical aspects. Tick-borne infections in Siberian region*. Novosibirsk: Publishing Office "Science", SB RAS;2011. p. 279–95. [in Russian].

27. Loginova NV, Efanova TN, Kovalchuk AV, et al. Effectiveness of virazol, realdiron and interferon inductors in experimental Omsk hemorrhagic fever. *Vopr Virusol* 2002;**47**(6):27–30 [in Russian].

8.2.1.2 Powassan Virus

1. McLean DM, Donohue WL. Powassan virus: isolation of virus from a fatal case of encephalitis. *Can Med Assoc J* 1959;**80**(9):708–11.

2. Artsob H. Powassan encephalitis //. In: Monath TP, editor. *The arboviruses: epidemiology and ecology*, 3. Boca Raton, FL: CRC Press;1988. p. 29–31.

3. McLean DM, McQueen EJ, Petite HE, et al. Powassan virus: field investigations in Northern Ontario, 1959 to 1961. *Can Med Assoc J* 1962;**86**(21):971–4.

4. Thomas LA, Kennedy RC, Eklund CM. Isolation of a virus closely related to Powassan virus from *Dermacentor andersoni* collected along North Cache la Poudre River, Colo. *Proc Soc Exp Biol Med* 1960;**104**:355–9.

5. Lvov DK, Leonova GN, Gromashevsky VL, et al. Isolation of the Powassan virus from *Haemaphysalis neumanni* Donitz, 1905 ticks in the Maritime Territory. *Vopr Virusol* 1974;(5):538–41 [in Russian].

6. Mandl CW, Holzmann H, Kunz C, et al. Complete genomic sequence of Powassan virus: evaluation of genetic elements in tick-borne versus mosquito-borne flaviviruses. *Virology* 1993;**194**:173–84.

7. Simmonds P, Becker P, Collet MS, et al. Flaviviridae. In: King AMQ, Adams MY, Carsdens ED, Lefirowitz EY, editors. *Virus taxonomy. Ninth Report of the International Committee on Taxonomy of Viruses.* London: Elsevier Academic Press;2012. p. 1003–20.

8. Ebel GD, Spielman A, Telford SR. Phylogeny of North American Powassan virus. *J Gen Virol* 2001;**82**(7):1657–65.

9. Leonova GN, Kondratov IG, Ternovoi VA, et al. Characterization of Powassan viruses from Far Eastern Russia. *Arch Virol* 2009;**154**(5):811–20.

10. Hardy JI. Arboviral zoonoses of North America. In: Beran GW, editor. *Handbook of zoonoses. Section B: Viral.* Boca Raton, FL: CRC Press;1994. p. 185–200.

11. McLean DM, Cobb C, Gooderham SE, et al. Powassan virus: persistence of virus activity during 1966. *Can Med Assoc J* 1967;**96**(11):660–4.

12. McLean DM, Crawford MA, Ladyman SR, et al. California encephalitis and Powassan virus activity in British Columbia, 1969. *Am J Epidemiol* 1970;**92**(4):266–72.

13. Kruglyak SP, Leonova GN. The significance of Ixodes ticks in the southern Far East in the circulation of Powassan virus. *Vopr Virusol* 1989;**34**(3):358–62 [in Russian].

14. Lvov DK, Gromashevsky VL, Skvortsova TM, et al. Isolation of viruses from natural sources in USSR. In: Lvov DK, editor. *Actual problems of common and medical virology.* Moscow: USSR Academy of Medical Sciences;1986. p. 28–37.

15. Main AJ, Carey AB, Downs WG. Powassan virus in Ixodes cookei and Mustelidae in New England. *J Wildl Dis* 1979;**15**(4):585–91.

16. Leonova GN, Kruglyak SP, Lozovskaya SA, et al. The role of wild murine rodents in the selection of different strains of tick-borne encephalitis and Powassan viruses. *Vopr Virusol* 1987;**32**(5):591–5 [in Russian].

17. Leonova GN, Isachkova LM, Baranov NI, et al. Role of Powassan virus in the etiological structure of tick-borne encephalitis in the Primorsky krai. *Vopr Virusol* 1980;(2):173–6 [in Russian].

18. Leonova GN, Sorokina MN, Kruglyak SP. The clinico-epidemiological characteristics of Powassan encephalitis in the southern Soviet Far East. *Zh Mikrobiol Epidemiol Immunobiol* 1991;(3):35–9.

19. Isachkova LM, Frolova MP, Leonova GN, et al. Pathomorphology of experimental infection caused by Powassan virus isolated in the Primorsky Territory. *Vopr Virusol* 1978;(1):68–74 [in Russian].

8.2.1.3 Tick-Borne Encephalitis Virus and Alma-Arasan Virus (var. Tick-borne encephalitis virus)

1. Pavlovsky EN. *Natural foci of transmissible infections.* Moscow/Leningrad: Nauka;1964 [in Russian].

2. Zilber LA. Spring (spring-summer) endemic tick-borne encephalitis. *Archiv Biologicheskikh Nauk* 1939;**56**(2):9–37 [in Russian].

3. Calisher CH. Antigenic classification and taxonomy of flaviviruses (family Flaviviridae) emphasizing a universal system for the taxonomy of viruses causing tick-borne encephalitis. *Acta Virol* 1988;**32**(5):469–78.

4. Calisher CH, Karabatsos N, Dalrymple JM, et al. Antigenic relationships between flaviviruses as determined by cross-neutralization tests with polyclonal antisera. *J Gen Virol* 1989;**70**(1):37–43.

5. Ecker M, Allison SL, Meixner T, et al. Sequence analysis and genetic classification of tick-borne encephalitis viruses from Europe and Asia. *J Gen Virol* 1999;**80**(1):179–85.

6. Emek E, Kozuch O, Nosek J, et al. Arboviruses in birds captured in Slovakia. *J Hyg Epidemiol Microbiol Immunol* 1977;**21**:353–9.

7. Holzmann H, Vorobyova MS, Ladyzhenskaya IP, et al. Molecular epidemiology of tick-borne encephalitis virus: cross-protection between European and Far Eastern subtypes. *Vaccine* 1992;**10**(5):345–9.

8. Hubalek Z, Pow I, Reid HW, et al. Antigenic similarity of central European encephalitis and louping-ill viruses. *Acta Virol* 1995;**39**(5–6):251–6.

9. Kopecky J, Tomkova E, Grubhoffer L, et al. Monoclonal antibodies to tick-borne encephalitis (TBE) virus: their use for differentiation of the TBE complex viruses. *Acta Virol* 1991;**35**(4):365–72.

10. Lvov DK, Il'ichev VD. *Migration of birds and the transfer of the infectious agents.* Moscow: Nauka;1979 [in Russian].

11. Karimov SK. Arboviruses in Kazakhstan region [Doctor medical sciences dissertation]. Alma-Ata/Moscow; 1983 [in Russian].

12. Lvov DK, Klimenko SM, Gaidamovich SY. *Arboviruses and arbovirus infections.* Moscow: Meditsina;1989 [in Russian].

13. Lvov DK, Alkhovsky SV, Shchelkanov MYu, et al. Genetic characteristics of Powassan virus (POWV) isolated from *Haemaphysalis longicornis* ticks in Primorsky krai and two strains of Tick-borne encephalitis virus (TBEV) (Flaviviridae, Flavivirus): Alma-Arasan virus (AAV) isolated from *Ixodes persulcatus* in Kazakhstan and Malyshevo virus (MALOV) isolated from *Aedes vexans* mosquitoes in Khabarovsk krai. *Vopr Virusol* 2014;**59**(5) [in Russian].

14. Karganova GG. Genetic variability of tick-borne encephalitis virus: fundamental and applied aspects. In: Lvov DK, Uryvaev LV, editors. *Investigation of virus evolution in the limits of the problems of biosafety and socially significant infections.* Moscow; 2011. p. 190–9 [in Russian].

15. Pogodina VV, Bochkova NG, Karan LS, et al. The Siberian and Far-Eastern subtypes of tick-borne

encephalitis virus registered in Russia's Asian regions: genetic and antigen characteristics of the strains. *Vopr Virusol* 2004;**49**(4):20−5 [in Russian].

16. Zlobin VI, Belikov SI, Dzhioev YuP, et al. Molecular epidemiology of tick-borne encephalitis. Irkutsk; 2003 [in Russian].

17. Korenberg AI. *Biochorological structure of species (on the example of taiga tick)*. Moscow: Nauka;1979.

18. Chumkov MP, Zeitlenok NA. Tick-borne spring-summer encephalitis in the wider Ural area. *Archiv Biologicheskikh Nauk* 1939;**56**(3):11−17 [in Russian].

19. Ierusalimsky AP. Tick-borne encephalitis. Novosibirsk, 2001. [in Russian]

20. Jaaskeläinen AE, Sironen T, Murueva GB, et al. Tick-borne encephalitis virus in ticks in Finland, Russian Karelia and Buryatia. *J Gen Virol* 2010;**91**(11): 2706−12.

21. Kuklin VV, Povalyagina NS. Infection rate of the ticks in different landscapes of Altai krai. In: *Geography of natural foci diseases of Altai krai*. Leningrad; 1976. p. 16−19 [in Russian].

22. Levkovich EN, Karpovich LG. Studies on biological properties of viruses of tick-borne encephalitis complex in tissue culture. In: Libikova H, editor. *Biology of viruses of the tick-borne encephalitis complex*. Prague: Publ House Czech Acad Sci;1962. p. 161−5 [in Russian].

23. Smorodintsev AA, Alekseev BP, Gulamova VP, et al. Epidemiological features of biphasic virus of meningoencephalitis. *Zh Mikrobiol* 1953;**5**:54−9 [in Russian].

24. Zlobin VI, Shamanin VA, Drokin DA, et al. The geographical distribution of genetic variants of the tick-borne encephalitis virus. *Vopr Virusol* 1992;**37** (5−6):252−6 [in Russian].

25. Brummer-Korvenkontio M, Saikku P, Korhonen P, et al. Arboviruses in Finland. I. Isolation of tick-borne encephalitis (TBE) virus from arthropods, vertebrates, and patients. *Am J Trop Med Hyg* 1973;**22**(3):382−9.

26. Tonteri E, Jaaskeläinen AE, Tikkakoski T, et al. Tick-borne encephalitis virus in wild rodents in winter, Finland, 2008−2009. *Emerg Infect Dis* 2011;**17**(1): 72−5.

27. Holmgren H, Forsgren M. Epidemiology of tick-borne encephalitis in Sweden 1956−1989: a study of 1116 cases. *Scand J Infect Dis* 1990;**22**:287−95.

28. Pettersson JH, Golovljova I, Vene S, et al. Prevalence of tick-borne encephalitis virus in *Ixodes ricinus* ticks in northern Europe with particular reference to Southern Sweden. *Parasit Vectors* 2014;(7):102.

29. Lindquist L, Vapalahti O. Tick-borne encephalitis. *Lancet* 2008;**371**(9627):1861−71.

30. Larsen AL, Kanestrom A, Bjorland M, et al. Detection of specific IgG antibodies in blood donors and tick-borne encephalitis virus in ticks within a non-endemic area in southeast Norway. *Scand J Infect Dis* 2014;**46** (3):181−4.

31. Skarpaas T, Golovljova I, Vene S, et al. Tick-borne encephalitis virus, Norway and Denmark. *Emerg Infect Dis* 2006;**12**(7):1136−8.

32. Gresikova M. Recovery of the tick-borne encephalitis virus from the blood and milk of subcutaneously infected sheep. *Acta Virol* 1958;**2**(2):113−19.

33. Frey S, Essbauer S, Zoller G, et al. Full genome sequences and preliminary molecular characterization of three tick-borne encephalitis virus strains isolated from ticks and a bank vole in Slovak Republic. *Virus Genes* 2014;**48**(1):184−8.

34. Hudopisk N, Korva M, Janet E, et al. Tick-borne encephalitis associated with consumption of raw goat milk, Slovenia, 2012. *Emerg Infect Dis* 2013;**19**(5): 806−8.

35. Georgiev B, Rosicky B, Pavlov P, et al. The ticks of the natural focus of tick-borne encephalitis of sheep and man in the Rhodope Mountains (Bulgaria). *Folia Parasitol (Praha)* 1971;**18**(3):267−73.

36. Gerzsenyi K, Gresikova M, Kuti V, et al. Seroepidemiological and ecological investigations of natural foci of tick-borne encephalitis in Hungary. In: Labuda M, Calisher CH, editors. *New aspects in ecology of arboviruses*. Bratislava: Inst Virol SAS;1980. p. 427−44.

37. Molnar E, Gulyas MS, Kubinyi L, et al. Studies on the occurrence of tick-borne encephalitis in Hungary. *Acta Vet Acta Sci Hung* 1976;**26**(4):419−37.

38. Stefanoff P, Orlikova H, Príkazsky V, et al. Cross-border surveillance differences: tick-borne encephalitis and lyme borreliosis in the Czech Republic and Poland, 1999−2008. *Cent Eur J Public Health* 2014;**22**(1):54−9.

39. Zajkowska J, Moniuszko A, Czupryna P, et al. Chorea and tick-borne encephalitis, Poland. *Emerg Infect Dis* 2013;**19**(9):1544−5.

40. Jemersic L, Dezðek D, Brnic D, et al. Detection and genetic characterization of tick-borne encephalitis virus (TBEV) derived from ticks removed from red foxes (*Vulpes vulpes*) and isolated from spleen samples of red deer (*Cervus elaphus*) in Croatia. *Ticks Tick Borne Dis* 2014;**5**(1):7−13.

41. Bormane A, Lucenko I, Duks A, et al. Vectors of tick-borne diseases and epidemiological situation in Latvia in 1993−2002. *Int J Med Microbiol* 2004;(Suppl. 37): 36−47.

42. Han X, Juceviciene A, Uzcategui NY, et al. Molecular epidemiology of tick-borne encephalitis virus in *Ixodes ricinus* ticks in Lithuania. *J Med Virol* 2005;**77**(2): 249−56.

43. Golovljova I, Vene S, Sjolander KB, et al. Characterization of tick-borne encephalitis virus from Estonia. *J Med Virol* 2004;**74**(4):580–8.

44. Ryltseva EV, Kulova NM, Bochkova MYu. Natural focies of tick-borne encephalitis in Baltic. IV. Further investigation of the areal of fleas. Deponent of VINITI. 1978; N 2696 [in Russian].

45. Ackermann R, Rehse-Kupper B. Central European encephalitis in the Federal Republic of Germany. *Fortschr Neurol Psychiatr Grenzgeb* 1979;**47**(3):103–22 [in German].

46. Kiffner C, Vor T, Hagedorn P, et al. Factors affecting patterns of tick parasitism on forest rodents in tick-borne encephalitis risk areas, Germany. *Parasitol Res* 2011;**108**(2):323–35.

47. Suss J, Schrader C, Falk U, et al. Tick-borne encephalitis (TBE) in Germany-epidemiological data, development of risk areas and virus prevalence in field-collected ticks and in ticks removed from humans. *Int J Med Microbiol* 2004;(Suppl. 37):69–79.

48. Suss J. Tick-borne encephalitis in Europe and beyond—the epidemiological situation as of 2007. *Euro Surveill* 2008;**13**(26):18916.

49. Frey S, Essbauer S, Zoller G, et al. Complete Genome Sequence of Tick-Borne Encephalitis Virus Strain A104 Isolated from a Yellow-Necked Mouse (*Apodemus flavicollis*) in Austria. *Genome Announc* 2013;**1**(4): e00564–613.

50. Knap N, Korva M, Dolinsek V, et al. Patterns of tick-borne encephalitis virus infection in rodents in Slovenia. *Vector Borne Zoonotic Dis* 2012;**12**(3):236–42.

51. Herpe B, Schuffenecker I, Pillot J, et al. Tickborne encephalitis, southwestern France. *Emerg Infect Dis* 2007;**13**(7):1114–46.

52. Beltrame A, Cruciatti B, Ruscio M, et al. Tick-borne encephalitis in Friuli Venezia Giulia, northeastern Italy. *Infection* 2005;**33**(3):158–9.

53. Beltrame A, Ruscio M, Cruciatti B, et al. Tickborne encephalitis virus, northeastern Italy. *Emerg Infect Dis* 2006;**12**(10):1617–19.

54. Barandika JF, Hurtado A, Juste RA, et al. Seasonal dynamics of *Ixodes ricinus* in a 3-year period in northern Spain: first survey on the presence of tick-borne encephalitis virus. *Vector Borne Zoonotic Dis* 2010;**10**(10):1027–35.

55. Hayasaka D, Suzuki Y, Kariwa H, et al. Phylogenetic and virulence analysis of tick-borne encephalitis viruses from Japan and far-Eastern Russia. *J Gen Virol* 1999;**80**(12):3127–35.

56. Kim SY, Jeong YE, Yun SM, et al. Molecular evidence for tick-borne encephalitis virus in ticks in South Korea. *Med Vet Entomol* 2009;**23**(1):15–20.

57. Yun SM, Kim SY, Ju YR, et al. First complete genomic characterization of two tick-borne encephalitis virus isolates obtained from wild rodents in South Korea. *Virus Genes* 2011;**42**(3):307–16.

58. Zhang Y, Si BY, Liu BH, et al. Complete genomic characterization of two tick-borne encephalitis viruses isolated from China. *Virus Res* 2012;**167**(2): 310–13.

59. Khasnatinov MA, Danchinova GA, Kulakova NV, et al. Genetic characteristics of the causative agent of tick-borne encephalitis in Mongolia. *Vopr Virusol* 2010;**55**(3):27–32 [in Russian].

60. Briggs BJ, Atkinson B, Czechowski DM, et al. Tick-borne encephalitis virus, Kyrgyzstan. *Emerg Infect Dis* 2011;**17**(5):876–9.

61. Krivanec K, Kopecký J, Tomková E, et al. Isolation of TBE virus from the tick *Ixodes hexagonus*. *Folia Parasitol (Praha)* 1988;**35**(3):273–6.

62. Alekseev AN. System tick-infection agent and its emergent properties. St. Petersburg; 1993 [in Russian].

63. Kozuch O, Nosek J. Transmission of tick-borne encephalitis (TBE) virus by *Dermacentor marginatus* and *D. reticulatus* ticks. *Acta Virol* 1971;**15**(4):334.

64. Naumov RL, Gutova VP, Chunikhin SP. Ixodid ticks and the causative agent of tick-borne encephalitis. 2. The genera Dermacentor and Haemaphysalis. *Med Parazitol (Moscow)* 1980;**49**(3):66–9 [in Russian].

65. Nosek J, Kozuch O. Replication of tick-borne encephalitis (TBE) virus in ticks *Dermacentor marginatus*. *Angew Parasitol* 1985;**26**(2):97–101.

66. Riedl H, Kozuch O, Sixl W, et al. Isolation of the tick-borne encephalitis virus (TBE-virus) from the tick *Haemaphysalis concinna* Koch. *Arch Hyg Bakteriol* 1971;**154**(6):610–11 [in German].

67. Inokuma H, Ohashi M, Jilintai, et al. Prevalence of tick-borne Rickettsia and *Ehrlichia in Ixodes persulcatus* and *Ixodes ovatus* in Tokachi district, Eastern Hokkaido, Japan. *J Vet Med Sci* 2007;**69**(6):661–4.

68. Takeda T, Ito T, Chiba M, Takahashi K, et al. Isolation of tick-borne encephalitis virus from *Ixodes ovatus* (Acari: Ixodidae) in Japan. *J Med Entomol* 1998;**35**(3):227–31.

69. Yun SM, Song BG, Choi W, et al. Prevalence of tick-borne encephalitis virus in ixodid ticks collected from the republic of Korea during 2011–2012. *Osong Public Health Res Perspect* 2012;**3**(4):213–21.

70. Ivanova LM. Contemporary epidemiology of infections with natural focality in Russia. *Med Parasitol* 1984;**62**(2):17–21 [in Russian].

71. Karpov SP, Fedorov YuV. Epidemiology and prophylaxis of tick-borne encephalitis. Tomsk; 1963 [in Russian].

72. Zlobin VI, Gorin OZ. *Tick-borne encephalitis: etiology, epidemiology and prophylaxis in Siberia.* Novosibirsk: Nauka;1996 [in Russian].

73. Balashov YuS. *Ixodidae ticks—parasites and vectors of infectious agents.* St. Petersburg: Nauka;1998 [in Russian].

74 Alekseev AN, Chunikhin SP. The experimental transmission of the tick-borne encephalitis virus by ixodid ticks (the mechanisms, time periods, species and sex differences). *Parazitologiya* 1990;**24**(3):177–85 [in Russian].

75. Alekseev AN. The present knowledge of tick-borne encephalitis vectors. *Vopr Virusol* 2007;**52**(5):21–6 [in Russian].

76. Bakhvalova VN, Potapova OF, Panov VV, et al. Vertical transmission of tick-borne encephalitis virus between generations of adapted reservoir small rodents. *Virus Res* 2009;**140**(1–2):172–8.

77. Alekseev AN. Ecology of tick-borne encephalitis virus: part of Ixodidae tick males in its circulation. In: *Ecological parasitology.* Leningrad-Petrozavodsk; 1992. p. 48–58 [in Russian].

78. Chunikhin SP, Stefuktina LF, Korolev MB, et al. Sexual transmission of the tick-borne encephalitis virus in ixodid ticks (Ixodidae). *Parazitologiya* 1983;**17**(3):214–17 [in Russian].

79. Chunikhin SP, Alekseev AN, Reshetnikov IA. Experimental study of the role of male ixodid ticks in the circulation of the tick-borne encephalitis virus. *Med Parazitol (Moscow)* 1989;**3**:86–7 [in Russian].

80. Seropolko AA. Isolation of tick-borne encephalitis virus from *Anopheles hyrcanus* mosquitoes in Chuysky valley of Kyrgyzstan. In: *Proceedings of Kyrgyz Institute for Epidemiology, Microbiology and Hygiene*, Frunze, vol. 15; 1975. p. 146–78 [in Russian].

81. Volynets LV, Bogdanov II, Fedorov TN. Recurring isolation of virus belonging to tick-borne encephalitis group from blood-sucking mosquitoes in forest-steppe focies of Omsk hemorrhagic fever in Western Siberia. In: *Arboviruses.* Moscow; 1969. p. 184 [in Russian].

82. Gaidamovich SY, Demenev VA, Roslan IG, et al. Malishevo strain of Negishi virus. Deposition sertificate of Russian State Collection of viruses N 682; 1978.

83. Vorobyeva RN, Ivanov LI, Arutyan NI, et al. To the question of the distribution of arboviruses in Priamurye and Sachalin. In: Lvov DK, editor. *Itogi nauki i techniki. Virology: Arboviruses and arboviral infections.* Moscow: Russian Academy of Sciences;1991p. 8–9 [in Russian].

84. Vorobyeva RN, Ivanov LI, Volkov VI, et al. Arboviruses in Priamurye and Sachalin. In: Lvov DK, editor. *Itogi nauki i techniki. Virology: arboviruses and arboviral infections, part 1.* Moscow: Russian Academy of Sciences;1992p. 10–16 [in Russian].

85. Negishi (NEGV). In: Karabatsos N, editor. *International catalogue of arboviruses including certain other viruses of vertebrates.* San Antonio, TX: American Society of Tropical Medicine and Hygiene;1975. p. 733–4.

86. Chipanina VM, Kozlovskaya OL. Experimental investigation of circulation of tick-borne encephalitis virus in the nest of bank vole at means of fleas Frontopsylla elata botis Jord., 1929. Reports of Irkutsk anti-plaque station, vol. 9; 1971. p. 237–8 [in Russian].

87. Chipanina VM, Kozlovskaya OL. Experimental study of the role of fleas in the circulation of tick-borne encephalitis virus. Natural focies of the Far East. Khabarovsk; 1976. p. 86–7 [in Russian].

88. Tagiltsev AA, Tarasevich LN. *Arthropods of shelter complex in the natural focies of arboviral infections.* Novosibirsk: Nauka;1982.

89. Tarasevich LN. Virological investigation of the role of blood-sucking arthropods belonging to shelter complex in the circulation of arboviruses. In: *Arboviruses.* Sverdlovsk; 1977. p. 40-5.

90. Labuda M, Jones LD, Williams T, et al. Efficient transmission of tick-borne encephalitis virus between cofeeding ticks. *J Med Entomol* 1993;**30**(1):295–9.

91. Safronov VG, Leonov GN, Sorochenko SA. Isolation of tick-borne encephalitis virus complex from birds in Chukotka. In: Akhundov VYu, editor. Ecology ofviruses. Baku; 1976. p. 128 [in Russian].

92. Rostasy K. Tick-borne encephalitis in children. *Wien Med Wochenschr* 2012;**162**(11–12):244–7.

93. Il'enko VI, Platonov VG, Smorodintsev AA. Biological variants of the tick-borne encephalitis virus isolated in various foci of the disease. *Vopr Virusol* 1974;**4**:414–18 [in Russian].

94. Korenberg EI, Kovalevskii YV. Zoning of the areal of tick-borne encephalitis. Moscow; 1981 [in Russian].

95. Korenberg EI, Kovalevskii YV. Main features of tick-borne encephalitis eco-epidemiology in Russia. *Zentralbl Bakt* 1999;**289**(5–7):525–39 [in German].

96. Petri E, Gniel D, Zent O. Tick-borne encephalitis (TBE) trends in epidemiology and current and future management. *Travel Med Infect Dis* 2010;**8**(4):233–45.

97. Mantke OD, Schadler R, Niegrig M. A survey on cases of tick-borne encephalitis in European countries. *Ero Survell* 2008;**13**(17) pii: 18848.

98. Lvov DK. Immunoprophylaxis of Tick-borne encephalitis [Thesis of Doctor of Medical Sciences]. Moscow; 1965 [in Russian].

99. Lvov DK, Bolshev LN, Rudik AP, et al. Calculation of the intensity of infection by Tick-borne encephalitis virus. *Med Parasitol Parasit Dis* 1968;**3**:274–5 [in Russian].

100. Lvov DK, Bolshev LN, Kruopis YuI, et al. Calculation of the intensity of the infection and morbidity of

arboviral infections. *Med Parasitol Parasit Dis* 1969;**4**:410–15 [in Russian].

101. Lvov DK, Lebedev AD. *Ecology of arboviruses..* Moscow: Meditsina;1974 [in Russian].

102. Lvov DK, Zlobin VI. Prevention of tick-borne encephalitis at the present stage: strategy and tactics. *Vopr Virusol* 2007;**52**(5):26–30 [in Russian].

103. Aitov KA, Tarbeev AK, Borisov VA, et al. Current aspects of the clinical picture of tick-borne encephalitis. *Vopr Virusol* 2007;**52**(5):33–7 [in Russian].

104. Borisov VA, Malov IV, Yushchyuk ND. *Tick-borne encephalitis*. Novosibirsk: Nauka;2002 [in Russian].

105. Asher DM. Persistent tick-borne encephalitis infection in man and monkeys: relation to chronic neurologic disease. In: Kurstak K, editor. *Arctic and tropical arboviruses*. NY: Academic Press;1979.

106. Pogodina VV, Frolova MP, Erman BA. *Chronic tick-borne encephalitis*. Novosibirsk: Nauka;1986 [in Russian].

107. Heinz FX, Stiasny K, Holzmann H, et al. Vaccination and tick-borne encephalitis, central Europe. *Emerg Infect Dis* 2013;**19**(1):69–76.

108. Pogodina VV, Levina LS, Skrynnik SM, et al. Tick-borne encephalitis with fulminant course and lethal outcome in patients after plural vaccination. *Vopr Virusol* 2013;**58**(2):33–7 [in Russian].

109. Loew-Baselli A, Poellabauer EM, Pavlova BG, et al. Prevention of tick-borne encephalitis by FSME-IMMUN vaccines: review of a clinical development programme. *Vaccine* 2011;**29**(43):7307–19.

110. Loew-Baselli A, Pavlova BG, Fritsch S, et al. A non-adjuvanted whole-virus H1N1 pandemic vaccine is well tolerated and highly immunogenic in children and adolescents and induces substantial immunological memory. *Vaccine* 2012;**30**(41):5956–66.

111. Romanenko VV, Prokhorova OG, Zlobin VI. New strategy for specific prophylaxis of tick-borne encephalitis: an experience mass vaccination organization in Sverdlovsk region. *Epidemiol Vaktsinoprofil* 2005;**3**:24–7 [in Russian].

8.2.1.4 Japanese Encephalitis Virus

1. Huang JH, Lin TH, Teng HJ, et al. Molecular epidemiology of Japanese encephalitis virus, Taiwan. *Emerg Infect Dis* 2011;**16**:876–8.

2. Mitamura T, Kitaoka M, Watanabe K. Study on Japanese encephalitis virus. Animal experiments and mosquito transmission experiments. *Kansai Iji* 1936;**1**:260.

3. Hoke CH, Gingrich JB. Japanese encephalitis. In: Beran GW, Steel JH, editors. *Handbook of zoonoses. Section B: Viral*. Boca Raton, FL: CRC Press;1994. p. 59–69.

4. Chen WR, Tesh RB, Rico-Hesse R. Genetic variation of Japanese encephalitis virus in nature. *J Gen Virol* 1990;**71** (12):2915–22.

5. Schuh AJ, Tesh RB, Barrett AD. Genetic characterization of Japanese encephalitis virus genotype II strains isolated from 1951 to 1978. *J Gen Virol* 2011;**92**(3):516–27.

6. Solomon T, Ni H, Beasley DW, et al. Origin and evolution of Japanese encephalitis virus in southeast Asia. *J Virol* 2003;**77**(5):3091–8.

7. Shimoda H, Ohno Y, Mochizuki M, et al. Dogs as sentinels for human infection with Japanese encephalitis. *Emerg Infect Dis* 2010;**16**:1137–9.

8. Leonova GN, Kruglyak SP, Maistrovskaya OS, et al. Results of serological surveillance for Japanese encephalitis in 1980 in Primorsky krai. In: Lvov DK, editor. *Arboviral infections*. Moscow: All-Union Institute of Scientific-Technical Information;1989 (24), 41–5 [in Russian].

9. Rybachuk VN, Leonova GN, Kruglyak SP, et al. Circulation of arboviruses in contemporary ecological conditions of the Southern Far East. In: Lvov DK, editor. *Arboviruses and arboviral infections. Proceedings of International Symposium*, Moscow; 1989. p. 46 [in Russian].

10. Lvov DK, Il'ichev VD. *Migration of birds and the transfer of the infectious agents*. Moscow: Nauka;1979 [in Russian].

11. Shchelkanov MYu, Ananyev VYu, Lvov DN, et al. Complex environmental and virological monitoring in the Primorje Territory in 2003–2006. *Vopr Virusol* 2007;**52**(5):37–48 [in Russian].

12. Hanna J, Ritchie S, Philips DA, et al. An outbreak of Japanese encephalitis in the Torres Strait, Australia, 1995. *Med J Austral* 1996;**165**(5):256–60.

13. Mackenzie JS. Emerging viral diseases: an Australian perspective. *Emerg Infect Dis* 1999;**5**(1):1–8.

14. Ritchie SA, Phillips D, Broom A, et al. Isolation of Japanese encephalitis virus from *Culex annulirostris* mosquitoes in Australia. *Am J Trop Med Hyg* 1997;**56**:80–4.

15. Wang J-L, Pan X-L, Zhang H-L, et al. Japanese encephalitis viruses from bats in Yunnan, China. *Emerg Infect Dis* 2009;**15**(6):939–42.

16. Cui J, Counor D, Sheen D, et al. Detection of Japanese encephalitis virus 2008—antibodies in bats in southern China. *Amer J Trop Med Hyg* 2008;**18**:1007–11.

17. Lvov DK. Importance of emerging–reemerging infections for biosafety. *Vopr Virusol* 2002;**47**(5):4–7 [in Russian].

18. Loginova NV, Karpova EF. The biological properties and variants of the Japanese encephalitis virus from a Russian collection. *Vopr Virusol* 1994;**39**(4):163–5 [in Russian].

19. Erlanger TE, Weiss S, Keiser J, et al. Past, present and future of Japanese encephalitis. *Emerg Infect Dis* 2009;**15**:1–7.

20. Hills SL, Phillips DC. Past, present and future of Japanese encephalitis. *Emerg Infect Dis* 2009;**15**(8): 11333.

21. Van den Hurk AF, Ritchie SA, Mackenzie JS. Ecology and geographical expansion of Japanese encephalitis virus. *Annu Rev Entomol* 2009;**54**:17–35.

22. Morita K. Molecular epidemiology of Japanese encephalitis in East Asia. *Vaccine* 2009;**27**:7131–2.

23. Yun SM, Cho JE, Ju YR, et al. Molecular epidemiology of Japanese encephalitis virus circulating in South Korea, 1983–2005. *Virol J* 2010;**7**:127.

24. Nabeshima T, Loan HT, Inoue S, et al. Evidence of frequent introductions of Japanese encephalitis virus from South-East Asia and continental East Asia to Japan. *J Gen Virol* 2009;**90**:827–32.

25. Wang H, Li Y, Liang G. Japanese encephalitis in mainland China. *Jpn J Infect Dis* 2009;**62**:331–6.

26. Chen Y-Y, Fan Y-C, Tu W-C, et al. Japanese encephalitis virus genotype replacement, Taiwan, 2009–2010. *Emerg Infect Dis* 2011;**17**(12):2354–6.

27. Li Y-X, Li M-H, Fu S-H, et al. Japanese encephalitis, Tibet, China. *Emerg Infect Dis* 2011;**17**(5):934–6.

28. Fulmali PV, Sapkal GN, Athawale S, et al. Introduction of Japanese encephalitis virus genotype I, India. *Emerg Infect Dis* 2011;**17**(2):319–21.

29. Loginova NV, Deryabin PG, Tikhomirov EE, et al. Tissue culture inactivated vaccine for the prevention of Japanese encephalitis: experimental and laboratory process and control layout. *Vopr Virusol* 2007;**3**:26–9 [in Russian].

30. Loginova NV, Karpova EF. Genetic analysis of Japanese encephalitis virus strains and variants from Russian collection. *Vopr Virusol* 2001;**46**(2):46–7 [in Russian].

31. Konishi F, Pincus S, Paoletti E, et al. A highly attenuated host range—restricted vaccinia virus strain, NUVAC encoding the PRM, E, and NS1 agents of Japanese encephalitis virus prevent JEV viremia in swine. *Virology* 1992;**190**(1):454–8.

32. Lvov DK, Deryabin PG. Japanese encephalitis. In: Lvov DK, editor. *Handbook of Virology. Viruses and viral infection of human and animals.* Moscow: MIA;2013p. 719–21 [in Russian].

33. Tauber E, Dewasthaly S. Japanese encephalitis vaccines-needs, flaws and achievements. *Biol Chem* 2008;**389**:547–50.

34. Halstead SB. Japanese encephalitis. In: Artenstein AW, editor. *Vaccines: a biography.* NY: Springer;2010. p. 317–34.

35. Beasley DWC, Lewthwaite P, Solomon T. Current use and development of vaccines for Japanese encephalitis. *Expert Opin Biol Ther* 2008;**8**:95–106.

36. Elias C, Okwo-Bele J, Fisher M. A strategic plan for Japanese encephalitis control by 2015. *Lancet Infect Dis* 2009;(9):7.

37. Tandan JB, Ohrr H, Sohn YM, et al. Single dose of Sa 14-14-2 vaccine provides long-tern protection against Japanese encephalitis; a case-control study in Nepal children 5 years after immunization. *Vaccine* 2007;**25**:5041–5.

8.2.1.5 Tyuleniy Virus and Kama Virus

1. Timofeeva AA, Pogrebenko AG, Gromashevsky VL, et al. Natural foci infections on the Iona island in Okhotsk Sea. *Zool J (Moscow)* 1974;**53**(6):906–11 [in Russian].

2. Lvov SD. *Natural virus foci in high latitudes of Eurasia. Soviet medical reviews. Section E: Virol reviews,* 3. USA: Harwood Academic Publishers GmbH;1993. p. 137–85..

3. Saumarez Reef virus. In: Karabatsos N, editor. *International catalogue of arboviruses and some others viruses of vertebrates.* San Antonio, TX: American Society of Tropical Medicine and Hygiene;1985. p. 913–14.

4. Simmonds P, Becher P, Collett MS, et al. Flaviviridae. In: King AMQ, Adams MJ, Carstens EB, Lefkowitz EJ, editors. *Virus taxonomy.* 9th ed. UK/USA: Elsevier Academic Press;2012. p. 1003–20.

5. Clifford CM, Yunker CE, Thomas LA, et al. Isolation of a group B arbovirus from *Ixodes uriae* collected on Three Arch Rocks national wildlife refuge, Oregon. *Am J Trop Med Hyg* 1971;**20**(3):461–8.

6. Efremova GA. The role of parasites of swallow diggings in the preservation of natural foci infections. Materials of 12-th all-Union conference on the natural foci diseases. Novosibirsk; 1989. p. 79–80.

7. Marin MS, Zanotto PM, Gritsun TS, et al. Phylogeny of TYU, SRE, and CFA virus: different evolutionary rates in the genus Flavivirus. *Virology* 1995;**206**(2):1133–9.

8. Tyuleniy virus. In: Karabatsos N, editor. *International catalogue of arboviruses and some others viruses of vertebrates.* San Antonio, TX: American Society of Tropical Medicine and Hygiene;1985. p. 1045–6.

9. Votyakov VI, Voinov IN, Samoilova TI, et al. Isolation of arboviruses from colonial seabirds of Barents Sea. In: *Materials of symposium on the ecology of viruses associated with birds.* Minsk; 1974. p. 42–4.

10. Zanotto PM, Gao GF, Gritsun T, et al. An arbovirus cline across the northern hemisphere. *Virology* 1995;**210**(1):152–9.

11. Grard G, Moureau G, Charrel RN, et al. Genetic characterization of tick-borne flaviviruses: new insights into evolution, pathogenetic determinants and taxonomy. *Virology* 2007;**361**(1):80–92.

12. Gushchin BV, Tsilinsky YaYa, Aristova VA, et al. Electron microscopic study of chicken fibroblasts infected with the Tyuleniy virus. *Vopr Virusol* 1975;(5):540–5 [in Russian].

13. Chastel C, Main AJ, Guiguen C, et al. The isolation of Meaban virus, a new Flavivirus from the seabird tick *Ornithodoros (Alectorobius) maritimus* in France. *Arch Virol* 1985;**83**(3–4):129–40.

14. Meaban virus. In: Karabatsos N, editor. *International catalogue of arboviruses and some others viruses of vertebrates*. San Antonio, TX: American Society of Tropical Medicine and Hygiene;1985. p. 675–6.

15. Lvov SD. Concept of circumpolar distribution of arboviruses. In: *Materials of 18-th Session of the Society of microbiologists, epidemiologists and parasitologists*. Alma-Ata; 1989. p. 224–5.

16. Lvov DK, Timopheeva AA, Smirnov VA, et al. Ecology of tick-borne viruses in colonies of birds in the USSR. *Med Biol* 1975;**53**(5):325–30.

17. Babenko LV. *Ixodes lividus* Koch as the members of burrow-shelter complex. *Studies of Fauna and Flora of USSR* 1956;**3**:21–105.

18. Lvov DK, Aristova VA, Gromashevsky VL, et al. Kama, a new virus (Flaviviridae, Flavivirus, Tiulenii antigenic group), isolated from *Ixodes lividus* ticks. *Vopr Virusol* 1998;**43**(2):71–4.

19. Lvov DK, Chervonski VI, Gostinshchikova IN, et al. Isolation of Tyuleniy virus from ticks *Ixodes (Ceratixodes) putus* Pick.-Camb. 1878 collected on Commodore Islands. *Arch Gesamte Virusforsch* 1972;**38**(2):139–42.

20. Lvov DK, Deryabin PG, Aristova VA, et al. *Atlas of distribution of natural foci virus infections on the territory of Russian Federation*. Moscow: SMC MPH RF Publ;2001 [in Russian].

21. Balashov YuS. Bloodsucking ticks (Ixodoidae)—vectors of disease of man and animals. *Miscel Publicat Entomol Soc Amer* 1972;**8**(5):1–376.

22. Bekleshova AYa, Terskikh II, Smirnov YuA. Arboviruses isolated in the areas of the extreme North. *Vopr Virusol* 1970;**15**:436–40 [in Russian].

23. Berezina LK, Smirnov VA, Zelensky VA. Experimental infection of birds by Tyuleniy virus. In: Lvov DK, editor. *Ecology of virus*. Moscow: Russian Academy of Medical Sciences;1974. p. 13–17.

24. Lvov DK, Gromashevsky VL, Skvortsova TM, et al. Arboviruses of high latitudes in the USSR. In: Kurstak E, editor. *Arctic and tropical arboviruses*. NY/San-Francisco/London: Harcourt Brace Jovanovich Publ.;1979. p. 21–38 [Chapter 3].

25. Thomas LA, Clifford CM, Yunker CE, et al. Tick-borne viruses in western North America. I. Viruses isolated from *Ixodes uriae* in coastal Oregon in 1970. *J Med Entomol* 1973;**10**(2):165–8.

26. Lvov DK, Il'ichev VD. *Migration of birds and the transfer of the infectious agents*. Moscow: Nauka;1979 [in Russian].

27. Lvov DK, Timofeeva AA, Chervonsky VI, et al. Tyuleniy virus: possibly new arbovirus from group B. *Vopr Virusol* 1971;**16**(2):180–4 [in Russian].

28. Lvov SD. Arboviruses in high latitudes. In: Lvov DK, Klimenko SM, Gaidamovich SY, editors. *Arboviruses and arbovirus infections*. Moscow: Meditsina;1989p. 269–89 [in Russian].

29. Filippova NA. *Fauna of USSR. Acariformes*, Moscow/Leningrad: Academy of Sciences of USSR;1966vol 4(3). Ixodoidea. Argas ticks (Argasidae). [in Russian].

30. Filippova NA. *Fauna of USSR. Moscow-Leningrad: Academy of Sciences of USSR*, Moscow/Leningrad: Academy of Sciences of USSR;19774(4). Ixodoidea. Ixodinae subfamily ticks [in Russian].

31. Lvov DK, Timopheeva AA, Chervonsky VI, et al. Tuleniy virus. A new group B arbovirus isolated from *Ixodes (Ceratixodes) putus* Pick.-Camb. 1878 collected on Tuleniy Island, Sea of Okhotsk. *Am J Trop Med Hyg* 1971;**20**(3):456–60.

32. Kempf F, Boulinier T, De Meeûs T, et al. Recent evolution of host-associated divergence in the seabird tick *Ixodes uriae*. *Mol Ecol* 2009;**18**(21):4450–62.

33. Gadget's Gully virus. In: Karabatsos N, editor. *International catalogue of arboviruses and some others viruses of vertebrates*. San Antonio, TX: American Society of Tropical Medicine and Hygiene;1985. p. 407–8.

34. Shchelkanov MYu, Gromashevsky VL, Lvov DK. The role of ecologo-virological zoning in prediction of the influence of climatic changes on arbovirus habitats. *Vestnik Rossiiskoi Akademii Meditsinskikh Nauk* 2006;**2**:22–5 [in Russian].

35. Lvov DK, Timofeeva AA, Gromashevsky VL, et al. Isolation of arboviruses from *Ixodes (Ceratixodes) putus* Pick.-Camb. 1878 ticks collected in the bird colony on the Tyuleniy island, Okhotsk Sea. *Vopr Virusol* 1970;**15**(4):440–4 [in Russian].

36. Chastel C, Guiguen C, Le Lay G, et al. Arbovirus serological survey among marine and non-marine birds of Brittany. *Bull Soc Pathol Exot Filiales* 1985;**78**(5):594–605 [in French].

37. St George TD, Standfast HA, Doherty RL, et al. The isolation of Saumarez Reef virus, a new flavivirus, from bird ticks *Ornithodoros capensis* and *Ixodes eudyptidis* in Australia. *Aust J Exp Biol Med Sci* 1977;**55**(5):493–9.

38. Lvov DK, Alkhovsky SV, Shchelkanov MYu, et al. Genetic characterization of viruses fromantigenic complex Tyuleniy (Flaviviridae, Flavivirus): Tyuleniy virus (TYUV) isolated from ectoparasites of colonial seabirds— *Ixodes (Ceratixodes) uriae* White, 1852 ticks collected in the hagh latitudes of Northern Eurasia—and Kama virus (KAMV) isolated from *Ixodes lividus* Roch, 1844 collected in the digging colonies of the middle part of Russian Plane. *Vopr Virusol* 2014;**59**(1):18–24 [in Russian].

8.2.1.6 Dengue Virus (imported)

1. Baaten GG, Sonder GJB, Zaayer HL, et al. Travel-related Dengue virus infection, the Netherlands, 2006–2007. *Emerg Infect Dis* 2011;**17**(5):821–8.
2. Calisher CH, Karabatsos N, Dalrymple JM, et al. Antigenic relationships between flaviviruses as determined by cross-neutralization tests with polyclonal antisera. *J Gen Virol* 1989;**70**(1):37–43.
3. Ooi EE, Goh KT, Gubler DJ. Dengue prevention and 35 years of vector control in Singapore. *Emerg Infect Dis* 2006;**12**(8):887–93.
4. Gibbons RV, Streitz M, Babina T, et al. Dengue and US military operations from the Spanish–American war through today. *Emerg Infect Dis* 2012;**18**(4):623–30.
5. Rovida F, Percivalla E, Campanini G, et al. Viremic degue virus infections in travelers: potencial for local outbreak in Northern Italy. *J Clin Virol* 2011;**50**: 76–9.
6. Butenko AM. Arbovirus circulation in the Republic of Guinea. *Med Parazitol (Moscow)* 1996;**2**:40–5 [in Russian].
7. Larichev VF, Saifullin MA, Akinshina YuA, et al. Imported cases of arboviral infections in Russian Federation. *Epidemiol Infektbolezni* 2012;(1):35–9 [in Russian].
8. Tissera HA, Ooc EE, Gubler DJ, et al. New Dengue virus type 1 genotype in Colombo, Sri Lanka. *Emerg Infect Dis* 2011;**17**(4):2053–5.
9. Calisher CH. Persistent emergence of Dengue. *Emerg Infect Dis* 2005;**11**(5):738–9.
10. Sergiev VP. Problems of medical parasitology. *Zh Mikrobiol Epidemiol Immunobiol* 2013;**1**:102–4 [in Russian].
11. Shu P-Y, Yang C-F, Kao JF, et al. Application of the Dengue virus NS1 antigen rapid test for on-site detection of imported Dengue cases at airports. *Clin Vaccine Immonol* 2009;**16**:589–91.
12. Sigueira JB, Martelli CM, Coelho GE, et al. Dengue and Dengue hemorrhagic fever, Brazil, 1981–2002. *Emerg Infect Dis* 2005;**11**(1):41–53.
13. Butenko AM. *Modern status of the problem of Crimean hemorrhagic fever, West Nile fever and other arboviral infections in Russian Federation. Investigation of virus evolution in the limits of the problems of biosafety and socially significant problems. Materials of scientific conference (Moscow, Russia; December 24, 2011).* Moscow: Russian Academy of Medical Sciences;2011. p. 176–90 [in Russian].
14. Naidenova EV, Kuklev VE, Yashechkin YuI, et al. Modern status of laboratory diagnostics of denge fever (a review). *Probl Specially Dangerous Infect* 2013; (4):89–94 [in Russian].
15. Araujo F, Nogueira R, de Souza Arajo M, et al. Dengue in patients with central nervous system manifestations, Brazil. *Emerg Infect Dis* 2012;**18**(4):677–9.
16. Vasilakis N, Tesh RB, Weaver SC. Sylvatic Dengue virus type 2 activity in humans, Nigeria, 1968. *Emerg Infect Dis* 2008;**14**(3):502–4.
17. Chahar IIS, Bharaj P, Dar L, et al. Co-infection with Chikungunya and Dengue virus in Delhi, India. *Emerg Infect Dis* 2009;**15**(7):1077–80.
18. Ganushkina LA, Dremova VP. *Aedes aegypti* L. and *Aedes albopictus* Skuse mosquitoes are a new biological threat to the south of Russia. *Med Parazitol (Moscow)* 2012;**3**:49–55 [in Russian].
19. Ashburn PM, Craig CF. Experimental investigations regarding the etiology of dengue fever. *J Infect Dis* 1907;**4**:440–75.
20. Halstead SB, Deen J. The future of dengue vaccines. *Lancet* 2002;**360**:1243–5.
21. Ganushkina LA, Bezzhonova OV, Patraman IV, et al. Distribution of *Aedes (Stegomyia) aegypti* L. and *Aedes (Stegomyia) albopictus* Skus. mosquitoes on the Black Sea coast of the Caucasus. *Med Parazitol (Moscow)* 2013;**1**:45–6 [in Russian].
22. Sabin AB, Schlesinger RW. Production of immunity to denge with virus modified by propagation in mice. *Science* 1945;**101**(2634):640–2.
23. Premartna R, Pathmeswaran A, Amarasekara ND, et al. A clinical guide for early detection of Dengue fever and timing of investigations to detect patients likely to develop complications. *Trans R Soc Trop Med Hyg* 2009;**103**:127–31.
24. Pelaez O, Guzman MG, Kouri G, et al. Dengue 3 epidemic, Havana 2001. *Emerg Infect Dis* 2004;**10**(4):719–22.
25. Schmidt-Chanasi J, Haditsch N, Schoneberg I, et al. Dengue virus infection in traveler returning from Groatia to Germany. *Euro Suveill* 2010;**15**:19677.
26. Holmes EC, Twiddy SS. The origin, emergence and evolutionary genetics of dengue virus. *Infect Gen Evol* 2003;**3**(1):19–28.
27. Twiddy SS, Farrar JJ, Vinh Chau N, et al. Phylogenetic relationships and differential selection pressures among genotypes of dengue-2 virus. *Virology* 2002;**298** (1):63–72.
28. Kanukarathe N, Wahala WMP, Messer WB, et al. Severe Dengue epidemics in Sri Lanka, 2003–2006. *Emerg Infect Dis* 2009;**15**(2):192–8.
29. La Ruche G, Souares Y, Armengaud A, et al. First two autochtonous Dengue virus infections in metropolitan France, September 2010. *Eur Surveill* 2010;**15**:19676.
30. Hammon WM, Rudnick A, Sather GE. Viruses associated with epidemic hemorrhagic fevers of the Philippines and Thailand. *Science* 1960;**131**(3407): 1102–3.

31. Nogueira RMR, Schatzmay HG, De Filippis AMB, et al. Dengue virus type 3, Brazil. *Emerg Infect Dis* 2005;**11** (9):1376–81.

32. Reller ME, Bodinayake C, Nagahawatte A, et al. Unsuspected Dengue and acute febrile illness in rural and semi-urban southern Sri Lanka. *Emerg Infect Dis* 2012;**18**(2):256–63.

33. Centers for Disease Control and Prevention. Locally acquired Dengue—Key West, Florida, 2009–2010. *MMWR* 2010;**59**:577–81.

34. Rosen L. Dengue in Greece in 1927 and 1928 and the pathogenesis of dengue hemorrhagic fever: new data and a different conclusion. *Am J Trop Med Hyg* 1986;**35**(3):642–53.

35. Thu HM, Lowry K, Myint TT, et al. Myanmar Dengue outbreak associated with displacement of serotypes 2, 3, and 4 by dengue 1. *Emerg Infect Dis* 2004;**10** (4):593–7.

36. Vaughn DW, Green SW, Kalayanarooj S, et al. Dengue viremia titer and body response pattern, and virus serotype correlate with disease severity. *J Infect Dis* 2000;**181**:2–9.

37. Hotta S. *Dengue and related arboviruses*. Kobe, Japan: Yukosha Printing House;1995.

38. Lanciotti RS, Lewis JG, Gubler DJ, Trent DW. Molecular evolution and epidemiology of dengue-3 viruses. *J Gen Virol* 1994;**75**(1):65–75.

39. Gubler DJ. Dengue/Dengue hemorrhagic fever: history and current status. *Novartis Found Symp* 2006;**277**:3–16.

40. Eritja R, Escosa R, Lucientes J, et al. Worlwide invasion of vector mosquitoes: present European distribution and challenges for Spain. *Biol Invasions* 1994;**7**:87–97.

41. Gubler DJ, Clark GG. Dengue/Dengue hemorrhagic fever: the emergence of a global health problem. *Emerg Infect Dis* 1995;**1**(2):55–7.

42. Moi ML, Takasaki T, Kotaki A, et al. Importation of Dengue virus type 3 to Japan from Tanzania and Cote d'Ivoire. *Emerg Infect Dis* 2010;**16**(11):1770–2.

43. WHO. Dengue guidelines for diagnosis, treatment, prevention and control. 3rd ed. Geneva; 2009. p. 3–17.

8.2.1.7 Sokuluk Virus

1. Sulkin SE, Allen R. Bats as reservoir hosts for arboviruses. In: *Proc 8-th Cong Trop Med Malar*, Teheran; 1968. p. 694.

2. Lvov DK. *Ecological sounding of the USSR territory for natural foci of arboviruses*. Soviet medical reviews. Section E: Virology reviews., 5. USA: Harwood Academic Publishers GmbH;1993. p. 1–47.

3. Shepherd RC, Williams MC. Studies on viruses in East African bats (Chiroptera). 1. Haemagglutination inhibition and circulation of arboviruses. *Zoonoses Res* 1964;**3** (3):125–39.

4. Gershtein VI, Vargina SG, Kuchuk LA, et al. *About ecology of arboviruses, associated with shelter biocenosis. Progress in Science and Technics. Series « Virology. Arboviruses and arboviral infection »*, 24. Moscow: Russian Academy of Sciences;1991. p. 39–40 [in Russian].

5. Johnson HN. Ecological implications of antigenically related mammalian viruses for which arthropod vectors are unknown and avian associated soft tick viruses. *Jpn J Med Sci Biol* 1967;**20**:160–6.

6. Simmonds P, Becher P, Collett MS, et al. Family Flaviviridae. In: King AMQ, Adams MJ, Carstens EB, Lefkowitz EJ, editors. *Virus Taxonomy: Ninth Report of the International Committee on taxonomy of viruses*. UK/USA: Elsevier Science;2011. p. 1003–20.

7. Skvortsova TM, Gromashevsky VL, Sidorova GL, et al. Results of virological survey of the vectors in Turkmenia. In: Lvov DK, editor. *Ecology of the Viruses*, 2. Moscow: Acad Med Nauk USSR;1974. p. 139–44 [in Russian].

8. Sulkin SE, Allen R, Miura T, et al. Studies of arthropod-borne virus infections in chiroptera. VI. Isolation of Japanese B encephalitis virus from naturally infected bats. *Am J Trop Med Hyg* 1970;**19**(1): 77–87.

9. Varelas-Wesley I, Calisher CH. Antigenic relationships of flaviviruses with undetermined arthropod-borne status. *Am J Trop Med Hyg* 1982;**31**(6):1273–84.

10. Salaun JJ, Klein JM, Hebrard G. A new virus, Phnom-Penh bat virus, isolated in Cambodia from a short-nosed fruit bat, "*Cynopterus brachyotis angulatus*" Miller, 1898.. *Ann Microbiol (Paris)* 1974;**125**(4):485–95 [in French].

11. Vargina SG, Breinger IG. *Monitoring of natural foci of arboviruses in Kyrgyzia. Progress in Science and Technics. Series « Virology. Arboviruses and arboviral infection »*, 24. Moscow: Russian Academy of Science;1991:15–16 [in Russian].

12. Vargina SG, Karas FR, Gershtein VI, et al. About ecology of Sokuluk virus. In: Gaydamovich SYa, Priimyagi LS, editors. *Arboviruses*. Tallin; 1984. p. 13–14 [in Russian].

13. Vargina SG, Kuchuk LA, Breinger IG. *About relations between arboviruses, vectors and hosts. Progress in Science and Technics. Series « Virology. Arboviruses and arboviral infection »*, 24. Moscow: Russian Academy of Science;1991. p. 40–41 [in Russian].

14. Bell JF, Thomas LA. A new virus, « MML », enzootic in bats (*Myotis lucifugus*) of Montana. *Am J Trop Med Hyg* 1964;**13**:607–12.

15. Johnson HN. The Rio Bravo virus: virus identified with Group B arthropod-borne viruses by hemagglutination inhibition and complement fixation test. In: *Proc 9-th Pacific Sci Cong*; 1957. p. 39.

16. Karabatsos N. Characterization of viruses isolated from bats. *Am J Trop Med Hyg* 1969;**18**:803–9.

17. Bres P, Chambon L. Isolation at Dakar of a strain of arbovirus from the salivary glands of the bat (preliminary note). *Ann Inst Pasteur (Paris)* 1963;**104**:705–12 [in French].

18. Tajima S, Takasaki T, Matsuno S, et al. Genetic characterization of Yokose virus, a flavivirus isolated from the bat in Japan. *Virology* 2005;**332**(1):38–44.

19. Lvov DK. Arboviral zoonoses of Northern Eurasia (Eastern Europe and the Commonwealth of Independent States). In: Beran GW, editor. *Handbook of zoonoses. Section B: Viral*. Boca Raton/London/Tokyo: CRC Press;1994. p. 237–60.

20. Lvov DK. *Natural foci of arboviruses in the USSR. Soviet medical reviews. Virology*, 1. UK: Harwood Academic Publishers GmbH;1987153–96.

21. Vargina SG, Steblenko SK, Karas FR, et al. Investigation of viruses, associated with the birds in Chuiskaya Valley in Kyrgyz SSR. In: Lvov DK, editor. *Ecology of the Viruses*. Moscow: USSR Acad Med Nauk;1973. p. 74–80 [in Russian].

22. Lvov DK. Arbovirus infection in sub-tropic and in southern of tempered latitudes in USSR. In: Lvov DK, Klimenko SM, Gaidamovich SY, editors. *Arboviruses and arboviral infection*. Moscow: Meditsina;1989p. 235–49 [Chapter 8] [in Russian].

23. Lvov DK, Alkhovsky SV, Shchelkanov MYu, et al. Taxonomy of Sokuluk virus (SOKV) (Flaviviridae, Flavivirus, Entebbe bat virus group), isolated from bats (*Vespertilio pipistrellus* Schreber, 1774), ticks (Argasidae Koch, 1844), and birds in Kyrgyzstan. *Vopr Virusol* 2014;**59**(1):30–4 [in Russian].

24. Timofeev EM. Foci of arboviruses in southern-western part of Osh district in Kyrgyz SSR [PhD thesis]. Moscow; 2008 [in Russian].

25. Burns KF, Fabinacei CJ. Virus of bats antigenically related to St. Louis encephalitis. *Science* 1956;**123**(3189):227.

26. Lvov DK, Tsyrkin YM, Karas FR, et al. "Sokuluk" virus, a new group B arbovirus isolated from *Vespertilio pipistrellus* Schreber, 1775, bat in the Kirghiz SSR. *Arch Gesamte Virusforsch* 1973;**41**(3):170–4 [in German].

27. Karas FR, Vargina SG, Gershtein VI, et al. About a vertebrate hosts of arboviruses in natural foci in Kirgizia. In: *Proceeding of 10-th all-union conference for natural-focal illness*, Dushanbe; 1979. p. 97–9 [in Russian].

28. Williams MC, Simpson DI, Shepherd RC. Studies on viruses in East African bats. (Chiroptera). 2. Virus isolation. *Zoonoses Res* 1964;**3**(3):141–53.

29. Lumsden WH, Williams MC, Mason PJ. A virus from insectivorous bats in Uganda. *Ann Trop Med Parasitol* 1961;**55**:389–97.

30. Entebbe bat virus. In: Karabatsos N, editor. *International catalogue of arboviruses and some others viruses of vertebrates*. San Antonio, TX: American Society of Tropical Medicine and Hygiene;1985. p. 385–6.

31. Lvov DK, Il'ichev VD. *Migration of birds and the transfer of the infectious agents*. Moscow: Nauka;1979 [in Russian].

8.2.1.8 West Nile Virus

1. Smithburn KC, Hughes TP, Burke AW, et al. A neurotropic virus isolated from the blood of a native of Uganda. *Am J Trop Med* 1940;**20**(1):471–92.

2. Melnick JL, Paul JR, Riordan JT, et al. Isolation from human sera in Egypt of a virus apparently identical to West Nile virus. *Proc Soc Exp Biol Med* 1951;**77**(4):661–5.

3. Kolobukhina LV, Lvov DN. West Nile fever. In: Lvov DK, editor. *Handbook of Virology. Viruses and viral infection of human and animals*. Moscow: MIA;2013. p. 721–31 [in Russian].

4. Clarke DH, Casals J. Arboviruses group. In: Horsfall B, Tamm FL, editors. *Viral and rickettsial infections of man*. Philadelphia; 1965. p. 606–58.

5. Lanciotti RS, Ebel GD, Deubel V, et al. Complete genome sequences and phylogenetic analysis of West Nile virus strains isolated from the United States, Europe, and the Middle East. *Virology* 2002;**298**(1):96–105.

6. Anishchenko M, Shchelkanov MYu, Alekseyev VV, et al. Pathogenecity of West Nile virus: molecular markers. *Vopr Virusol* 2010;**55**(1):4–10 [in Russian].

7. Bakonyi T, Ivanics E, Erdelyi K, et al. Lineage 1 and 2 strains of encephalitic West Nile virus, central Europe. *Emerg Infect Dis* 2006;**12**(4):618–23.

8. Lvov DK, Butenko AM, Gromashevsky VL, et al. West Nile and other zoonotic viruses as examples of emerging–reemerging situations in Russia. *Arch Virol* 2004; (Suppl. 18):85–96.

9. Bondre VP, Jadi RS, Mishra AC, et al. West Nile virus isolates from India: evidence for a distinct genetic lineage. *J Gen Virol* 2007;**88**(3):875–84.

10. Vazquez A, Sanchez-Seco MP, Ruiz S, et al. Putative new lineage of West Nile virus, Spain. *Emerg Infect Dis* 2010;**16**(3):549–52.

11. Lvov DK. West Nile fever. *Vopr Virusol* 2000;**2**:4–9 [in Russian].

12. Figuerola J, Baouab RE, Soriguer R, et al. West Nile virus antibodies in wild birds, Morocco. *Emerg Infect Dis* 2009;**15**(10):1651–3.

13. Venter M, Human S, VanNiekerk S, et al. Fatal neurologic disease and abortion in Mare infected with lineage 1 West Nile virus, South Africa. *Emerg Infect Dis* 2011;**17**(8):1534–6.

14. Lvov DK, Il'ichev VD. *Migration of birds and the transfer of the infectious agents.* Moscow: Nauka;1979 [in Russian].

15. Chowers MY, Lang R, Nassar F, et al. Clinical characteristics of the West Nile fever outbreak, Israel, 2000. *Emerg Infect Dis* 2001;**7**:675−8.

16. Yen J-Y, Kim HJ, Nah J-J, et al. Surveillance for West Nile virus in dead wild birds, South Korea, 2005−2008. *Emerg Infect Dis* 2010;**17**(2):299−301.

17. Li S, Li X, Qin E, et al. Kunjin virus replicon—a novel viral vector. *Sheng Wu Gong Cheng Xue Bao* 2011;**27** (2):141−6 [in Chinese].

18. Mann RA, Fegan M, O'Riley K, et al. Molecular characterization and phylogenetic analysis of Murray Valley encephalitis virus and West Nile virus (Kunjin subtype) from an arbovirus disease outbreak in horses in Victoria, Australia, in 2011. *J Vet Diagn Invest* 2013;**25** (1):35−44.

19. Prow NA. The changing epidemiology of Kunjin virus in Australia. *Int J Environ Res Public Health* 2013;**10** (12):6255−72.

20. Frost MJ, Zhang J, Edmonds JH, et al. Characterization of virulent West Nile virus Kunjin strain, Australia, 2011. *Emerg Infect Dis* 2012;**18**(5):792−800.

21. Hubalek Z, Halouzka J, Juricova Z, et al. First isolation of mosquito-borne West Nile virus in the Czech Republic. *Acta Virol* 1998;**42**(2):119−20.

22. Hubalek Z, Rudolf I, Bakonyi T, et al. Mosquito (Diptera: Culicidae) surveillance for arboviruses in an area endemic for West Nile (Lineage Rabensburg) and Tahyna viruses in Central Europe. *J Med Entomol* 2010;**47**(3):466−72.

23. Hubalek Z, Savage HM, Halouzka J, et al. West Nile virus investigations in South Moravia, Czechland. *Viral Immunol* 2000;**13**(4):427−33.

24. Weissenbock H, Bakonyi T, Rossi G, et al. Usutu virus, Italy, 1996. *Emerg Infect Dis* 2013;**19**(2):274−7.

25. Hubalek Z, Halouzka J, Jurikova Z, et al. Serologic survey of birds for West Nile Flavivirus in Southern Moravia (Czech Republic). *Vector Borne Zoonotic Dis* 2008;**8**(5):659−66.

26. Hubalek Z, Wegner E, Halouzka J, et al. Serologic survey of potential vertebrate hosts for West Nile virus in Poland. *Viral Immunol* 2008;**21**(2):247−53.

27. Hayes EB, Komar N, Nasci RS, et al. Epidemiology and transmission dynamics of West Nile virus disease. *Emerg Infect Dis* 2005;**11**(8):1167−73.

28. Gobbi F, Barzon L, Capelli G, et al. Surveillance for West Nile, Dengue and Chikungunya virus infections, Veneto Region, Italy, 2010. *Emerg Infect Dis* 2012;**18**(4):671−3.

29. Rossini G, Carletti F, Bordi L, et al. Phylogenetic analysis of West Nile virus isolates, Italy, 2008−2009. *Emerg Infect Dis* 2011;**17**(5):903−6.

30. Papa A, Bakonyi T, Xanthopoulou K, et al. Genetic characterization of West Nile virus lineage 2, Greece, 2010. *Emerg Infect Dis* 2011;**17**(5):920−2.

31. Papa A, Politis C, Tsoukala A, et al. West Nile virus lineage 2 from blood donor, Greece. *Emerg Infect Dis* 2012;**18**(4):688−9.

32. Brown EBE, Adkin A, Fooks AR, et al. Assesing risk of West Nile virus infected mosquitoes from transatlantic aircraft implications for disease emergence in the United Kingdom. *Vector Borne Zoonotic Dis* 2012;**12**(4):310−20.

33. Brinton MA. The molecular biology of West Nile virus: a new invader of the Western Hemisphere. *Ann Rev Microbiol* 2002;**56**:371−402.

34. Diaz LA, Komar N, Visintin A, et al. West Nile virus in birds, Argentina. *Emerg Infect Dis* 2008;**14**(4):689−91.

35. Lvov DK, Deryabin PG, Aristova VA, et al. *Atlas of distribution of natural foci virus infections on the territory of Russian Federation.* Moscow: SMC MPH RF Publ;2001 [in Russian].

36. Bell JA, Brewer CM, Mickelson NJ, et al. West Nile virus epizootology, Central Red river valley, North Dakota and Minnesota, 2002−2005. *Emerg Infect Dis* 2006;**12**(8):1245−7.

37. Carson PJ, Borchardt SM, Custer B, et al. Neuroinvasive disease and West Nile virus infection, North Dakota, USA, 1999−2008. *Emerg Infect Dis* 2012;**18**(4):684−6.

38. Carney RM, Ahearn SC, McConchie A, et al. Early warning system for West Nile virus risk areas, California, USA. *Emerg Infect Dis* 2011;**17**(8):1445−53.

39. Lvov DK, Kovtunov AI, Yashkulov KB, et al. Circulation of West Nile virus (Flaviviridae, Flavivirus) and some other arboviruses in the ecosystems of Volga delta, Volga-Akhtuba flood-lands and adjoining arid regions (2000−2002). *Vopr Virusol* 2004;**49**(3):45−51 [in Russian].

40. Danis K, Papa A, Theocharopoulos G, et al. Outbreak of West Nile virus infection in Greece, 2010. *Emerg Infect Dis* 2011;**17**(10):1868−72.

41. Johnson G, Nemeth N, Hale K, et al. Surveillance for West Nile virus in American white pelicans, Montana, USA, 2006−2007. *Emerg Infect Dis* 2010;**16**(3):1406−11.

42. St Leger J, Wu G, Anderson M, et al. West Nile infection in killer whale, Texas, USA, 2007. *Emerg Infect Dis* 2011;**17**(8):1531−3.

43. McMullen AR, May FJ, Guzman H, et al. Evolution of new genotype of West Nile virus in North America. *Emerg Infect Dis* 2011;**17**(5):785−93.

44. Lvov DK. Arboviral zoonoses on Northern Eurasia (Eastern Europe and Commonwealth of Independent States. In: Beran GW, Steel JH, editors. *Handbook of zoonoses. Section B: Viral.* Boca Raton, FL: CRC Press;1994. p. 237−68.

45. Bushkieva BTs, Shchelkanov MYu, Teldzhiev SB, et al. Ecologo-virological classification of arid landscapes on the territory of Kalmyk Republic. *J Infect Pathol* 2005;**12** (3–4):83–4 [in Russian].

46. Butenko AM. Modern status of the problem of Crimean hemorrhagic fever, West Nile fever and other arboviral infections in Russia. In: Lvov DK, Uryvaev LV, editors. *Investigation of virus evolution in the limits of the problems of biosafety and socially significant infections.* Moscow; 2011. p. 175–9 [in Russian].

47. Dzharkenov AF, Shchelkanov MYu, Lvov DN, et al. The experience of the supporting of field investigations in the natural foci of West Nile virus (Flaviviridae, Flavivirus) on the territory of Astrakhan region. *J Infect Pathol* 2005;**12**(3–4):87–8 [in Russian].

48. Kovtunov AI, Kolobukhina LV, Moskvina TM, et al. West Nile fever among residents of the Astrakhan region in 2002. *Vopr Virusol* 2003;**48**(5):9–11 [in Russian].

49. Lvov DK, Butenko AM, Gromashevsky VL, et al. *West Nile and other emerging–reemerging viruses in Russia. Emerging biological threat. NATO science series. Series I. life and behavior sciences*, vol. 370. Amsterdam: IOS Press;2005. p. 33–42.

50. Lvov DK, Kolobukhina LV, Shchelkanov MYu, et al. Clinical profile and diagnostics algorithm of Crimean-Congo hemorrhagic fever and West Nile fever. In: *Methodical handbook.* Moscow; 2006 [in Russian].

51. Lvov DK, Kovtunov AI, Yashkulov KB, et al. Safety issues in new and emerging infections. *Vestnik Rossiiskoi Akademii Meditsinskikh Nauk* 2004;**5**:20–5 [in Russian].

52. Lvov DK, Shchelkanov MYu, Dzharkenov AF, et al. Mosquito- and tick-transmitted infections in the Northern Part of Caspian Region (1999–2007). In: *Materials of XIV International Congress of Virology*, Istanbul, Turkey, August 10–15. Turkey: IUMS; 2008. VOP-116.

53. Lvov DN. Population interactions between WNV and other arboviruses with artropod vectors, vertebrate animals and humans in the middle and low belts of Volga delta [PhD thesis]. Moscow: D.I. Ivanovsky Institute of Virology; 2008 [in Russian].

54. Lvov DN, Dzharkenov AF, Aristova VA, et al. The isolation of Dhori viruses (Orthomyxoviridae, Thogotovirus) and Crimean-Congo hemorrhagic fever virus (Bunyaviridae, Nairovirus) from the hare (*Lepus europaeus*) and its ticks *Hyalomma marginatum* in the middle zone of the Volga delta, Astrakhan region, 2001. *Vopr Virusol* 2002;**47**(4):32–6 [in Russian].

55. Lvov DN, Shchelkanov MYu, Dzharkenov AF, et al. Population interactions of West Nile virus (Flaviviridae, Flavivirus) with arthropod vectors, vertebrates, humans in the middle and low belts of Volga

delta in 2001–2006. *Vopr Virusol* 2009;**54**(2):36–43 [in Russian].

56. Shchelkanov MYu, Alkhovsky SV. *Arbovirus infections in ecosystems of Volga-Akhtuba and Volga delta high zone. III Conference of Russian scientists with foreign participation "Fundamental sciences and progress in clinical medicine"* (Moscow, Russia; January, 20–24, 2004). Moscow: MMA Publ;2004. p. 460 [in Russian].

57. Shchelkanov MYu, Aristova VA, Kulikova LN, et al. Ecological and physiological peculiarities of Ixodidae ticks *Rhipicephalus pumilio* Sch., 1935 on the territory of north-western part of Caspian region. In: *Modern problems of epizootology. Materials of International Scientific Conference*, Krasnoobsk, Novosibirsk region, Russia, June 29; 2004. p. 290–3 [in Russian].

58. Shchelkanov MYu, Finogenov OV, Sapronov BN, et al. Circulation of West Nile virus (Flaviviridae, Flavivirus) among wild animal populations on the territory of North-Western Caspian Region. In: *Wildlife infections. Materials of international scientific and practical conference*, Pokrov, Vladimir region, Russia, September 28–30; 2004. p. 80–5 [in Russian].

59. Shchelkanov MYu, Gromashevsky VL, Aristova VA, et al. Distribution of some arboviruses on the territory of Kalmyk Republic according to investigation of human and domestic animal blood sera (2001–2002 data). In: *Modern problems of epizootology. Materials of international scientific conference*, Krasnoobsk, Novosibirsk region, Russia, June, 29; 2004. p. 293–7 [in Russian].

60. Shchelkanov MYu, Gromashevsky VL, Lvov DK. The role of ecologo-virological zoning in prediction of the influence of climatic changes on arbovirus habitats. *Vestnik Rossiiskoi Akademii Meditsinskikh Nauk* 2006; (2):22–5 [in Russian].

61. Shchelkanov MYu, Lvov DK. Role of paleogeographical reconstruction in the prognosis of arbovirus natural habitats. In: *Abstracts of International Conference "Development of International Collaboration in Infectious Disease Research"*, Koltsovo, Novosibirsk region, Russia; September 8–10. Novosibirsk: CERIS; 2004. p. 279.

62. Lvov DK, Butenko AM, Gromashevsky VL, et al. Isolation of two strains of West Nile virus during an outbreak in southern Russia, 1999. *Emerg Infect Dis* 2000;**6**(4):373–6.

8.3 Family Orthomyxoviridae

1. McCauley JW, Hongo S, Kaverin NV, et al. Orthomyxoviridae. In: King AMQ, Adams MJ, Carstens EB, Lefkowitz EJ, editors. *Virus taxonomy. 9th Rep Intern Comm Taxonomy of Viruses.* UK/USA: Elsevier Academic Press;2012. p. 749–61.

2. Shchelkanov MYu, Fedyakina IT, Proshina ES, et al. Taxonomic structure of Orthomyxoviridae: current

views and immediate prospects. *Vestnik Rossiiskoi Akademii Meditsinskikh Nauk* 2011;**5**:12–19 [in Russian].

8.3.1 Genus *Influenza A Virus*

1. Jagger BW, Wise HM, Kash JC, et al. An overlapping protein-coding region in influenza A virus segment 3 modulates the host response. *Science* 2012;**337** (6091):199–204.
2. Muramoto Y, Noda T, Kawakami E, et al. Identification of novel influenza A virus proteins translated from PA mRNA. *J Virol* 2013;**87**(5):2455–62.
3. Shchelkanov MYu, Lvov DK. Genotypic structure of the genus influenza A virus. *Vestnik Rossiiskoi Akademii Meditsinskikh Nauk* 2011;**5**:19–23 [in Russian].
4. Tong S, Zhu X, Li Y, et al. New world bats harbor diverse influenza A viruses. *PLoS Path* 2013;**9**(10):e1003657.
5. Wise HM, Foeglein A, Sun J, et al. A complicated message: identification of a novel PB1-related protein translated from influenza A virus segment 2 mRNA. *J Virol* 2009;**83**(16):8021–31.
6. Tong S, Li Y, Rivailler P, et al. A distinct lineage of influenza A virus from bats. *Proc Nat Acad Sci USA* 2012;**109**(11):4269–74.
7. Shchelkanov Myu, Lvov DK. New subtype of influenza A virus from bats and new tasks for ecologo-virological monitoring. *Vopr Virusol* 2012;(Suppl. 1):159–68 [in Russian].

8.3.1.1 Influenza A Viruses (H1–H18)

1. Meynell GG. John Locke and the preface to Thomas Sydenham's Observationes medicae. *Med Hist* 2006;**50** (1):93–110.
2. Supotnitsky MV. Microorganisms, toxins, and epidemics. Moscow; 2000 [in Russian].
3. Gezer G. History of mass diseases. St. Petersburg; 1867 [in Russian].
4. Shchelkanov MYu. *Evolution of highly pathogenic avian influenza virus (H5N1) in ecosystems of Northern Eurasia (2005–2009)* [Doctor dissertation thesis]. Moscow: D.I. Ivanovsky Institute of Virology; 2010.
5. Beveridge WIB. *Influenza: the last great plague. An unfinished story of discovery.* London: Heinemann;1977.
6. Hampson AW. Surveillance for pandemic influenza. *J Infect Dis* 1997;**176**(Suppl. 1):8–13.
7. Patterson DK. *Pandemic influenza 1700–1900, a study of historical epidemiology.* New Jersey: Rowman & Littlefield;1986.
8. Supotnitsky MV. Pandemic "Spanish" flu (1918–1920) in the context of other influenza pandemic and "avian flu". *Meditsinskaya Kartoteka* 2006;**11**:31–4 [in Russian].
9. Vogralik GF. Doctrine of epidemic diseases. Tomsk; 1935 [in Russian].
10. Shope RE. Swine influenza: III. Filtration experiments and etiology. *J Exp Med* 1931;**54**:373–85.
11. Shope RE. The incidence of neutralizing antibodies for swine influenza virus in the sera of human beings of different ages. *J Exp Med* 1936;**63**:669–84.
12. Andrewes CH, Laidlaw PP, Smith W. Experiments on the immunization of ferrets and mice. *Br J Exp Path* 1935;**16**:291–302.
13. Smith W, Andrewes CH, Laidlaw PP. A virus obtained from influenza patients. *Lancet* 1933;**2**:66–8.
14. Efremenko AA, Levtova AA. M.I. Afanasiev, the founder of St. Petersburg microbiological school. *Zh Mikrobiol Epidemiol Immunobiol* 1961;**11**:145–7.
15. Pfeiffer RF. Vorläufige Mittheilungen über den Erreger der Influenza. *Deutsche Med Wochensch (Berlin)* 1892;**18**:28 [in German].
16. Pfeiffer RF. Die Aetiologie der Influenza. *Zeitschr Hygiene Infektions* 1893;**13**:357–86 [in German].
17. Pertseva TA, Plekhanova OV, Dmitrichenko VV. Clinically significant agents of respiratory infections. Conspectus for clinical physician. Part 3. Haemophilus. Moraxella. *Kliniches Immunol Allergol Infektol* 2007;**6**:7–15 [in Russian].
18. Smorodintsev AA, Korovin AA. *Influenza.* Leningrad: Nauka;1961 [in Russian].
19. Payne AM. Symposium on the Asian influenza epidemic. *Proc Roy Soc Med* 1957;**51**:1009–15.
20. Potter CW. A history of Influenza. *J App Microbiol* 2001;**91**(4):572–9.
21. Smorodintsev AA. *Influenza and its prophylaxis.* Moscow: Meditsina;1984 [in Russian].
22. Frolov AF, Shablovskaya EA, Shevchenko LF, et al. *Influenza.* Kiev: Zdorovie;1985 [in Russian].
23. Lvov DK. *Molecular ecology of influenza and other emerging viruses in Northern Eurasia: global consequences. Materials of International Conference "Emerging influenza viruses (H5N1, H1N1)", February 15–16, 2010, Marburg, Germany).* Marburg: Koch-Mechnikov Forum;2010. p. 23.
24. Lvov DK, Burtseva EI, Shchelkanov MYu, et al. Spread of new pandemic influenza A(H1N1)v virus in Russia. *Vopr Virusol* 2010;**55**(3):4–9 [in Russian].
25. Lvov DK, Malyshev NA, Kolobukhina LV, et al. Influenza provoked by new pandemic virus A/H1N1 swl: clinics, diagnostics, treatment. Methodological recommendations. Moscow; 2009 [in Russian].
26. Shchelkanov MYu, Lvov DN, Fedyakina IT, et al. Trends in the spread of pandemic influenza A(H1N1) swl in the Far East in 2009. *Vopr Virusol* 2010;**55** (3):10–15 [in Russian].
27. Shchelkanov MYu, Prilipov AG, Lvov DK. *Evolution of emerging influenza viruses in Northern Eurasia. Materials of*

International Conference "Emerging influenza viruses (H5N1, H1N1)" (February 15–16, 2010, Marburg, Germany). Marburg: Koch-Mechnikov Forum;2010. p. 31.

28. Zimmer SM, Burke DS. Historical perspective-emergence of Influenza A (H1N1) viruses. *N Engl J Med* 2009;**361**:279–85.

29. Smorodintsev AA, Luzyanina AA, Ivanova NA. Returning of Influenza A-prim and the problem of pandemic strains emergence. *Vopr Virusol* 1979;**1**:87–90 [in Russian].

30. Gorbunova AS, Pysina TV. *Animal flu (of mammals and birds)*. Moscow: Kolos;1973 [in Russian].

31. Sopikov PM. *Diseases of the birds*. Moscow/Leningrad: Selkhozgiz;1953 [in Russian].

32. Kohler M, Kohler W. Zentralblatt für Bakteriologie—100 years ago an outbreak of fowl plague in Tyrol in 1901. *Int J Med Microbiol* 2001;**291**(5):319–21 [in German].

33. Mancini GC. Eugenio Centanni and the rise of immunology in Italy. *Med Secoli* 2004;**16**(3):603–12.

34. Perroncito E. Epizoozia tifoide nei gallinacei. *Annali Accademia Agricoltura (Torino)* 1878;**21**:87–126 [in Italian].

35. Schafer W. Sero-immunologic studies on incomplete forms of the virus of classical fowl plague. *Arch Exp Vet Med* 1955;**9**:218–30.

36. Schafer W. Vergleichende sero-immunologische Untersuchungen über die Viren der Influenza und klassischen Geflügelpest. *Zeitschr Naturforschung* 1955;**10B**:81–91 [in German].

37. Becker WB. The isolation and classification of Tern virus: influenza A-Tern South Africa-1961. *J Hyg (London)* 1966;**64**(3):309–20.

38. Lvov DK. Possible significance of natural biocenoses in the variability of influenza A viruses. *Vopr Virusol* 1974;**6**:740–4 [in Russian].

39. Laver WG, Webster RG. Ecology of influenza viruses in lower mammals and birds. *Br Med Bull* 1979;**35**(1):29–33.

40. Lvov DK. Influenza A viruses—a sum of populations with a common protected gene pool. In: Zhdanov VM, editor. *Soviet medical reviews. Section E. Virology reviews*, 2. Glasgow: Bell and Bain Ltd.;1987. p. 15–37.

41. Lvov DK. Populational interactions in biological system: influenza virus A—wild and domestic animals—humans; relations and consequences of introduction of high pathogenic influenza virus A/H5N1 on Russian territory. *Zh Mikrobiol Epidemiol Immunobiol* 2006;**3**:96–100 [in Russian].

42. Lvov DK, Kaverin NV. Avian influenza in Northern Eurasia. In: Klenk H-D, Matrosovich MN, Stech J, editors. *Avian influenza*. Basel, Switzerland: Karger;2008. p. 41–58.

43. Tsimokh PF. Hemagglutination reaction in infectious sinusitis of ducklings. *Veterinariya* 1961;**38**(12):63–5 [in Russian].

44. Prokofieva MT, Gurova EI, Tsimokh PF. Virus influenza of ducklings. *Veterinariya* 1963;**40**(10):33–5 [in Russian].

45. Prokofieva MT, Tsimokh PF. *Virus influenza of ducks. Poultry diseases*. Moscow: Kolos;1966 [in Russian].

46. Smirnova GA, Stakhanova VM. Biological properties of animal influenza viruses. *Veterinariya* 1964;**41**(4):7–12 [in Russian].

47. Zakstelskaja LY, Evstigneeva NA, Isachenko VA, et al. Influenza in the USSR: new antigenic variant A2/Hong Kong/1/68 and its possible precursors. *Am J Epidemiol* 1969;**90**:400–5.

48. Zakstelskaya LY, Antonova IV, Evstigneeva NA, et al. A new variant of influenza A2 (Hong Kong) 1–68 virus and its connection with the outbreaks of influenza observed in the USSR in 1968. *Vopr Virusol* 1969;**14**(3):320–6 [in Russian].

49. Osidze NG, Tkachenko AV, Bogautdinov ZF, et al. Study of some biological properties of influenza virus strains of poultry and horses isolated in the USSR. *Vopr Virusol* 1973;**18**(6):705–9 [in Russian].

50. Isachenko VA, Molibog EV, Zakstelskaya LY. Antigenic characteristics of the influenza viruses isolated from domestic animals and birds in the USSR. *Vopr Virusol* 1973;**18**(6):700–5 [in Russian].

51. Zakstelskaya LY, Isachenko VA, Osidze NG, et al. Some observations on the circulation of influenza-viruses in domestic and wild birds. *Bull WHO* 1972;**47**:497–501.

52. Osidze NG, Lvov DK, Syurin VN, et al. A new variety of chicken influenza virus. *Veterinariya* 1979;**56**(9):29–31 [in Russian].

53. Lvov DK, Il'ichev VD. *Migration of birds and the transfer of the infectious agents*. Moscow: Nauka;1979 [in Russian].

54. Zhezmer VYu, Lvov DK, Isachenko VA, et al. Isolation of the Hong Kong variant of type A influenza virus from sick chickens in the Kamchatka region. *Vopr Virusol* 1973;**18**(6):94–9 [in Russian].

55. Zakstelskaya LY, Isachenko VA, Timofeeva SS, et al. Serological evidence of the circulation of various variants of the influenza type A viruses among migratory birds in northern and western zones of the USSR. *Vopr Virusol* 1973;**18**(6):760–4 [in Russian].

56. Lvov DK, Sidorova GA, Eminov AE, et al. Properties of Hav6Neq2 and Hsw1(H0)Nav2 influenza viruses isolated from waterfowl in southern Turkmenia. *Vopr Virusol* 1980;**25**(4):415–19 [in Russian].

57. Sayatov MKh, Beisembaeva RU, Lvov DK, et al. Influenza viruses isolated from wild birds. *Vopr Virusol* 1981;**26**(4):466–71 [in Russian].

58. Lvov DK, Yamnikova SS, Shemyakin IG, et al. Persistence of the genes of epidemic influenza viruses (H1N1) in natural populations. *Vopr Virusol* 1982;**27**(4):401–5 [in Russian].

59. Pysina TV, Lvov DK, Braude NA, et al. Influenza virus A/Anas Acuta/Primorie/695/76 isolated from wild ducks in the USSR. *Vopr Virusol* 1978;**23**(3):300–4 [in Russian].

60. Lvov DK, Sazonov AA, Chernetsov YuV, et al. Comprehensive study of the ecology of the influenza viruses in Komandory Islands. *Vopr Virusol* 1973;**18**(6):747–50 [in Russian].

61. Sazonov AA, Lvov DK, Webster RG, et al. Isolation of an influenza virus, similar to A/Port Chalmers/1/73 (H3N2) from a common murre at Sakhalin Island in USSR (strain A/common murre/Sakhalin/1/74). *Arch Virol* 1977;**53**(1–2):1–7.

62. Pysina TV, Lvov DK, Braude NA, et al. Influenza virus A/Anas acuta/Primorie/730/76(H3N2) isolated from wild ducks in the Maritime Territory. *Vopr Virusol* 1979;**24**(5):489–93 [in Russian].

63. Shablovskaya EA, Lvov DK, Sazonov AA, et al. Isolation of influenza strains identical to influenza virus A/Anglia/42/72 from semisynanthropic bird species in Rovno Province, the Ukrainian SSR. *Vopr Virusol* 1977;**22**(4):414–18 [in Russian].

64. Zhdanov VM, Yamnikova SS, Isachenko VA, et al. Human and avian viruses of the Hong Kong series. *Vopr Virusol* 1981;**26**(6):657–64 [in Russian].

65. Sidorenko EV, Ignatenko TA, Yamnikova SS, et al. Isolation of an influenza virus from a tree sparrow and the infection rate of the virus in wild birds in the mid-Dnieper Region. *Vopr Virusol* 1985;**30**(6):657–61 [in Russian].

66. Webster RG, Isachenko VA, Carter M. A new avian influenza virus from feral birds in the USSR: recombination in nature? *Bull WHO* 1974;**51**(4):325–32.

67. Zakstelskaya LY, Timofeeva SS, Yakhno MA, et al. Isolation of influenza A viruses from wild migratory waterfowl in the north of Europian part of the USSR. In: Lvov DK, editor. *Ecology of viruses*, vol. 3. Moscow; 1975. p. 58–63 [in Russian].

68. Roslaya IG, Lvov DK, Yamnikova SS. Incidence of influenza virus infection in black-headed gulls. *Vopr Virusol* 1984;**29**(2):155–7 [in Russian].

69. Lvov DK, Andreev VP, Braude NA, et al. Isolation of influenza virus with the antigenic formula Hav4 Nav2 and Hav5 Nav2 during epizootic infection among sea gulls in the Astrakhan district in the summer of 1976. *Vopr Virusol* 1978;**23**(4):399–403 [in Russian].

70. Chernetsov YuV, Slepushkin AN, Lvov DK, et al. Isolation of influenza virus from *Chlidonias nigra* and serologic examination of the birds for antibodies to influenza virus. *Vopr Virusol* 1980;**25**(1):35–40 [in Russian].

71. Lvov DK, Yamnikova SS, Fedyakina IT, et al. Ecology and evolution of influenza viruses in Russia (1979–2002). *Vopr Virusol* 2004;**49**(3):17–24 [in Russian].

72. Hinshaw VS, Air GM, Gibbs AJ, et al. Antigenic and genetic characterization of a novel hemagglutinin subtype of influenza A viruses from gulls. *J Virol* 1982;**42**:865–72.

73. Yamnikova SS, Kovtun TO, Dmitriev GA, et al. Circulation of the influenza A virus of H13 serosubtype among seagulls in the Northern Caspian (1979–1985). *Vopr Virusol* 1989;**34**(4):426–30 [in Russian].

74. Chambers TM, Yamnikova S, Lvov DK, et al. Antigenic and molecular characterization of subtype H13 hemagglutinin of influenza virus. *Virology* 1989;**172**(1):180–8.

75. Yamnikova SS, Gambaryan AS, Fedyakina IT, et al. Monitoring of the circulation of influenza A viruses in the populations of wild birds of the North Caspian. *Vopr Virusol* 2001;**46**(4):39–43 [in Russian].

76. Yashkulov KB, Shchelkanov MYu, Lvov SS, et al. Isolation of influenza virus A (Orthomyxoviridae, Influenza A virus), Dhori virus (Orthomyxoviridae, Thogotovirus), and Newcastle's disease virus (Paramyxoviridae, Avulavirus) on the Malyi Zhemchuzhnyi Island in the north-western area of the Caspian Sea. *Vopr Virusol* 2008;**53**(3):34–8 [in Russian].

77. Kawaoka Y, Yamnikova S, Chambers TM, et al. Molecular characterization of a new hemagglutinin, subtype H14, of influenza A virus. *Virology* 1990;**179**: 759–67.

78. Okazaki K, Takada A, Ito T, et al. Precursor genes of future pandemic influenza viruses are perpetuated in ducks nesting in Siberia. *Arch Virol* 2000;**145**:885–93.

79. Shchelkanov MYu, Kirillov IM, Kolobukhina LV, et al. *Highly pathogenic Influenza A (H5N1) virus: epidemic threat is remaining. Infectious diseases and antivirals. Materials of X-th scientific-practice conference (Moscow, Russia; October, 02–03, 2012).* Moscow: Infomedfarm-Dialog;2012. p. 84–8 [in Russian].

80. Lvov DK, Gorin OZ, Yamnikova SS, et al. Isolation of influenza A viruses from wild birds and a muskrat in the western part of the East Asia migration route. *Vopr Virusol* 2001;**46**(4):35–9 [in Russian].

81. Lvov DK, Yamnikova SS, Fedyakina IT, et al. Evolution of H4, H5 influenza A viruses in natural ecosystems in Northern Eurasia (2000–2002). In: Kawaoka Y, editor.

Options for the control of influenza V. International congress series, 1263. Elsevier;2004. p. 169—83.

82. Razumova YuV, Shchelkanov MYu, Durymanova AA, et al. Genetic diversity of influenza A virus in the populations of wild birds in the south of Western Siberia. *Vopr Virusol* 2005;50(4):31—5 [in Russian].

83. Razumova YuV, Shchelkanov MYu, Zolotykh SI, et al. The 2003 results of monitoring of influenza A virus in the populations of wild birds in the south of Western Siberia. *Vopr Virusol* 2006;51(3):32—7 [in Russian].

84. Shestopalov AM, Tserennorov D, Shchelkanov MYu, et al. The study of circulation of influenza and West Nile viruses in the natural biocenoses of Mongolia. In: *Current Topics of Virology. Abstract Book of X National Mongolian Conference* (November 25, 2004, Mongolia). Ulaanbaatar; 2004. p. 122—3.

85. Lvov DK. Virus ecology. In: Lvov DK, editor. *Medical virology*. Moscow: Guide. Medical information agency;2008. p. 101—18.

86. Kaverin NV, Gambaryan AS, Bovin NV, et al. Postreassortment changes in influenza A virus hemagglutinin restoring HA-NA functional match. *Virology* 1998;244:315—21.

87. Kaverin NV, Matrosovich MN, Gambaryan AS, et al. Intergenic HA-NA interactions in influenza A virus: postreassortment substitutions of charged amino acid in the hemagglutinin of different subtypes. *Virus Res* 2000;66:123—9.

88. Zhang G, Shoham D, Gilichinsky D, et al. Evidence of influenza A virus RNA in Siberian lake ice. *J Virol* 2006;80:12229—35.

89. Lvov DK, Shchelkanov MYu, Deryabin PG, et al. Isolation of influenza A/H5N1 virus strains from poultry and wild birds in West Siberia during epizooty (July 2005) and their depositing to the state collection of viruses (August 8, 2005). *Vopr Virusol* 2006;51 (1):11—14 [in Russian].

90. Shchelkanov MYu, Vlasov NA, Kireev DE, et al. Clinical symptoms of bird disease provoked by highly pathogenic variants of influenza A/H5N1 virus in the epicenter of epizooty on the south of Western Siberia. *J Infect Path (Moscow)* 2005;12(3—4):121—4 [in Russian].

91. Lvov DK, Prilipov AG, Shchelkanov MYu, et al. Molecular genetic analysis of the biological properties of highly pathogenic influenza A/H5N1 virus strains isolated from wild birds and poultry during epizooty in Western Siberia (July 2005). *Vopr Virusol* 2006;51 (2):15—19 [in Russian].

92. Chen H, Smith GJD, Zhang SY, et al. Avian flu: H5N1 virus outbreak in migratory waterfowl. *Nature* 2005;436 (7048):191—2.

93. Lipatov AS, Govorkova EA, Webby RJ, et al. Influenza: emergence and control. *J Virol* 2004;78:8951—9.

94. Shestopalov AM, Durimanov AG, Evseenko VA, et al. H5N1 influenza virus, domestic birds, Western Siberia, Russia. *Emerg Infect Dis* 2006;12:1167—8.

95. Onishchenko GG, Shestopalov AM, Ternovoi VA, et al. Study of highly pathogenic H5N1 influenza virus isolated from sick and dead birds in Western Siberia. *Zh Mikrobiol Epidemiol Immunobiol* 2006;(5):47—54 [in Russian].

96. Al-Azemi A, Bahl J, Al-Zenki S, et al. Avian Influenza A virus (H5N1) outbreaks. Kuwait, 20. *Emerg Infect Dis* 2008;14(6):958—61.

97. Alexander DJ. Summary of avian influenza activity in Europe, Asia, Africa, and Australasia, 2002—2006. *Avian Dis* 2007;(Suppl. 1):161—6.

98. Cattoli G, Monne I, Fusaro A, et al. Highly pathogenic avian influenza virus subtype H5N1 in Africa: a comprehensive phylogenetic analysis and molecular characterization of isolates. *PLoS One* 2009;4(3):e4842.

99. Joannis T, Lombin LH, De Benedictis P, et al. Confirmation of H5N1 avian influenza in Africa. *Vet Rec* 2006;158(9):309—10.

100. Leslie T, Billaud J, Mofleh J, et al. Knowledge, attitudes, and practices regarding avian influenza (H5N1), Afghanistan. *Emerg Infect Dis* 2008;14 (9):1459—61.

101. Lvov DK, Shchelkanov MYu, Deryabin PG, et al. Highly pathogenic influenza A/H5N1virus-caused epizooty among mute swans (*Cygnus olor*) in the low estuary of the Volga River (November 2005). *Vopr Virusol* 2006;51(3):10—16 [in Russian].

102. Tosh C, Murugkar HV, Nagarajan S, et al. Outbreak of avian influenza virus H5N1 in India. *Vet Rec* 2007;161(8):279.

103. Lvov DK, Shchelkanov MYu, Deryabin PG, et al. Isolation of highly pathogenic avian influenza (HPAI) A/H5N1 strains from wild birds in the epizootic outbreak on the Uvs-Nur Lake (June 2006) and their incorporation to the Russian Federation State Collection of viruses (July 3, 2006). *Vopr Virusol* 2006;51(6):14—18 [in Russian].

104. Lvov DK, Shchelkanov MYu, Prilipov AG, et al. Molecular genetic characteristics of the strain A/chicken/Moscow/2/2007 (H5N1) strain from a epizootic focus of highly pathogenic influenza A among agricultural birds in the near-Moscow region (February 2007). *Vopr Virusol* 2007;52(6):40—7 [in Russia].

105. Lvov DK, Shchelkanov MYu, Deryabin PG, et al. Epizooty caused by high-virulent influenza virus A/H5N1 of genotype 2.2 (Qinghai-Siberian) among wild and domestic birds on the paths of fall migrations to the north-eastern part of the Azov Sea basin (Krasnodar Territory). *Vopr Virusol* 2008;53(2):14—19 [in Russian].

106. Lvov DK, Shchelkanov MYu, Prilipov AG, et al. Interpretation of the epizootic outbreak among wild and domestic birds in the south of the European part of Russia in December 2007. *Vopr Virusol* 2008;**53** (4):13−18 [in Russian].

107. Lvov DK, Shchelkanov MYu, Prilipov AG, et al. Evolution of HPAI H5N1 virus in Natural ecosystems of Northern Eurasia (2005−2008). *Avian Dis* 2010;**54** (1):483−95.

108. Li M, Wang B. Homology modeling and examination of the effect of the D92E mutation on the H5N1 non-structural protein NS1 effector domain. *J Mol Model* 2007;**13**(12):1237−44.

109. Long JX, Peng DX, Liu YL, et al. Virulence of H5N1 avian influenza virus enhanced by a 15-nucleotide deletion in the viral nonstructural gene. *Virus Genes* 2008;**36**(3):471−8.

110. Deryabin PG, Lvov DK, Isaeva EI, et al. The spectrum of vertebrate cell lines sensitive to highly pathogenic influenza A/tern/SA/61 (H5N3) and A/duck/ Novosibirsk/56/05 (H5N1) viruses. *Vopr Virusol* 2007;**52**(1):45−7 [in Russian].

111. Lu J, Zhang D, Wang G. Highlight the significance of genetic evolution of H5N1 avian flu. *Chin Med J* 2006;**119**(17):1458−64.

112. Russell CJ, Webster RG. The genesis of a pandemic influenza virus. *Cell* 2005;**123**(3):368−71.

113. Lvov DK, Fedyakina IT, Shchelkanov MYu, et al. *In vitro* effects of antiviral drugs on the reproduction of highly pathogenic influenza A/H5N1 virus strains that induced epizooty among poultry in the summer of 2005. *Vopr Virusol* 2006;**51**(2):20−2 [in Russian].

114. Shchelkanov MYu, Prilipov AG, Lvov DK, et al. Dynamics of virulence for highly virulent influenza A/H5N1 strains of genotype 2.2 isolated on the territory of Russia during 2005−2007. *Vopr Virusol* 2009;**54** (2):8−17 [in Russian].

115. Lvov DK, Shchelkanov MYu, Vlasov NA, et al. The first break-trough of the genotype 2.3.2.1 of highly virulence influenza A/H5N1 virus, which is new for Russia, in the Far East. *Vopr Virusol* 2008;**53**(5):4−8 [in Russian].

116. Kang HM, Batchuluun D, Kim MC, et al. Genetic analyses of H5N1 avian influenza virus in Mongolia, 2009 and its relationship with those of eastern Asia. *Vet Microbiol* 2011;**147**(1−2):170−5.

117. Sakoda Y, Sugar S, Batchluun D, et al. Characterization of H5N1 highly pathogenic avian influenza virus strains isolated from migratory waterfowl in Mongolia on the way back from the southern Asia to their northern territory. *Virology* 2010;**406**(1):88−94.

118. Köbe K. Die Aetiologie der Ferkelgrippe (enzootische Pneumonie des Ferkels). *Zentralbl Bakt* 1933;**129**:161−8 [in German].

119. Gulrajani TS, Beveridge WI. Infectious pneumonia of pigs. *Nature* 1951;**167**(4256):856−7.

120. Gulrajani TS, Beveridge WI. Studies on respiratory diseases of pigs. IV. Transmission of infectious pneumonia and its differentiation from swine influenza. *J Comp Pathol* 1951;**61**(2):118−39.

121. Betts AO, Beveridge WI. Investigations on a virus pneumonia of long duration prevalent in pigs. *J Pathol Bacteriol* 1952;**64**(1):247−8.

122. Sasahara J. Studies on the HVJ (Hemagglutinating virus of Japan) newly isolated from the swine. *Rep Natl Inst Anim Health (Tokyo)* 1955;**30**:1−32.

123. Shimizu T, Ishizaki R, Kono Y, et al. Multiplication and cytopathogenic effect of the hemagglutinating virus of Japan (HVJ) in swine kidney tissue culture. *Jpn J Exp Med* 1955;**25**(6):211−22.

124. Andrewes CH, Stuart-Harris CH. Discussion on virus infections of the upper respiratory tract. *Proc R Soc Med* 1958;**51**(7):469−74.

125. Francis TJr, Quilligan Jr. JJ, Minuse E. Resemblance of a strain of swine influenza virus to human A-prime strains. *Proc Soc Exp Biol Med* 1949;**71**(2): 216−20.

126. Labutina A. Investigation of swine respiratory diseases in Lithuania SSR. In: *Proceedings of Estonia Academy of farm industry*, Tartu; 1965. p. 18−24 [in Russian].

127. Aaver EA. *About etiology of swine influenza. Bulletin of scientific-technical information.* Tallinn: Estonian Institute of Agriculture and Melioration;1957. p. 73−7. [in Russian].

128. Parnas J, Lorkiewicz Z, Szczygielska J. Результаты сравнительного исследования вируса G1, выделенного от животных, и его отношение в вирусам гриппа человека. *Bull Acad Polon Sci* 1957;**5**(3):89−95.

129. Osidze DF. Virological and serological investigation of swine influenza. *Veterinariya* 1964;**1**:19−22 [in Russia].

130. Gois M, Mensik J, Davidova M, et al. Attempt to standardize techniques used in isolating influenza virus from pig lungs. *Acta Virol* 1963;**7**:455−64.

131. Harnach R, Hubik R, Chivatal O. Isolation of influenza virus in Czechoslovakia. *Cas Cesk Vet* 1950;**5** (13):289 [in Czech].

132. Barb K, Farkas E, Romvary J, et al. Comparative investigation of influena virus strains isolated from domestic animals in Hungary. *Acta Virol* 1962;**6**(3):207−13.

133. Kaplan MM, Payne AM. Serological survey in animals for type A influenza in relation to the 1957 pandemic. *Bull WHO* 1959;**20**:465−7.

134. Mensik J. Experimental infection of pregnant sows with swine influenza virus. *Ved Prace Ustavi Vet (Brno)* 1962;**6**:409−15 [in Czech].

135. Gambaryan AS. Receptor specificity of Influenza A viruses from different hosts [Doctor dissertation

thesis]. Moscow: Institute of Poliomyelitis and Virus Encephalitis; 2007 [in Russian].

136. Gambaryan AS. *Three receptors of Influenza A virus. Medical virology.* Moscow: Institute of Poliomyelitis and Virus Encephalitis;2007. p. 245—54. [in Russian].

137. Lvov DK, Zaberezhny AD, Aliper TI. Influenza viruses: events and prognosis. *Nature (Moscow)* 2006;**6**:3—13 [in Russian].

138. Marinina VP, Gambaryan AS, Tuzikov AB, et al. Evolution of the receptor specificity of influenza viruses hemagglutinin in its transfer from duck to pig and man. *Vopr Virusol* 2004;**49**(3):25—30 [in Russian].

139. Gray GC, McCarthy T, Capuano AW, et al. Swine workers and swine influenza virus infections. *Emerg Infect Dis* 2007;**13**(12):1871—8.

140. Ito T, Couceiro JN, Kelm S, et al. Molecular basis for the generation in pigs of influenza A viruses with pandemic potential. *J Virol* 1998;**72**(9):7367—73.

141. Robinson JL, Lee BE, Patel J, et al. Swine influenza (H3N2) infection in a child and possible community transmission, Canada. *Emerg Infect Dis* 2007;**13**(12):1865—70.

142. Yu H, Hua RH, Zhang Q, et al. Genetic evolution of swine influenza A (H3N2) viruses in China from 1970 to 2006. *J Clin Microbiol* 2008;**46**(3):1067—75.

143. Sergeev VA, Nepoklonov EA, Aliper TI. *Viruses and antivirus vaccines.* Moscow: Biblionika;2007 [in Russian].

144. Smetanin MA, Gumerov NK. Circulation of Influenza A viruses among domestic animals. *Veterinariya* 1982;**8**:25—6 [in Russian].

145. Syurin VN, Samuilenko AYa, Soloviev BV, et al. Viral diseases of animals. Moscow; 1998 [in Russian].

146. Yamnikova SS, Kurinov GV, Lomakina NF, et al. Infection of pigs with influenza A/H4 and A/H5 viruses isolated from wild birds on the territory of Russia. *Vopr Virusol* 2008;**53**(6):30—4 [in Russian].

147. Brown IH, Alexander DJ, Chakraverty P, et al. Isolation of an influenza A virus of unusual subtype (H1N7) from pigs in England, and the subsequent experimental transmission from pig to pig. *Vet Microbiol* 1994;**39**(1—2):125—34.

148. Brown IH, Hill ML, Harris PA, et al. Genetic characterisation of an influenza A virus of unusual subtype (H1N7) isolated from pigs in England. *Arch Virol* 1997;**142**(5):1045—50.

149. Yasuhara H, Hirahara T, Nakai M, et al. Further isolation of a recombinant virus (H1N2, formerly Hsw1N2) from a pig in Japan in 1980. *Microbiol Immunol* 1983;**27**(1):43—50.

150. Saito T, Suzuki H, Maeda K, et al. Molecular characterization of an H1N2 swine influenza virus isolated in Miyazaki, Japan, in 2006. *J Vet Med Sci* 2008;**70** (4):423—7.

151. Starikov NS, Shchelkanov My, Bovin NV, et al. Statistical approaches to the analysis of receptor specificity spectra of Influenza A virus. In: *Proceedings of IX-th scientific-practice conference "Infectious disease and antivirals" (Moscow, Russia; October 6—7, 2011).* Moscow; 2011. p. 80—1 [in Russian].

152. Lvov DK, Yashkulov KB, Prilipov AG, et al. Detection of amino acid substitutions of asparaginic acid for glycine and asparagine at the receptor-binding site of hemagglutinin in the variants of pandemic influenza A/H1N1 virus from patients with fatal outcome and moderate form of the disease. *Vopr Virusol* 2010;**55** (3):15—18 [in Russian].

153. Lvov DK, Burtseva EI, Prilipov AG, et al. A possible association of fatal pneumonia with mutations of pandemic influenza A/H1N1 swl virus in the receptor-binding site of HA1 subunit. *Vopr Virusol* 2010;**55** (4):4—9 [in Russian].

154. Kolobukhina LV, Merkulova LN, Malyshev NA, et al. Strategy of early antiviral therapy of Influenza A virus as prophylaxis of severe complexities. *Pulmonology (Moscow)* 2010;(Suppl. 1):9—14 [in Russian].

155. Kolobukhina LV, Malyshev NA, Shchelkanov My, et al. *Peculiarities of epidemiological season of influenza 2010—2011—the first postpandemic season. Information-analytical report on the materials of Moscow infection clinical hospital N 1. Infectious diseases and antivirals. Materials of IX-th scientific-practice conference (Moscow, Russia; October 06—07, 2011).* Moscow: Infomedfarm-Dialog;2011. p. 40—2. [in Russian].

156. Lvov DK, Shchelkanov MYu, Bovin NV, et al. Correlation between the receptor specificities of pandemic influenza A (H1N1) pdm09 virus strains isolated in 29—211 and the structure of the receptor-binding site and the probabilities of fatal primary virus pneumonia. *Vopr Virusol* 2012;**57**(1):14—20 [in Russian].

157. Kolobukhina LV, Shchelkanov MYu, Proshina ES, et al. Clinic and pathogenetic peculiarities and optimization of antiviral therapy of pandemic influenza A (H1N1) pdm09. *Vopr Virusol* 2012;(Suppl. 1):189—98 [in Russian].

158. Proshina ES, Starikov NS, Kirillov IM, et al. *Interrelations between the receptor specificity coefficient of Influenza A (H1N1) pdm09 virus strains during 2010—2011 epidemiological season. Infectious diseases and antivirals. Materials of X-th scientific-practice conference (Moscow, Russia; October 02—03, 2012).* Moscow: Infomedfarm-Dialog;2012:67—9 [in Russian].

159. Lavrischeva VV, Burtseva EI, Khomyakov YuN, et al. Etiology of fatal pneumonia cause by influenza A

(H1N1) pdm09 virus during the pandemic in Russia. *Vopr Virusol* 2013;**58**(3):17–21 [in Russian].

160. Agapov SI. About pathogenic properties of Influenza A viruses from pigs and humans. *Zh Mikrobiol Epidemiol Immunobiol* 1936;**XVII**(4):543–7 [in Russian].

161. Rosocha J, Newbert J. Influenza of swine influenza virus from rats and study of its characteristics. *Folia Vet (Kosice)* 1956;**1**:179–87.

162. Romvary J, Takatsy G, Barb K, et al. Isolation of influenza virus strains from animals. *Nature* 1962;**193** (4818):907–8.

163. Hoyle L. Studies of pneumonia in mice. *J Patal Bact* 1935;**41**:163–71.

164. Lvov DK, Shchelkanov MYu, Fedyakina IT, et al. Strain of influenza A/IIV-Anadyr/177-ma/2009 (H1N1) pdm09 adapted for laboratory mice lung tissues. Patent of Russian Federation N 2487936. Priority of the invention 02.02.2012 [in Russian].

165. Lu X, Tumpey TM, Morken T, et al. A mouse model for the evaluation of pathogenesis and immunity to influenza A (H5N1) viruses isolated from humans. *J Virol* 1999;**73**(7):5903–11.

166. Shchelkanov MYu, Fedyakina IT, Kirillov IM, et al. *Biological model of lethal primary viral pneumonia in laboratory mice provoked by specially adapted Influenza A/IIV-Anadyr/177-ma/2009 (H1N1) pdm09 strain. Infectious diseases and antivirals. Materials of X-th scientific-practice conference (Moscow, Russia; October 02–03, 2012)*. Moscow: Infomedfarm-Dialog;2012. p. 83–4. [in Russian].

167. Wright PF, Neumann G, Kawaoka Y. *Orthomyxoviruses*. 5th ed. *Fields Virology*, 2. Philadelphia/Baltimore/NY/London/Buenos Aires/Hong Kong/Sydney/Tokyo: Walter Kluwer, Lippincott Williams & Wilkins;2007. p. 1691–40. [Chapter 48].

168. Matsuura Y, Yanagawa R, Noda H. Experimental infection of mink with influenza A viruses. *Arch Virol* 1979;**62**:71–6.

169. Klingeborn B, Englund L, Rott R, et al. An avian influenza A virus killing a mammalian species—the mink. Brief report. *Arch Virol* 1985;**63**:347–51.

170. Klopfleisch R, Wolf PU, Uhl W, et al. Distribution of lesions and antigen of highly pathogenic avian influenza virus A/Swan/Germany/R65/06 (H5N1) in domestic cats after presumptive infection by wild birds. *Vet Pathol* 2007;**44**(3):261–8.

171. Klopfleisch R, Wolf PU, Wolf C, et al. Encephalitis in a stone marten (*Martes foina*) after natural infection with highly Pathogenic Avian Influenza Virus Subtype H5N1. *J Comp Pathol* 2007;**137**(2–3):155–9.

172. Nakamura J, Iwasa T. On the fowl-pest infection in cat. *Jpn J Vet Sci* 1942;**4**:511–23 [in Japanese].

173. Paniker CK, Nair CM. Infection with A2 Hong Kong influenza virus in domestic cats. *Bull WHO* 1970;**43** (6):859–62.

174. Xian-zhu X, Yu-Wei G, Rong-liang H. The first finding of tiger influenza by virus isolation and specific gene amplification. *Chin J Vet Sci* 2003;**23**(2):107–10 [in Chinese].

175. Keawcharoen J, Oraveerakul K, Kuiken T, et al. Avian influenza H5N1 in tigers and leopards. *Emerg Infect Dis* 2004;**10**(12):2189–91.

176. Thanawongnuwech R, Amonsin A, Tantilertcharoen R, et al. Probable tiger-to-tiger transmission of avian influenza H5N1. *Emerg Infect Dis* 2005;**11**:699–701.

177. Amonsin A, Songserm T, Chutinimitkul S, et al. Genetic analysis of influenza A virus (H5N1) derived from domestic cat and dog in Thailand. *Arch Virol* 2007;**152**(10):1925–33.

178. Kuiken T, Rimmelzwaan G, Van Amerongen G, et al. Avian H5N1 influenza in cats. *Science* 2004;**306**:241.

179. Weber S, Harder T, Starick E, et al. Molecular analysis of highly pathogenic avian influenza virus of subtype H5N1 isolated from wild birds and mammals in northern Germany. *J Gen Virol* 2007;**88**(2):554–8.

180. Titiva SM. Experimental Influenza A virus infection among dogs. *Vopr Med Virusol* 1954;**4**:114–21 [in Russian].

181. Ado AD, Titova SM. Investigation of experimental influenza in dogs. *Vopr Virusol* 1959;**2**:165–9 [in Russian].

182. Todd JD, Cohen D. Studies of influenza in dogs. I. Susceptibility of dogs to natural and experimental infection with human A2 and B strains of influenza virus. *Am J Epidemiol* 1968;**87**(2):426–39.

183. Pysina TV, Syurin NG. Isolation of Influenza A virus related to A2 (Hong Kong) from dogs. *Vopr Virusol* 1972;**17**(2):245–8 [in Russian].

184. Nikitin T, Cohen D, Todd JD, et al. Epidemiological studies of A/Hong Kong/68 virus infection in dogs. *Bull WHO* 1972;**47**:471–9.

185. Thiry E, Zicola A, Addie D, et al. Highly pathogenic avian influenza H5N1 virus in cats and other carnivores. *Vet Microbiol* 2007;**122**(1–2):25–31.

186. Maas R, Tacken M, Ruuls L, et al. Avian influenza (H5N1) susceptibility and receptors in dogs. *Emerg Infect Dis* 2007;**13**(8):1219–21.

187. Zini E, Glaus TM, Bussadori C, et al. Evaluation of the presence of selected viral and bacterial nucleic acids in pericardial samples from dogs with or without idiopathic pericardial effusion. *Vet J* 2007;**179** (2):225–9.

188. Sovinova O, Tumova B, Pouska F, et al. Isolation of a virus causing respiratory disease in horses. *Acta Virol* 1958;**2**:52–61.

189. Gibson CA, Daniels RS, Oxford JS, et al. Sequence analysis of the equine H7 influenza virus haemagglutinin gene. *Virus Res* 1992;**22**(2):93–106.

190. Lief FS, Cohen D. Equine influenza. Studies of the virus and of antibody patterns in convalescent, interepidemic and postvaccination sera. *Am J Epidemiol* 1965;**82**(3):225–46.

191. Domracheva ZV. Outbreak of Influenza A2 among humans and horses (preliminary report). *Zh Mikrobiol Epidemiol Immunobiol* 1961;**7**:31–6 [in Russian].

192. Waddell GH, Teigland MB, Sigel MM. A new influenza virus associated with equine respiratory disease. *J Am Vet Med Assoc* 1963;**143**:587–90.

193. Scholtens RG, Steele JH, Dowdle JH, et al. U.S. epizootic of equine influenza, 1963. *Public Health Rep* 1964;**79**:393–402.

194. Sommamoreira RE, Tosi HC, Vallone EF, et al. Evolution of the antibody curve in animals experimentally inoculated with influenza virus Aequi/Uruguay/540/1963. *An Fac Med Univ Repub Montev Urug* 1964;**49**:436–9.

195. Wilson JC, Bryans JT, Doll ER. Recovery of influenza virus from horses in the equine influenza epizootic of 1963. *Am J Vet Res* 1965;**26**(115):1466–8.

196. Yamnikova SS, Mandler J, Bekh-Ochir ZH, et al. A reassortant H1N1 influenza A virus caused fatal epizootics among camels in Mongolia. *Virology* 1993;**197**(2):558–63.

197. Molibog EV, Smetanin MA, Yakhno MA. Investigation of Influenza A viruses isolated from cattle. In: *Ecology of viruses*. Moscow; 1975 (Issue 3). p. 63–6 [in Russian].

198. Farkhutdinova M, Kiryanova AI, Isachenko VA, et al. Isolation and identification of Influenza A/Hong Kong (H3N2) during respiratory diseases of cattle. *Vopr Virusol* 1973;**4**:474–8 [in Russian].

199. Labengarts YaZ, Rybakova AM, Orlova AV. Investigation of sera from cattle for the presence of anti-flu antibodies. *Zh Mikrobiol Epidemiol Immunobiol* 1983;**3**:100–3 [in Russian].

200. Maksimovich MB, Shablovskaya EA, Kozlovsky MM. Investigation of ecology of Influenza A virus in the Western part of Ukraine SSR. In: *Ecology of viruses*. Moscow; 1982. p. 82–6 [in Russian].

201. Sidorenko EV. Investigation of Influenza A virus infection among farm animals according to serological data. In: *Influenza and acute respiratory diseases*. Kiev; 1966. p. 45–9 [in Russian].

202. Sidorenko EV, Ignatenko TA, Zelenskaya TP. About longitudinal circulation of Influenza A virus strains of humans and in the nature. In: *Ecology of viruses*. Moscow; 1982. p. 86–90 [in Russian].

203. Bronitki A, Gabriella I, Malian A. L'incidence des anticorps antirippaux chez l'homme et chez les animaux domestiques. Le recueil des VIII Congres international de pathologie comparee (September 7–22, 1966, Beyrouth, Liban). Tomme 11 « La grippe »: 230 [in French].

204. McQueen JL, Davenport FM. Experimental influenza in sheep. *Proc Soc Exp Biol Med* 1963;**112**:1004–6.

205. Pysina TV, Lvov DK, Braude NA, et al. Isolation of Influenza A virus from reindeer. *Veterinariya* 1979;**9**:32.

206. Ehrengut W, Sarateanu DE, Rutter G. Influenza A antibodies in deer and elk. *Dtsch Med Wochenschr* 1979;**104**(31):1112 [in German].

207. Ehrengut W, Sarateanu DE, Rutter G. Influenza A antibodies in Cervine animals. *Infection* 1980;**8**(2):66–9.

208. Li SQ. Serological study of influenza virus antibody in deer. *Zhonghua Yu Fang Yi Xue Za Zhi* 1983;**17**(4):237–9 [in Chinese].

209. Lvov DK. Influenza. *Health (Moscow)* 1979;**1**:7–11 [in Russian].

210. Lvov DK, Zhdanov VM, Sazonov AA, et al. Comparison of influenza viruses isolated from man and from whales. *Bull WHO* 1978;**56**(6):923–30.

211. Hinshaw VS, Bean WJ, Geraci JR, et al. Characterization of two influenza A viruses from a pilot whale. *J Virol* 1986;**58**:655–6.

212. Hinshaw VS, Bean WJ, Webster RG, et al. Are seals frequently infected with avian influenza viruses? *J Virol* 1984;**51**:863–5.

213. Callan RJ, Early G, Kida H, et al. The appearance of H3 influenza viruses in seals. *J Gen Virol* 1995;**76**:199–203.

214. Webster RG, Geraci J, Petursson G, et al. Conjunctivitis in human beings caused by influenza A virus of seals. *N Engl J Med* 1981;**304**:911.

215. Webster RG, Hinshaw VS, Bean WJ, et al. Characterization of an influenza A virus from seals. *Virology* 1981;**113**:712–14.

216. Isakov VA. Severe forms of influenza (clinics and step treatment system) [Doctor dissertation thesis]. St. Petersburg; 1996 [in Russian].

217. Kiselev OI, Isakov VA, Sharonov BP, et al. Pathogenesis of the severe forms of influenza. *Vestnik Rossiiskoi Academii Meditsinskikh Nauk* 1994;**9**:32–6 [in Russia].

218. Kolobukhina L, Shchelkanov MYu. Virus infections of respiratory tract. In: Chuchalin AG, editor. *Pulmonology. National guidance*. Moscow: GEOTAR-Media;2013. p. 143–70 [Chapter 6]. [in Russian].

219. Chuchalin AG. Syndrom of acute lesion of lungs. *Rus Med J (Moscow)* 2006;**14**(22):1582 [in Russian].

220. Kolobukhina LV. Viral infections of respiratory system. *Rus Med J (Moscow)* 2000;**8**(13–14):559–64 [in Russian].

221. Kolobukhina LV. Clinics and treatment of influenza. *Rus Med J (Moscow)* 2001;**9**(16–17):710–13 [in Russian].

222. Chen J, Skehel JJ, Wiley DC. N- and C-terminal residues combine in the fusion-pH influenza hemagglutinin HA(2) subunit to form an N cap that terminates the triple-stranded coiled coil. *Proc Natl Acad Sci USA* 1999;**96**(16):8967–72.

223. Cross KJ, Burleigh LM, Steinhauer DA. Mechanism of cell entry by influenza virus. *Expert Rev Mol Med* 2001;**1**:1–18.

224. Korte T, Ludwig K, Booy FP, et al. Conformational intermediates and fusion activity of influenza virus hemagglutinin. *J Virol* 1999;**73**(6):4567–74.

225. Palese P, Shaw ML. *Orthomyxoviridae: the viruses and their replication*. 5th ed. *Fields virology*, 2. Philadelphia/Baltimore/NY/London/Buenos Aires/Hong Kong/Sydney/Tokyo: Walter Kluwer/Lippincott Williams & Wilkins;2007. p. 1647–89 [Chapter 47].

226. Skehel JJ, Bizebard T, Bullough PA, et al. Membrane fusion by influenza hemagglutinin. *Cold Spring Harb Symp Quant Biol* 1995;**60**:573–80.

227. Horimoto T, Kawaoka Y. Influenza: lessons from past pandemics, warning from current incidents. *Nature Rev Microbiol* 2005;**3**:591–600.

228. Kobayashi Y, Horimoto T, Kawaoka Y, et al. Pathological studies of chickens experimentally infected with two highly pathogenic avian influenza viruses. *Avian Pathol* 1996;**25**:285–304.

229. Mo IP, Brugh M, Fletcher OJ, et al. Comparative pathology of chickens experimentally inoculated with avian influenza viruses of low and high pathogenecity. *Avian Dis* 1997;**41**:125–36.

230. Kawaoka Y, Naeve CW, Webster RG. Is virulence of H5N2 influenza viruses in chicken associated with loss of carbohydrate from the hemagglutinin ? *Virology* 1984;**139**:303–16.

231. Kawaoka Y, Webster RG. Sequence requirements for cleavage activation of influenza virus hemagglutinin expressed in mammalian cells. *Proc Natl Acad Sci USA* 1988;**85**:324–8.

232. Chen H, Deng G, Li Z, et al. The evolution of H5N1 influenza viruses in ducks in southern China. *Proc Natl Acad Sci USA* 2004;**101**:10452–7.

233. Gabriel G, Dauber B, Wolff T. The viral polymerase mediates adaptation of an avian influenza virus to a mammalian host. *Proc Natl Acad Sci USA* 2005;**102**:18590–5.

234. Syurin VN, Osidze NG, Chistova ZYa, et al. Epizootic potential of avian influenza virus. *Veterinariya* 1972;**49**(8):41–3 [in Russian].

235. Bakulin VA. Mycotoxicoses of birds. *Zooindustriya* 2006;**7**:4–10 [in Russian].

236. Obukhov IL. Properties and the cycle of Chlamydia development. *Farm Biology (Moscow)* 1999;**4**:12–27 [in Russian].

237. Spesivtseva NA. *Mycoses and mycotoxicoses*. Moscow: Kolos;1964 [in Russian].

238. Lvov DK, Zhdanov VM. Persistence of the genes of epidemic influenza A viruses in natural populations. *Uspekhi Sovremennoi Biologii* 1982;**93**(3):323–37 [in Russian].

239. Slepushkin AN, Pysina TV, Gonsovsky FK, et al. Haemagglutination-inhibiting activity to type A influenzaviruses in the sera of wild birds from the far east of the USSR. *Bull WHO* 1972;**47**:527–30.

8.3.2 Genus *Quaranjavirus*

1. McCauley JW, Hongo S, Kaverin NV, et al. Orthomyxoviridae. In: King AMQ, Adams MJ, Carstens EB, Lefkowitz EJ, editors. *Virus taxonomy. IX-th Report of Internationl Committee of Taxonomy of Viruses*. UK/USA: Elsevier Academic Press;2012. p. 749–61.

2. Austin FJ. Johnston Atoll virus (Quaranfil group) from Ornithodoros capensis (Ixodoidea: Argasidae) infesting a gannet colony in New Zealand. *Am J Trop Med Hyg* 1978;**27**(5):1045–8.

3. Quaranfil (QRFV). In: Karabatsos N, editor. *International catalogue of arboviruses and certain other viruses of vertebrates*. San Antonio, TX: American Society of Tropical Medicine and Hygiene;1985. p. 849–50.

4. Presti RM, Zhao G, Beatty WL, et al. Quaranfil, Johnston Atoll, and Lake Chad viruses are novel members of the family Orthomyxoviridae. *J Virol* 2009;**83**(22):11599–606.

5. Taylor RM, Hurlbut HS, Work TH, et al. Arboviruses isolated from Argas ticks in Egypt: Quaranfil, Chenuda, and Nyamanini. *Am J Trop Med Hyg* 1966;**15**(1):76–86.

6. Kemp GE, Lee VH, Moore DL. Isolation of Nyamanini and Quaranfil viruses from *Argas (Persicargas) arboreus* ticks in Nigeria. *J Med Entomol* 1975;**12**(5):535–7.

7. Jupp PG, McIntosh BM. Identity of argasid ticks yielding isolations of Chenuda, Quaranfil and Nyamanini viruses in South Africa. *Entomol Soc South Afr* 1986;**49**:392–5.

8.3.2.1 Tyulek Virus

1. Karas FR. Ecology of arboviruses of mountain system in Central Asian region of USSR [Doctor dissertation thesis]. Moscow: D.I. Ivanosky institute of Virology; 1979 [in Russian].

2. Gershtein VI, Vargina SG, Kuchuk LA, et al. About ecology of arbovirusses, associated with shelter biocenosis. In: Lvov DK, editor. *Uspekhi nauki i tekhniki. Ch «Virology. Arboviruses and arboviral infection»*. Moscow: VINITI;1992. p. 45–9.

3. Karas FR. Arboviruses in Kirgizia. In: Gaidamovich SY, editor. *Arboviruses*. Moscow: D.I. Ivanosky institute of Virology of USSR;1978p. 40–4 [in Russian].

4. McCauley JW, Hongo S, Kaverin NV, et al. Orthomyxoviridae. In: King AMQ, Adams MJ, Carstens EB, Lefkowitz EJ, editors. *Virus taxonomy. IX-th Report of International Committee of Taxonomy of Viruses*. UK/USA: Elsevier Academic Press;2012. p. 749–61.

5. Quaranfil (QRFV). In: Karabatsos N, editor. *International catalogue of arboviruses and certain other viruses of vertebrates*. San Antonio, TX: American Society of Tropical Medicine and Hygiene;1985. p. 849–50.

6. Presti RM, Zhao G, Beatty WL, et al. Quaranfil, Johnston Atoll, and Lake Chad viruses are novel members of the family Orthomyxoviridae. *J Virol* 2009;**83** (22):11599–606.

7. Lvov DK, Il'ichev VD. *Migration of birds and the transfer of the infectious agents*. Moscow: Nauka;1979 [in Russian].

8. Lvov DK. Isolation of the viruses from natural sources in USSR. In: Lvov DK, Klimenko SM, Gaidamovich SY, editors. *Arboviruses and arboviral infections*. Moscow: Meditsina;1989p. 220–35 [Chapter 7].

9. Lvov DK. Quaranfil fever. In: Lvov DK, editor. *Viruses and viral infection*. Moscow: MIA;2013. p. 789.

10. Austin FJ. Johnston Atoll virus (Quaranfil group) from *Ornithodoros capensis* (Ixodoidea: Argasidae) infesting a gannet colony in New Zealand. *Am J Trop Med Hyg* 1978;**27**(5):1045–8.

11. Taylor RM, Hurlbut HS, Work TH, et al. Arboviruses isolated from Argas ticks in Egypt: Quaranfil, Chenuda, and Nyamanini. *Am J Trop Med Hyg* 1966;**15** (1):76–86.

12. Kemp GE, Lee VH, Moore DL. Isolation of Nyamanini and Quaranfil viruses from *Argas (Persicargas) arboreus* ticks in Nigeria. *J Med Entomol* 1975;**12**(5):535–7.

13. Lvov DK, Alkhovsky SV, Shchelkanov MYu, et al. Taxonomic status of Tyulek virus (TLKV) (Orthomyxoviridae, Quaranjavirus, Quaranfil group) isolated from ticks *Argas vulgaris* Filippova, 1961 (Argasidae) from the birds burrow nest biotopes in the Kyrgyzstan. *Vopr Virusol* 2014;**59**(2):28–32 [in Russian].

14. Jupp PG, McIntosh BM. Identity of argasid ticks yielding isolations of Chenuda, Quaranfil and Nyamanini viruses in South Africa. *Entomol Soc South Afr* 1986;**49**:392–5.

15. Filippova NA. *Fauna of USSR. Acariformes*, Moscow/ Leningrad: Academy of Sciences of USSR;1966. vol. 4 (3). Ixodoidea. Argas ticks (Argasidae) [in Russian].

16. Gershtein VI. Current issue of arbovirus ecology in Kirgizia. *Frunze* 1981;:38–9 [in Russian].

17. Vargina SG, Breininger IG. Monitoring of natural foci of arboviruses in Kirgizia. In: Lvov DK, editor. *Uspekhi nauki i tekhniki. Ch* «*Virology. Arboviruses and arboviral infection*». Moscow: Academy of Science of USSR;1991. p. 15–16 [in Russian].

18. Vargina SG, Breininger IG, Gershtein VI. Dynamics of circulation of arboviruses in Kirgizia. In: Lvov DK, editor. *Uspekhi nauki i tekhniki. Ch* «*Virology. Arboviruses and arboviral infection*». Moscow: Academy of Science of USSR;1992. p. 38–45 [in Russian].

8.3.3 Genus *Thogotovirus*

1. McCauley JW, Hongo S, Kaverin NV, et al. Orthomyxoviridae. In: King AMQ, Adams MJ, Carstens EB, Lefkowitz EJ, editors. *Virus taxonomy. IX-th Report of International Committee of Taxonomy of Viruses*. UK/USA: Elsevier Academic Press;2012. p. 749–61.

2. Bussetti AV, Palacios G, Travassos da Rosa A, et al. Genomic and antigenic characterization of Jos virus. *J Gen Virol* 2012;**93**(2):293–8.

3. Williams RE, Hoogstraal H, Casals J, et al. Isolation of Wanowrie, Thogoto, and Dhori viruses from Hyalomma ticks infesting camels in Egypt. *J Med Entomol* 1973;**10** (2):143–6.

4. Jupp PG. Arboviral zoonoses in Africa. In: Beran GW, Steel JH, editors. *Handbook of zoonoses Section B: Viral*. Boca Raton, FL: CRC Press;1994. p. 262–73.

8.3.3.1 Dhori Virus and Batken Virus (var. Dhori virus)

1. Anderson CR, Casals J. Dhori virus, a new agent isolated from *Hyalomma dromedarii* in India. *Ind J Med Res* 1973;**61**(10):1416–20.

2. Williams RE, Hoogstraal H, Casals J, et al. Isolation of Wanowrie, Thogoto, and Dhori viruses from Hyalomma ticks infesting camels in Egypt. *J Med Entomol* 1973;**10** (2):143–6.

3. Filipe AR, Casals J. Isolation of Dhori virus from *Hyalomma marginatum* ticks in Portugal. *Intervirology* 1979;**11**(2):124–7.

4. Yashkulov KB, Shchelkanov M, Lvov SS, et al. Isolation of influenza A virus (Orthomyxoviridae, Influenza A virus), Dhori virus (Orthomyxoviridae, Thogotovirus), and Newcastle's disease virus (Paromyxoviridae, Avulavirus) on the Malyi Zhemchuzhnyi Island in the north-western area of the Caspian Sea. *Vopr Virusol* 2008;**53**(3):34–8 [in Russian].

5. Lvov DN, Dzharkenov AF, Aristova VA, et al. The isolation of Dhori viruses (Orthomyxoviridae, Thogotovirus) and Crimean-Congo hemorrhagic fever virus (Bunyaviridae, Nairovirus) from the hare (*Lepus europaeus*) and its ticks *Hyalomma marginatum* in the middle zone of the Volga delta, Astrakhan region, 2001. *Vopr Virusol* 2002;**47**(4):32–6 [in Russian].

6. Lvov DK. Arboviral zoonoses on Northern Eurasia (Easter Europe and the Commonwealth of Independent States). In: Beran GW, Steel JH, editors. *Handbook of zoonoses. Section B: Viral.* Boca Raton, FL: CRC Press;1994. p. 237–60.

7. Butenko AM, Chumakov MP. Isolation of a new for USSR arbovirus "Astra" from ticks *H. plumbeum* and mosquitos *An. hyrcanus* in Astrakhan district. *Vopr Med Virusol* 1971;(2):11–12 [in Russian].

8. Lvov DK, Karas FR, Tsyrkin YM, et al. Batken virus, a new arbovirus isolated from ticks and mosquitoes in Kirghiz S.S.R. *Arch Ges Virusf* 1974;**44**(1):70–3 [in German].

9. Karas FZ. Arboviruses in Kirgizia. *Arboviruses* 1978;:40–4 [in Russian].

10. Batken. In: Karabatsos N, editor. *International catalogue of arboviruses including certain other viruses of vertebrates.* San Antonio, TX: American Society of Tropical Medicine and Hygiene;1985. p. 223–4.

11. Butenko AM, Leshchinskaia EV, Semashko IV, et al. Dhori virus—a causative agent of human disease. 5 cases of laboratory infection. *Vopr Virusol* 1987;**32** (6):724–9 [in Russian].

12. Filipe AR, Calisher CH, Lazuick J. Antibodies to Congo-Crimean haemorrhagic fever, Dhori, Thogoto and Bhanja viruses in southern Portugal. *Acta Virol* 1985;**29**(4):324–8.

8.4 Family Togaviridae

1. Kuhn R. Togaviridae: the viruses and their replication. In: Knipe DM, Howley PM, editors. *Fields virology.* 5th ed. Philadelphia, PA: Lippincott Williams & Wilkins;2007. p. 1001–22.

2. Westaway EG, Brinton MA, Gaidamovich S, et al. Togaviridae. *Intervirology* 1985;**24**(3):125–39.

3. Schmaljohn AL, McClain D. Alphaviruses (Togaviridae) and Flaviviruses (Flaviviridae). In: Baron S, editor. *Medical microbiology,* 4th ed. Galveston, TX; 1996.

4. Banatvala JE, Brown DW. Rubella. *Lancet* 2004;**363** (9415):1127–37.

5. Powers A, Huang H, Roehrig J, et al. Family Togaviridae. In: King AM, Adams MJ, Carstens EB, Lefkowitz EJ, editors. *Virus taxonomy. Classification and nomenclature of viruses: IX report of the International Committee on Taxonomy of Viruses, 1-st edition.* London: Elsevier;2012. p. 1103–10.

8.4.1 Genus *Alphavirus*

1. Kuhn R. Togaviridae: the viruses and their replication. In: Knipe DM, Howley PM, editors. *Fields virology.* Philadelphia, PA: Lippincott Williams & Wilkins;2007. p. 1001–22.

2. Gould EA, Coutard B, Malet H, et al. Understanding the alphaviruses: recent research on important emerging pathogens and progress towards their control. *Antiviral Res* 2010;**87**(2):111–24.

3. Hollidge BS, Gonzalez-Scarano F, Soldan SS. Arboviral encephalitides: transmission, emergence, and pathogenesis. *J Neuroimmune Pharmacol* 2010;**5**(3):428–42.

4. Schmaljohn AL, McClain D. Alphaviruses (Togaviridae) and Flaviviruses (Flaviviridae). In: Baron S, editor. *Medical microbiology.* Galveston, TX; 1996.

5. Powers A, Huang H, Roehrig J, et al. Family Togaviridae. In: King AM, Adams MJ, Carstens EB, Lefkowitz EJ, editors. *Virus taxonomy. Classification and nomenclature of viruses: Ninth Report of the International Committee on Taxonomy of Viruses.* London: Elsevier;2012. p. 1103–10.

6. McLoughlin MF, Graham DA. Alphavirus infections in salmonids—a review. *J Fis Dis* 2007;**30**(9):511–31.

7. Burt FJ, Rolph MS, Rulli NE, et al. Chikungunya: a re-emerging virus. *Lancet* 2012;**379**(9816):662–71.

8. Devaux CA. Emerging and re-emerging viruses: a global challenge illustrated by Chikungunya virus outbreaks. *World J Virol* 2012;**1**(1):11–22.

8.4.1.1 Chikungunya Virus (imported)

1. Ross RW. The Newala epidemic. III. The virus: isolation, pathogenic properties and relationship to the epidemic. *J Hyg (Lond)* 1956;**54**(2):177–91.

2. Robinson MC. An epidemic of virus disease in Southern Province, Tanganyika Territory, in 1952–53. I. Clinical features. *Trans R Soc Trop Med Hyg* 1955;**49** (1):28–32.

3. Mason PJ, Haddow AJ. An epidemic of virus disease in Southern Province, Tanganyika Territory, in 1952–53; an additional note on Chikungunya virus isolations and serum antibodies. *Trans R Soc Trop Med Hyg* 1957;**51** (3):238–40.

4. Casals J, Whitman L. Mayaro virus: a new human disease agent. I. Relationship to other arbor viruses. *Am J Trop Med Hyg* 1957;**6**(6):1004–11.

5. Porterfield JC. Cross-neutralization studies with group A arthropod-borne viruses. *Bull WHO* 1961;**24**:735–41.

6. Chastel C. Human infections in Cambodia by the Chikungunya virus or an apparently closely related agent. I. Clinical aspects. Isolations and identification of the viruses. Serology. *Bull Soc Pathol Exot Filiales* 1963;**56**:892–915 [in French].

7. Myers RM, Carey DE, Reuben R, et al. The 1964 epidemic of dengue-like fever in South India: isolation of Chikungunya virus from human sera and from mosquitoes. *Indian J Med Res* 1965;**53**(8):694–701.

8. Jadhav M, Namboodripad M, Carman RH, et al. Chikungunya disease in infants and children in Vellore: a report of clinical and haematological features of virologically proved cases. *Indian J Med Res* 1965;**53**(8):764−76.

9. Horwood PF, Reimer LJ, Dagina R, et al. Outbreak of Chikungunya virus infection, Vanimo, Papua New Guinea. *Emerg Infect Dis* 2013;**19**(9):1535−8.

10. Gubler DJ. Human arbovirus infections worldwide. *Ann NY Acad Sci* 2001;**951**:13−24.

11. Powers AM, Brault AC, Tesh RB, et al. Re-emergence of chikungunya and o'nyong-nyong viruses: evidence for distinct geographical lineages and distant evolutionary relationships. *J Gen Virol* 2000;**81**:471−9.

12. Powers AM, Christopher HL. Changing patterns of Chikungunya virus: re-emergence of a zoonotic arbovirus. *J Gen Virol* 2007;**88**:2363−77.

13. Staples JE, Breiman RF, Powers AM. Chikungunya fever: an epidemiological review of a re-emerging infectious disease. *Clin Inf Dis* 2009;**49**(6):942−8.

14. Cavrini F, Gaibani P, Pierro AM, et al. Chikungunya: an emerging and spreading arthropod-born viral disease. *J Infect Dev Ctzies* 2009;**3**:744−52.

15. Malinoski F. Chikungunya fever. In: Beran GW, editor. *Handbook of zoonoses. Section B: Viral*. Boca Raton, FL: CRC Press;1994. p. 101−9.

16. Pastorino B, Muyembe-Tamfum JJ, Bessaud M, et al. Epidemic resurgence of Chikungunya virus in Democratic Republic of the Congo: identification of a new central Asian strain. *J Med Virol* 2004;**74**:277−82.

17. Kalunda M, Lwanga-Ssozi C, Lule M, et al. Isolation of Chikungunya and Pongola viruses from patients in Uganda. *Trans R Soc Trop Med Hyg* 1985;**79**(4):567.

18. CDC. Chikungunya fever among U.S. Peace Corps volunteers-Republic of the Philippines. *Morb Mortal Wkly Rep* 1986;**35**(36):573−4.

19. Van den Bosch C, Lloyd G. Chikungunya fever as a risk factor for endemic Burkitt's lymphoma in Malawi. *Trans R Soc Trop Med Hyg* 2000;**94**(6):704−5.

20. Wiwanitkit S, Wiwanitkit V. Ckikungunya virus infection and relationship to rainfall, the relationship study from southern Thailand. *J Arthropod Borne Dis* 2013;**7**(2):185−7.

21. Harnett GB, Bucens MR. Isolation of Chikungunya virus in Australia. *Med J Aust* 1990;**152**(6):328−9.

22. Ivanov AP, Ivanova OE, Lomonosov NN, et al. Serological investigations of Chikungunya virus in the Republic of Guinea. *Ann Soc Belg Med Trop* 1992;**72**(1):73−4.

23. Thonnon J, Spiegel A, Diallo M, et al. Chikungunya virus outbreak in Senegal in 1996 and 1997. *Bull Soc Pathol Exot* 1999;**92**(2):79−82 [in French].

24. Kosasih H, Widjaja S, Surya E, et al. Evaluation of two IgM rapid immunochromatographic tests during circulation of Asian lineage Chikungunya virus. *Southeast Asian J Trop Med Public Health* 2012;**43**(1):55−61.

25. Lam SK, Chua KB, Hooi PS, et al. Chikungunya infection—an emerging disease in Malaysia. *Southeast Asian J Trop Med Public Health* 2001;**32**(3):447−51.

26. Muyembe-Tamfum JJ, Peyrefitte CN, Yogolelo R, et al. Epidemic of Chikungunya virus in 1999 and 2000 in the Democratic Republic of the Congo. *Med Trop (Mars)* 2003;**63**(6):637−8 [in French].

27. Chretien JP, Anyamba A, Bedno SA, et al. Drought-associated chikungunya emergence along coastal East Africa. *Am J Trop Med Hyg* 2007;**76**(3):405−7.

28. CDC. Chikungunya fever diagnosed among international travelers-United States, 2005−2006. *Morb Mortal Wkly Rep* 2006;**55**(38):1040−2.

29. Savini H, Gautret P, Gaudart J, et al. Travel-associated diseases, Indian Ocean Islands, 1997−2010. *Emerg Infect Dis* 2013;**19**(8):1297−301.

30. Demanou M, Antonio-Nkondjio C, Ngapana E, et al. Chikungunya outbreak in a rural area of Western Cameroon in 2006: a retrospective serological and entomological survey. *BMC Res Notes* 2010;(3):128.

31. Dutta SK, Pal T, Saha B, et al. Copy number variation of chikungunya CESA virus with disease symptoms among Indian patients. *J Med Virol* 2013;**86**(8):1386−92.

32. Johnson DF, Druce JD, Chapman S, et al. Chikungunya virus infection in travellers to *Australia. Med J Aust* 2008;**188**(1):41−3.

33. Soon YY, Junaidi I, Kumarasamy V, et al. Chikungunya virus of Central/East African genotype detected in Malaysia. *Med J Malaysia* 2007;**62**(3):214−17.

34. Panning M, Grywna K, van Esbroeck M, et al. Chikungunya fever in travelers returning to Europe from the Indian Ocean region, 2006. *Emerg Infect Dis* 2008;**14**(3):416−22.

35. Lee N, Wong CK, Lam WY, et al. Chikungunya fever, Hong Kong. *Emerg Infect Dis* 2006;**12**(11):1790−2.

36. Mizuno Y, Kato Y, Kudo K, et al. First case of chikungunya fever in Japan with persistent arthralgia. *Kansenshogaku Zasshi* 2007;**81**(5):600−1 [in Japan].

37. Rezza G, Nicoletti L, Angelini R, et al. Infection with Chikungunya virus in Italy: an outbreak in a temperate region. *Lancet* 2007;**370**(9602):1840−6.

38. Amador Prous C, López-Perezagua MM, Arjona Zaragozí FJ, et al. Chikungunya fever in a Spanish traveller. *Med Clin (Barc)* 2007;**129**(3):118−19 [in Spanish].

39. Cha GW, Cho JE, Lee EJ, et al. Travel-Associated Chikungunya Cases in South Korea during 2009−2010. *Osong Public Health Res Perspect* 2013;**4**(3):170−5.

40. Apandi Y, Lau SK, Izmawati N, et al. Identification of Chikungunya virus strains circulating in Kelantan, Malaysia in 2009. *Southeast Asian J Trop Med Public Health* 2010;**41**(6):1374−80.

II. ZOONOTIC VIRUSES OF NORTHERN EURASIA: TAXONOMY AND ECOLOGY

41. Lim CK, Nishibori T, Watanabe K, et al. Chikungunya virus isolated from a returnee to Japan from Sri Lanka: isolation of two sub-strains with different characteristics. *Am J Trop Med Hyg* 2009;**81**(5):865–8.

42. Grandadam M, Caro V, Plumet S, et al. Chikungunya virus, southeastern France. *Emerg Infect Dis* 2011;**17**(5):910–13.

43. Chaves TS, Pellini AC, Mascheretti M, et al. Travelers as sentinels for chikungunya fever, Brazil. *Emerg Infect Dis* 2012;**18**(3):529–30.

44. Mizuno Y, Kato Y, Takeshita N, et al. Clinical and radiological features of imported chikungunya fever in Japan: a study of six cases at the National Center for Global Health and Medicine. *J Infect Chemother* 2011;**17**(3):419–23.

45. Inoue S, Morita K, Matias RR, et al. Distribution of three arbovirus antibodies among monkeys (*Macaca fascicularis*) in the Philippines. *J Med Primatol* 2003;**32**(2):89–94.

46. Weinbren MP, Haddow AG, Williams MC. The occurrence of Chikungunya virus in Uganda. *Trans Roy Soc Trop Med Hyg* 1958;**52**:253–62.

47. Osterrieth PM, Deleplanque-Liegloris PS. Presence d'anticorp vis-a-visdes visus fran mis pour arthropods chez le chimpanzee (*Pan troglodytes*). Comparison de leur etalimmnitai a celui de l'homme. *Ann Soc Belg Med Trop* 1961;**41**:63–72 [in French].

48. McIntosh BM. Antibodies against Chikungunya virus in wild primates in Southern Africa. *S Afr J Med Sci* 1970;**35**:65–7.

49. Bodekar SD, Pavri KM. Studies with Chikungunya virus. I. Susceptibility of birds and small mammals. *Indian J Med Res* 1969;**57**:1181–92.

50. Shah KV, Daniel RW. Attempts of experimental infection of the Indian fruit-bat Pteropus giganteus with Chikungunya and denge-2 viruses and antibody survey of bat sera for the same viruses. *Indian J Med Res* 1966;**54**:714–22.

51. Chippaux-Hippolixe C, Chippaux A. Contribution a l'etude d'un reservoir de virus anitses dans le cycle de certains abovirus en Centrafrique. I. Etude immunologique chez divers animaux domestiques et savages. *Bull Soc Pathol Exot* 1969;**62**:1034–45 [in French].

52. McIntosh BM, Paterson HE, Donaldson JM, et al. Chikungunya virus: viral susceptibility and transmission studies with some vertebrates and mosquitoes. *S Afr J Med Sci* 1969;**28**:45–52.

53. Abu Bakar S, Sam IC, Wong PF, et al. Vectors of Chikungunya, Malaysia. *Emerg Infect Dis* 2007;**13**:1264–6.

54. Diallo M, Thonnon J, Traole-Lamizana M, et al. Vectors of Chikungunya virus in Senegal: current data and transmission cycles. *Am J Trop Med Hyg* 1999;**60**:281–6.

55. Pavri KM. Presence of chikungunya antibodies in human sera collected from Calcutta and Jamshedpur before 1963. *Indian J Med Res* 1964;**52**:698–702.

56. Sarkar JK, Chatterjee SN, Chakravarty SK, et al. The causative agent of Calcutta haemorrhagic fever: chikungunya or dengue. *Bull Calcutta Sch Trop Med* 1965;**13**:53–4.

57. Halstead SB, Scanlon JE, Umpaivit P, et al. Dengue and Chikungunya virus infection in man in Thailand, 1962–1964. IV. Epidemiologic studies in the Bangkok metropolitan area. *Am J Trop Med Hyg* 1969;**18**(6):997–1021.

58. Mavalankar D, Shastri P, Bandyopadhyay T, et al. Increased mortality rate associated with Chikungunya epidemic, Ahmadabad, India. *Emerg Infect Dis* 2008;**14**:412–15.

59. Beltrame A, Angheben A, Bisoffi Z, et al. Imported Chikungunya infection, Italy. *Emerg Infect Dis* 2007;**13**:1264–6.

60. Bonilauri P, Bellini R, Calzolari M, et al. Chikungunya virus in *Aedes albopictus*, Italy. *Emerg Infect Dis* 2008;**14**:852–4.

61. Gould EA, Gallian P, Lambalerie C, et al. First cases of autochthonous Dengue fever and Chikungunya fever in France: from bad dream to reality. *Clin Microbiol Infect* 2010;**16**:1702–4.

62. Lanciotti RS. Chikungunya virus in US travelers returning from India, 2006. *Emerg Infect Dis* 2007;**13**:764–7.

63. Kobayashi M, Nibei N, Kurihara T. Analysis of Northern distribution of *Aedes albopictus* (Diptera, Culicidae)in Japan by GIS. *J Med Entomol* 2002;**39**:4–11.

64. Shchelkanov MYu, Lvov DK, Kolobukhina LV, et al. Isolation of Chikungunya virus in Moscow from Indonesian visitor (September, 2013). *Vopr Virusol* 2014;**59**(3):28–34.

65. Larichev VF, Saifullin MA, Akinshin YuA, et al. Introduced cases of arbovirus infections in Russian Federation. *Epidemiol Infekcion bolezni* 2012;**1**:35–8 [in Russian].

66. Ganushkina LA, Dremova VP. *Aedes aegypti* L. and *Aedes albopictus* Skuse mosquitoes are a new biological threat to the south of Russia. *Med Parazitol (Moscow)* 2012;**3**:49–55 [in Russian].

67. Ganushkina LA, Bezzhonova OV, Patraman IV, et al. Distribution of *Aedes (Stegomyia) aegypti* L. and *Aedes (Stegomyia) albopictus* Skus. mosquitoes on the Black Sea coast of the Caucasus. *Med Parazitol (Moscow)* 2013;**1**:45–6 [in Russian].

68. Lvov DK. Chikungunya fever. In: Lvov DK, editor. *Guide for Virology. Viruses and viral infections of humans and animals.* Moscow: MIA;2013p. 707–10 [in Russian].

8.4.1.2 Getah Virus

1. Aaskov JC, Yoherty RI. Arboviral zoonoses of Australasia. In: Beran GW, editor. *Handbook of zoonoses. Section B: Viral.* Boca Raton, FL: CRC Press;1994. p. 289–304.

2. Doherty RL, Whitehead RH, Gorman BM, et al. The isolation of a third group A arbovirus in Australia with preliminary observations on its relationships to epidemic polyarthritis. *Austral J Sci* 1963;**26**:185.

3. La Ruche G, Souares Y, Armengaud A, et al. First two autochthonous Dengue virus infections in metropolitan France, September 2010. *Europ Surveill* 2010;**15**: 19676.

4. Bowen EJM, Simpson DIH, Platt GS, et al. Arbovirus infection in Sarawak, October 1968—February 1970. Human serological studies in a Land Dyak village. *Trans Roy Soc Trip Med Hyg* 1975;**69**:182.

5. Ksiazek TG, Trosper JH, Ross JH. Isolation of Getah virus from Ecija province, Republic of Pilippines. *Trans Roy Soc Trip Med Hyg* 1981;**75**:312.

6. Chumakov MP, Moskvin AV, Andreeva EB. Isolation of 5 strains of Getah virus from mosquitoes on the South of Amur region of the USSR. In: *Proceedings of the Institute of poliomyelitis and viral encephalitis*, vol. 22. Moscow; 1974. p. 65−71 [in Russian].

7. Calisher CH, Shope RE, Brandt W, et al. Proposed antigenic classification of registered arboviruses. I. Togaviridae, Alphavirus. *Intervirology* 1980;**14**:229−32.

8. Wekesa SN, Inoshima Y, Murakami K, et al. Genomic analysis of some Japanese isolates of Getah virus. *Vet Microbiol* 2001;**83**(2):137−46.

9. Wen JS, Zhao WZ, Liu JW, et al. Genomic analysis of a Chinese isolate of Getah-like virus and its phylogenetic relationship with other Alphaviruses. *Virus Genes* 2007;**35**(3):597−603.

10. Gur'ev EL, Gromashevsky VL, Prilipov AG, et al. Analysis of the genome of two Getah virus strains (LEIV-16275Mar and LEIV-17741MPR) isolated from mosquitoes in the North-Eastern Asia. *Vopr Virusol* 2008;**53**(5):27−31 [in Russian].

11. Seo HJ, Kim HC, Klein TA, et al. Characterization of recent Getah virus isolates from South Korea. *Acta Virol* 2012;**56**(3):265−7.

12. Gurjev EL, Gromashevsky VL, Prilipov AG, et al. Molecular genetic analysis of a diversity of Getah virus strains isolated from mosquitoes in the North-Eastern Asia. *Vopr Virusol* 2008;**53**(3):9−12 [in Russian].

13. Andreev VP, Ahmed D, Lvov SD. Isolation of Getah-like virus (Togaviridae, Alphavirus, Semliki Forest antigenic complex) from mosquitoes collected in Mongolia. *Arch Virol* 1990;(Suppl. 1):346.

14. Andreeva EB. *Studies of biological properties of Getah viruses isolated in Amur region. Voprosy Meditsinskoi Virusologii*. Moscow: Abstract Book;1975255−56. [in Russian].

15. Bockova NG, Pogodina VV. Ecological and differential-diagnostics aspects of investigation of mixed arboviral natural focies in Primorsky krai. In: *Proceedings of 10-th all-Russian conference on natural foci diseases*. Dushanbe; 1999. p. 27−8 [in Russian].

16. Kruglyak SP, Leonova GN. Chronic Getah infection among horses as the main factor determining the circulation of virus in natural foci. In: *Itogi Nauki i Techniki. Virologiya*, vol. 27. Moscow; 1992. p. 16−22 [in Russian].

17. Lvov DK, Il'ichev VD. *Migration of birds and the transfer of the infectious agents*. Moscow: Nauka;1979 [in Russian].

18. Lvov DK, Deryabin PG, Aristova VA, et al. *Atlas of distribution of natural foci virus infections on the territory of Russian Federation*. Moscow: SMC MPH RF Publ;2001 [in Russian].

19. Lvov SD. *Natural virus foci in high latitudes of Eurasia. Soviet medical reviews. Section E: Virol reviews*, 3. USA: Harwood Academic Publishers GmbH;1993137−85.

20. Lvov SD, Gromashevsky VL, Andreev VP, et al. Natural foci of arboviruses in far northern latitudes of Eurasia. *Arch Virol* 1999;(Suppl. 1):267−75.

21. Lvov SD, Gromashevsky VL, Aristova VA, et al. Isolation of Getah virus (Togaviridae, Alphavirus) in the North-Eastern Asia. *Vopr Virusol* 2000;**45**(5):14−18 [in Russian].

22. Shchelkanov MYu, Gromashevsky VL, Lvov DK. The role of ecologo-virological zoning in prediction of the influence of climatic changes on arbovirus habitats. *Vestnik Rossiiskoi Akademii Meditsinskikh Nauk* 2006;**2**:22−5 [in Russian].

23. McClure HE. Migration and survival of the birds in Asia. Bangkok, 1974.

24. Li XD, Qui FX, Yang H. Isolation of Getah virus from mosquitoes collected on Hainan Island, China, and results of serosurvey. *S Asian J Trop Med Publ Health* 1992;**23**(4):730−4.

25. Shibata I, Katano Y, Nishimura Y, et al. Isolation of Getah virus from dead fetuses extracted from naturally infected sow in Japan. *Vet Microbiol* 1991;**27**:385−97.

26. Shirako Y, Yamaguchi Y. Genome structure of Sagayama virus and its relatedness to other alphaviruses. *J Gen Virol* 2000;**81**:1353−60.

27. Takashima I, Hashimoto N. Getah virus in several species of mosquitoes. *Trans Roy Soc Trop Med Hyg* 1985;**79**:546−50.

28. Theiler M, Casals J, Mouthouhes C. Etiology of the 1927−28 epidemic of Dengue in Greece. *Proc Soc Exp Biol Med* 1960;**103**:244−6.

29. Theiler M, Downs WG. *The arthropod-born viruses of vertebrates*. New Haven/London: Vale University Press;1973.

30. Fukunaga Y, Kunamomido T, Kamoda M. Getah virus as equine pathogen. *Vet Clin N Am Equin Pract* 2000;**16** (3):605−17.

31. Wekesaa SN, Inoshimaa Y, Murakamia K, et al. Genomic analysis of some Japanese isolates of Getah virus. *Vet Microbiol* 2001;**83**:137—46.

32. Tesh RB, Gujduzek DC, Baeruto RM, et al. The distribution and prevalence of group A arbovirus neutralizing antibodies among human populations in Southern Asia and the Pacific Islands. *Am J Trop Med Hyg* 1975;**24**:664—9.

33. Yago K, Hagiwara S, Kawamura H, et al. A fatal case in newborn piglets with Getah virus infection. Isolation of the virus. *Jpn J Vet Sci* 1987;**49**:899—903.

34. Brown CM, Timoney PJ. Getah virus infection of Indian horses. *Trop Anim Hlth Product* 1998;**30**(4):241—52.

35. Kono Y. Getah virus disease. In: Monath TP, editor. *The Arboviruses: epidemiology and ecology.* Boca Raton, FL: CRC Press;1988. p. 21—36 [Chapter 25].

8.4.1.3 Sindbis Virus and a Set of Var. Sindbis Virus: Karelian Fever Virus, Kyzylagach Virus ()

1. Taylor RM, Hurlbut HS, Work TH, et al. Sindbis virus: a newly recognized arthropod transmitted virus. *Am J Trop Med Hyg* 1955;**4**(5):844—62.

2. Sindbis. In: Karabatsos N, editor. *International catalogue of arboviruses including certain other viruses of vertebrates.* San Antonio, TX: American Society of Tropical Medicine and Hygiene;1985. p. 1146.

3. Jost H, Bialonski A, Storch V, et al. Isolation and phylogenetic analysis of Sindbis viruses from mosquitoes in Germany. *J Clin Microbiol* 2010;**48**(5):1900—3.

4. Weaver SC, Kang W, Shirako Y, et al. Recombinational history and molecular evolution of western equine encephalomyelitis complex alphaviruses. *J Virol* 1997;**71**(1):613—23.

5. Hubalek Z. Mosquito-borne viruses in Europe. *Parasitol Res* 2008;**103**(Suppl. 1):S29—43.

6. Sane J, Kurkela S, Putkuri N, et al. Complete coding sequence and molecular epidemiological analysis of Sindbis virus isolates from mosquitoes and humans, Finland. *J Gen Virol* 2012;**93**(Pt 9):1984—90.

7. Marshall ID, Woodroofe GM, Hirsch S. Viruses recovered from mosquitoes and wildlife serum collected in the Murray Valley of South-eastern Australia, February 1974, during an epidemic of encephalitis. *Am J Exp Biol Med Sci* 1982;**60**(Pt 5):457—70.

8. Bardos V, Sixl W, Wisidagama CL, et al. Prevalence of arbovirus antibodies in sera of animals in Sri Lanka. *Bull World Health Organ* 1983;**61**(6):987—90.

9. Lundstrom JO, Pfeffer M. Phylogeographic structure and evolutionary history of Sindbis virus. *Vector Borne Zoonotic Dis* 2010;**10**(9):889—907.

10. Simpson DA, Davis NL, Lin SC, et al. Complete nucleotide sequence and full-length cDNA clone of S.A. AR86 a South African alphavirus related to Sindbis. *Virology* 1996;**222**(2):464—9.

11. Shirako Y, Niklasson B, Dalrymple JM, et al. Structure of the Ockelbo virus genome and its relationship to other Sindbis viruses. *Virology* 1991;**182**(2):753—64.

12. Traore-Lamizana M, Fontenille D, Diallo M, et al. Arbovirus surveillance from 1990 to 1995 in the Barkedji area (Ferlo) of Senegal, a possible natural focus of Rift Valley fever virus. *J Med Entimil* 2001;**38**(4):480—92.

13. Calisher CH, Karabatsos N, Lazuick JS, et al. Reevaluation of the western equine encephalitis antigenic complex of alphaviruses (family Togaviridae) as determined by neutralization tests. *Am J Trop Med Hyg* 1988;**38**(2):447—52.

14. Lvov DK, Vladimirtseva EA, Butenko AM, et al. Identity of Karelian fever and Ockelbo viruses determined by serum dilution-plaque reduction neutralization tests and oligonucleotide mapping. *Am J Trop Med Hyg* 1988;**39**(6):607—10.

15. Lvov DK, Gromashevskii VL, Skvortsova TM, et al. Kyzylagach virus (family Togaviridae, genus Alphavirus), a new arbovirus isolated from *Culex modestus* mosquitoes trapped in the Azerbaijani SSR. *Vopr Virusol* 1979;**5**:519—23.

16. Liang GD, Li L, Zhou GL, et al. Isolation and complete nucleotide sequence of a Chinese Sindbis-like virus. *J Gen Virol* 2000;**81**(Pt 5):1347—51.

17. Diseases transmitted from animals to men. In: Hubbert WT, McCulloch WF, Schnurrenberger WF, editors. Springer Fields: Charles Thoma Publ; 1975.

18. Weaver SC, Frey TK, Huang HV, et al. Family Togaviridae. In: Fauquet CM, Mayo MA, Maniloff J, et al., editors. *Virus taxonomy: Eight Report of the International Committee on taxonomy of viruses.* Elsevier Academic Press;2005. p. 999—1008.

19. Lvov DK, Deryabin PG, Aristova VA, et al. *Atlas of distribution of natural foci virus infections on the territory of Russian Federation.* Moscow: SMC MPH RF Publ;2001 [in Russian].

20. Lvov DK, Il'ichev VD. *Migration of birds and the transfer of the infectious agents.* Moscow: Nauka;1979 [in Russian].

21. Voinov IN, Rytik PG, Grigorieva AI. *Arboviral infections in Belarus. Viruses and viral infections of humans.* Moscow: Nauka;1987:86—87 [in Russian].

22. Page LA. *Wildlife diseases.* NY: Plenum Press;1976.

23. Rosen L. Arthropods as hosts and vectors of Alphaviruses and Flaviviruses—experimental infections. In: Schlesinger W, editor. *Togaviruses: biology, structure, replication.* NY/London: Academic Press, Hacourt Brace Jovanovich Publ.;1980. p. 123—239.

Double-Stranded DNA Viruses

9.1 FAMILY ASFARVIRIDAE

The Asfarviridae family comprises only one genus, *Asfivirus*, with species African swine fever virus (ASFV). The family belongs to the clade of the nucleocytoplasmic large DNA viruses (NCLDV) of the proposed order Megavirales, which brings together diverse eukaryotic DNA viruses that reproduce mainly in the cytoplasm of infected cells. The order includes at least 10 families, among which are the Poxviridae, Iridoviridae, and Phycodnaviridae, as well as several newly formed families—the Mimiviridae, Marseilleviridae, Pandoraviridae, and Pithoviridae—that contain a growing number of giant viruses. Metagenomic study has revealed that giant viruses are a vast group of viruses that infect a myriad of phagocytic protozoa in the oceans, freshwaters, permafrost, soil, etc. Genome fragments related to ASFV sequences were also found in sewage and human sera.[1,2]

9.1.1 Genus *Asfivirus*

As mentioned in the previous section, there is but a single genus in this family: *Asfivirus*. Moreover, only a single species has been described to date in this genus: the Prototype strain, ASFV/Behin97/1.

The virion is enveloped; pleomorphic, including spherical; and 175–215 nm in diameter. The capsid is round and exhibits icosahedral symmetry ($T = 189-217$) corresponding to 1,892–2,172 capsomers.

The genome of the virus is a linear dsDNA genome 170–190 kbp in length. The end sequences are complementary. The genome encodes for at least 150 open reading frames (ORFs). Attachment of the viral proteins to host receptors mediates endocytosis of the virus into the host cell. The virus fuses with the membrane of the endocytic vesicle, and the genome's DNA is released into the host cytoplasm. Transcription of early viral genes occurs, and replication of the DNA genome in the cytoplasm begins about 6 h post infection. Following DNA replication, late viral genes, including structural proteins, are transcribed.

New virions are assembled in cytoplasmic viral factories. Endoplasmic reticulum cisternae are recruited and transformed , giving rise to precursor viral membranes, which represent the first identifiable viral structures. The viral membranes become icosahedral particles by the gradual assembly of the outer capsid layer, formed by protein p72. At the same time, the matrix shell is formed underneath the viral envelope and the viral DNA and nucleoproteins are packaged and condensed. Virions migrate to the plasma membrane on microtubules and buds.[1,2]

9.1.1.1 *African Swine Fever Virus*

ASF is a highly contagious disease of domestic and wild pigs (family Suidae, order Artiodactyla) that is characterized by high mortality rates among domestic animals, often reaching 100%. ASF causes high economic losses due to the necessity of depopulating pig farms in affected areas, implementing sanitary measures, imposing trade restrictions, etc. For instance, economic losses were as large as $US267 million in 2001 alone.[1]

History. The first records of ASF date back to the period between 1903 and 1905, when the disease occurred in domestic pigs exported from Europe to southern African countries. Once imported into the African countries, the animals would develop signs of illness and die at an almost 100% rate, with symptoms and pathoanatomical changes highly evocative of classical swine fever, sometimes also called hog cholera. Until the 1950s, ASF was present only in Africa.[2-4]

ASFV was first isolated in Kenya by R.E. Montgomery (1880–1932) (Figure 9.1) in 1921 when filtrates from a desert warthog (*Phacochoerus aethiopicus*) were injected into a healthy domestic pig.[5]

The disease first penetrated into Europe when it appeared in Portugal in 1957. It then appeared in Latin America, entering Cuba in 1971. Between 1957 and 2000, ASF was reported in a total of 12 European and Latin American countries.[4]

Domestic pigs of all breeds and ages are susceptible to ASF, irrespective of the season, as are wild boars and wild African suids, such as the common warthog (*Phacochoerus africanus*), considered the main reservoir of the disease, the red river hog (*Potamochoerus porcus*), and the giant forest hog (*Hylochoerus meinertzhageni*).[6] Other animal species and humans are not susceptible to this virus.[7]

In Russia, ASF is studied by Aleksey Dmitrievich Zaberezhny (1960) (Figure 9.2), Taras Ivanovich Aliper (1957) (Figure 9.3), Evgeny Anatolyevich Nepoklonov (1956) (Figure 9.4), and other virology research teams.

Taxonomy. As has been mentioned, the causative agent of ASF is a DNA virus classified into a genus and a family of its own.[8]

ASFV is highly resistant to environmental influences and may remain infectious for 5–7 years at 5°C, for 18 months at room temperature, and for 10–30 days at 37°C. The virus stays viable for 5–6 months in soil and dung, and for over 30 months in frozen meat. It is stable at a pH between 4 and 13, is sensitive to lipid solvents such as ether and chloroform, and can be inactivated at 60°C within 30 min.

FIGURE 9.1   Robert Eustace Montgomery (1880–1932).

FIGURE 9.2   Aleksey Dmitrievich Zaberezhny (1960).

FIGURE 9.3   Taras Ivanovich Aliper (1957).

FIGURE 9.4   Evgeny Anatolyevich Nepoklonov (1956).

Under laboratory conditions, ASFV replicates in pig leukocytes, pig bone marrow cells, and pig and monkey kidney passaged cell cultures. If the virus is grown in pig leukocytes or bone marrow cells, erythrocyte adsorption on the cell surface can be observed. Only some of the virus strains cause hemadsorption, while others, isolated from lymph nodes of chronically sick animals, do not.

The pathogenicity of ASFV varies greatly. All known isolates have been classified into strains that are highly virulent, moderately virulent, low virulent, or nonvirulent, causing corresponding acute, subacute, chronic, and no infection.

Around 10 serological types of ASFV have been identified on the basis of their performance in a hemadsorption inhibition test. Antibodies inhibiting hemadsorption are type specific; however, they exert no noticeable influence on virus accumulation and have no cytopathic effect. A certain amount of genetic diversity has also been described for ASFV. Sequence analysis of the central variable region in the gene encoding the main capsid protein p72 has enabled researchers to classify the 42 known African isolates of ASFV into 22 genotypes.[9]

ASFV is the only virus of the *Asfrivirus* genus in the Asfarviridae family.[9] In the course of genomic analysis, ASF was shown to be phylogenetically related to the "giant" DNA virus.[10] Unidentified genome fragments distantly related to ASFV have been also found in human blood serum and sewage samples, a finding that is suggestive of the existence of genetically related viruses.[11] The ASFV virion has a complex internal organization: The particles have an average size of 200 nm and are surrounded by a hexagonal outer membrane, the formation of which is dependent on the membrane protein known as p54. The outer membrane is visualized by electron microscopy methods at early stages of viral particle assembly, whereas mature icosahedral particles are formed with active participation on the part of the p17 transmembrane protein residing on the inner surface of the virion membrane.[12–14] The ASFV genome is a double-stranded DNA molecule 170,000–190,000 bp long, depending on the strain. Both ends of the genome carry inverted repeats; the central region, 125,000 bp long, is flanked by two variable regions. The BA7v strain, for instance, has a total of 151 ORFs.[15] Over 100 proteins synthesized as a result of the viral infection were detected in infected macrophages, with 50 of these proteins interacting with blood sera of sick animals. Some of these proteins, such as p73, p54, p30, and p12, are powerful antigens and can thus be used for serological diagnosis. ASFV is well adapted for replication in a number of passaged

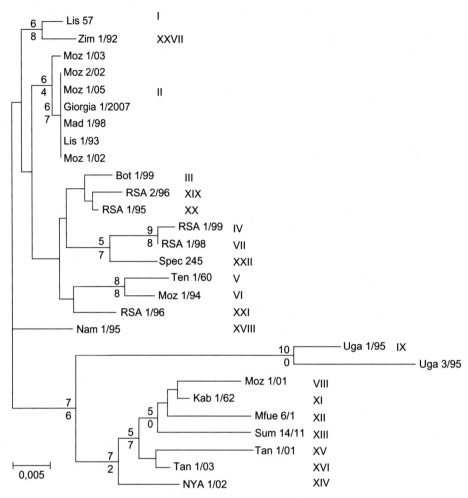

FIGURE 9.5   A phylogenetic dendrogram based on D646L gene nucleotide sequences and representing evolutionary relationships between a number of ASFV isolates of different genotypes. The genotypes are designated by Roman numerals. The Georgian isolates all belong to genotype II.

cell lines, including Vero, CV, and COS-1.[16,17] The virus propagates mostly in monocytes and macrophages,[18] and, to a lesser extent, in endothelium cells, renal tubular epithelial cells,[19] hepatocytes,[20] and neutrophils.[15] ASFV also propagates in certain tick species, mainly *Ornithodoros moubata*[21] and *Ornithodoros erraticus*.[4] More recently, *Ornithodoros porcinus* was also shown to contribute to the circulation of ASF.[7,22]

Twenty-two ASFV genotypes have been described.[23] The strain isolated in Georgia in 2007 belongs to genotype II, which also circulates in Mozambique, Madagascar, and Zambia (Figure 9.5).[4,24−26] Its genome is 189,344 bp long and contains 166 ORFs. Phylogenetic analysis on the basis of 125 conservative cistrons revealed maximum similarity with the Mkuzi/1979 isolate.[8]

Distribution. Natural foci of the disease are mainly equatorial and located in subtropical pastoral ecosystems. ASF is enzootic in many African countries, including. Angola, Mozambique, the Republic of South Africa, Senegal, Uganda, and Zimbabwe. Beyond Africa, ASF was first reported in Portugal in 1957, where it resulted in a lethal outcome in almost 100% of domestic pigs.[27] Spain and Portugal remained affected by ASF until as late as 1995. ASF appeared in France in 1964; in Malta, Sardinia, Brazil, and the Dominican Republic in 1978; and in Italy in 1967 and 1980; and entered Haiti in 1975 and Cuba in 1980.[28,29] Cases have also been reported in Belgium and the Netherlands. ASF first appeared in Georgia and the Russian Federation in 2007 (Figure 9.5).[30]

Epizootiology. Pigs are the only domestic species susceptible to ASF. The disease poses no threat to humans. European wild boars (*Sus scrofa*) are the only wild suid species to display the same symptoms and the same mortality rates as domestic pigs.[6] By contrast, three African wild suid species—the desert warthog (*Phacochoerus aethiopicus*), the giant forest hog (*Hylochoerus meinertzhageni*), and the red river hog *Potamochoerus porcus*—are normally only asymptomatically infected and constitute a natural ASFV reservoir alongside arthropod vectors: ticks (Argasidae) of the species *Ornithodoros moubata* and *O. porcinus* in Africa and *O. erraticus* in Europe.

Infection rates among warthogs (*Phacochoerus africanus*), considered the main ASF reservoir host, amount to 21—71%. Ticks of the *Ornithodoros* genus have a life span of 10—12 years, in some cases as long as 25 years. In ticks, the virus can be transmitted sexually, transphasially, and transovarially and is thus capable of circulating for long periods in the tick population in the absence of vertebrate hosts.

Upon penetration of ASFV into a population of domestic pigs, the emergence of carrier animals is inevitable at some point. Identifying these animals by serological methods is crucial to ASF eradication, as it was in Spain.[31]

The ASF virus is transmitted via the oronasal route, via injection by any route, and via tick bites.[21,32] The virus is highly resistant to adverse environmental conditions and can even be isolated from blood serum after 18 months of storage at room temperature. At 60°C, ASFV can be inactivated within 30 min[33]; common disinfectants can also be used for inactivation.[34] The virus can be preserved in meat products for months and persists indefinitely in frozen meat.[35,36] In the course of production of Spanish ham from contaminated meat, ASFV is inactivated over a period of 112—140 days, according to different sources.[37]

Clinical Picture. In the event of an outbreak of ASF caused by a virulent strain in a naive commercial pig population, symptoms include fever, cough, anorexia, skin cyanosis, ataxia, diarrhea, lymphoreticular tissue damage, vasculitis, profuse hemorrhages, thrombosis, infractions, and abortion. The mortality rate is almost as high as 100%. Infection with other strains can result in a mild form of the infection or a carrier status.

Cattle, sheep, goats, dogs, and rabbits are not susceptible to ASF, although the virus has sometimes been detected in rabbits and goats. Human disease has not been reported.

The incubation period for ASF is 4—19 days. Primary replication of the virus takes place in monocytes and macrophages of the lymph nodes closest to the site of penetration of the virus. The virus then migrates via blood and lymph across the lymph nodes, bone marrow, spleen, lungs, liver, and kidneys. Viremia occurs 4—8 days after the infection and lasts for several days or even months, with the host unable to produce virus-neutralizing antibodies. In the acute form, ASF causes the formation of multiple hemorrhages, attributed to phagocyte activation in endothelial cells and ASFV propagation therein. The resulting lymphopenia is supposedly caused by lymphocyte apoptosis, despite a lack of evidence that the virus propagates in T- or B-lymphocytes.[19,38]

Lung edema developing due to alveolar macrophage activation is the main reason for the lethal outcome.[32,39]

The nature of the lesions that develop depends on the ASFV strain, the exposure dose, and the infection route. The disease can develop in a hyperacute, acute, subacute, chronic, or latent (asymptomatic) form. In the hyperacute form, ASFV causes sudden death or symptoms such as hyperthermia (41−42°C), accelerated breathing, and reddening of the skin. Animals die within 1−3 days after the first symptoms appear, and the death rate equals nearly 100%. The acute and subacute forms of infection develop upon exposure to highly virulent and moderately virulent ASFV isolates, respectively, the symptoms being fever (body temperature rising to 42°C), leucopenia, and anorexia, as well as exhaustion, diarrhea, conjunctivitis accompanied by serosanguineous discharge, and symptoms of pneumonia and lung edema. Reddish purple spots can be observed on the skin. One to two days before death, signs of neurodegenerative lesions, such as cramps, limb paresis, and paralysis, can be observed. The abortion rate in pregnant sows is almost 100%. The duration of the disease is 4−10 days for the acute form, with a mortality rate of nearly 100%, and 15−25 days in the subacute form. In the latter, some of the animals survive but remain ASFV carriers. The acute form is predominant in Africa,[38,40] while elsewhere the disease often takes on the chronic form, of which respiratory disorders, abortions, and low mortality rates are typical.[29]

The mechanism behind the immunity response triggered by infection with ASFV has been very poorly studied, and attempts to design a vaccine have so far proven fruitless. ASFV has a pronounced antigenic activity, with the production of IgM starting on the fourth day of infection and IgG appearing 6−8 days post exposure.[41] Antibodies remain present for long periods, a feature that is attributed to a slowdown in the progression of the disease, a drop in the level of viremia, and a decline in mortality rates.[42,43] Early studies showed that virus-neutralizing antibodies were not produced, even though animals that recovered from ASF remained capable of synthesizing antibodies in response to other pathogens.[44] Neutralization of some ASFV strains with blood sera from convalescent pigs has been demonstrated, but approximately 10% of the virus population preserve their infectious properties. It can thus be said that antibodies induced by ASFV do not possess a neutralizing activity in the conventional sense. By contrast, T-killer cells were shown to be capable of destroying ASFV-affected macrophages, suggesting an important role for cellular immunity.[35,45]

The main issue in diagnosing ASF is that the results of laboratory testing often end up overdue. At this moment, state-of-the-art ASF diagnostic tools are in use in a number of countries, including Russia.[5,46−48] Isolating on the virus and carrying out a hemadsorption test are specific and highly sensitive, but require a lot of time and effort on the part of laboratory personnel in order to be routine methods of diagnosis. The design and implementation of a polymerase chain reaction (PCR) test with diagnostic oligonucleotides complementary to a conservative fragment of the ASFV genome allows all strains of ASFV to be detected with a high degree of sensitivity.[49−51] Immunofluorescence assays have also been developed; however, because of the presence of antibodies, the sensitivity of these tests can drop to as low as 40% in diagnosing the subacute and chronic forms of the disease.[52]

Another important objective is the detection of ASFV-specific antibodies. First and foremost, their presence confirms infection—an important diagnostic event because vaccines against ASF are unavailable. Second, antibody production occurs at an early stage of the infection, and antibodies persist for a long time. Accordingly, serological surveillance is of vital importance. Serological methods currently in use are indirect immunofluorescence, indirect solid-phase enzyme-linked immunosorbent assay (ELISA), and immunoblotting.[50]

In Russia, a real-time PCR assay for diagnosing ASF recently was designed by a Russian research-and-development team, and an immunoenzymatic assay for the detection of antibodies in porcine blood sera has been created in collaboration with Ingenasa, a Spanish biotechnology company.

The PCR and real-time PCR assays that have been designed are highly sensitive and specific, and enable diagnosticians to rapidly detect ASFV in the blood, spleen, liver, and muscular tissue of latently infected, sick, and fallen animals.[46,53]

There is no efficient strategy for treating ASF.

Efforts have been aimed at developing a vaccine ever since 1963, when the first live modified vaccine was used in Portugal. It ensured partial protection from clinical signs of ASF in the event of infection with a homologous strain, although some animals developed a chronic infection and became virus carriers.[4] Cross-strain protection was observed, albeit only in some of the experimental animals, upon immunization with a vaccine developed on the basis of the Benin 97/1 strain of genotype I and upon subsequent inoculation with the Uganda 1965 strain of genotype X. This protection correlated with the activation of cellular immune response mechanisms.[2,54] Immunization with vaccines developed on the basis of the most immunogenic group-specific peptides did not ensure protection, but resulted in a significant delay of the lethal outcome in experimentally inoculated animals.[55] The use of inactivated vaccines resulted in no noticeable protection.

For the time being, preventing ASF consists in protecting an ASF-free herd from contact with the virus. The most crucial aspects are thus strict adherence to veterinary and sanitary security standards, prevention of the animals' exposure to food waste and to the environment whenever contact with the virus is possible, and regular serological surveillance.

ASF in the Russian Federation Six months after ASF emerged in Georgia, outbreaks were reported for the first time in the region of Shatoy in the Republic of Chechnya in Russia (Figure 9.6). The disease was found in North Ossetia in mid-2008[29] and spread rapidly into adjacent regions,[56,57] with cases reported in four federal subjects (jurisdictions) of Russia. Over 2009 and 2010, Russia acquired a permanent ASF-affected status, with as many as 24 federal subjects reporting cases of the disease in both wild and domestic animals from 2007 to 2011. More than 250 affected sites and 37 farming units were affected by the virus, and some 440,000 pigs were slaughtered in the course of eradicating the infection from the sites.

ASFV circulation among wild boars was confirmed in Southern and North Caucasian Federal Districts, as well as in Tver Oblast. Wild boars constitute an ASFV reservoir and pose a permanent threat of the disease penetrating into other regions.

Direct and indirect losses brought about by ASF in Russia over the period between 2008 and 2011 totaled over 10 billion rubles ($US290 million). Compensation paid from the federal subject budgets to farm owners for confiscated animals exceeded 1.3 billion rubles. If drastic measures are not taken to eradicate ASF, economic losses could reach as much as 100 billion rubles ($US2.9 billion).

Since the first case of ASF was reported in Georgia, the World Organisation for Animal Health (Office International des Epizooties, or OIE) has received over 230 reports of outbreaks, mostly from southern Russia and the countries between the Black Sea and the Caspian Sea (Figure 9.6).

Some of the outbreaks occurred at large distances from the main site—for instance, in the

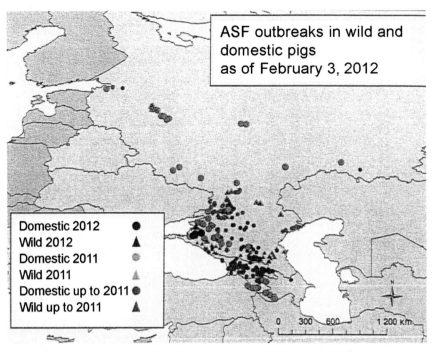

FIGURE 9.6   ASF outbreaks in Russia in wild and domestic pigs as of February 3, 2012.

region of Leningrad in 2009, 2010, and 2011 (among domestic pigs) and in the region of Tver in 2011 (among both wild and domestic pigs). These outbreaks were successfully dealt with, although repeated penetration of the virus into the same region suggests that trade routes exist via which contaminated pig-breeding products are brought from an affected region to a region free of ASF.[30,53]

Social and economic aspects such as back-yard pig breeding and the human factor thwart the effort invested in ASF control. The majority of outbreaks in Russia started in small farms, where swill feeding and free-range pig breeding are customary. Both of these practices increase the risk of infection via feed and via contact with wild animals, respectively. Unsurveyed migration and the transport of infected animals and pig-breeding products further increase the risk of ASF outbreaks, even in remote regions.

The wild boar (Sus scrofa) also constitutes a significant risk factor; however, recent studies demonstrate that, at this time, like domestic pigs, wild animals experience ASF in the acute form, diminishing their role in the spreading of ASF.[12,19,58]

The role of yet another risk factor in ASFV circulation— argasid ticks of the Ornithodoros genus—has been well described; at the moment, however, no evidence has arisen indicating that ticks contribute to ASF circulation in Russia.

Over the past 40 years, Russian experts have accumulated a considerable amount of experience in successfully controlling acute ASF. Three major outbreaks took place in 1977 following penetration of ASFV through the port of Odessa: one in the region of Odessa itself and, later, two more, in the region of Kiev and in the town of Tavda in the region of Sverdlovsk.[3,30,59–61]

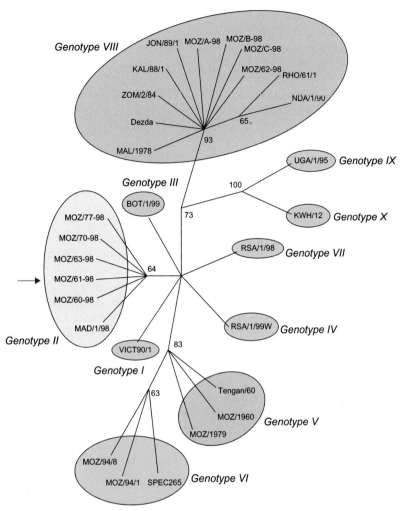

FIGURE 9.7 Dendrogram illustrating phylogenetic relationships among different ASFV genotypes. Genotype II is widespread in southeastern Africa (Zambia, Mozambique, and Madagascar) as well as in Georgia, Armenia, Azerbaijan, and Russia. *(Courtesy of Professor Trevor Drew, Veterinary Laboratories Agency, UK.)*

Eradication in the context of enzootic stability is quite feasible, as is suggested by the experience of Spain and Portugal in the 1980s and 1990s. Despite the extreme complexity of the situation, aggravated by the presence of ticks and wild boars, and by backyard free-range farming, eradication was effectuated in two stages, first at industrial pig-breeding units between 1985 and 1987 and then at small farms and in backyards between 1987 and 1995. Because no vaccine was available, the program was based on the detection of infected animals with the use of laboratory diagnostic tools, as well as on strict adherence to veterinary and sanitary security standards. The program's crucial aspects were (i) creating a network of mobile veterinary specialist teams that were charged with surveillance of pig-breeding units and early detection of ASF; (ii) conducting serological testing of pigs;

(iii) introducing stricter sanitation and hygiene standards for pig breeding; (iv) eliminating outbreaks of ASF by slaughtering and securely disposing of all pigs at an affected farm, by introducing a ban on pig movements within a 3-km zone, and by conducting serological surveillance of every herd and disposal of carrier animals within a 10-km zone; and (v) tracking the movements of pigs, including testing every animal before fattening or breeding it.

Genetic typification of the ASFV that penetrated into Europe revealed that the most probable origin of the virus is the west coast of Africa, from Angola in the south all the way to Senegal in the north.[37] The 22 currently known genotypes of ASFV are represented in the dendrogram in Figure 9.7, which shows the phylogenetic relationships among different genotypes.

It is currently unrealistic to pinpoint the origin of the broadly spread genotype I, which caused numerous outbreaks over a significant period between 1959 and 2000.[62] Nine genotypes are present in African countries. ASF outbreaks that occurred in the west of the continent between 1996 and 2000 were caused by genetically close viruses, all belonging to genotype I. Meanwhile, in Madagascar, the Republic of South Africa, and Botswana, the infection was brought about by genotypes II,

III, and VII, respectively. Genome sequences of 11 ASFV isolates collected in Africa at different points in time and analyzed in 2010 revealed that the most variable domain of the ASFV genome is its 5'-region.[23]

The virus isolated in Georgia belongs to genotype II, which circulates in Mozambique, Madagascar, and Zambia (Figure 9.7).[4,25] Its genome is 189,344 bp long and has 166 ORFs. Phylogenetic analysis on the basis of 125 conservative cistrons revealed maximum similarity with the Mkuzi 1979 isolate.[8] On the basis of nucleotide sequences of a p72 fragment and the entire p54 gene, genetic typification was completed on nine ASFV isolates collected at outbreak sites in Abkhazia, South Ossetia, Armenia, and different regions of Russia.[25]

Diagnostic confirmation of ASF by laboratory methods is imperative to the analysis not only of blood treated with anticoagulants (e.g., ethylenediaminetetraacetic acid, or EDTA), but also of sera and the spleen, kidney, lungs, and lymphatic nodes. Many diagnostic methods are currently in use, including virological, serological, and molecular biology tools. A number of the most notable methods are listed in Table 9.1 in accordance with the *Manual of Diagnostic Test and Vaccines for Terrestrial Animals*, published by OIE.

TABLE 9.1   African Swine Fever Laboratory Diagnostic Methods

Virus detection method	Description	
Hemadsorption assay	ASFV is detected in pig macrophage primary cultures. The virus is capable of causing infection and replicating naturally in peripheral blood leukocyte cultures, where, apart from causing cytopathic effect in infected macrophages, it displays a recognizable hemadsorbing effect and cell lysis. Under a microscope typical erythrocyte rosettes anchored to leukocytes can be observed. The hemadsorption assay remains the most specific and sensitive method of ASFV detection, since no other porcine virus displays a similar	Hemadsorption tests are currently only used at reference laboratories. The assay takes 3–10 days

(Continued)

TABLE 9.1 (Continued)

Virus detection method	Description	
	activity. Even though the hemadsorption assay is extremely laborious and time consuming as opposed to other diagnostic methods (results can only be obtained over a period of 5–10 days), it still remains preferable to other techniques. It is also noteworthy that some ASFV strains are incapable of hemadsorption, in which case confirmation of ASFVs presence requires additional tests like PCR or immunofluorescence	
Immunofluorescence assay	The immunofluorescence assay is used to detect viral antigens in the course of dyeing cryostatic slices or tissue tracing with antibodies to ASFV with the addition of fluorescein isothiocyanate. It is a very rapid, simple and sensitive method that can also be used on cell cultures infected with homogenized organs and tissues from suspicious animals. Under a microscope cytoplasmic inclusion bodies can be observed. At late stages of infection fluorescence can appear grainy. If the assay is performed over 10 post infection, antibodies that appear at this point can block conjugate binding. The immunofluorescence assay is therefore used alongside antibody detecting techniques (indirect immunofluorescence, ELISA, immunoblotting)	It is only advisable to run immunofluorescence tests when PCR cannot be performed or the facility has insufficient experience to run it reliably. It is important to remember that every negative result must be confirmed by, for example, antibody detection methods. The immunofluorescence assay requires 75 min
Polymerase chain reaction (PCR)	PCR is a highly sensitive and specific technique that allows to confirm the presence of ASFV by way of amplification of the viral DNA in the sample. PCR is reliant on nucleotide primers complementary to a highly conservative ASFV genome fragment, and most known ASFV strains can thus be detected, both hemadsorbing and non-hemadsorbing. This method is currently used at reference laboratories both to diagnose ASF and confirm the presence of ASFV. Both tissue and serum samples from animals displaying clinical signs can be tested, since viremia is still occurring. PCR can be used for detection of ASFV in blood starting from the second day of infection to several weeks post exposure	PCR is currently very widely used for ASF diagnosing, but requires highly thorough personnel training. The assay can be completed in 5–6 h
ELISA	Sandwich ELISA has been adapted for ASF diagnosis. However, despite a superb sensitivity at early stages of infection, it drops drastically after 9–10 days post infection due to antibody blockage like in the case of immunofluorescence assays	ELISA for antigen detection did not come to be in demand. The assay requires 3 h

Antibody detection	Description	
Indirect immunofluorescence assay	Indirect immunofluorescence assays are rapid, highly sensitive and specific. The assay relies on specific antibodies present in the serum or exudate on the surface of a cell layer infected with ASFV. The results are visualized upon addition of iodized protein A or antibodies to porcine IgG conjugated with a fluorescent marker. If the cell layer contains positive samples in areas close to the nucleus, fluorescence can be observed, indicating to ASFV replication sites	Indirect immunofluorescence tests are not very widely used. Commercial reagents are not available for sale. The assay requires 2 h
ELISA	ELISA is a method used for mass prophylactic and epizootic studies. At this moment a soluble antigen is used that contains most of the viral proteins. ELISA is highly sensitive and specific, as well as rapid and easy to perform. An ELISA assay with non-infectious reagents has also been designed with recombinant proteins p32, p54 and pp72 as viral antigens. The sensitivity and specificity of ELISA assays allow one to use them to test poorly preserved serum samples	ELISA is currently widely used. Commercial diagnostic kits are available on the market. The assay requires 2 h
Immunoblotting	During immunoblotting the ASFV proteins are transferred onto nitrocellulose membranes acting as antigen strips. The suspicious sample reacts with peroxidase-conjugated protein A to detect the possible antibodies to ASFV. Immunoblotting is used to measure antibody reactivity in the serum upon contact with proteins unique to ASFV. Specificity, sensitivity and minimum impact of the human factor make immunoblotting a perfect serological diagnostic technique	Diagnostic kits are currently not available for sale. Reagents are manufactured at a number of reference laboratories in the EU and at OIE. Immunoblotting is a superb technique if serological confirmation of spurious results is required. The assay requires 3 h

References

9.1 Family Asfarviridae

1. Dixon LK, Alonso C, Esciribano JM, et al. Family asfarviridae. In: King AMQ, Adams MJ, Carstens EB, et al., editors. *Virus Taxonomy: Ninth report of the International Committee on Taxonomy of Viruses*. London, UK: International Union of Microbiological Societies/ Elsevier Academic Press;2012. p. 153−62.
2. Tulman ER, Delhon EA, Ku BK, et al. African swine fever virus. In: Van Ethen JL, editor. *Lesser known large DNA viruses*. Current Topics in Microbiology and Immunology. 2009. p. 43−87.

9.1.1.1 African Swine Fever Virus

1. United States Department of Agriculture (USDA). Russian Federation. Global Agricultural Information Network 2011; Report RS 1144.
2. King K, Chapman DA, Argilaguet JM. Protection of European domestic pigs from virulent African isolates of African swine fever virus by experimental immunization. *Vaccine* 2011;**29**(28):4593−600.
3. KovalenkoYaR, Sidorov MA, Burba LG. African Swine Fever. M. 1972 [in Russian].
4. Sanchez-Botija C. Reservorios del virus de la Peste Porcina Africana. Investigation del virus de la PPA en los artropodos mediante la prueba de la hemoadsorcion. *Bull OIE* 1963;:895−9.

5. Montgomery RE. On a form of swine fever occurring in British East Africa (Kenya colony). *J Comp Pathol* 1921;**34**:159–91.

6. Jori F. Bastos ADS. Role of wild suids in the epidemiology of African swine fever. *Ecohealth* 2009;**6** (2):296–310.

7. Sanchez-Vizcaino JM. *African swine fever. Diseases of swine, ninth edition*. Ames, Iowa: Blackwell Publishing; 2006:291–98.

8. Chapman DA, Darby AC, Da Silva M, et al. Genomic analysis of highly virulent Georgia 2007/1 Isolate of African swine fever virus. *Emerg Infect Dis* 2011;**17** (4):599–605.

9. Dixon LK, Alonso C, Eseribano JM. Asfarviridae. In: King AMQ, Carstens EB, Lefkowitz E, et al., editors. *Virus taxonomy. Ninth report of the International Committee on the Taxonomy of Viruses. Elsevier;*2011. p. 153–62.

10. Ogata H, Toyoda K, Tomaru Y. Remarkable sequence similarity between the dinoflagellate-infecting marine virus and the terrestrial pathogen African swine fever virus. *Virol J* 2009;**6**:178.

11. Loh J, Zhao G, Presti RM. Detection of novel sequences related to African swine fever virus in human serum and sewage. *J Virol* 2009;**83**(24):13019–25.

12. Hess WR. *African swine fever virus. Virology monographs*, 9. Vienna: Springer Verlag;1971:1–33.

13. Suarez C, Gutierrez-Berzal J, Andres G. African swine fever virus protein p17 is essential for the progression of viral membrane precursors toward icosahedral intermediates. *J Virol* 2010;**84**(15):7489–99.

14. Tulman ER, Delhon GA, Ku BR. African swine fever virus. *Curr Top Microbiol Immunol* 2009;**328**:43–87.

15. Yanez RJ, Rodriguez JM, Nogal ML. Analysis of the complete nucleotide sequence of African swine fever virus. *Virology* 1995;**208**(1):249–78.

16. Carrascosa AL, Bustos MJ, de Leon P. Methods for growing and titrating African swine fever virus: field and. *Curr Protoc Cell Biol* 2011; [chapter 26. Unit 26.14].

17. Hurtado C, Bustos MJ, Carrascosa AL. The use of COS-1 cells for studies of field and laboratory African swine fever virus samples. *J Virol* 2014;**164**(1–2):131–4.

18. Malmquist WA, Hay D. Hemadsorption and cytopathic effect produced by African swine fever virus in swine bone marrow and buffy coat cultures. *Am J Vet Res* 1960;**21**:104–8.

19. Gomez-Villamandos JC, Hervas J, Mendez A. Experimental African swine fever: apoptosis of lymphocytes and virus replication in other cells. *J Gen Virol* 1995;**76**(9):2399–405.

20. Sierra MA, Bernabe A, Mozos E. Ultrastructure of the liver in pigs with experimental African swine fever. *Vet Pathol* 1987;**24**(5):460–2.

21. Plowright W, Perry CT, Peirce MA, et al. Experimental infection of the argasid tick, *Ornithodoros moubata porcinus*, with African swine fever virus. *Arch Ges Virusforsch* 1970;**31**(1):33–50.

22. Ravaomanana J, Michaud V, Jori F. First detections of African swine fever virus of *Ornithodoros porcinus* in Madagascar and new insights into tick distribution and taxonomy. *Parasit Vectors* 2010;**3**:115.

23. Villiers EP, Gallardo C, Arias M, et al. Phylogenomic analysis of 11 complete African swine fever virus genome sequences. *Virology* 2010;**400**:128–36.

24. Gabriel C, Blome S, Malogolovkin A. Characterization of African swine fever virus Caucasus isolate in European wild boars. *Emerg Infect Dis* 2011;**17**:2342–54.

25. Kalabekov IM, Elsunova AA, Shendrick AG. Phylogenetic analysis of field isolates of African swine fever virus. *Veterinariya* 2010;**5**:31–3 [in Russian].

26. Swaney LM, Lyburt F, Mebus CA. Genome analysis of African swine fever virus isolated in Italy in 1983. *Vet Microbiol* 1987;**14**:101–4.

27. Manso Ribeiro J, Azevedo R, Teixeira J. An atypical strain of swine fever virus in Portugal. *Bull OIE* 1963;**50**:516–34.

28. Manelli A, Sotgia S, Patta C. Effect of husbandry methods on seropositivity to African swine fever virus in Sardinian swine herds. *Prev Vet Med* 1997;**32**:235–41.

29. Arias M, Escribano JM, Rueda A, et al. La Peste Porcina Africana. *Med Veter* 1986;**3**:333–50 [in Spanish].

30. Aliper TI, Zaberezhny AD, Grebennikova TV. African swine fever in the Russian Federation. *Vop Virusol* 2012;**1**:127–36 [in Russian].

31. Arias M, Sanchez-Vizcaino JM. *African swine fever eradication: The Spanish model. Trends in emerging viral infections of swine—1st edition*. 2000133–9.

32. Colgrov G, Haelterman EO, Coggins L. Pathogenesis of the African swine fever virus in young pigs. *Am J Vet Res* 1969;**30**:1343–59.

33. Plowright W, Parker J. The stability of African swine fever virus with particular reference to heat and pH inactivation. *Arch Ges Virusforsch* 1967;**21**:383–402.

34. Krug PW, Larson CR, Eslami AC, et al. Disinfection of foot-and-mouth disease and African swine fever virus with citric acid and sodium hypochlorite on birch wood carriers. *Vet Microbiol* 2012;**156**(1–2):96–101.

35. Pharo H, Cobb SP. The spread of pathogens through trade in pig meat: overview and recent developments. *Rev Sci Tech* 2011;**30**(1):139–48.

36. Wieringa-Jelsma T, Wijnker JJ, Zijlstra-Willems EM. Virus inactivation by salt (NaCl) and phosphate supplemented salt in a 3D collagen matrix model or natural sausage casings. *Int J Food Microbiol* 2011;**148** (2):128–34.

37. Mebus C, House C, Ruiz F. Survival of foot-and-mouth disease, African swine fever and hog cholera virus in Spanish serrano cured hams and Iberian cured hams, shoulder and loin. *Food Microbiol* 1993;**10**:133–43.

38. Mebus C, McVicar JW, Dardiri AH. Comparison of the pathology of high and low virulence African swine fever infections. In: Wilkinson PJ, editor. African Swine Fever. EUR 8466 EN: Proceedings of CEC/FAO Research Seminar, Sardinia. 1983. p. 183–94.

39. Sierra MA, Carrasco L, Gomez-Villamandos JC. Pulmonary intravascular macrophages in lungs of pigs inoculated with African swine fever virus of differing virulence. *J Comp Pathol* 1990;**102**(3):323–34.

40. Gomez-Villamandos JC, Bautista MJ, Carrasco L. African swine fever virus infection of bone marrow: lesions and pathogenesis. *Vet Pathol* 1997;**34**(2):97–107.

41. Sanchez-Vizcaino JM, Slauson DO, Ruiz-Gonzalvo F, et al. Lymphocyte function and cell-mediated immunity in pigs with experimentally induced African swine fever. *Am J Vet Res* 1981;**42**:1335–41.

42. Onisk D, Borca M, Kutish G. Passively transferred African swine fever virus antibodies protect swine against lethal infection. *Virology* 1994;**198**:350–4.

43. Schlafer DH, Mebus CA, McVicar JW. African swine fever in neonatal pigs: Passive acquired protection from colostrum or serum from recovered pigs. *Am J Vet Res* 1984;**45**:1367–72.

44. De Boer CJ. Studies to determine neutralizing antibody in sera from animals recovered from African swine fever and laboratory animals inoculated with African virus with adjuvants. *Arch Ges Virusforsch* 1967;**20**(2):164–79 [in German].

45. Martins C, Leitao A. Porcine immune response to African swine fever virus (ASFV) infection. *Vet Immunol Immunopathol* 1994;**34**:99–106.

46. Grebennikova TV, Zaberezhny AD, Aliper TI, et al. Diagnostics of African swine fever in Russian Federation. *Vop Virusol* 2013;**1**:64–80 [in Russian].

47. Ronish B, Hakhverdyan M, Stahl K. Design and verification of a highly reliable Linear-After-The-Exponential PCR (LATE-PCR) assay for the detection of African swine fever virus. *J Virol Methods* 2011;**172**(1–2):8–15.

48. Tignon M, Gallardo C, Iscaro C, et al. Development and interlaboratory validation study of an improved new real-time PCR assay with internal control for detection and laboratory diagnosis of African swine fever virus. *J Virol Methods* 2011;**178**(1–2):161–70.

49. Aguero M, Fernandes J, Romero L, et al. Highly sensitive PCR assay for routine diagnosis of African swine fever virus in clinical samples. *J Clin Microbiol* 2003;**41**(9):4431–4.

50. Eberling AJ, Bieker-Stefanelli J, Reising MM, et al. Development, optimization, and validation of a Classical swine fever virus real-time reverse

transcription polymerase chain reaction assay. *J Vet Diagn Invest* 2001;**23**(5):994–8.

51. Zsak L, Borca MV, Risatti GR. Preclinical diagnosis of African swine fever in contact-exposed swine by a real-time PCR assay. *J Clin Microbiol* 2005;**43**:112–19.

52. Thibezov VV, Terehova JO, Balandina MV, et al. Development of immunoferment methods of diagnosis African swine fever. *Veterinariya Kubani* 2012;**1**:20–3 [in Russian].

53. Zaberezhny AD, Aliper TI, Grebennikova TV, et al. African swine fever in Russian Federation. *Vop Virusol* 2012;**5**:4–10 [in Russian].

54. Neilan JG, Zsak L, Lu Z. Neutralizing antibodies to African swine fever virus proteins p30, p54 and p72 are not sufficient for antibody-mediated protection. *Virology* 2004;**319**:337–42.

55. Ivanov V, Efremov EE, Novikov BV. Vaccination with viral protein-mimicking peptides mortality in domestic pigs infected by African swine fever virus. *Mol Med Rep* 2011;**4**(3):395–401 [in Russian].

56. Gulenkin VM, Korennoy FI, Karaulov AK, et al. Cartographical analysis of African swine fever outbreaks in the territory of Russian Federation and computer modeling of the basic reproduction ratio. *Prev Vet Med* 2011;**102**(3):167–74.

57. Kurinnov VV, Kolbasov DV, Tsybanov SD, et al. Diagnosis and monitoring of African swine fever outbreaks in Caucasus Repablics. *Veterinariya* 2008;**10**:20–5 [in Russian].

58. Balyshev VM, Kurinov VV, Tsybanov SZh. Biological properties of Russian isolates of the African swine fever virus. *Veterinarnyi Zhurnal* 2010;**7**:25–7 [in Russian].

59. Kovalenko YR, Bourba LG, Sidorov MA. African swine fever infection routes in swine. In: *Proceedings of Russian Institute of Experimental Veterinary Medicine*, vol. 31. Moscow; 1965. p. 336–42 [in Russian].

60. Kovalenko YR, Bourba LG, Sidorov MA. Persistence potential of the African swine fever virus in the environment. *The Agricultural Bulletin (Moscow)* 1964;**3**:62–5 [in Russian].

61. Kovalenko YR, Ivanov BG. Experimental inoculation of pigs with the African swine fever virus. In: *Proceedings of Russian Institute of Experimental Veterinary Medicine*, vol. 24. Moscow; 1961. p. 53–61 [in Russian].

62. Boshoff CI, Bastos ADS, Gerber LJ. Genetic characterization of African swine fever virus from outbreaks in southern Africa (1973–1999). *Vet Microbiol* 2007;**121**:45–55.

Prions

... It's a last story about Mowgli ... *R. Kipling*

History. The transmissible spongiform encephalopathies (TSEs) make up a unique group of zoonotic neurodegenerative diseases affecting both animals and humans. Among the TSEs is scrapie in sheep; bovine spongiform encephalopathy (BSE) in cattle; chronic wasting disease (CWD) in cervids; and Creutzfeldt–Jakob disease (CJD), fatal familial insomnia (FFI), Gerstmann–Sträussler–Scheinker syndrome (GSS), and kuru in humans. Pathogenic clinical features typical of these conditions are progressive dementia, cerebral ataxia, and 100% mortality.

TSEs attracted unprecedented interest on the part of the scientific community over 20 years ago after Stanley Prusiner (Figure 10.1)[1] proposed the hypothesis that the only etiological agent of the infections listed in the previous paragraph is a pathogenic isoform (or isoforms) of the otherwise normal cellular protein, a hypothesis which meant that no DNA or RNA molecules were involved in the infections. The basic dogma of molecular biology was thus questioned for the first time. Nevertheless, a large proportion of experimental data that were subsequently obtained supported this groundbreaking, unconventional hypothesis, and in 1997 Prusiner won the second Nobel Prize awarded for TSE research

after the 1976 award to Dr. D.C. Gajdusek (Figure 10.2),[2] who, in 1966, had discovered and described the lethal kuru disease circulating in some tribes of Papua New Guinea, where ritual cannibalism was a common practice. The infectious agent was purified, described, and named "prion," an abbreviation for "proteinaceous" and "infectious" followed by the common suffix "-on." The term "prion diseases" then became synonymous with "TSEs."

Prions define a new class of infectious agents, all or nearly all of which are abnormal isoforms of the host-encoded prion protein, PrP. In strict accordance with Prusiner's "no DNA or RNA" (also called "protein-only") hypothesis, no nucleic acid has been specifically associated with infectivity so far. However, hypotheses such as the viral, or "virino," hypothesis, which states that the infectious agent is composed of a small amount of nucleic acid protected by the prion protein, cannot be dismissed yet, because they are supported by some pathophysiological data.[1,3–5]

Thus, despite a very long history of research into TSEs (clinical signs of ovine scrapie were reported at the beginning of the twentieth century), the group of lethal prion infections is still referred to as "mystery diseases," compelling scientists to conduct further research.

Zoonotic Viruses of Northern Eurasia.
DOI: http://dx.doi.org/10.1016/B978-0-12-801742-5.00010-6

FIGURE 10.1   Stanley Prusiner.

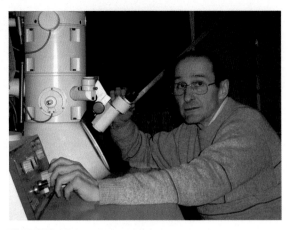

FIGURE 10.3   Vladimir Grigoriev.

FIGURE 10.2   Daniel Carleton Gajdusek.

FIGURE 10.4   Sergey Kalnov.

In Russia, prion diseases are actively studied especially by Vladimir Grigoriev (Figure 10.3), Sergey Kalnov (Figure 10.4), and Oleg Verkhovsky (Figure 10.5), results of whose research are presented in this section.

Taxonomy. The standard scientific approaches used for the taxonomic classification of bacteria and viruses are insufficient for a characterization and systematic description of prions. One hypothesis that is current is the hypothesis that the only infectious agent causing TSEs is a protease-resistant isoform of the normal cellular protein PrP^c. The hypothesis has been confirmed by a large amount of experimental data.

FIGURE 10.5   Oleg Verkhovsky.

The gene that encodes the prion protein can be found in nearly all vertebrate genomes assessed, including domestic chickens, turtles, and fish, as well as in the fruit fly *Drosophila melanogaster* and low eukaryotes such as the common yeast *Saccharomyces cerevisiae*. PrP^c, encoded by that gene, is a normal membrane glycoprotein produced in all types of cells, except monocytes. As of now, what triggers an infection, what happens at the very outset, and which factors have an impact are still unknown; however, the conformational change in the PrP's molecular structure has been confirmed.[4,6] The amino acid sequences of PrP^c and its pathogenic counterpart are identical, and it is the conformational change in the protein structure that results in altered biochemical and physical properties that, in their turn, usher in the subsequent pathogenic process(es). The abnormal prion protein isoform considered to be pathogenic is designated variously, depending on the type of prion disease: PrP^{Sc} (scrapie), PrP^{BSE} (BSE), PrP^{CJD} (Creutzfeld–Jacob disease), PrP^{CWD} (chronic wasting disease), PrP^{RES} (Proteinase K resistant) or, more generally, PrP^d ("d" for the "disease").

Prion strains. One issue undermining the "protein only" hypothesis (i.e., the postulated noninvolvement of nucleic acids in the transformation of PrP^c into PrP^d) is the proposed existence of so-called TSE prion strains.[7] To understand this issue, note that prion strains vary in their biological properties, such as their incubation period, the time that clinical signs of the disease first appear, and the distribution of PrP^d deposits in brain regions and tissues beyond the central nervous system (CNS).[8] These strains have different glycosylation sites, and their protein patterns (peptide fingerprints) also vary after standard treatment with Proteinase K. Sick animals with identical genotypes have displayed symptoms suggesting that they could be affected by different strains.[8] In accordance with the protein-only hypothesis, the maturation and differentiation of these strains should be determined just by the degree of conformational changes in their three-dimensional PrP^c molecular structure, not by mutations in the corresponding nucleic acid. On the basis of this assumption and some of the experimental data, then, one must conclude that the three-dimensional molecular structure of the newly formed infectious PrP^d isoform is transmitted to the strain-specific phenotype via a nongenetic mechanism, which comes across as an extremely puzzling and previously unknown genetic phenomenon.[3] One explanation is that additional cofactors that are of a proteinaceous nature, such as chaperones, could be at work converting PrP^c to pathogenicity.[9]

Three types of TSEs can be distinguished: sporadic (as in CDJ and scrapie), genetic/familial (as in FFI and GSS), and transmissible/infectious (as in kuru, iatrogenic CJD, BSE, and CWD) (Table 10.1). The common pathogenic clinical features for these variants are progressive dementia, cerebral ataxia, and a 100%-lethal outcome.

Native prion protein PrP^c. The term "PrP^c" denotes the prion protein isoform that is normally present in cells, with "c" abbreviating "cell." Prions are distinct from viruses, insofar as both of the PrP isoforms (PrP^c and the pathogenic PrP^d) are encoded by a chromosomal gene designated as PRNP in humans and Prnp in mice and located, respectively, on the short arm of the 20th chromosome and the synthetic region of the 2nd chromosome.[1,2]

The expression product of the PRNP gene varies insignificantly in molecular mass, depending on the animal species. The PrP^c polypeptide chain is 253 aa long in humans, 265 aa long in cattle, etc., with the overall homology in vertebrates amounting to 85–95%. PrP is a glycoprotein located on the cell surface, where it is bound to a glycosyl–phosphatidylinositol anchor.[1] The processing of the PrP^c glycoprotein starts in

TABLE 10.1 Prion Diseases of Humans and Animals

Disease	Host	Transmission routes
Kuru	Human	Alimentary infection *via* ritualistic cannibalism
Infectious Creutzfeldt-Jakob disease	Human	Infection through injection of growth hormone, transplantation of brain envelope, etc.
Hereditary CJD	Human	Mutations in the PRNP gene
Gerstmann-Straussler-Schneiker syndrome	Human	Mutations in the PRNP gene
Fatal familial insomnia	Human	Mutations in the PRNP gene (D 178N, M 129)
Sporadic Creutzfeldt-Jakob diseases	Human	Somatic mutation or spontaneous PrP^c/PrP^d conversion
New variant CJD	Human	Alimentary infection through by-products
Scrapie (Sc)	Sheep, goats	Genetic factory
Bovine spongiform encephalopathy (mad cow disease)	Cattle	Alimentary infection through by-products/meet-and-bone meal
Atypical Bovine spongiform encephalopathy	Cattle	Alimentary infection through by-products/meet-and-bone meal. Infectious agent PrP^d non-resistant to Proteinase K
Transmissive mink encephalopathy	Mink	Alimentary infection through by-products, bovine, sheep, goat meet and meet-and-bone meal
Chronic Wasting Disease	Deer, Moose	Unknown
Spongiform encephalopathy cats (*Felidae*)	Cats, Leopard, Tiger	Alimentary infection through by-products/bovine meet-and-bone meal
Exotic encephalopathy of ungulates	Antelope: Kudu, Nyala, Orix	Alimentary infection through meet-and-bone meal in Zoos and national parks

the endoplasmic reticulum, after which one or two sugars are added to asparagines in positions N180 and N196 (sites of normal glycosylation) in the Golgi complex, followed by further saccharide modification. Isolated PrP^c protein is thus visualized in three distinct bands, corresponding to the nonglycosylated (25 kD), partially glycosylated (30 kD), and fully glycosylated (32–40 kD) forms, if SDS electrophoresis is performed.

As soon as processing and folding are completed, the PrP^c protein installs itself on the cellular membrane surface in clusters, or "rafts." The PrP^c protein also acts as a receptor molecule: Despite binding to the outer cell membrane surface via a hydrophobic anchor, it is essentially a mobile molecule, able to move and contact membranes of adjacent cells. In the course of pathogenesis, this same mechanism effectuates transmission of the infectious agent from neuron to neuron, enabling interaction with the abnormal isoform PrP^d. Over the past several years, data shedding light on the possible three-dimensional conformation of the PrP^c molecule were obtained for at least the recombinant PrP^c protein of mouse, hamster, and human being, whereas amino acid sequences were established for 29 species.[3]

Neurons supposedly contain over 5,000 PrP^c molecules per cell, whereas in other types of cells, including lymphocytes, their proportion does not exceed 3,000 molecules per cell. The protein can be found in a variety of tissues and is highly expressed in the CNS. Although the precise physiological function of PrP remains to be established, a wealth of experimental data demonstrates that protein's essential role in the pathogenesis of TSE and susceptibility thereto.[10]

PrP^c function. The physiological role of PrP^c is still unclear. One study showed that PrP^c knockout mice (PrP$^{0/0}$) did not exhibit any differences in phenotype compared with wild mice. However, another study found that PrP$^{0/0}$ mice had lower numbers of Purkinje neurons, displayed electrophysiological abnormalities, and had sleep patterns different from those of other mice.[7] Also, the PrP$^{0/0}$ mice were resistant to the TSE etiological agent (i.e., they did not show clinical signs of pathology after experimental intracerebral infection with a thousand-fold infectious dose of PrP^d, a result very much in line with the assumption that PrP^c is necessary and sufficient for the development and transmission of pathology).

The pathogenic infectious isoform PrP^d. As a TSE infection progresses, the level of mRNA does not rise. The prion isoforms PrP^c and PrP^d both have a molecular mass of 33–35 kD, but differ significantly when it comes to their three-dimensional structure, which is shown for PrP^{Sc} in Figure 10.6. The abnormal isoform contains a heightened proportion of beta sheets and exhibits partial resistance to Proteinase K (PK) proteolysis.[7] Infectious prionic agents correspond entirely or at least partially to misfolded PrP^d protein or its N-terminal truncated version (PrP27–30) produced following PrP^d proteolysis.[1,11] Most evidence suggests that PrP^d appears as a result of a conformational change from PrP^c to PrP^d. Alpha-helical regions and unstructured regions are most likely converted to beta sheets, the alpha-helix content being 42% in PrP^c, but as low as 30% in PrP^d and 21% in PrP27–30. In the same vein, the proportion of beta sheets is 3% in PrP^c, 43% in PrP^d, and 54% in PrP.[11–14] The transformation of Proteinase-K-sensitive PrP into Proteinase-K-resistant PrP has been demonstrated in cell-free systems.[15] However, the "direct and pure" generation of infectious PrP^d from PrP^c has not been experimentally reproduced. To this day, it is still believed that the presence of cofactors such as molecular chaperones may aid PrP^c-to-PrP^d conversion.[15]

The conformational change of PrP^c is believed to take place in endolysosomes or lysosomes. Freshly generated PrP^d molecules stimulate the further conversion of more and

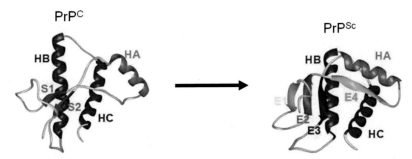

FIGURE 10.6 The scheme of conformational change (conversion, misfolding) of the normal cellular protein PrP^c to its pathogenic isoform PrP^{Sc}. Conformational change (misfolding) of PrP^c is characterized by breakage of disulfide bonds, a drop in the proportion of alpha-helices (HA, HC, HA), and an increased proportion of beta sheets (E1, E2, E3).

more PrP^c molecules to PrP^d, which accumulates and destroys nerve cells. As a result, vacuoles form in the brain, amyloidosis takes place, and a fatal outcome becomes inevitable.[15,16]

Because the amino acid sequences of different infectious PrP^d variants are identical to the original native PrP^c molecule, those isoforms are referred to as "conformers."

The PrP^c–PrP^d conformational transition leads to a radical change in physical properties of the protein. Unlike PrP^c, the infectious isoform PrP^d is insoluble in detergents, is stable against inactivation by ionizing irradiation or UV treatment and is partially resistant to Proteinase K treatment, which only results in the formation of a protein fragment PrP^{RES} with a molecular weight of 27–30 kD and that is also infectious.

Evidence has recently been obtained that a proteinase-sensitive (PrP^{sens}) infectious isoform might exist that is different from PrP^d [17]. Apart from this finding, a number of short proteinase-resistant polypeptides with a molecular weight of 12–13 kD were discovered and their infectivity was demonstrated in laboratory animals. The new proteinase-sensitive PrP isoform was isolated from so-called mad cows, (i.e., cows displaying symptoms of BSE similar to symptoms of the prion disease (Table 10.1)). This BSE variant was named atypical bovine spongiform encephalopathy or a BSE.[1]

The "protein-only" hypothesis suggests that the conversion of endogenous PrP^c to PrP^d is an autocatalytic process, triggered by an initial portion of PrP^d in the matrix and supposedly further stimulated by an interaction between both isoforms, also known as "seeding" contact.

In accordance with this model, thermodynamic equilibrium is in place between the two isoforms while an additional portion of PrP^d breaks the equilibrium and initiates an intensive accumulation of ever more PrP^d.

As a result of the initial interaction between the "seeding agent" and native cellular membrane PrP^c, a polymer chain consisting of PrP^d molecules is formed. The chain grows in a "geometric progression" as new freshly transformed PrP monomers are added.[1] This model is believed to describe the hereditary spongiform encephalopathy scenario.

Interspecies barrier. Interspecies transmission of spongiform encephalopathies normally correlates with longer incubation periods for the prion disease. Numerous data obtained in experiments on both transgenic cell lines and animals indicate that the degree of resistance to prion infectious agents is determined by the level of homology between the prion proteins of the donor and recipient,[18,19] as well as by the degree of congruence between three-dimensional molecules in different species (Figure 10.7).

Distribution. BSE was first identified in 1986 in the United Kingdom, where it quickly spread to reach epidemic proportions, affecting over 30,000 head of cattle per year by 1992. Through continual export of live cattle as well as meat-and-bone meal obtained from carcasses of slaughtered animals, the disease

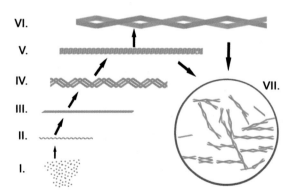

FIGURE 10.7 The scheme of amyloidlike fibril formation following conversion of recombinant prion PrP^c protein *in vitro*.[20,21] I – PrP^c soluble monomer, II – PrP^{Sc} polypeptide, III – PrP^{Sc} protofibrils, IV–VII – variants of transformation of PrP^{Sc} protofibrils to mature amyloidlike pathogenic fibrils.

spread throughout most of Europe and some non-European countries. By 2006, 20 years after its first appearance in the United Kingdom, the disease had been reported in 24 more countries. Numerous regulations were introduced both in the United Kingdom in the late 1980s and elsewhere in the 1990s, with a view to minimizing the entry of contaminated tissues into animal and human food chains in order to eventually prevent the disease from spreading internationally.

These measures proved to be successful to the extent that disease did not penetrate into any of the BSE-free countries over the past year, and incidence rates fell significantly in most countries where BSE had been reported. In other countries, where surveillance is insufficient, the situation remains unclear. Although the general opinion is that the epidemic was spawned by cross-species contamination with ovine scrapie as a result of feeding meat-and-bone meal derived from ovine carcasses to cattle, an alternative hypothesis suggested that the disease occurred in cattle spontaneously and spread via cattle-derived meat-and-bone meal. Facts speaking both against and in favor of these two hypotheses have been debated, and neither hypothesis is likely to be convincingly proven or refuted. The spontaneous-disease hypothesis was kept in limbo until the recent discovery of "atypical" strains of BSE, a discovery that reopened the question. Prion research teams across the world are now considering the importance of atypical BSE from the perspective of the entire picture of spontaneous disease and whether such cases, if they exist, could account for at least some cases of apparently spontaneous CJD in humans.

The discovery of Proteinase-K-sensitive PrP^d isoforms raised questions about the standard BSE, new-variant CJD (nvCJD), CWD, and all other TSE immunoassay variants, which are based on the detection of PrP^d (27–30 kD) protein fragment antigen, which forms after controlled exposure to Proteinase K *in vitro*. From 1986, when BSE was discovered in the United Kingdom, to today, confirmed cases of BSE in animals or in humans (nvCJD) have been reported in over 29 countries. The list of affected countries lengthens, with new cases of TSE being reported in the United States, Europe, and Asia (Japan).

Economic losses from BSE outbreaks have amounted to $US13 billion in Great Britain alone. The number of young people infected with nvCJD, whose incubation period is short, also keeps increasing.[3,6]

Vertebrate hosts. Since the discovery of BSE in November 1986 at the Central Veterinary Laboratory in Weybridge, UK, there has been a great deal of public concern about the safety of meat and meat products. BSE has a broad experimental host range and has possibly been transmitted to domestic cats and to a variety of nondomestic felids, ruminants, and nonhuman primates in zoological parks. PrP^d accumulation (CWD) has been demonstrated in captive and free-ranging mule deer, white-tailed deer, and elk in limited areas of the United States.[22–32]

Animal pathology. Following a long incubation period, the CNS that is affected by prions degenerates progressively and death ensues a few months after the onset of clinical symptoms. TSEs can be experimentally transmitted to laboratory animals, nonhuman primates, mice, and hamsters. Progression of the disease is currently believed to correlate with the accumulation of the abnormal PrP^c isoform.[1,9,11,12]

For scrapie, both naturally occurring in sheep and experimentally reproduced in rodents, infectivity has been demonstrated in a number of tissues, which fall into four categories based on the infectivity level in accordance with the WHO classification:

I. High infectivity—brain, spinal cord
II. Medium infectivity—spleen, lymph nodes, tonsils, ileum, proximal colon

III. Low infectivity—sciatic nerve, distal colon, adrenal, nasal mucosa, hypophysis

IV. Low detectable infectivity (undetectable by enzyme-linked immunosorbent assay (ELISA), immunohistochemistry (IHC), etc.)—blood clot, serum, milk, colostrum, mammary gland, skeletal muscle, heart, kidney, saliva, ovaries, uterus, seminal glands, feces.

For TSEs, the only way of measuring infectivity is a bioassay that relies on intracerebral inoculation into mice. Such experiments were performed for BSE, and it was shown that the infectivity level in peripheral tissues, aside from the CNS, is relatively lower than that for ovine scrapie. Infectivity had previously not been observed in blood; bone marrow; tallow; the gastrointestinal tract; the heart, kidney, liver, lungs, or lymph nodes; muscle tissue; peripheral nerves; or the pancreas, gonads, spleen, skin, trachea, or tonsils in field case studies of BSE.[13]

Chronic Wasting Disease. There are currently two principal views on the threat of further geographic spreading of TSEs, especially bovine TSE (BSE), and on the prospect of their eradication. On the one hand, statistical data obtained by the European Union and the World Organisation for Animal Health's Office International des Epizooties (OIE) indicate that, following the introduction of international regulations and implementation of control and surveillance measures with regard to BSE (1976–2001), cases of the disease arose only infrequently inside countries where it was originally found (the United Kingdom, Ireland, etc.). These low rates suggest lower risks of TSE outbreaks globally, so hopes of complete eradication of BSE are not unduly harbored. On the other hand, new cases have been reported lately in countries where a ban on recirculating meat by-products was introduced a long time ago. Cases of bovine TSE have recently been reported in eastern Europe, and a new "strain" of bovine TSE causative agent has been observed in the United States and a number of European countries. On this basis, it is obvious that diagnostic tools need to be enhanced drastically in order for surveillance and control to be successful in Mongolia and the former Soviet Union with regard to bovine and other TSEs, including scrapie and CWD in wild ungulates. Even more crucial to the surveillance of TSEs in domestic and wild animals is the availability of methods for detecting infectious prion forms (i.e., PrP^d) in blood. Indeed, designing TSE diagnostic tools in Russia and former Soviet republics could be further thwarted not only by the variability of prion isoforms—especially with new ones continually being discovered—but also by the genetic diversity of economically significant animal species.[14,33]

CWD was first described as a TSE, or a prion disease, occurring in wild ungulates, both free and captive.[34] Three species in North America were shown to be susceptible: the mule deer (*Odocoileus hemionus*), the white-tailed deer (*Odocoileus virginianus*) and the Rocky Mountain elk (*Cervus elaphus nelsoni*).[32,35–38] The moose (*Alces alces*) has recently been added to this list.[17,39–41] It has been shown that CWD can be freely transmitted horizontally to susceptible animals upon direct or indirect contact despite a high genetic variability within and among species.[42,43] The genetic polymorphism of the Prnp gene, encoding the normal prion protein PrP^c, is known to determine the incubation period and the animal's susceptibility to two strains causing classical ovine scrapie[44] and also to contribute to the contagiousness of the recently discovered prion "strain" Nor98.[45]

Sparse distribution and long incubation periods were shown to correlate with codon 132 polymorphism of the Prnp gene in the Rocky Mountain elk.[46,47] Low susceptibility to CWD was demonstrated to correlate with codon 225 polymorphism in the mule deer and with codon 96 polymorphism of the same gene in the white-tailed deer.[48–50]

Other species susceptible to CWD have not yet been identified, nor are their genotypes amenable to the disease. Although the CWD agent has not been proven to infect cattle or humans, the geographic range of CWD carriers and vectors among other ruminants remains unstudied and thus unknown. Equally unknown is the number of CWD "strains" in circulation and the exact geographic distribution of the disease outside North America. In addition, certainty has not been attained regarding the extent of variability of the PrP^c amino acid sequence in wild ungulates, and this lack of knowledge could dramatically complicate the detection of infectious conformers by immunology tools that are reliant on monoclonal antibodies, which are designed to bind to protein epitopes.

Clinical signs of CWD include severe exhaustion, weight loss, profuse salivation, ataxia, deafness, overall listlessness, and changes in behavior and coordination, such as walking in circles.

Histological studies invariably reveal spongiform lesions in the brain (in the area of the obex), vacuoles forming in neurons, and neuronal degeneration. The disease was experimentally transmitted from deer to other deer, minks, ferrets, goats, and squirrel monkeys. In the United States, there was one recorded case of lethal infection with CJD: The patients (twins) were hunters who had consumed homemade deer-meat sausage.

Prnp gene polymorphism in wild ungulates in Russia. Estimating the risk of TSEs emerging and spreading in Russia, former Soviet Republics, Mongolia, and neighboring countries is much complicated by a great species diversity of wild ungulates, which populate an immense territory. The emergence and spread of CWD cannot be detected without basic data on prnp genotype frequencies in these species and without a technique for assessing CWD's interspecies transmission potential *in vitro*. Besides, although the lethality of prion CWD "strains" present in North American wild ungulates remains

unconfirmed, there is absolutely no guarantee that previously unknown CWD causative agents capable of triggering infection in cattle and humans will not be detected in Russia, former Soviet Republics, and adjacent regions.[33,51,52]

In Russia, comprehensive molecular genetic studies of wild ungulates' phylogenetic characteristics have been carried out only for the past decade. Previous studies of cervids' morphology have recently been supplemented by gene sequence analysis of their cDNA and polymerase chain reaction (PCR) analysis of their satellite DNA at Severtsev Institute for Ecology and Evolution Research in Moscow. To understand the kinship among the Cervidae, Bovidae, Moschidae, and Antilocapridae families at the molecular level, taxonometric studies were carried out on the basis of comparative analyses of ribosomal mitochondrial RNA nucleotide sequences and amino acid sequences of a number of proteins, such as ribonucleases and caseins.[5,39,41,53−55] Similar approaches were used to reveal the kinship between the North American moose (*Alces alces*), which is susceptible to CWD, and deer (Cervidae) in Russia and the former Soviet Union. The data obtained lay the foundation for comprehensive studies of Prnp genotypes of closely related species.

As of now, no data are available as to whether CWD is present in Russia or in any of the former Soviet Republics. This absence of data could be due to insufficient and sporadic diagnostic studies performed or to inadequately sensitive diagnostic tools for use with wild ungulates. There is a risk of CWD penetrating into Russia via imported cervids and other animals that were in contact with infected wild ungulates. Cases of CWD were reported in Korea among deer imported from Canada.[11] Breeding between CWD-infected imported animals and local closely related cervid species creates a full-scale threat that the disease will emerge in Russia. Ninety-three deer, including CWD-infected animals, were imported into Korea in 1997, and prion surveillance tests carried out between 2001 and 2004 among Canadian and local animals revealed new cases of CWD.

There is thus an array of reasons for why the threat of CWD penetrating and spreading in Russia cannot as yet be discarded: (i) Endemic species have not been identified; (ii) sporadic cases of TSE in ruminants and wild ungulates have not been recorded; and (iii) deer and elks might get, or might have already gotten, infected as a result of contact with imported animals or migrant animals of closely related species from countries such as Poland, Korea, China, and Mongolia. Prion protein gene polymorphism is believed to be paramount to assessing the risk of CWD emerging in Russia.

BSE has a broad spectrum of hosts and might have been transmitted to domestic cats, other feline species, ruminants, and primates in zoos. The abnormal prion protein isoform (PrP^d, PrP^{CWD}), as well as the expansion of CWD toward Canada, Asia, and Europe, was originally demonstrated for mule deer, white-tailed deer, and American elks in certain regions of the United States. The following goals were therefore set for Russian research teams as part of the international TSE initiative:

1. to expand tissue sample collections from cattle and wild ungulates from different regions of Russia and the former Soviet Republics; to use these samples to carry out molecular typing of cervid prion genes;
2. to identify the CWD risk group on the basis of the genotype for moose, deer, and other wild ungulates in Russia and the former Soviet Republics; to identify genotypes sensitive to infection with the CWD causative agent; to determine the variability of the prion protein gene in cervid subpopulations;
3. to test cervids for CWD, using immunohistochemistry and ELISA;

4. to synthesize recombinant prion proteins of cervids; to design antigens and antibodies for specific detection of wild ungulates' prion proteins.

In Russia, the genotyping of prnp in wild ungulates was conducted alongside sequencing and DNA analysis.[41,54] These studies consisted in sample collection, tissue pretreatment before IHC testing, PCR analysis, sequencing, phylogenetic analysis, and ultrastructure analysis by electron microscopy. In the course of this research, 350 tissue samples from deer and elk species dominant in the Russian Federation were added to the cervid tissue bank. Among these were samples from the red deer (*Cervus elaphus*), the reindeer (*Rangifer tarandus*), and the moose (*Alces alces*) (Figure 10.8). The sika deer (*Cervus nippon*) and the Altai maral (*Cervus elaphus sibiricus*) also were tested.

The data acquired were eventually in agreement with preliminary results obtained earlier that implied a higher variability of a specific fragment in the PrP gene in moose from Siberia compared with that estimated for moose from the European part of Russia. Apart from this difference, the variability of the mtDNA control region was studied in the same group of animals and further confirmed a greater subspecies diversity among Siberian moose. These data also reliably showed that the majority of the moose population in Russia falls into the so called high-risk CWD susceptibility group. Furthermore, it was demonstrated that animals homozygous for a mutation at amino acid position 132 in the PrP gene predominate among the Altai maral subspecies (the Altai maral being the most closely related species to the North American elk), a finding that also places animals with that mutation in the high-risk group for CWD.

A certain degree of correlation among PrP gene polymorphism, susceptibility to the infection, and the duration of the incubation period

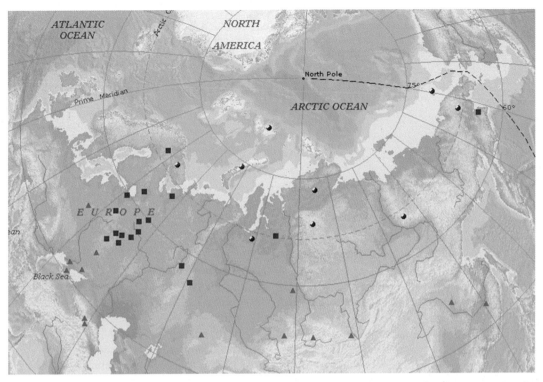

FIGURE 10.8 Geography of Cervidae (deer and moose) samples, collected in Russia for chronic wasting disease analysis (triangles – Red deer, *Cervus elaphus*; circles – Northern deer, *Rangifer tarandus*; quadrates – Moose, *Alces alces*).

was found for prion diseases of ruminants.[56] Also, mutation positions related to susceptibility were located: first and foremost, position 132 for the North American elk[47,49] and position 209 for the moose.[34] Numerous studies are being carried out at present to reveal the nature of PrP gene polymorphism in different wild ungulate species. Until recently, ungulates inhabiting Russia were not studied from this perspective, and it remained unknown to what extent variability in the PrP gene was related to variability elsewhere in the genome.

On the basis of the correlations found, the chief goal of these studies was identified as the performance of research into the previously unexplored PrP gene variability in the moose and Altai maral in Russia and a comparison thereof with the variability of the mtDNA control region, a fragment of mitochondrial DNA

known to possess the highest variability. Analysis of PrP gene polymorphism in moose from different parts of their geographic range revealed a number of region-specific differences in this parameter. As previously mentioned , the PrP gene variability in moose is significantly higher in Siberia than in regions west of the Urals (Table 10.2).

Research data obtained both overseas[57] and in Russia suggest that, on the basis of mtDNA variability, moose inhabiting eastern Eurasia are genetically close to North American moose. Assumptions have been made that animal migration was occurring via Beringia around the Pleistocene—Holocene boundary and gene flow between the two regions did take place. If CWD existed among the moose of America and, potentially, Beringia during that time, then natural selection might have favored the

TABLE 10.2   Variability of the PrP Gene and the mtDNA Control Region in Moose in Different Regions of Russia

Region	No. of samples	No. of alleles	Number of synonymous/ nonsynonymous substitutions	Length of PrP gene segment amplified with a forward and reverse primer, in base pairs (bp)	Number of mtDNA control region (464 bp) haplotypes (nVS)
West of Ural	40	1(0)	0/0	800	5 (9 vs)
Western Siberia	19	2(1)	1/0	(790) 500	18 (30 vs)
Eastern Siberia	36/21	3(5)	5/1	600—700	8 (27 vs)

positive I209 mutation and led to its fixation in moose of these regions. The west Siberian moose population falls into an intermediate position between the European and east Siberian populations. Thus, with the moose as an example, a correlation was demonstrated between mtDNA and PrP gene variability.

The majority of the Russian moose population falls into the high-risk group for CWD infection.

PrP gene polymorphism assessment in Altai marals from different regions in Russia. Data obtained to this day have ushered in the assumption that deer which are homozygous for L132 in the PrP gene could be less susceptible to CWD and have a longer incubation period than that of M132 homozygous and heterozygous animals.[36,46,47] In a group of 43 animals, both wild and from a farm in Altai, all animals but one wild deer were shown to carry M132. Thirty variable sites were described, and 18 haplotypes identified, for a D-loop fragment of mtDNA 464 bases long in a group of 19 Altai marals from west Siberia. Results obtained in Russia suggest that the variability of the mtDNA control region significantly exceeds that of the PrP gene. Overall, this difference corresponds to that in the mutation speed between these two genes, as the noncoding region of the mtDNA used as the control region is one of the fastest mutating genome fragments in mammals (Figure 10.9). At the same time, the PrP-T allele was observed in two samples

from animals falling into the same haplogroup for the mtDNA control region in the west Siberian study group. A predominance of animals homozygous for the PrP 132 position mutation was thus demonstrated for Altai marals in Russia, a fact that can be translated into a heightened risk of CWD infection.

High susceptibility to CWD is determined by a number of genetic factors, among which are (i) the genetic variability of the prnp coding region and associated genotypes, (ii) the genetic variability of prnp regulatory regions, (iii) a combination of the quantitative and qualitative impact of genetic variability inside the coding and regulatory regions of the prnp gene, and (iv) the variability of other genes. Mutations at critical positions do not always result in an increase in susceptibility to CWD.[58,59] The aforementioned factors have been poorly studied in wild ungulates, particularly in cervids; understanding is likely to be attained only as data on the prnp gene variability in different deer populations accumulate. Special attention will have to be paid to statistics on the proportion of sick animals for different genotypes. The variability pattern for the mtDNA control region in Altai marals corresponds mostly to that for the allelic variants of the PrP gene. *Inter alia*, a tendency for a number of mtDNA haplotypes to correlate with certain alleles was observed. In other words, in Altai marals, a higher variability of

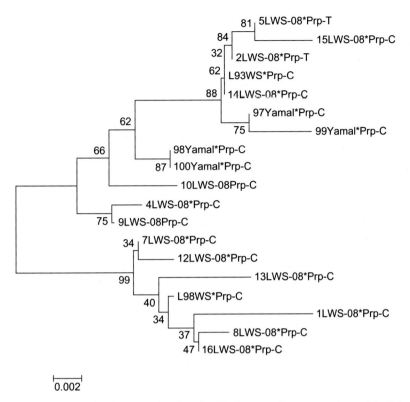

FIGURE 10.9 Neighbor-joining dendrogram, developed with the use of a parametric model of the mtDNA D-loop sequence (464 bp) for Maral deer, *Cervus elaphu sibiricus*, (464 bp gene fragment) located in western Siberia of the Russian Federation.

the mitochondrial genome is bound up with a higher polymorphism of the PrP gene.

In conclusion, it is worth remarking that tens of thousands of deer infected with CWD are identified in the United States annually and are "safely disposed of" (i.e., their carcasses are burned). Pastures are regularly treated with disinfectants, such as sodium hypochlorite. Despite all of these efforts, attempts to eradicate CWD have so far been unsuccessful.

References

1. Prusiner SB. Molecular biology of prion diseases. *Science* 1991;**252**:1515–22.
2. Gajdusek DC, Gibbs Jun CJ, Alpers M. Experimental transmission of a kuru-like syndrome to chimpanzees. *Nature* 1966;**209**:794–6.
3. Grigoriev VB. Prions. *Vopr Virusol* 2004;**6**:4–12 [in Russian].
4. Grigoriev VB, Lvov DK. Prion infections. In: Lvov DK, editor. *Handbook of virology: viruses and virus infections of human and animals*. Moscow: MIA;2013, p. 818–21 [in Russian].
5. Ribakov SS. Scrapie and other prion diseases of animals and human. Moscow; 2003 [in Russian].
6. Kalnov SL, Verkhovsky OA, Aliper TI. Prion diseases of animals. In: Lvov DK, editor. *Handbook of virology: viruses and virus infections of human and animals*. Moscow: MIA;2013, p. 910–21 [in Russian].
7. Weissmann C. PrP effect clarified. *Curr Biol* 1996;**6**:1359.
8. Safar J, Wille H, Itri V, et al. Eight prion strains have PrPSc molecules with different conformations. *Nat Med* 1998;**4**:1157–65.
9. Qin K, Yang D-S, Yang Y, et al. Copper (II)-induced conformational changes and protease resistance in recombinant and cellular PrP. *J Biol Chem* 2000;**275**: 19121–31.

10. Lasmezas CI, Weiss S. Molecular biology of prion diseases. In: Cary JW, Linz JE, Bhatnagar D, editors. *Microbial foodborne diseases: mechanisms of pathogenesis and toxin synthesis*. Lancaster, PA: Techn. Publishing Co.;2000.

11. Prusiner SB. Novel proteinaceous infection particles cause Scrapie. *Science* 1982;**216**:136–44.

12. Hadlow WJ. Differing neurohistologic images of scrapie, transmissible mink encephalopathy, and chronic wasting disease of mule deer and elk. In: Gibbs Jr. CJ, editor. *Bovine spongiform encephalopathy. The BSE dilemma*. NY: Springer;1996. p. 122–37.

13. Fraser H. Diversity in the neuropathology of scrapie-like diseases in animals. *Br Med Bull* 1993;**49**:792–809.

14. Tcherkasskaya O, Davidson EA, Schmerr MJ, et al. Conformational biosensor for diagnosis of prion diseases. *Biotech Lett* 2005;**27**:671–5.

15. McBride PA, Eickelenboom H, Kraal G, et al. PrP protein is associated with follicular dendritic cells of spleens and lymph nodes in uninfected and scrapie-infected mice. *J Pathol* 1992;**168**:413–18.

16. Korth C, Stierli B, Streit P, et al. Prion (PrP-Sc)-specific epitope defined by a monoclonal antibody. *Nature* 1997;**390**:74–7.

17. Sigurdson CJ, Aguzzi A. Chronic wasting disease. *Biochim Biophys Acta* 2007;**1772**:610–18.

18. Wadsworth JDF, Joiner S, Hill AF, et al. Tissue distribution of protease resistant prion protein in variant Creutzfeldt-Jakob disease using a highly sensitive immunoblotting assay. *Lancet* 2001;**358**:171–80.

19. Trieschmann L, Santos N, Kaschig K, et al. Ultra-sensitive detection of prion protein fibrils by flow cytometry in blood from cattle affected with bovine spongiform encephalopathy. *BioMed Central Biotechnol* 2005;**5**:26.

20. Kalnov SL, Grigoriev VB, Alekseev KP, et al. Production and characterization of bovine full-length recombinant PrPc. *Bull Exp Biol Med* 2006;**141**(1):68–71 [in Russian].

21. Reshetnikov GG, Petrova ON, Ribakov SS, et al. Usage of polypeptide antibodies mixture, generated against PrP for diagnostics of BSE by IHC. *Ann ARRIAH* 2006;**IV**:116–23 [in Russian].

22. Aldhous P. Spongiform encephalopathy found in cat. *Nature* 1990;**345**:194.

23. Bons N, Mestre-Frances N, Belli P, et al. Natural and experimental oral infection of nonhuman primates by bovine spongiform encephalopathy agents. *PNAS* 1999;**96**:4046–51.

24. Guiroy DC, Williams ES, Yanagihara R, et al. Immunolocalization of scrapie amyloid (PrP27–30) in chronic wasting disease of Rocky Mountain elk and hybrids of captive mule deer and white-tailed deer. *Neurosci Lett* 1991;**126**:195–8.

25. O'Rourke KI, Baszler TV, Besser TE, et al. Preclinical diagnosis of scrapie by immunohistochemistry of third Eyelid Lymphoid tissue. *J Clin Microbiol* 2000;**38**:3254–9.

26. O'Rourke KI, Holyoak GR, Clark WW, et al. PrP genotypes and experimental scrapie in orally inoculated Suffolk sheep in the United States. *J Gen Virol* 1997;**78**:975–8.

27. O'Rourke KI, Melco RP, Mickelson JR. Allelic frequencies of an ovine scrapie susceptibility gene. *Anim Biotech* 1996;**7**:155–62.

28. O'Rourke KI, Baszler TV, Miller JM, et al. Monoclonal antibody F89/160.1.5 defines a conserved epitope on the ruminant prion protein. *J Clin Microbiol* 1998;**36**:1750–5.

29. O'Rourke KI, Baszler TV, Parish SM, et al. Preclinical detection of PrP-Sc in nictitating membrane lymphoid tissue of sheep. *Vet Rec* 1998;**142**:489–91.

30. Wells GAH, Scott AC, Johnson CT, et al. A novel progressive spongiform encephalopathy in cattle. *Vet Rec* 1987;**121**:419–20.

31. Wells GAH, McGill IS. Recently described scrapie-like encephalopathies of animals: case definitions. *Res Vet Sci* 1992;**53**:1–10.

32. Wells GAH, Simmons MM. The essential lesion profile of bovine spongiform encephalopathy (BSE) in cattle is unaffected by breed or route of infection. *Neuropathol Appl Neurobiol* 1996;**22**:453.

33. Williams ES. Chronic wasting disease. *Vet Pathol* 2005;**42**:530–49.

34. Baylis M, Goldmann W. The genetics of scrapie in sheep and goats. *Curr Mol Med* 2004;**4**:385–96.

35. Jewell JE, Conner MM, Wolfe LL, et al. Low frequency of PrP genotype 225SF among free-ranging mule deer (*Odocoileus hemionus*) with chronic wasting disease. *J Gen Virol* 2005;**86**:2127–34.

36. Spraker TR, Balachandran A, Zhuang D, et al. Variable patterns of distribution of PrPCWD in the obex and cranial lymphoid tissues of Rocky Mountain elk (*Cervus elaphus nelsoni*) with subclinical chronic wasting disease. *Vet Rec* 2004;**155**:295–302.

37. Spraker TR, Miller MW, Williams ES, et al. Spongiform encephalopathy in free-ranging mule deer (*Odocoileus hemionus*), white-tailed deer (*Odocoileus virginianus*) and Rocky Mountain elk (*Cervus elaphus nelsoni*) in northcentral Colorado. *J Wildlife Dis* 1997;**33**:1–6.

38. Spraker TR, O'Rourke KI, Gidlewski T, et al. Detection of the abnormal isoform of the prion protein associated with chronic wasting disease in the optic pathways of the brain and retina of Rocky Mountain Elk (*Cervus elaphus nelsoni*). *Vet Pathol* 2010;**47**:536–46.

39. Kholodova MV, Kolpatshikov LA, Kuznetsova LA, et al. Genetic diversity of wild Northern deer (*Rangifer tarandus*) from Taimir region: analysis of control region mtDNA polymorphism. *Izvestiya*

Rossiiskoi Academii Nauk, Biology 2011;**1**:52—60 [in Russian].

40. Kreeger TJ, Montgomery DL, Jewell JE, et al. Oral transmission of chronic wasting disease in captive Shira's moose. *J Wildlife Dis* 2006;**42**:640—2.

41. Zvichainaya EU, Kholodova MV. Distribution of haplotypes and alleles of mt-DNA control region and PrP gene alleles in isolated Maral (*Cervus elaphus sibiricus*) population. In: *Proceedings of International Conference "Zoo—and phylogenic problems of Vertebrata"*. Moscow; 2009. p. 35 [in Russian].

42. Miller MW, Williams ES, Hobbs NT, et al. Environmental sources of prion transmission in mule deer. *Emerg Infect Dis* 2004;**10**:1003—6.

43. Raymond GJ, Bosser A, Raymond LD, et al. Evidence of a molecular barrier limiting susceptibility of humans, cattle and sheep to chronic wasting disease. *EMBO J* 2000;**19**:4425—30.

44. Hunter N. PrP genetics in sheep and implications for scrapie and BSE. *Trends Microbiol* 1997;**5**:331—4.

45. Moum T, Olsaker I, Hopp P, et al. Polymorphisms at codons 141 and 154 in the ovine prion protein gene are associated with scrapie Nor98 cases. *J Gen Virol* 2005;**86**:231—5.

46. Hamir AN, Gidlewski T, Spraker TR, et al. Preliminary observations of genetic susceptibility of elk (*Cervus elaphus nelsoni*) to chronic wasting disease by experimental oral inoculation. *J Vet Diagn Invest* 2006;**18**:110—14.

47. O'Rourke KI, Besser TE, Miller MW, et al. PrP genotypes of captive and free-ranging Rocky Mountain elk (*Cervus elaphus nelsoni*) with chronic wasting disease. *J Gen Virol* 1999;**80**:2765—79.

48. Johnson C, Johnson J, Vanderloo JP, et al. Prion protein polymorphisms in white-tailed deer influence susceptibility to chronic wasting disease. *J Gen Virol* 2006;**87**:2109—14.

49. O'Rourke KI, Spraker TR, Hamburg LK, et al. Polymorphisms in the prion precursor functional gene but not the pseudogene are associated with susceptibility to chronic wasting disease in white-tailed deer. *J Gen Virol* 2004;**85**:1339—46.

50. Vanik DL, Surewicz KA, Surewicz WK. Molecular basis of barriers for interspecies transmissibility of mammalian prions. *Mol Cell* 2004;**14**:139—45.

51. Williams ES, Young S. Spongiform encephalopathy of Rocky Mountain Elk. *J Wildlife Dis* 1982;**18**: 465—71.

52. Williams ES, Young S. Chronic wasting disease of captive mule deer: a spongiform encephalopathy. *J Wildlife Dis* 1980;**16**:89—98.

53. Kholodova MV. Comparable phylogeografic: molecular methods, ecological analysis. *Mol Boil* 2009;**43** (5):910—17 [in Russian].

54. Kholodova MV, Prikhodko VI. Molecular genetic diversity of Musk Deer *Moschus moschiferus* L., 1758 (Ruminantia, Artiodactyla) from the Northern Subspecies Group. *Rus J Gen* 2006;**42**(7):783—9 [in Russian].

55. Pokidishev AN. Characterization of brec-PrP (*Bos taurus*) and development of infectious prion isoform detection method [Ph.D. dissertation]. Moscow; 2009 [in Russian].

56. Kim TY, Shon HJ, Joo YS, et al. Additional cases of chronic wasting disease in imported deer in Korea. *J Vet Med Sci* 2005;**67**(8):753.

57. Huson HJ, Happ GM. Polymorphisms of the prion protein gene (PRNP) in Alaskan moose (*Alces alces gigas*). *Anim Genet* 2006;**37**:425—6.

58. Hundertmark KJ, Shields GF, Udina IG, et al. Mitochondrial phylogeography of moose (*Alces alces*): Late Pleistocene divergence and population expansion. *Mol Phylogen Evol* 2002;**22**(3):375—87.

59. Baeten LA, Powers BE, Jewell JE, et al. A natural case of chronic wasting disease in a free-ranging moose (*Alces alces shirasi*). *J Wildlife Dis* 2007;**43**: 309—14.

Index

Note: Page numbers followed by *"f"*, and *"t"* refers to figures and tables respectively.